Springer Geography

The Springer Geography series seeks to publish a broad portfolio of scientific books, aiming at researchers, students, and everyone interested in geographical research.

The series includes peer-reviewed monographs, edited volumes, textbooks, and conference proceedings. It covers the major topics in geography and geographical sciences including, but not limited to; Economic Geography, Landscape and Urban Planning, Urban Geography, Physical Geography and Environmental Geography.

Springer Geography—now indexed in Scopus

More information about this series at http://www.springer.com/series/10180

Assefa M. Melesse · Wossenu Abtew ·
Semu A. Moges
Editors

Nile and Grand Ethiopian Renaissance Dam

Past, Present and Future

Springer

Editors
Assefa M. Melesse
Department of Earth and Environment
Florida International University
Miami, FL, USA

Wossenu Abtew
Water and Environment Consulting LLC
Boynton Beach, FL, USA

Semu A. Moges
Senior Hydrologist
Dallas, TX, USA

ISSN 2194-315X ISSN 2194-3168 (electronic)
Springer Geography
ISBN 978-3-030-76436-4 ISBN 978-3-030-76437-1 (eBook)
https://doi.org/10.1007/978-3-030-76437-1

This Springer imprint is published by the registered company Springer Nature Switzerland AG
The registered company address is: Gewerbestrasse 11, 6330 Cham, Switzerland

Preface

Global freshwater availability is projected to be 40% short of the demand by 2030. The value of water has been increasing including becoming a commodity on the stock market. Rising number of water conflicts between communities, regions in a country, and between countries is a sign of the freshwater access challenges of the future. Transboundary basins in this regard become points of water conflict as transboundary rivers cross international boundaries. There are several international experiences of transboundary water and benefit sharing between riparian countries through basin management agreements and institutional arrangements. The Nile Basin is one of the large transboundary basins where basin management agreement has neither been realized nor institutional mechanism established. The Nile Basin Initiative, a partnership organization of the 10 Nile riparian countries, was initiated in 1999. It is an intergovernmental partnership for the objective of achieving sustainable socio-economic development through the equitable resource and benefit utilization. The participants are Burundi, DR Congo, Egypt, Ethiopia, Kenya, Rwanda, South Sudan, Sudan, Tanzania, and Uganda. Eritrea is an observer. Parallel to the NBI, a political process was initiated between the riparian countries in 1997 to formulate a Cooperative Framework Agreement (CFA) that was intended to be the guiding legal regime for the basin. The process of developing CFA was completed in 2010 and opened for member countries to sign the agreement. The objective of the agreement is to promote integrated management, sustainable development, and harmonious utilization of the water resources of the Nile Basin with conservation and protection of resources. Seven countries agreed while Egypt and Sudan showed reservation. Burundi, Ethiopia, Kenya, Rwanda, Tanzania, and Uganda signed the agreement. Ethiopia, Rwanda, Tanzania, and Uganda ratified the agreement by their parliaments. The CFA requires ratification by six riparian countries for implementation as the basin law. With this in the background, Ethiopia started building a large hydropower dam, the Grand Ethiopian Renaissance Dam (GERD), on the Blue Nile River on its territory near the Sudan. Concern was raised by downstream countries of Sudan and Egypt on the filling and operation of the dam for fear of losing their current use of 100% of the Nile water. Population growth in upstream countries has made it clear that it is time to share the Nile water for irrigation and hydropower from the Nile tributaries that originate or flow through their territories. Series of negotiations were conducted on the filling and operation of the GERD between Egypt, Ethiopia, and Sudan since the launching of the dam in 2011. In 2015, a milestone agreement known as the Declaration of Principles (DoP) was reached. The main features of the DoP included principles of cooperation, development, regional integration, and sustainability, not causing significant harm, equitable and reasonable utilization, to cooperate in the first filling and operation of the dam, confidence building, exchange of information and data, dam safety, sovereignty and territorial integrity, and principle of peaceful settlement of dispute conflicts. The specific negotiations, however, are at impasse with signs of conflict in many forms. The time has passed for upstream countries to survive on rainfed agriculture, and downstream countries unchecked use of the Nile water as a result of population explosion in the basin, economic development, and climatic factors. The need for energy to revamp the economies of the upstream countries calls for hydroelectric power projects like GERD. Freshwater and electricity access in Ethiopia is many folds lower than Egypt. Ethiopia has experience of recurrent drought and food insecurity, a major challenge for many years. The level of land degradation and poor agricultural practices in the headwater of the Blue Nile River is a major impediment for boosting agricultural productivity. Rainfed agriculture has been suffering for a long time due to the less reliability of the timing and volume of rainfall. Ethiopia is emerging and determined to be self-sufficient in food production through irrigated agriculture and also increase in energy accessibility through GERD and other projects. The need for cooperation and basin-wide water and benefit sharing agreement is timely. Conflict between downstream and upstream countries will make the situation worse. The longer the wait to agree, the lesser the chance to have an amicable agreement mainly due to the new demands induced by population growth and climate change impacts on the water sector. This book is a contribution by the presenters of the 2020 International Conference on the Nile and GERD. The book covers Nile water claims past and present, international transboundary basin cooperation, Nile water supply and demand management, Blue Nile/Abbay and Grand Ethiopian Renaissance Dam, land and water degradation and watershed management, emerging threats of the Lakes Region in the Nile Basin, and hydrologic variation and monitoring. This book is beneficial for students, researchers, sociologists, engineers, policy-makers, water resources and environmental managers, the people, and governments of the Nile Basin.

Boynton Beach, USA — Wossenu Abtew
Miami, USA — Assefa M. Melesse
Rowlette, USA — Semu A. Moges

Contents

Chapter 1
Introduction

Semu A. Moges, Wossenu Abtew, and Assefa M. Melesse

Abstract Many countries in the world are connected by rivers and groundwater aquifers that cross the political boundaries. In many of these cases, transboundary countries have benefited from this and equitably share the common water resources and forge collaboration, boost economic development and also promote cultural exchange. Even though Nile River is the longest river crossing eleven countries, sometimes its history and fame are overshadowed by the transboundary water use conflict between Egypt and Sudan on one side and Ethiopia and the remaining upstream Nile countries on the other side. This conflict and lack of collaboration to use the common good are exacerbated after the construction of the Grand Ethiopian Renaissance Dam (GERD), a hydroelectric dam for hydropower production. This non-consumptive water use project has multiple benefits for Ethiopia, Sudan and Egypt. Regardless of the various efforts to reach a negotiated agreement on the filling and operation of the dam, the three countries have yet to build trust, narrow their differences and use this unique project to their benefit and learn from other successful transboundary agreements to work together and tackle the common challenges of increased water demand, climate change and watershed degradation. The time is now to deescalate the issue and reach to an agreement driven by science and good data. This chapter looks into the history of the Nile, colonial era non-inclusive agreements, the efforts of the Nile Basin Initiative (NBI) to forge a Cooperative Framework Agreement and also the recent negotiations on GERD. It also recommends some directions in resolving the conflict for a sustainable trans-boundary water management.

S. A. Moges
School of Civil and Environmental Engineering, University of Connecticut, Mansfield, USA
e-mail: semu.moges@uconn.edu

W. Abtew
Water and Environment Consulting LLC, 6153 Terra Rosa Circle,
Boynton Beach, FL 33472, USA
e-mail: wabtew@gmail.com

A. M. Melesse (✉)
Department of Earth and Environment, Institute of Environment,
Florida International University, Miami, USA
e-mail: melessea@fiu.edu

A. M. Melesse et al. (eds.), *Nile and Grand Ethiopian Renaissance Dam*,
Springer Geography, https://doi.org/10.1007/978-3-030-76437-1_1

Keywords Nile river · GERD · Ethiopia · Egypt · Sudan · Transboundary water management · Nile basin initiative · Cooperative framework agreement

Overview

Nile is the world's longest river, traversing more than 6700 km and covers eleven African countries. The Basin extends for more than 3 million square kilometers. Nile is one of the world's basins shared by more than ten countries that include Burundi, Democratic Congo, Egypt, Eritrea, Ethiopia, Kenya, Rwanda, Tanzania, South Sudan, Sudan and Uganda (Fig. 1.1). The total population of the Nile basin countries exceeds 550 million by the end of 2020 and is predicted to grow to more than a billion people by the end of 2050 (UNDESA 2019). According to NBI estimate, 54% of the total population lives in the basin part of the riparian countries (NBI, www.nilebasin.org). Four of the eleven countries that share the Nile River are among the world's ten poorest countries, and poverty is widespread with millions living on less than a dollar a day (NBI, www.nilebasin.org).

The Nile receives its flows mainly from two basin sources, the Equatorial Lakes plateau (Burundi, DRC, Kenya, Rwanda, Tanzania and Uganda), and the Ethiopian highland plateau. More than 85% of the annual flow to the main Nile is contributed from the Ethiopian Highlands. The remaining 15% of the flow is contributed from the equatorial plateau. The total annual flow at the border between Sudan and Egypt has historically been taken (before any significant abstraction) as 84 km^3 (1901–1959). However, the Nile flow in recent years (1965–2009) exhibited markedly higher variability between 42 and 105 BCM (Ahmed et al. 2019). Compared to the land area, the Nile has the lowest specific discharge (0.98 $L/s/km^2$) of all world rivers having a basin area exceeding 1 million km^2. This is almost one-tenth of the estimated specific discharge of the Congo basin, which is geographically the closest to the Nile basin (Shahin 1985).

The most important salient characteristics of the Nile hydrology today are its high evaporation and conveyance losses. Blackmore and Whittington (2008) summarized that about 19 BCM is lost annually from man-made reservoirs. A major source of evaporation is High Aswan Dam (HAD) in Egypt. Secondly, approximately 20 to 30 BCM of water is lost as conveyance loss in the channels, and significant losses occur from irrigation systems operated throughout the Nile basin.

All the Nile basin countries have predominantly agricultural economies, and irrigation has become essential for food security in the basin. Of the total irrigation in the Nile basin, 98.7% is in Egypt and northern Sudan. The legacy of conflict, civil war, hydrological variability and limited capacity on the one hand and over protection of the bilateral colonial water use treaties and hydrohegemony by the lower riparian countries on the other hand contributed to the low level of irrigation development in the upper riparian countries of the basin. Most importantly, the lack of cooperation framework and legal regime in the Nile basin contributed to skewed

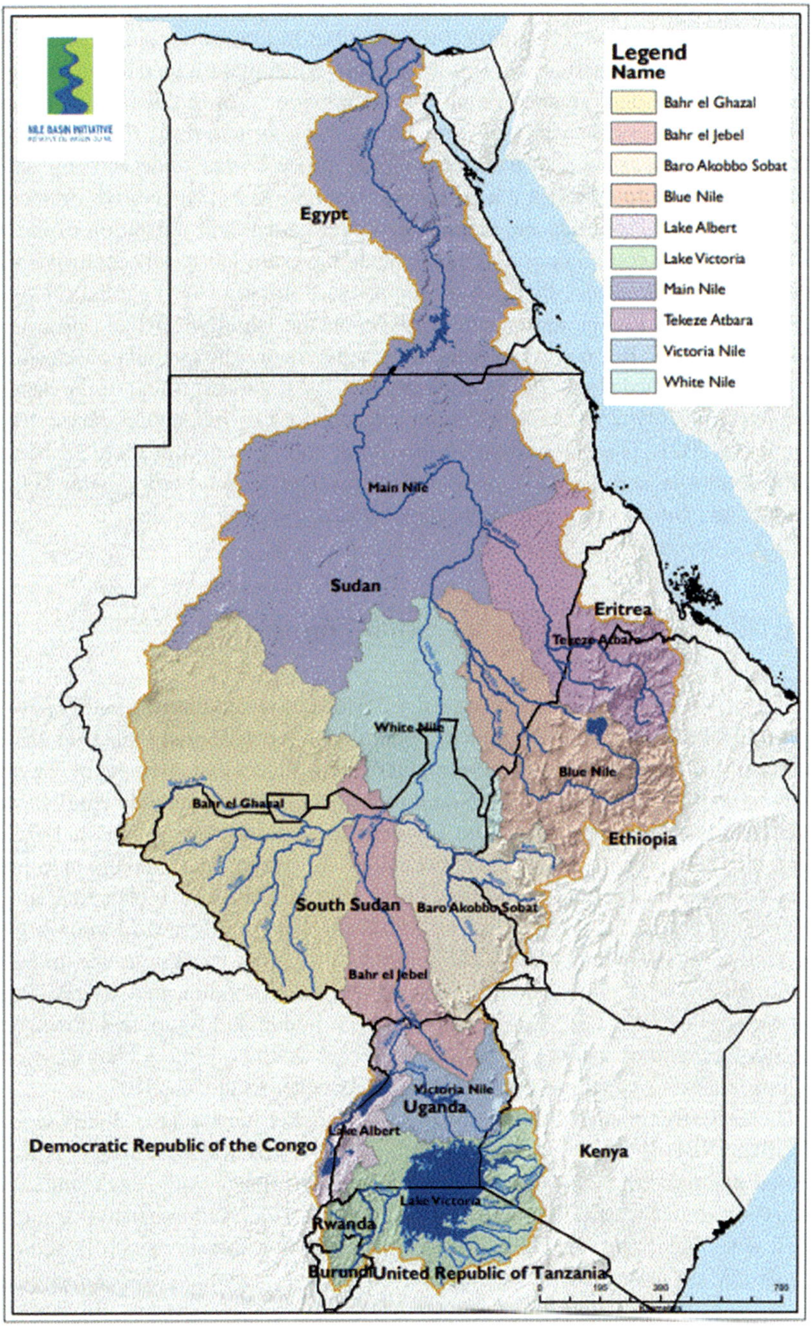

Fig. 1.1 Nile is one of the world's river basins shared by eleven countries

development in the basin. But now, population pressure and social, economic and technological growth are changing the statuesque in upstream countries.

The Nile Basin Initiative (NBI) is an intergovernmental partnership organization established with the shared vision of "*achieving sustainable socio-economic development through the equitable utilization of, and benefit from, the common Nile Basin water resources.*" A parallel political process was underway by all the riparian countries to establish a legal mechanism known as Cooperative Framework Agreement (CFA) to guide the sustainable development and utilization of the Nile resources. These comprehensive engagements have not led to any inclusive water agreement acceptable to all riparian countries. Similarly, the negotiation on the Grand Ethiopian Renaissance Dam (GERD) on the Blue Nile River, close to the Sudan border in 2011, could not reach to an agreement. The overall construction of the dam has reached about 78.3% as of January 2021, the first filling of the dam was completed in July 2020, and Ethiopia expects to fill the second phase by the summer of 2021. The negotiation on the filling and operation of the dam between Egypt, Ethiopia and Sudan is still underway. The recent border issue between Ethiopia and Sudan complicates the negotiation.

The Nile Conflict and Cooperation Status

The history of Nile can be characterized as history of water conflict mainly between Egypt and Ethiopia. The documented conflict of the Nile dates back to the last quarter of the nineteenth century, when there were successive attempts by Egyptian rulers to control the source of the Nile through war. The two notable conflicts were successive wars between Egypt and Ethiopia at the Battles of Gundet in 1875 and Gura in 1876 (Arsano 2007). The signature of Egypt and Sudan a bilateral agreement for total control of the flow of the Nile in 1959, which cleared a way for the construction of the largest High Aswan Dam (HAD), infuriated Ethiopia. Since the Second World War, Ethiopia has continually asserted its right to use the waters of the Blue Nile for the benefit of its people and communicated its position of non-recognition of the 1959 agreement between Sudan and Egypt and downstream water control and expanding water use. As recent as 1997, Ethiopia has transmitted her concerns on expanding the use of the Nile water (Salman 2010).

The first all-inclusive cooperation attempt was the launching of the Nile Basin Initiative (NBI) in 1999 by ten riparian countries. NBI was established as a transitional intergovernmental mechanism to facilitate basin-wide engagement and discussion, build trust and confidence for a basin-wide comprehensive agreement (NBI, www.Nilebasin.org). NBI was established with a shared vision of "*achieving sustainable socio-economic development through the equitable utilization of, and benefit from, the common Nile Basin water resources.*" Over the last 20 years, NBI has managed to maintain a platform for cooperation, communication, generate knowledge, and contribute to capacity building.

A parallel political process was underway by all the riparian countries to establish a legal mechanism known as Cooperative Framework Agreement (CFA) to guide the sustainable development and utilization of the Nile resources. After 13 years of inclusive negotiation, the countries agreed with all 45 articles of CFA except one article (the 14b sub-article) and concluded the negotiation in 2010 without full agreement. The downstream countries of Egypt and Sudan insisted to include sub-article that protects the current water utilization, which is 100%. While the upstream riparian countries pushed for a compromise deal that does not foreclose their right to utilize the Nile water equitably and reasonably. Salman (2013) described the unfortunate stalling of the CFA negotiation to be used as a tool to consolidate the two known positions of upstream and downstream riparian countries. Egypt and Sudan consolidated their position on the *colonial treaties and acquired uses and rights of lower riparian countries,* while the majority of upstream riparian countries consolidated on the position of *equitable and reasonable utilization of the shred water resources.* The CFA is currently signed by six countries and ratified by four countries. It will be the law of the Nile basin countries if the document is ratified by six countries.

Grand Ethiopian Renaissance Dam Facts

At the backdrop of the failure of CFA process and the growing energy demand in the country, Ethiopia launched the construction of Grand Ethiopian Renaissance Dam (GERD) on Blue Nile River close to the Sudan border in 2011. The GERD is the first major dam on the Blue Nile (Abbay) River in Ethiopia. The dam is a roller-compacted concrete dam with maximum height of 155 m and width of 1800 m. On the side, a 5-km saddle dam of 50 m height was built to raise a lower land feature to the desired elevation. It has initial storage capacity of 74 billion cubic meters with power generation of 6000 MW. Currently, the dam is in the final phase of construction, and it is self-financed. The overall construction of the dam has reached about 78.5% as of January 2021. The first filling of the dam was done in July 2020, and Ethiopia expects to fill the second phase by the summer of 2021.

Ethiopia cooperated on sharing of documents and data pertaining to the design of the dam. Ethiopia supplied over 150 design reports to Egypt and Sudan, in good phase, and agreed to establish the International Panel of Experts (IPoE) in September 2011 to study potential impacts of the GERD. A National Technical Committee (NTC) composed of the three countries was established in 2014 to follow the IPoE studies and coordinate the implementation of the recommendations.

Egypt, Ethiopia and Sudan conducted several meetings concerning the impact of GERD on downstream countries of Egypt and Sudan. In 2015, the series of negotiations culminated with the signature of the Principles of Declaration in 2015, by the head of states of the three countries of Egypt, Ethiopia and Sudan. The three states agree to cooperate implementation of outcomes of joint studies on the GERD. Specifically, they agreed on guidelines and rules for filling and annual operation of

GERD, and to inform downstream states of any unforeseen or urgent circumstances. Priority is also given to downstream states to purchase power generated by GERD. They agreed as well to work together on the dam safety. Since then, the countries were undergoing technical negotiation on the filling and operation of the GERD. It is said that in September 2018, the National Independent Scientific Research Group (NISRG) presented their consensus report on the filling and operation of the dam to the Water Affair Ministers for signature and was declined by Egypt. The next negotiations that took place in Washington DC from November 2019 to February 2020 were also unsuccessful (Egypt 2020). This time Ethiopia refused to sign the agreement document stating that it fringes on its right to use the Blue Nile River for irrigation upstream of the GERD. Ethiopia did not accept the clause that demands Ethiopia guarantee quotas of annual releases from its dam during drought periods and normal operations. While hydropower operation requires maximum flexibility to adapt to changing weather and power demand conditions, Egypt and Sudan are insisting Ethiopia should be bound by legal agreement on the operation of the dam. Currently, this is a major sticking point.

The current negotiation process under the auspicious of African Union is going slowly. The sticking point is similar to the previous Washington DC negotiation. Putting it in different wording, Ethiopia believes the sticking points implicitly carry protection of prior water quota of downstream countries. The continuously surfacing Washington DC proposed negotiation document embeds water quota in the name of drought mitigation and minimum flows with legally binding agreement. Ethiopia believes that this provision undermines the performance of power generating capacity of the GERD and also affects Ethiopia from using its share of water for irrigation upstream of the dam. Ethiopia's growing population is in dire need of irrigation expansion to combat food insecurity. The GERD negotiation is beyond the filling and operation issue for the three countries and may need to be viewed as an issue requiring a more comprehensive and cooperative agreement based on equitable and reasonable principles of transboundary basin norms.

The Way Forward

We believe the negotiation of the shared resources should be dictated by reason, science and above all shared human destiny. The population of the Nile basin countries was about 86 million in 1950. The basin population has grown almost fivefold in 2010 (417 million) by the time of the CFA negotiation was completed. The population stands at more than 550 million in 2020 and will grow to about a billion people by 2050. There is no other option than cooperative management of the Nile water. Specifically,

i. As a first principle, the three governments of Egypt, Ethiopia and Sudan need to depoliticize and denationalize water and advocate water as greater social and economic good endowed to all of us by nature, operationalize the equitable and reasonable utilization principles of the shared resources.
ii. Secondly, water-related negotiation requires a paradigm shift toward a progressive and adaptive negotiation and agreement style. The three countries must capitalize on agreed part of the negotiation (e.g., the filling) and sign as a way of confidence building and continued working together. Proceed with the long-term operation and comprehensive legal arrangements including water sharing modalities.
iii. GERD is a hydropower dam, and operation of a hydropower dam needs flexible operation to accommodate weather and climate fluctuations, dam safety and maintenance and meeting power demands. It is not amenable to a rigid legally binding agreement.
iv. Steps should immediately begin on basin-wide agreement on joint development management, cooperation and water and benefit sharing.
v. The time to agree on fair and feasible dam operation and water sharing agreements is now. The waiting game and dragging negotiations are not in the best interest of all parties.
vi. Every year the water demand is increasing; the uncertainty of available freshwater in the basin is growing adding additional layer of complicity to negotiation. Countries are advised to take a more cooperative and human-centered approach in dealing with precious natural resources like water.

References

Ahmed Y, Al-Faraj F, Sholtz M (2019) Assessment of upstream human intervention coupled with climate change impact for a transboundary river flow regime: Nile River basin. Water Resour Manage. https://doi.org/10.1007/s11269-019-02256-1

Arsano Y (2007) Ethiopia and the Nile: dilemmas of national and regional hydro-politics. Center for Security Studies, Swiss Federal Institute of Technology, Zurich

Blackmore D, Whittington D (2008) Opportunities for cooperative water resources development on the Eastern Nile: risks and rewards. In: An independent report of the scoping study team to the Eastern Nile Council of Ministers. The World Bank, Washington, D.C., 102 pp

Egypt (2020) Permanent Mission of the Arab Republic of Egypt to the United Nations, June 19, 2020. Letter to the United Nations Security Council on GERD. https://www.securitycouncilreport.org/atf/cf/%7B65BFCF9B-6D27-4E9C-8CD3-CF6E4FF96FF9%7D/S_2020_566%20Egypt%20letter%20of%2019%20June.pdf. Accessed 4 Feb 2021

NBI Nile Basin Initiative (2021) www.nilebasin.org. Accessed on 22 Feb 2021

Permanent Mission of the Arab Republic of Egypt to the United Nations (2020) Letter to the United Nations Security Council on GERD. https://www.securitycouncilreport.org/atf/cf/%7B65BFCF9B-6D27-4E9C-8CD3-CF6E4FF96FF9%7D/S_2020_566%20Egypt%20letter%20of%2019%20June.pdf. Accessed 31 Dec 2020

Salman SMA (2010) Downstream riparian can also harm upstream riparian: the concept of foreclosure of future uses. Water Int 35(4):350–364

Salman SMA (2013) The Nile Basin cooperative framework agreement: a peacefully unfolding African spring? Water Int 38(1):17–29
Shahin M (1985) Hydrology of the Nile basin, Developments in water science 21. Elsevier
UNDESA United Nations, Department of Economic and Social Affairs, Population Division (2019) World population prospects 2019, Online Edition Rev 1

Chapter 2
A Look into the Historical Depths of the Nile Waters: What to Learn from History

Valentina Acquafredda

Abstract The recent events concerning the GERD's construction lead to retrace the crucial moments in the history of the Nile, inserting them in the dynamics that have shaped the management of the river's water through the centuries. In the pre-colonial time, the turning point is identified in the Egyptian policy of cotton cultivation and the construction of the Suez Canal. The British conquest of Egypt and the competition among European powers within the Scramble of Africa had a strong impact on the governance of the Nile. The river became an object of hegemonic claim by downstream countries to the detriment of upstream ones, sanctioned by an unfair system of colonial agreements and treaties, which continued until the 1980s. Since the Nineties, efforts of cooperation between the countries of the basin have been carried out, which despite being steps forward are still poorly performative.

Keywords History of Nile water · Transboundary waters and power · Colonial Nile water treaties · Regional cooperation in Nile Basin · GERD

Introduction

The similarities between the flowing of the water river and history are common and suggestive, accumulated by the concept of the inexorability of time, of the change inherent in cyclicity and infinity. But about the latter one is in error, history is "an argument without end", according to Pieter Geyl's well-known definition (Schölch 1976, 773), but not water. On the contrary, its scarcity is increasing, making the risk of conflicts higher, as historians and scholars from different fields have warned for decades in the Nile Basin, a particular complex case of transboundary river, for number of countries and lives of people involved, among which there is no multilateral agreement recognized and in effect. It is a matter of solving the problem of

V. Acquafredda (✉)
Università Degli Studi Di Urbino Carlo Bo, Urbino, Italy
e-mail: v.acquafredda@campus.uniurb.it

A. M. Melesse et al. (eds.), *Nile and Grand Ethiopian Renaissance Dam*, Springer Geography, https://doi.org/10.1007/978-3-030-76437-1_2

the finiteness of water through the infinity of history, its ability to teach us that irreversibility is the condition only of the past, but not of the present and even less of the future. New approaches and new paths can be tried.

This chapter aims to investigate historically the relationships of power that have determined the management of the Nile waters. First, it is about placing the position of Egypt and of Ethiopia in the dynamics that there are usually between upstream and downstream riparian countries where the hydrologically weaker state is the stronger state. Secondly, it is intended to stress that, even before we could technically talk about water right claims, it was thought to control floods along the Nile. Thirdly, with the colonial regime, the river ceased to be the prerogative of the countries of its basin and became a matter of transboundary water that had to be regulated and bent to the will of colonial powers. In this space, it is intended to restore centrality to all agreements signed between 1891 and 1993, questioning the official sources. On the one hand, we analyse how colonial age agreements are a colonially imposed legal regime, so not valid; on the other hand, as well as in the post-colonial age, none of the treaties signed involved all the riparian countries. In this, the River Nile differs from other cases of transboundary water, but it did not prevent its waters from being the scene of a scramble. On the contrary, this has happened and continues to happen for more than a century with the construction and planning of dams, which are unilateral projects, not opportunities to be seized at basin level, of which GERD is only the last and most fiery act financed by Ethiopia (Abtew and Dessu 2019b). The asymmetric control of the Nile system by downstream countries, which ignored upstream riparian's rights, could not last forever. Indeed, it had to be redefined in favour of equitably sharing and utilizing.

The need for a rethinking of the status quo had already become necessary with the dramatic events of the 1980s, providing for the involvement of upstream countries, putting aside the aim of sovereignty over waters to a more comprehensive development dialogue, with the help of aid agencies and donors. But despite considerable progress, such as the establishment of the NBI, and the rectification of the UNWC, (United Nation Watercourses Convention), there is now a vacuum: there are no treaties that are being treated in practice as valid by all riparian countries. Finally, this chapter seeks to shed light on the fact that the colonial treaties, although not legally binding, continue to affect the present. This is not only because they are invoked by Egypt's nationalist policies and press as natural and historical rights, but because all the actors in the field are still moving in a logic of power relations, old and new.

The effects of climate change and population growth make it urgently necessary to find solutions not only to resolve *hic et nunc* the GERD impasse between Egypt, Ethiopia and Sudan, but specially to build regional cooperation between all the countries of the Nile Basin.

Structural Unbalanced Condition of Basins

The power relations are of two types: one of structural nature and the other specific of the Nile's history, which we will see in detail in the following paragraphs. The first refers to the relationship between Egypt as the "gift of the Nile" (Colonna and Bevilacqua 1996, 287; Griffiths 1966) and Ethiopia as the "water tower of Africa" (Swain 1997, 688), but also the "great unknown" (Swain 1997, 686). It is common in the dynamics between the upstream and downstream countries where the ideologically weaker state is the stronger state (Wouters 1992). Unless certain exceptions, in any dispute an upper basin state would appear to have a significant advantage with a lower basin state, because it could stop the flow before it enters a lower state, making the latter particularly vulnerable. But from several examples in the space and in the time, it is known to be the opposite. On the other side, for lower down countries it is easier to start development activities: irrigation, farming, urbanization and navigation. That lets lower basin to develop earlier and faster than the upper basin countries, for who the water use has been extremely difficult, if not impossible, in terms of technology and money. For millennia, upper basin state did not have the technology to achieve anything more than a short-term interruption of the water flow to lower basin states (Dellapenna 1996, 1997). The Nile Basin illustrates this pattern of water resources development.

There is an example in the history of Egypt–Ethiopia relationship, in the early decades of the fourteenth century during the reign of the Christian emperor "Amdä Seyon" (1312–1342) and the Mamluk sultan of Egypt Al-Nasir Muhammäd ibn Qala'un (1285–1341). The Sultan began to persecute Egyptian Copts and to destroy their churches, actions to which the emperor reacted by sending a dispatch to Cairo in 1321–1322, with which he threatened to persecute Muslims in his lands and to divert the course of the Nile River preventing the water from reaching Egypt. Al-Nasir Muhammäd ibn Qala'un was not at all shaken by these words, knowing that the enemy did not possess the necessary technological ability to carry out his plan, but "Fear that the Ethiopian might tamper with the Nile was nevertheless to remain with Egyptians for many centuries" (Pankhurst 1997, 41).

A similar episode is recounted by Jean de Lastic who had reported to Charles VII, in a letter of 1448, that, since Sultan continued to harass Christians, Prester John had intimated that the course of the Nile should be diverted. Such as this, there are many other historical episodes and legends, which became very frequent in the Middle Ages, concerning threats and dangers of damming the river narrated by Langer (1968), documented with bibliographic richness, precision and abundance of details. The first evidence of such a danger is contained in the Chronicles of the Saracenic Empire, written by El Macin in the thirteenth century, and tells how, at the end of the eleventh century, "the king of Abyssinia put aside the waters of the Blue Nile, until the Sultan of Egypt sent the Coptic patriarch with gifts". Also, Friar Jordanus in 1330 narrates the use of donations, as a deterrent, by the Sultan of Egypt to keep the Abyssinians quiet. Marignoli specifies that the latter "have the power to shut off the water, and then Egypt would perish". The same story is also told by Simone Sigoli in 1348, who repeats it adding that “every time this Prester Giovanni (i.e., the king of Abyssinia) chooses to open certain river locks he can drown Cairo, Alexandria and the whole country”. The same kind of fear is found in the 1422 tales of Guillebert de Lannoy and in Bertrandon de la Broquière’s one of 1432, in which it is written that Prester John could make the river change direction if he wanted to and refrained from doing so because Christians would starve to death along with the Muslims in Egypt. In 1450 is King Alphonse of Aragon to send a message to Abyssinia to attack Egypt and block the Nile to support the Aragonese naval campaign in the Holy Land. In the Commentaries of the Great Alfonso D’alboquerque is written that in the sixteenth century, at the time of his explorations in Sudan in the areas of Keneh and Kossair, the Alavers used the Nile to take revenge on the Sultan, taking advantage of the high water level to break through the bank and cause flooding in the valley. The theme enters also in the epic–chivalric poem by Ludovico Ariosto Orlando Furioso of 1516 (canto XXXIII, 106 octave): “The Soldan, King of the Egyptian Land,/Pays tribute to this sovereign, as his head,/They say; since having Nile at his command/ He may divert the stream to other bed./Hence, with its district upon either hand,/ Forthwith might Cairo lack its daily bread”. Another testimony goes back to James Bruce’s account of a letter written by the King of Abyssinia to the Pasha of Egypt in 1704 and expresses himself thus “The Nile would be sufficient to punish you, for God has placed its source, outlet and rise in our power, and we can dispose of it to do you harm” (Langer 1968, 103–105).

Therefore, this sense of insecurity was perceived since ancient times by the Egyptians, in turn used, manipulated and acted as an instrument of power by Ethiopia and will come until the end of the nineteenth century, but it was a feeling of threat, rather than a real danger. Indeed, it should be noted that, if the recent Amharic songs on the Nile convey that Ethiopians did not benefit from the Nile as well as they should (Getahun 2014), the old ones complained about the flood period of the river, the difficulty in managing them, because this was for a long time the main concern of all countries, downstream and upstream (Knobelsdorf 2006, 625–626).

From the end of the nineteenth century, however, other components, the technological and technical possibilities and the European powers, engaged in the Scramble for Africa, made the Egyptian’s ancestral and legendary scares real and made the course of the Nile’s history irreversible, at least until today. It is the irreversibility from which we must now emerge.

The Construction of Historical Unbalanced Condition of Basins

Pre-colonial Time

The foundations for change were laid by Muhammad Ali, ruler of Egypt 1805–1848, a key figure in the history of modern country. He was the commander of the

Ottoman Albanian forces who had led the conquest of Egypt on behalf of the Sultan, after Napoleon's French invasion (1798–1801), and was awarded the title of Wali of Egypt in 1805. He was determined to make the most of his power, especially considering the weakening of the Mamluks' central power, and seven years later, he had swept away the impediments to achieving his goals of expansion and modernization "to establish Egypt's effective independence of Constantinople" (Sayyid-Marsot 2001, 652).

Muhammad Ali's politics is indispensable to the analysis of Nile history for two reasons: the so-called Turkish–Egyptian conquest of Sudan and expansion of cotton production. At the base of the military campaign in Sudan from 1820 to 1824, there were several motivations, among others also the desire to "occupy the Sudanese ports of Sawakin (Suakin) and Massawa to strengthen Egyptian positions in the Red Sea" (Beška 2019, 33). Instead, Kendie (1999, 145) argues that there is a direct relationship between the conquest of Sudan and the intent to ensure Egypt's security and prosperity by conquering the provinces from which it receives its great reserves of water. And moreover, it was a stepping stone "to impose Egypt's will on Ethiopia, and either to occupy it or force it to give up the Lake Tana area" (Kendie 1999, 145). Native cotton was cultivated and used in Egypt since ancient times for clothing and for wrapping mummies, and during the Middle Ages people started to export it, but that had nothing to do with the production system inaugurated by Muhammad Ali. Whatever his real intentions were, whether he aimed specifically at controlling the waters of the Nile or whether the conquest was part of a broader policy of expansion aimed at the heart of the Ottoman Empire, the Wali had in fact extended the borders of Egypt to Sudan. This is by securing control of the waters of the Blue and White Nile, inaugurating a policy of expansion destined in later years to reach the territories of the Upper White Nile (Moore-Harell 2010, 11; Cleveland and Bunton 2009, 72) and beginning to put, at least in part, his country, whose entire existence depended entirely on those resources, away from ancient fears. Riaz Pasha, an Egyptian statesman, asserted in December 1888 how the Nile was life to the Egyptians and "Now the Nile means the Soudan, and no one will doubt that the ties and connections that unite Egypt to the Soudan are as inseparable as those that unite the soul to the body" (Langer 1968, 107). Okidi (2014, 178) traces the first agreement on the Nile, dated 12 October 1841, expressed in the form of a unilateral declaration, to make the river navigable, back to Muhammad Ali's government.

Thereby, it guaranteed access to water for irrigation, through dredging canals, constructing barrage storing system of water, the 600,000 acres of arable land and an army of peasants in a condition of semi-slavery, fellahin. It favoured the cultivation of varieties of long-staple cotton and mills operation, as a basis for Egypt industrialization, under the control and ownership of the state, following the meeting with the French entrepreneur of the cotton industry, Louis Jumel. The dimensions of his project, as well as in general his expansionist ambitions of conquer, since Muhammad Ali wanted to make Egypt a textile industrial power, were scaled down and condemned to failure by the European powers, who intervened against him twice. They were worried by the monopoly system introduced in Egypt and by a too weak Ottoman Empire on which loomed the threat of the

Russian Empire, and with the Anglo-Ottoman Treaty of Balta Liman in 1838 and the Treaty of London in 1841 (Cleveland and Bunton 2009, 74) integrated Egypt as a primary agricultural producer of row cotton for export.

Such policy was carried out by Sa'id, son of Muhammad Ali, who succeeded as Wali of Egypt from 1854 to 1863, to his brother Abbas (1848–1854), position that the conqueror of Albanian origin made hereditary giving birth to a dynasty that will remain in power until 1952. Ibrahim took advantage of the American Civil War and the consequent "cotton famine", which had struck among the European powers, especially England, to insert Egypt definitively within the global economy system, gaining a place in the front rank of cotton-producing countries.

In fact, between 1861 and 1863 there was a revolution in Egypt. While American cotton revenues had dropped to an insignificant fraction of normal imports, Egyptian cotton, preferred for its quality and also for socio-economic reasons to Indian one, starting from the Nile Valley ended up in the Lancashire cotton mills, which risked being completely paralysed without it. The level of exports grew feverishly, reaching about five hundred per cent between 1860 and 1865, and cotton exports increased from about thirty-six to about ninety-two per cent of total Egyptian exports (Earle 1926, 535). The relationship between England and Egypt became very close. The latter had become valuable for the European power, and this had heavy consequences on Egypt itself. From having a self-sufficient agriculture, it became a one-crop country dependent, contrary to the plans Muhammad Ali had for the region, forced to respect English qualitative and quantitative needs and to live up to their investments in the country.

The British were not the only ones to profit from the relations with Egypt, but also France. Since 1854, Sa'id entrusted the French engineer Ferdinand de Lesseps the concession to construct a canal across the Isthmus of Suez from the Mediterranean to the Red Sea, with a disastrous agreement, indeed fatal for Egypt. It contributed to increase the importance of the country in the eyes of the great powers and, internally, to exacerbate the social consequences of relations with the Europeans who became increasingly disliked by the Egyptians.

Those sentiments grew further under Khedive Ismail (1863–1879), a complex figure, who intended to make Egypt a European rather than an African power, obviously independent of the power of the Ottoman Empire, dedicated to greatness in terms of infrastructure, economic growth and cultural development, and in turn a conqueror. Under Ismail in 1869, work on the canal was completed in grand style. The huge revenues of cotton were useful, but not enough to support the costs of the project and the ambitions of the Khedive, for which he went into debt with the European powers, a choice that ended up becoming the Achilles' heel of his government and to mark irreparably the fate of Egypt (Cleveland and Bunton 2009, 81). Suez, meanwhile, quickly became an important international waterway benefiting all maritime states by reducing the distance and cost of transport of goods and passengers between the East and Europe, whose traffic tripled between the years 1876 and 1890, and almost all of those was British. In fact, it was Great Britain the first commercial nation at the time with possessions in India and the Far East, to draw the greatest benefit from the channel.

In addition, in 1875 Great Britain also ended up buying Egyptian 44 per cent interest in the Suez Canal Company for £4 million, necessary for the Khedive exploits, and a Dual Control (whereby two controllers, an Englishman and a Frenchman, came to control the Egyptian economy) was imposed in 1876. With a financial set-up already severely compromised, Ismail launched expeditions that seized the Red Sea coast to Cape Guardafui and conquered Kunama in 1869 in Northeast Ethiopia and Harare in the east in 1875. But in that same year between 14 and 16 November, Ismail was defeated by Emperor Johannes IV of Ethiopia in the Battle of Gundet and at Gura on 7–9 March 1876 with heavy losses by the Egyptians (Jesman 1959; Swain 1997, 676).

In the light of the considerations of the Khedive's army, Sir Samuel Baker and General Charles George Gordon (Langer 1968, 102–106) stated that one can see how Ismail's politics is more programmatic in securing full control and exclusive access to the Ethiopian great reserve of water. Sir Baker proved to be a great connoisseur of the Nile and its economic and political implications for the life of Egypt, to the point that he believed the biblical telling of the seven years of famine in Egypt (Genesis 41:1–7) referred to some diversion of the Nile water (Langer 1968, 105–106). General Gordon arrived at the upper reaches of the Nile, to Lake Albert and so-called Victoria Nile, and established posts in the provinces of Bahr-el-Ghazal and Equatoria, as well as in Unyoro and up to the very frontiers of Uganda, which, however, was difficult to penetrate from within. The latter was convinced that it was challenging to pass from the inside and suggested to the Khedive to take possession of the East African Coast and in particular of Mombasa to get to Uganda and for this reason Barawa and Kismayu were occupied, but then they stopped for internal reasons.

Moreover, the situation in Egypt was becoming increasingly worrying. From the financial point of view in 1876 there was a risk of bankruptcy, and loans had interest rates of 10% and in addition were asked to repay previous ones. The society, intolerant of European interference in Egyptian affairs and their privileges, was increasingly in conflict with itself, and Ismail tried to satisfy it with an extreme gesture, by firing the two French and English controllers (Cleveland and Bunton 2009, 99). That measure worried the European powers, Egypt's main creditors who, with the complicity of King Abdul Hamid II, decided to depose Ismail in 1879, in favour of his more malleable son Tawfiq (1879–1892) who guaranteed their full financial settlement through the Law of Liquidation. It was a move that led to the revolt headed by Ahmad Urabi, a colonel of peasant origin, (1879–1882), intent, to the claim of "Egypt to the Egyptians", to eliminate foreign control of Egypt's finances and to put a brake on the autocracy of the Khedive. Ismail, in turn, tolerated him and tried to integrate him into his government. In the eyes of European powers, all this further complicated the already precarious situation of the country stability. Security and free access to the canal could be jeopardized if the Urabi revolt took over the Khedive, as well as the payment of the huge debt incurred and in addition cotton production had declined and become qualitatively lower due to poor attention to agricultural management (Earle 1926, 539). Thus, after sending a joint Anglo-French naval force in May 1882, after the violence that broke out in Alexandria (June 1882), and the bombing of the city by the Royal Navy (11–13 July 1882), in the face of the refusal of intervention by the Sultan and the French, the British landed in Egypt, because of commercial interests and political considerations (Porter 2001, 632; Schölch 1976; Hopikins 1986).

They defeated the Egyptian nationalist forces in the Battle of Tel-el-Kebir (13 September) with an action that had to be momentary and aimed at re-establishing the power of the Khedive Tawfiq. "Despite the desire to restore and retire" (Collins 1967, 948), they departed from Egypt in 1952 and from that time "The dual watch on Egypt's accounts thus became a British watch on the Nile" (Newbury 2001, 634). One would be forgiven for wondering what this excursus on the history of Egypt from 1805 to 1882 has to do with the history of the Nile that has only been mentioned a few times so far? Here is the foundation for explaining how we go from the legends and threats of diversion of the Nile to the disputes over its control and to the crippling treaty system, that is, how the historical unbalanced condition of basins is constructed, starting from the appearance of some elements and their interconnections. Indeed, the security and control exerted by the expansion of the borders to present-day Sudan and South Sudan, and the importance of Muhammad Ali's cultivation of Jumel cotton for the British economy, as well as the Suez Canal, are the two key elements for fully understanding how, on the one hand, the British no longer left Egypt and how the Nile became their absolute priority, all while moving into the complex arena of colonialism in Africa.

Colonial Time

For Robinson and Gallagher, the Scramble for Africa began with the British occupation of Egypt, in "Africa and the Victorians. The Official Mind of Imperialism" (1961), Robinson and Gallagher, argue the Scramble for Africa began with the British occupation of Egypt. This position has been superseded by more recent historiographic analyses (see Louis 2006, 5). However, it was undoubtedly a significant episode and a turning point in the history of the Nile, even if not immediately.

As Napoleon had guessed, Suez was crucial to the British. "The immediate military objective of the French expedition was to strike at Britain's communications routes with India" (Cleveland and Bunton 2009, 65). The situation changed with the failure of the negotiations conducted for England by Sir Drummond Wolf (Hornik 1940), which led to the 1888 Convention of Constantinople to regulate the use of the canal. Due to the strong opposition of the other European powers, France and Russia, the English conditions, required by Articles IV and V, which formalized the English role of guardian for three years in Egypt and left open the possibility of their stay in case of internal or external danger, did not end up in the convention (Hornik 1940, 616–617). Instead, it was guaranteed the passage to all ships in times of both war and peace (Art.X), without any particular power to the Great Britain. "By then, the watch on the Nile had consequences for British partition elsewhere on the continent" (Newbury 2001, 635).

Though, before that, Egypt, and therefore England, had also lost Sudan, due to the Mahadist revolution that broke out in 1881. It started as a revolt with a strongly religious connotation, greatly underestimated, and faced with pale attempts at reconquest. And then, Maddiya became "a powerful and militant state that dominated Sudan for fourteen years" (Boahen 1985, 77), putting an end on 26 January 1885 to the Turkish–Egyptian rule in the Sudan.

By 1888, from the Egyptians, from the British on the spot, from Salisbury in London, there was a clear awareness of the danger: "The possibility that some power might control the Nile serves seemed to threaten Egypt's agricultural surplus, finances and debt servicing" (Newbury 2001, 639). But the British did not want to deal with it for economic and political reasons, and as long as the Mahdists controlled Sudan it was not a source of concern, because they did not have the necessary technology to worry Egypt.

Sir Baker wrote in 1888, although the evidence of his apprehension dates back to 1884, dreading that Sudan would become a land of conquest for others, "I can positively state the plan is feasible, and that, should any European be in command at the rebellious centre of the Sudan, his first strategic operation would be to deprive Egypt of the water that is necessary for her existence…If I were myself an enemy of Egypt I know the place where I should commence the fatal work upon River Atbara" (Langer 1968, 106). In 1892, a "graver anxiety behind" emerged in the words of Lord Miller, a British statesman, who spoke about the situation in the Soudan, "The savages of the Sudan may never themselves possess sufficient engineering skill to play tricks with the Nile, but for all that it is an uncomfortable thought that the regular supply of water by the great river, which is to Egypt not a question of convenience and prosperity but actually of life, must always be exposed to some risk, as long as the upper reaches of that river are not under Egyptian control. Who can say what might happen, if someday a civilized Power, or a Power commanding civilized skill, were to undertake great engineering works on the Upper Nile, and to divert for the artificial irrigation of that region the water which is essential for the artificial irrigation of Egypt?" (Langer 1968, 107).

Sir Colin Scott-Moncrieff, as head of work on the Nile, also wrote in 1895: "What the Mahdi could not do a civilised people do…I may be allowed to point out an evident enough fact, that the civilised possessor of the Upper Nile Valley holds Egypt in his grasp…A civilised nation on the Upper Nile would surely build regulating sluices across the outlet of the Victoria Nyanza and control the great sea as Manchester controls Thirlmere. This would probably be an easy operation. Once done, the Nile supply would be in their hands; and if poor little Egypt had the bad luck to be at war with this people on the upper waters, they might flood them, or cut off their water supply at their pleasure" (Langer 1968, 107–108).

Therefore, it was instead the competing imperialist European powers that needed to be kept at arm's length from Sudan. "Protection of the upper Nile-this became one of the keynotes of British policy in Africa, a policy which, from the very nature of the problem, had endless ramifications […]" (Langer 1968, 108).

Besides, the vision of the British Prime Minister, Salisbury, was clear: "We should insist on the command of all affluents on the Nile, so far as Egypt formerly possessed them" (Langer 1968, 110).

Thus, excluding momentarily the possibility of a military conquest of Sudan, while waiting to find the funds, the English weaved a system of agreements and treaties with both European powers and African rulers. In those official documents, they also included provisions and measures, in the form of articles, concerning the control of the waters of the Nile (Upper, White, Blue) and their use to define spheres of influence.

"The acquisition of a sphere of influence by treaty was usually the first stage in the occupation of an African state by a European power" (Boahen 1985, 33). The attacks British had to defend against could come from multiple directions, the Red Sea, the African East Coast and the Congo (Langer 1968, 108). For this reason, they started first of all by supporting the Italian conquest of Ethiopia, another land crucial for the security of the Nile and therefore of Egypt, so as to exclude the French, an operation that turned out to be more complex than expected. Italy under Crispi with the Uccialli Treaty of 1889 (Italian version) intended to reduce the entire territory of, that time, Abyssinia to an Italian protectorate, with an initiative that made the British fear expansionist aims on Sudan and the upper Nile. Thus, while Salisbury was dealing with the Italian Ambassador in London telling him "that Egypt had not given up their territorial claims, since England herself regarded the Nile Valley as vitally necessary to Egypt" (Langer 1968, 110), Crispi found no justification for such claims considering that "England had lost all prior claim to dominion in the Nile Valley when the Sudan was abandoned" (Langer 1968, 110).

However, in February 1891, the position of Italian Prime Minister passed into the hands of Antonio Starabba, Marchese di Rudinì, who was not interested in African affairs, more intent on not alienating the good relations with the Great Britain in view of the renewal of the Triple Alliance and obtaining Italian colonial recognition. Hence in Rome, on 15 April 1891, the protocol between the UK and Italy was signed to define their respective areas of influence in East Africa, without taking into consideration the interests of Ethiopia (Ferede and Abebe 2014). And its, specifically, Article 3 referred to the Nile: "The Italian Government engages to no construct on the Atbara, in view of irrigation, any works which might sensibly modify its flow into the Nile".

This is only the first of a succession of agreements, aimed to permit no work unless approved by the several interested states when water was relatively plentiful, enabling each state to trade off consent to projects in each state (Dellapenna 1996, 240). As the scholar goes on (to say), such agreements are common, and apparently of limited importance taken individually, "but cumulatively representing a high degree of joint development of a basin's water resources" (Dellapenna 1996, 240). And this is what happened in the Nile Basin. In the words of Mekonnen (2010, 423): "The advent of British colonialism in the basin brought with it a hegemonic plan geared towards controlling the entire basin with a view to ensuring the uninterrupted flow of the river downstream".

Authors usually make the first of the Nile treaties, designed to ensuring that there were no impacts and changes to the course and flow of the river's waters flowing into Egypt, and coincide with the Anglo-Italian treaty of 1891 (Garretson 1960; Kasimbazi 2010; Obengo 2016). But it is on 1 July 1890 the Anglo-German Treaty (Gray and Peters 1960), which "recognized the Upper Nile as falling within the British sphere of influence" (Uzoigwe 1985, 34). This treaty temporarily settled colonial disputes between Germany and Great Britain. It recognized Tanganyika as a German colony, and Heligoland, an island off the coast of Schleswig-Holstein in the North Sea, was ceded; in return, the Germans abstained from further encroaching into British Kenya and agreed the English colonial dominion on Uganda, Sultanate of Witu and their influence on Zanzibar.

Another treaty not brought into the common Nile's system framework is the Anglo-Congo Free State Treaty (12 May 1894). It "is equally significant because it settled the limits of the Congo Free State in such a way that it acted as a buffer between French territories and the Nile valley and provided for the British a Cape-to-Cairo corridor from Uganda via Lake Tanganyika; withdrawn in June because of German protest" (Uzoigwe 1985, 34). The treaty defined their spheres of influence in East and Central Africa along the Congo-Nile watershed. In Article II, the UK entrusted King Leopold with territories in the western drainage area of the Nile. "The Anglo-Congolese agreement […] was the culmination of the Foreign Office's efforts to seal off the Nile Valley by diplomacy" (Louis 2006, 65). It was immediately opposed by King Leopold and the French who were willing to be humiliated on the upper Nile, just as they had been humiliated on the lower Nile with the occupation of Egypt (Collins 1967, 949). Compared to what Hornik (1940, 623) wrote about the failed negotiations in Constantinople, the road to Fashoda was no longer dotted "by many roundabouts and detours", and it was now a straight line.

At the end of the nineteenth century, there was no more time for diplomatic manoeuvrers, and the British could no longer postpone their plans for supremacy over the Nile Valley and in 1896 the "savage and unnecessary bloodshed" (Uzoigwe 1985, 36). Anglo-Egyptian conquest or rather the reconquest of Sudan began. Having defeated the Sudanese forces, British met with the French in the summer of 1898 at Fashoda in Sudan, as the meeting point of the two different European directions of expansion. That "squalid, pestilential, and otherwise useless site of Fashoda" (Collins 1967, 948) was the symbol of the policy of control of the Nile. Whether we consider it a minor moment in the history of European colonialism, or whether we see it as the Waterloo of Africa, which inaugurated important cooperation between England and France, with the retreat of the latter from Fashoda, the English affirmed their hegemony over the Nile waters and Suez and guaranteed lifelines to India and the Orient (Collins 1967, 949). And about the conquered Sudan, it established the so-called dual flag policy and the Anglo-Egyptian Condominium, a solution particularly disliked by Egyptian nationalists, for whom the Sudan was more important to Egypt than Alexandria (Warburg 2007, 477) and which also did not delay in triggering a certain competition between the two countries downstream, especially for the cultivation of cotton (Swain 1997, 677).

In this new order of things, UK, now acting on behalf of Egypt and Sudan, entered into a colonial agreement involving the Nile, with Ethiopia on 15 May 1902. It aimed at establishing boundaries between Ethiopia and Sudan. Article 3 of the English version's Agreement prevented Ethiopia from using the waters of the Blue Nile or its tributaries that might impede the flow of the river to not damage Sudan. Also stated: "His Majesty the Emperor Menelik II, King of Kings of Ethiopia, engages himself towards the Government of His Britannic Majesty not to construct or allow to be constructed any work across the Blue Nile, Lake Tana, or the Sobat, which would arrest the flow of their waters in to the Nile except in agreement with His Britannic Majesty's Government and the Government of Sudan". As in the case of the Treaty of Uccialli, in Amharic there was a different agreement translation, or rather the omission of "the Government of the Sudan", mentioned however in Article IV alongside His Britannic Majesty's Government. "This omission might well become the source of potential disagreement between the contracting parties, and it is surprising that so obvious a discrepancy should have been allowed to pass" (Ullendorff 1967, 652). In addition, another important aspect concerns the fact that with the formula "not allow to be constructed", the British were referring to potential European competitors that could conquer Ethiopia, rather than Menelik II, underestimating the possibility that Ethiopia may acquire the skills to control the Nile with ancient Egyptian nightmares coming true (Ferede and Abebe 2014, 59).

Once the concerns about the Blue Nile were removed, the British moved on to securing the flow from the White Nile. Article 3 of the Treaty between the UK and Congo, (9 May 1906), amending the much-discussed Agreement of 12 May 1894, relating to the spheres of influence of Great Britain and the Independent Sate of Congo, intended to continue the British policy of guarantee and control over the waters. It eliminated the last competitor for control of the Southern Sudan, the Congo Free State itself, and provided: "The Government of the Independent State of the Congo undertake not to construct, or allow to be constructed, any work on or near the Semliki or Isango River, which would diminish the volume of water entering Lake Albert, except in agreement with the Soudanese Government". King Leopold renounced his claims to the territories of present-day South Sudan in exchange for maintaining the Lado Enclave for the duration of his reign and for certain commercial concessions, guaranteed interest on a railroad loan (Collins and Herzog 1961, 119).

The possibility for the French to build a railroad in the territories of Ethiopia that would reach their colony of Djibouti, at the confluence of the Red Sea and Gulf of Aden, was one of the hottest issues at the centre of a broader negotiation about the Ethiopia's fate after the supposed death of Menelik II. The negotiation involved the three European states with major interests in East Africa, France, England and Italy, committed to designing the future of the empire, excluding Ethiopia itself which rejected the Tripartite Treaty. In fact, after years of negotiations, on 13 December 1906, an agreement was reached between the UK, France and Italy, which, without altering Ethiopia's *status quo*, defined reciprocal spheres of influence in Article 4. It states: "[…] France, Great Britain, and Italy shall make every effort to preserve the integrity of Ethiopia. In any case, they shall concert together, […] in order to safeguard, among others: "a. The interests of Great Britain and Egypt in the Nile Basin, more especially as regards the regulation of the waters of that river and its tributaries (due consideration being paid to local interests), without prejudice to Italian interests […]".

The British, in the person of Harrington, Francophobe British minister in Ethiopia, initially feared that "France could hold Egypt hostage by the threat of diverting the flow of water from Lake Tana, the Ethiopian source of the Blue Nile's high season discharge" (Keefer 1981, 366), thus compromising British interests in the country. Eventually, "The British bought the good will of France extremely cheaply by accepting a railway which they did not have the financial resources to oppose and in the process staked out a future claim to a good part of Ethiopia including the British Empire's only primary strategic interest in Ethiopia, Lake Tana, the source of the Blue Nile" (Keefer 1981, 366, 380).

In the same direction, 19 years later in a context that had changed due to the consequences of the First World War, which had altered the structure of the colonies and protectorates, first of all making in 1914 Egypt a formal British protectorate, after the collapse of the Ottoman Empire, Great Britain continued "to pursue her interests in controlling the headwaters of the Blue Nile" (Kasimbazi 2010, 721) protecting the flow from any potential upstream diversions. For this reason, an agreement, produced in the form of an Exchange of Notes, was reached between Italy and the UK, on 14–20 December 1925, from which, violating its sovereignty, Ethiopia was once again excluded. The African country in the meantime, in September 1923, had become a member of the League of Nations to which it appealed to denounce the plots of the two colonial countries, which, to their great shame, were forced to acknowledge their actions. With such an agreement, Great Britain would have endorsed Italy in its attempt to obtain a concession from Ethiopia to construct a railway stretching from the border of Eritrea to the border of Italian Somaliland that was meant to intensify Italy's economic influence, and in return Italy made concessions to Great Britain and promised to help it obtain permission from Ethiopia to construct a dam on Lake Tana and her support in doing this and building a road from Sudan to Lake Tana for the transportation of goods and people (Ferede and Abebe 2014, 61).

The opening of the Note of the British Ambassador in Rome addressed to the Italian Prime Minister is in line with Great Britain's programmatic policy in Africa: "Your Excellency is well aware of the vital importance to Egypt and the Sudan of maintaining and, if possible, increasing the volume of water for irrigation purposes available in those from the Blue and White Niles and their tributary streams". But the aspect to underline is the Italian response: "The Italian Government, recognizing the prior hydraulic rights of Egypt and the Sudan engage not to construct on the head waters of the Blue Nile and the White Nile and their tributaries and effluents any work which might sensibly modify their flow into the main river".

For the first time in an official document, there is mention of hydraulic rights of Egypt and the Sudan, which will become Egypt's alone with the well-known Exchange of Notes between the UK and Egypt of 7 May 1929. It is the agreement constantly quoted next to that of 1959, when discussing the management of the Nile, the hydropower and in the summary history with which the Grand Ethiopian Renaissance Dam dispute is presented. Historically, its importance lies in the fact that it is first and foremost the first agreement that properly and entirely covers the use of waters of the Nile for irrigation, which does not end up related in a single article of a larger colonial agreement; moreover, it stood for 30 years, until 1959, and for the first time "it covered most of the Nile Basin" (Kasimbazi 2010, 722).

There was on the one hand the Egyptian government and on the other hand the British government that acted directly for the Sudan (Abdalla 1971), but in the meantime had extended its power in addition to Uganda, Kenya and Tanzania, the East African riparians to Lake Victoria, which had not been involved.

The Exchange of Notes of 1929 is in continuity with the first ones: it prohibits any kind of work, it continues to exclude Ethiopia, but it marks a turning point among the several treaties concluded between the colonial powers. It perpetuated the Egyptian ancient claims giving them a new configuration and legitimacy, thanks to the support of the British who consecrated definitively and officially Egypt as master of the Nile, investing it with the authority to exercise its hegemony over the entire basin.

The essence of the settlement is described in the paragraph 2(I): Egypt was in favour of supporting and complying with the requests of Sudan, in the increase of water needed for irrigation, "so far as this would not infringe on neither the natural and historical rights of Egypt on the waters of the Nile nor on its agricultural development needs subject to obtaining satisfactory assurances with regard to the protection of Egyptian interests". Echoing in paragraph 4, I letter: "no irrigation works shall be undertaken nor electric generators installed along the Nile and its branches nor on the lakes from which they flow if these lakes are situated in Sudan or in countries under British administration which could jeopardize the interests of Egypt either by reducing the quantity of water flowing into Egypt or appreciably changing the date of its flow or causing its level to drop".

And although Lord Lloyd, High Commissioner, on behalf of the British government, believed that the agreement "will certainly facilitate the development of Egypt and the Sudan and promote their prosperity" (Article 2, II letter), he was keen to confirm to Muhammed Mahmoud Pasha, Egyptian Prime Minister, as has been shown so far, that "H.M.G. in the United Kingdom considers the observance of these rights as a fundamental principle of the policy of Great Britain".

It was clearly an unbalanced agreement: Egypt was allocated 48 billion cubic metres of water per year, while Sudan was allocated only 4, with some 32 billion unallocated because the other riparian countries were not involved. The accord recognized Egypt's right to monitor upstream flows (a power of knowledge) and to undertake projects without the consent of upstream states and to veto any construction project that might adversely affect its interests (Kagwanja 2007, 324).

Nevertheless, beyond all this important and useful information to understand how the colonial legacy was built, a fundamental knowledge if we want to deconstruct and think and plan a different present and future from the point of view of cooperation and not of the claim of exclusive rights and powers, it is necessary to restore complexity to historical events, to probe memory not normalize it, to bring out its contradictions and oppositions so that history is not reduced only to an endless list of treaties.

In this regard, it should be reported what Crabitès (1929) wrote about the immediate Egyptian reactions to the agreement, which were not at all positive. Muhammed Mahmoud Pasha claimed his own choice, arguing "I, as an Egyptian, believe that the agreement on the waters of the Nile fully and completely safeguard Egypt's rights […] I am convinced that the agreement embodies the Egyptian point of view in regard to the waters of the Nile". But he was attacked first of all by the oppositions because the day in which the agreement was signed, the Parliament was not assembled and he did not have the authority to do so. The second reason for criticism was that the whole measure was based on the "good faith" of the British and the anti-European and, in particular, anti-British spirit was very strong in the country, especially on the part of the Wafd party. Far from grasping the gain in agricultural productivity and the advantageous political consequences, the agreement required an act of faith towards a power that, however benevolent, was a foreign and colonial one (Crabitès 1929, 147). An awareness, while Egypt's supremacy among African countries was being enshrined, was very strong among its inhabitants.

In addition, hydropolitics defined by the Exchange of Notes of 1929 included the enlargement of the Aswan Old Dam in 1933, built in 1889, and the construction of the Jebel Awliya Dam, near Khartoum on the White Nile, started in 1932 and completed in 1937, considered the counterpart of the Gezira project, for which the Sennar Dam on the Blue Nile was built in 1925.

The details of the construction of the Jebel Awliya Dam were entrusted to the Supplementary Agreement between the UK and Egypt, 1932. In fact, it was "an elaboration of the 1929 agreement on the Jebel Awliya" (Kasimbazi 2010, 723), but in Garretson's (1960, 140) opinion, alongside the 1959 agreement, one "of much greater importance in the creation of the conventional system of the Nile basin than the above-listed treaties to which the United Kingdom is party".

This Supplementary Agreement provided for the construction and maintenance of the Jebel Awliya Dam, for the storage of water for the use of Egypt, which for this reason financed it, and also provided for monetary compensation for the damage caused to Sudanese interests, which would be greater if it was also decided to enlarge it or carry out other constructions, but only after a coordinated and joint operation of Egypt and Sudan, in view of the recognition of their unity of purpose and strategy (Kasimbazi 2010, 723), reinforced by the exclusion of other countries.

An agreement that was intended to regulate the interests of countries "other" than Egypt and Sudan was the Agreement between Belgium and the UK, 22 November 1934, regarding the allocation of the waters of Kagera that is part of the upper Nile Basin and the main tributary of Lake Victoria, between Tanganyika and Rwanda–Burundi, which were once territories of German East Africa, placed under Belgian control by the Treaty of Versailles. Article 1 states "Water diverted from a part of a water course situated wholly within either territory shall be returned without substantial reduction to its natural bed at some point before such water course flow into the territory or at some point before such water course forms the common boundary" (Garretson 1960, 140). Words that according to Kasimbazi (2010, 723) show how "the agreement was aimed at ensuring that all riparian countries enjoy equal shares of the water and that there is no unreasonable interference in others' enjoyment of the water resource". If so, it would be a principle of fairness, which is a novelty in the Nile's system agreement. In any case, it helps to give an account of the breadth and scope of English interests and power to legislate in such a vast territory as the Nile Basin.

With Exchange of Notes between Egypt and UK, England returns to strengthen Egypt's hydro-hegemony on the Nile. This refers to four different Notes, contracted from 1949 to 1953 by Egypt and Great Britain acting for Uganda concerning the construction of the Owen Falls Dam on the outlet of Lake Victoria. This project was the closest to the British dream of the "century storage plan", designed to exploit the ideal position of Uganda, which being at the choke point in the Nile River represents strategic locations for large store facilities to smooth over the annual flow of the White Nile for the benefit of Egypt and the Sudan (Yohannes 2008, 121).

The Owen Falls Dam was intended to serve a dual purpose: irrigation purposes for Egypt and provide electric energy for Uganda. The construction, however, had to "not be prejudicial to the interests of Egypt in accordance with the 1929 Agreement on the waters of the Nile and shall not adversely affect the rate of the water supposed to pass through the dam in line with agreements which shall be concluded between the two Governments" (Notes of May 1949, paragraph 5). Such a condition had to be guaranteed by "an Egyptian irrigation expert so that he may be able to ascertain, through a personal on-the-spot examination, that the operation of this project is in line with the commitments undertaken [...]" (Notes of February 1949, paragraph 3), by Egyptian resident engineer of appropriate rank and his staff to track all construction activities implemented by the Ugandan Electricity Corporation (Notes of May 1949, paragraph 4) and by assurance that no action could be brought by the two parties without consultation (Notes of February 1949, paragraph 2; Notes of May 1949, paragraph 4). Moreover, it was foreseen, as in the case of Sudan, that Egypt would pay a compensation (Notes of July 1952, paragraph 2), as well as paying a fixed annual contribution to the expense of collecting and calculating the meteorological and hydrological data about the Equatorial Lakes by the Hydrological Department of Uganda (Notes of January 1950, paragraph 4).

Finally, the last aspect to emphasize regarding this agreement is that it is the last act of Egypt and Sudan under British control. Indeed, Egypt ended the monarchy and the formal protectorate in the fall of 1953 and then in 1956 it was the turn of Sudan. The latter did not agree to reconstitute itself into a single entity with Egypt, as it had been since the time of Khedive, and chose independence and with it the need to reformulate agreements with the Egyptian government concerning the Nile.

"In the 1950s, when the approach of the 'wind of change' made it evidently clear that the days of the British Nile empire were numbered, London tried to reach a kind of basin-wide agreement with Egypt based on a new and more equitable Nile Valley Plan, giving more attention to the water needs of the upstream countries still under British control, but they failed" (Tvedt 2009, 6). The agreement between Republic of the Sudan and United Arab Republic, of 8 November 1959, was "Ironically called the Agreement for the Full Utilization of the Nile" (Okidi 2014, 179).

But its full utilization, correcting the partial control guaranteed by the previous accord, concerned only the two now independent countries. It "was more equitable than the 1929 agreement" (Haynes and Whittington 1981, 18), reallocated 55.5 billion cubic metres to Egypt and 18.5 billion cubic metres to Sudan (Article 1). But still the agreement was not inclusive, the rest of the Nile Basin continued to be excluded, and it "assumes that upstream countries do not use any water" (McKenzie 2012, 582). It is taken into consideration in Article 5 that they can make "claim a share in the Nile waters", but they are not guaranteed any right; on the contrary, it reaffirms the centrality of Egypt and Sudan that have to express a common and shared vision about the possible requests. Their cooperation is further strengthened by the Permanent Joint Technical Commission (Article 4).

The symbol of this new and enhanced cooperative framework given the increase in available water resources was the construction of the Aswan Dam in Egypt and the Roseires Dam on the Blue Nile in Sudan, presented as "the first link of a series of projects on the Nile for over-years storage" (Article 2,1). These are impressive projects for the realization of which is expected to "the final transfer of the population of Halfa and all other Sudanese inhabitants whose lands shall be submerged by the stored water". This is the accord to which reference is always made, because of the consequences it has had in the basin, but now it should be banned in the policy and in the horizon of expectation of action of the countries as a reference: "The 1959 Agreement is virtually useless today considering the interconnected nature of contemporary water scarcity issues. Water degradation does not stop at political borders; therefore, the entire Basin's cooperation in sustaining the Nile's viability is imperative" (Wiebe 2001, 747).

The purpose of this lengthy excursus into colonial dynamics was twofold, on the one hand highlighting how "The British quickly acted on their excitement, creating a new reality that would have profound implications for inter-riparian relations long after their departure" (Yohannes 2008, 35) and on the other demonstrating how "the allocation of quantities of Nile water by Egypt and Sudan has been both recent and unilateral" (Okidi 2014, 179).

Post-colonial Time

The colonial legacy left on the management of the Nile waters was really heavy: the iniquitous system of agreements and the logic behind it, that is, the hegemonic control and the political competition.

Once also the other riparians gained independence between 1960 and 1963: "[…] a new era opened because suddenly, and for the first time in the long history of the river, nine sovereign states (ten, when Eritrea declared independence from Ethiopia (1991)) were responsible for using and sharing the Nile basin" (Tvedt 2009, 6–7). "The states have become eleven since South Sudan declared independence in 2011".

The upstream countries followed the example of Ethiopia, which had already denounced and unrecognized in the past the binding force of those bilateral treaties (Knobelsdorf 2006, 630). The riparians were not willing to accept the unfair conditions imposed by the colonial settlements, for which they were not consulted at all; indeed, they denied their validity outright. Nile Basin's countries have developed different positions and invoked different doctrines, from Harmon to the Nyerere Doctrine, regarding watercourses management and conflict by reference to the domestic and international law (Wiebe 2001, 747; Knobelsdorf 2006, 631–647; Kasimbazi 2010, 726–729; Salman 2013, 18; Ferede and Abebe 2014).

This debate was crucial in defining the governance of the Nile in the long run until today, although it did not come to produce a new and inclusive agreement that would include all the riverine countries, taking equal account of the needs and interests of each, basically because of the opposition of the downstream countries and especially Egypt.

Indeed, by exploiting the institutionalization through treaties of its own claims to "integrity of the river because of the priority of their use" (Dellapenna 1996, 247), still "Egypt was, throughout the post-colonial period, by far the most important actor on the Nile" (Tvedt 2009, 7), "Until recently, Egypt controlled Nile waters by retaining economic and military [...]" (Wiebe 2001, 747).

In order "to address the issue of water security", giving full effect to the Agreement of 1959, Gamal Abdel Nasser, when he was still the Egyptian Prime Minister, ordered the construction of the High Aswan Dam in 1960, with funds, at the end, of the Russians under Nikita Khrushchev. In the climate of the Cold War, the concerns of the Western powers for the Egyptian strategy of non-alignment induced them to withdraw funding for the monumental dam, a decision which determined the diplomatic fallout between Egypt and its western allies and the infamous Suez Canal crisis in 1956 (Abtew and Dessu 2019a, 160). After about two centuries, the nationalization of the canal for which everything had begun, it "was the death blow to Britain's Nile project" (Tvedt 2009, 5).

However, from the ashes of the British imperialist project, the Egyptian project of Nasser was born. His plans for greatness in his own country, to make it the "Japan of Africa" (Tvedt 2009, 5), met with the hydraulic mission of the Cold War period (Allan 1999, 2–3), in which the competition between the USA and the USSR also passed through the financing and construction of major works aimed at controlling nature, which in Egypt, as elsewhere, were also, among other things, in continuity with the broad legacy of engineering imperialism (Buchanan 1986, 513–514).

As was the case in colonial times, the management of the Nile Basin was being shaped by historical, political forces engaged in larger disputes and conflicts beyond its own waters. And the GERD also sees its origin in this conflicted period. At that time, the US Bureau of Reclamation revived the 1927 Ethiopian project to build a dam on Lake Tana, for which the J.G. White Engineering Corporation of New York conducted a feasibility study and estimated an expenditure of $20,000,000. The new study was completed in 1964 and had foreseen for the construction of 35 multipurpose projects, and 16 irrigation schemes for 439,440 hectares of land and 12 power projects could utilize 12 billion m^3, of water from the

Blue Nile, with a hydroelectric capacity three times the AHD (Kendie 1999, 149–150).

But the dream was destined to remain locked in the drawer because of the Nasser's aggressive expansionist policy to finally realize the dream of Ismail, to build the Unity of the Nile Valley. In fact, Jesman wrote in 1959 about the achievements of the last great Wazi and first Khedive of Egypt: "Just like the Egyptian extremists of today he was inflamed with the idea of the unity of the Nile valley from the great lakes to the delta under the green flag of Egypt" (Jesman 1959, 59).

Ethiopia was perfectly aware of this, as we can see from the words to Nasser in December 1956 by Melesse Andom, Ethiopian Ambassador in Sudan, suggested to him by Emperor Haile Selassie: "[…] We Ethiopians do not belong to your world, although like you we drink of the water of the Nile. You have military objectives. We do not know exactly what they may be, but we have no confidence in the strength of your armed forces" (Kendie 1999, 154).

The pax-egytiana, to ensure Egypt the monopoly of the waters of the Nile (Kagwanja 2007, 325), was also on the agenda of his successor, President Anwar Sadat, who had equally ambitious projects. In fact, first and foremost under his mandate was the construction of another major engineering work, Jonglei Canal in Sudan. This was developing, according to the 1959 Agreement, the Nile waters in Sudan and holding a claim on the anticipated increased flow and exploiting the good relations with Jaafar Nimeiri, the Sudanese President, whose regime he had saved twice. The canal, which would benefit Egypt and northern Sudan at the expense of Southern Sudan, was begun to be built in 1976, but they were suspended in 1984 due to attacks by the Sudan People's Liberation Army (SPLA). Secondly, Sadat designed the project to carry water through the Suez Canal to the Sinai Desert for irrigation (Kendie 1999, 159) and, if possible, to Israel (Kendie 1999, 157) after the Camp David negotiations (1978–9) had mended the rift with Western countries and inaugurated a friendly relationship with Israel. But it had undoubtedly complicated the situation between the countries bordering the Nile, first and foremost with Ethiopia.

In turn, Ethiopia was steadfast in defending its interests, as was evident in 1977, at the UN Water Conference in Mar de Plata, where its representatives declared "the sovereign right of any riparian state, in the absence of international agreement, to proceed unilaterally with the development of water resources in its territory" (Swain 1997, 687).

And in 1979, Mengistu Haile Marian, Dergue's leader, threatened to reduce the flow of the Blue Nile by building a dam on Lake Tana, in line with the secular power dynamics between the two upstream and downstream countries, illustrated in the previous pages. A danger related to Egypt's natural and structural vulnerability and to which the president responded with a warning, "If Ethiopia takes any action to block our right to the Nile water, there will be no alternative for us but to use force" (Swain 1997, 687; Tvedt 2009, 8).

The Nile continued to be the heart of Egyptian foreign policy or, as Kendie (1999, 158) defines it, the "sinister agenda" carried out from the 1960s until the early 1980s, aimed at further undermining Ethiopia's stability and economic resources through support to Eritrean and Somali fighters, and boycotting Ethiopian financial assistance from international and regional donors so that it would be definitively and irretrievably impossible to carry out the construction of a series of water projects along the tributaries of the Nile (Kendie 1999; Kagwanja 2007, 325).

Just during the 1980s, there were events, low rainfall contributed to an iconic global famine that dramatically affected Ethiopia in 1984–5, and an exceptional low flow on the Nile raised unprecedented concerns in Egypt in 1988 that contributed to the slow awareness of the problems of water scarcity and food insecurity, of their cross-cutting between downstream and upstream countries. Finally, the emergence of the new Ethiopian government after the fall of Mengistu's communist dictatorship created the conditions for fine-tuning a different perspective on water management (Nicol and Cascão 2011, 319–320).

The slogans on sovereignty over waters gave way to a more comprehensive and wider development dialogue, leading to a change at the end of the Nineties, at least in tones, postures and discourses, to efforts of cooperation.

As several authors point out (Kagwanja 2007; Mekonnen 2010; Salman 2013, 19), the first attempts by the Nile Basin countries to work together, going beyond the bilateral agreements that have shaped much of the history of river governance, date back as early as 1967 to the Hydromet project. It was followed by Kagera Basin Organization (KBO) in 1977; Undugu Project (UP) in 1983; TECCONILE Project in 1992; a new platform for dialogue, the Nile 2002 Conference in 1993; and Lake Victoria Environmental Management Project (LVEMP) in 1994. They were significant but not incisive moments, such as the Ethio-Sudanese Agreement regarding the Nile (1991) and the Ethio-Egyptian Framework Agreement (1993). These were the last act of the bilateral provisions, of whose performativity they had nothing. As much as they represented an important step forward, the fact that they were made separately invalidated both their meaning and their effectiveness, reflecting the different motivations between the countries and reduced to being more of a framework for future negotiation, rather than a binding document (Dellapenna 1997, 132; Arsano 2007, 102–103).

The foundation for the formal setting of cooperation among the Nile Basin riparians is laid in 1997 by the World Bank, United Nations Development Programme (UNDP), and other international donors aid agencies (Salman 2013, 19–20), also driven by the need to jointly address serious political threats such as the rise of Islamism, in Egypt the Muslim Brotherhood, in Sudan the more islamic orientation and in Somalia the state disintegration (Nicol and Cascão 2011, 320). Awareness of the seriousness of the problem can be seen in the words in 1995 of Ismail Serageldin, Vice-President of the World Bank: "The wars of the next century will be about water" (Kagwanja 2007, 322). A non-secondary role was played by the preparatory work of the Convention on the Law of the Non-navigational Uses of International Watercourses 1997 (UNWC), which, however, did not enter into force until 17 August 2014. Thus, Nile Basin Initiative (NBI) was born on 22 February 1999; for the first time, nine of the 10 countries of the Nile Basin, except Eritrea, were together and shared a "new orthodoxy" that leveraged the concept of equitable water and benefit sharing. The institutionalization of the NBI creates the basis for WB to finance "through lower and easier to achieve levels of agreement", two subsidiary action programmes (SAPs) at the subbasin level in Addis Ababa and in Kigali, joint water projects, until then inhibited by World Bank Policy when there was opposition from other countries. "Donors, particularly some bilateral, the World Bank and the African Development Bank- were encouraged to achieve 'quick win' development projects which would, they believed, provide examples of confidence building to strengthen the wider NBI" (Nicol and Cascão 2011, 320). But the goal was for these principles to be incorporated into a truly inclusive agreement, which was only achieved in 2010: Cooperative Framework Agreement (CFA), signed by 6 countries, if we consider the entry of South Sudan in 2012, alongside Ethiopia, Tanzania, Uganda, Rwanda, Kenya and Burundi, but not by Egypt, Sudan and the Democratic Republic of Congo. This had ended up drawing the same divisions between downstream and upstream countries that had prevented the rectification of the UNWC in 2012 (Salman 2007), further complicating the picture with the controversial concept of water security, despite the 15-year-long efforts of mediation and formal process of water and hydropolitical cooperation (Salman 2013, 19–20).

Actually, behind Nile cooperation, there were projects financed by bilateral agreements from China (Verhoeven 2020) or other Asian countries, such as Merowe Dam in Sudan and Tekeze Dam in Ethiopia, or domestic financing of the Toshka project in Egypt, national logics and interests that continued to prevail over a regional vision, aimed at maximizing hydropower and reducing water losses in the river and basin (Nicol and Cascão 2011, 321–322). Basically, they continued to build on the Nile with a plethora of projects smaller than in the past with less individual capital costs, the so-called third phase identified with foresight by Haynes and Whittington (1981), without reaching a total agreement, which instead, also by virtue of this moving trend, was increasingly necessary, but keeping to act in the unique context of the Nile, among the transboundary river in which "is the large number of agreements concluded, but not recognized, by one or more of the parties" (Salman 2016, 512).

Finally, the lack of a real "working jointly and cooperatively countries" (Nicol and Cascão 2011, 321; Chen and Swain 2014, 16) became evident with the three events of 2011, which changed the balance in the region. First, the secession referendum of South Sudan in January planned already in 2005, in the Sudanese Peace Agreement, rectified the separation of the two states and this meant for upstream states an additional partner in negotiations and agreements. The second event concerns the climate of political uncertainty in Egypt triggered by the Revolution and President Hosni Mubarak's deposition, also in January 2011. Third, following the "strategic" announcement in April of the construction of GERD by Ethiopia, the fact that Egypt and Sudan did not accept Ethiopia's proposal to make the dam a project owned, funded and operated by three countries. In words of Salman (2016, 525), "Absence of cooperation and the narrowly, and poorly, perceived national interests aborted that proposal. Lots of waters have flown over this proposal, and it is now clear that time has surpassed it".

Conclusion

What remains of this ride through the centuries? The umpteenth reconstruction through primary and secondary sources, which studies, deconstructs and reconstructs only a part of the immense bibliographic wealth on the subject stratified over time? History when dealing with technical or interdisciplinary topics serves first and foremost to create context. But the history of the Nile, now more than ever caught up in the search for technical solutions, cannot be reduced to acting as a backdrop for discourses and actions aimed at building cooperation among the countries of the Nile Basin and preventing rather than resolving conflicts in the region. When we seem to know everything about the succession of events and their connections, it is then that we begin to betray the sense of history, to which we must never stop asking questions and seeking advice. And from the analysis that has emerged in these pages, there are three lessons that must be drawn and kept in mind.

Firstly, the change as a primary natural and historical condition, in the words of Mekonnen (2010, 440): "The truth though must be told forcefully, that it is time for Egyptians to let go of the wrong belief 'that their country will have the right forever, *ad vitam aeternam*, to all of the water carried by the Nile, as at the time of the Pharaohs".

Secondly, the need to definitively abandon logics of supremacy and nationalism, so as not to fall into the populist trap of perpetrating revenge and reproducing asymmetrical power relations: "Upper-riparian states must not give in to the temptation of payback and resort to a version of the ancient Harmon Doctrine" (Okidi 2014, 178), because "Nothing is sadder than seeing history repeat itself" (Todorov 1992, 298).

And thirdly that GERD is an act still in the making and if despite the escalations of the last year, as is evident in Fantini's words (2020), "In the past Ethiopian politicians presented the GERD as an African project, benefiting the entire region. Nowadays the nationalistic pride trumps all alternative narratives, as exemplified by the hashtag #ItsMyDam that went viral on social media [...]" could finally represent the renaissance of the Nile Basin region, or rather its birth to all intents and purposes.

Notes: Official Documents

- The Anglo-Italian Protocol was signed in an attempt by Italy and Great Britain in the attempt of legitimizing their capture of parts of East Africa.
- A Treaty Between Ethiopia and Great Britain on the Delimitation of the Frontier between Ethiopia and Sudan.
- Agreement between Great Britain and the Independent State of the Congo, modifying the Agreement signed at Brussels, May 12, 1894, relating to the

Spheres of Influence of Great Britain and the Independent State of the Congo in East and Central Africa—Signed at London, May 9, 1906.
- Agreement between the United Kingdom, France, and Italy, Respecting Abyssinia, signed at London, December 13, 1906.
- Agreement between the Republic of the Sudan and the United Arab Republic for the full utilization of the Nile waters signed at Cairo, 8 November 1959.
- Convention between Great Britain, Germany, Austria-Hungary, Spain, France, Italy, The Netherlands, Russia and Turkey, respecting the free navigation of the Suez maritime canal signed at Constantinople, 29 October 1888.
- Exchange of Notes between Great Britain and Italy.
- Exchange of Notes between Her Majesty's Government in the United Kingdom and the Egyptian Government on the Use of Waters of the Nile for Irrigation.
- Exchange of Notes constituting an agreement between the Government of the United Kingdom of Great Britain and Northern Ireland (on behalf of Uganda) and the Government of Egypt regarding cooperation in meteorological, and Hydrological surveys in certain areas of the Nile Basin Parties: Egypt, Great Britain (Uganda).

References

Abdalla IH (1971) The 1959 Nile waters agreement in Sudanese-Egyptian Relations. Middle East Stud 7(3):329–341

Abtew W, Dessu SB (2019a) Chapter 9: Dialogue and diplomacy through the construction of the grand Ethiopian renaissance dam. In: The grand Ethiopian renaissance dam on the Blue Nile, pp 131–146

Abtew W, Dessu SB (2019b) Chapter 11: financing the grand Ethiopian renaissance dam. In: The grand Ethiopian renaissance dam on the Blue Nile, pp 161–169

Allan JA (1999) The Nile basin: evolving approaches to Nile waters management. SOAS water issues group occasional paper 20, June

Arsano Y (2007) Ethiopia and the Nile dilemmas of national and regional hydropolitics. ETH, Zurich

Beška E (2019) Muhammad Cali's conquest of Sudan (1820–1824). Asian Afr Stud 28(1):30–56

Boahen Adu A (1985) General history of Africa. Africa under colonial domination (1880–1935). First. Vol VII. VIII vol. Heinemann-California-UNESCO

Buchanan RA (1986) The diaspora of British engineering. Technol Cult 27(3):501–524

Chen H, Swain A (2014) The grand Ethiopian renaissance dam: evaluating its sustainability standard and geopolitical significance. Energy Dev Front 3(1):11–19

Cleveland WL, Bunton M (2009) A history of the modern Middle East, 4th edn. Westview Press, Philadelphia

Collins R, Herzog R (1961) Early British administration in the Southern Sudan. J Afr Hist 2 (1):119–135

Collins RO (1967) Sanderson (G. N.). England, Europe and the Upper Nile, 1882–1899. Revue belge de philologie et d'histoire 45(3):947–954

Colonna A, Bevilacqua F (1996) Le Storie di Erodoto. In. Vol. II. Libri V-IX. UTET, Torino

Crabitès P (1929) The Nile waters agreement. Foreign Aff 8(1):145–149

Dellapenna JW (1996) Rivers as legal structures: the examples of the Jordan and the Nile. Nat Resour Prog 32(2)

Dellapenna JW (1997) The Nile as a legal and political structure. In: The scarcity of water: emerging legal and policy responses, pp 121–134

Earle EM (1926) Egyptian cotton and the American civil war. Polit Sci Q 41(4):520–545

Fantini E (2020) The grand Ethiopian renaissance dam: from diplomatic deadlock to nationalistic lockdown, ISPI. https://www.ispionline.it/en/pubblicazione/grand-ethiopian-renaissance-dam-diplomatic-deadlock-nationalistic-lockdown-26040. Accessed 22 May 2020

Ferede W, Abebe S (2014) The efficacy of water treaties in the Eastern Nile Basin. Afr Spectr 49 (1):55–67

Garretson AH (1960) The Nile river system. In: Proceedings of the American society of international law at its annual meeting (1921–1969) 54(April):136–144

Getahun BE (2014) A Content Analysis of Amharic Songs on Nile River. Journal of Arts and Humanities (JAH) 3(7):92–104

Gray J, Peters C (1960) Anglo-German relations in Uganda, 1890–1892. J Afr Hist 1(2):281–297

Griffiths JG (1966) Hecataeus and herodotus on "A gift of the river." J near East Stud 25(1):57–61

Haynes KE, Whittington D (1981) International management of the Nile. Stage three? Geogr Rev 71(1):17–32

Hopikins AG (1986) The victorians and Africa: a reconsideration of the occupation of Egypt, 1882. J Afr Hist 27(2):363–391

Hornik MP (1940) The mission of Sir Henry Drummond-Wolff to constantinople, 1885–1887. Engl Hist Rev 55(220):598–623

Jesman C (1959) Egyptian invasion of Ethiopia. Afr Aff 58(230):75–81

Kagwanja P (2007) Calming the waters: the East African community and conflict over the Nile resources. J East Afr Stud 1(3):321–337

Kasimbazi EB (2010) The impact of colonial agreements on the regulation of the waters of the River Nile. Water Int 35(6):718–732

Keefer EC (1981) Great Britain, France, and the Ethiopian tripartite treaty of 1906. Q J Concerned Br Stud 13(4):364–380

Kendie D (1999) Egypt and the hydro-politics of the Blue Nile River. Northeast Afr Stud 6(1/2):141–169

Knobelsdorf V (2006) The Nile waters agreements: imposition and impacts of a transboundary legal system. Columbia J Transnational Law 44(2):622–648

Langer WL (1968) The diplomacy of imperialism, 1890–1902, 2nd ed. Alfred A. Knopf, New York, pp 100–118 (First published 1935)

Louis R (2006) Ends of British imperialism. In: The scramble for empire, suez and decolonization. I.B. TAURIS, New York

McKenzie SO (2012) Egypt's choice: from the Nile basin treaty to the cooperative framework agreement, an international legal analysis. Transnational Law Contemp Probl 21:571–599

Mekonnen DZ (2010) The Nile basin cooperative framework agreement negotiations and the adoption of a 'water security' paradigm: flight into obscurity or a logical cul-desac? Eur J Int Law 21(2):421–440

Moore-Harell A (2010) Egypt's African empire; samuel baker, charles gordon and the creation of equatoria. Sussex Academic Press, Eastbourne

Newbury C (2001) Great Britain and the partition of Africa, 1870–1914. In: The oxford history of the British empire. Oxford University Press, New York, pp 624–650

Nicol A, Cascão AE (2011) Against the flow new power dynamics and upstream mobilisation in the Nile Basin. Rev Afr Polit Econ 38(128):317–325

Obengo JO (2016) Hydropolitics of the Nile: the case of Ethiopia and Egypt. Afr Secur Rev 25 (1):95–103

Okidi CO (2014) Nile waters: the threat of war is not justifiable in modern times. Environ Policy Law 44(1/2):176–180

Pankhurst R (1997) The Ethiopian borderlands: essays in regional history from ancient times to the end of 18th century. The Red Sea Press Inc., Asmara

Porter A (2001) The oxford history of the British empire, Vol III. The nineteenth century, vol IV. Oxford University Press, New York

Salman SMA (2007) The United Nations watercourses convention ten years later: why has its entry into force proven difficult? Water Int 32(1):1–15

Salman SMA (2013) The Nile basin cooperative framework agreement: a peacefully unfolding African spring? Water Int 38(1):17–29

Salman SMA (2016) The grand Ethiopian renaissance dam: the road to the declaration of principles and the Khartoum document. Water Int 41(4):512–527

Sayyid-Marsot AL (2001) The British occupation of Egypt from 1882. In: The oxford history of the British empire. Oxford University Press, New York, pp 651–664

Schölch A (1976) The "Men on the Spot" and the English occupation of Egypt in 1882. Hist J 19 (3):773–785

Swain A (1997) Ethiopia, the Sudan, and Egypt: the Nile river dispute. J Mod Afr Stud 35(4):675–694

Todorov T (1992) La conquista dell'America. Il problema dell'<<altro>>, 2nd ed. Einaudi, Torino (First published 1982)

Tvedt T (2009) The river Nile in the post-colonial age conflict and cooperation among the Nile basin countries. I.B. TAURIS, New York

Ullendorff E (1967) The Anglo-Ethiopian treaty of 1902. Bull Sch Orient Afr Stud Univ Lond 30 (3):641–654

Uzoigwe GN (1985) European partition and conquest of Africa: an overview. In: General history of Africa. Heinemann-California-UNESCO, pp 19–44

Verhoeven H (2020) Bis is beautiful: Megadams, African water security, and China's role in the new global economy. Global Water Forum. https://globalwaterforum.org/2012/10/16/big-is-beautiful-megadams-african-water-security-and-chinas-role-in-the-new-global-political-economy/. Accessed 22 May 2020

Warburg GR (2007) The search for the sources of the White Nile and Egyptian-Sudanese relations. Middle East Stud 43(3):475–486

Wiebe K (2001) The Nile river: potential for conflict and cooperation in the face of water degradation. Nat Resour J 41(3):731–754

Wouters PK (1992) Allocation of the non-navigational uses of international watercourses: efforts at codification and the experience of Canada and the United States. Can YB Int Law 30:43–88

Yohannes O (2008) Water resources and inter-riparian relations in the Nile basin: the search for an integrative discourse. State University of New York, Albany

Chapter 3
The Grand Ethiopian Renaissance Dam and the Revival of the Dispute Over the Colonial Nile Water Treaties

Mahemud Eshtu Tekuya

Abstract For over five years, Ethiopia, Sudan, and Egypt have been conducting a series of negotiations over the filing and annual operation of the Grand Ethiopian Renaissance Dam (GERD) but failed to strike a way forward deal acceptable to all of them. The recent involvement of the USA and World Bank in the negotiation further complicated the dispute and resulted in diplomatic crises. Although international and regional institutions, such as the United Nations Security Council (UNSC), the African Union (AU), and the European Union (EU), have recently been involved in the negotiations, the three states have made no progress and yet to agree on several outstanding issues. This doctrinal legal research, using an evaluative legal research model, expounds the ramifications of the colonial and 1959 Nile Treaties (colonial Nile Water Treaties) and argues that they comprise the principal obstacle to the GERD negotiations. Building on the disagreement over the baseline for the GERD's impact studies, and examining the sticking points during the Washington, D.C., negotiations, the research indicates how the dispute over the colonial Nile Water Treaties is reviving and affecting the GERD negotiations. Finally, the research calls upon the three states to address the problems associated with the colonial Nile Water Treaties by limiting the scope of the current negotiations on the filling and annual operation of the GERD and leaving the long-term operation of the dam, water allocation, and dispute resolution for the Cooperative Framework Agreement (CFA).

Keywords Nile · GERD · Washington · D.C. · GERD negotiations · Declaration of principles · Cooperative framework agreement · Nile colonial treaties · Equitable utilization · International watercourses law

M. E. Tekuya (✉)
McGeorge School of Law, Sacramento, CA, USA

A. M. Melesse et al. (eds.), *Nile and Grand Ethiopian Renaissance Dam*, Springer Geography, https://doi.org/10.1007/978-3-030-76437-1_3

Introduction

Ethiopia contributes 85–90% of the Nile waters through three main tributaries: the Blue Nile, Sobat (Baro-Akobo), and the Atbara (Tekeze and Angereb) rivers (Nile Basin Initiative 2012). However, it has been unable to use the Nile waters for a long time due to a combination of factors, including a lack of capacity and resources (Yihdego 2017) and the Egyptian coercive hydro-hegemony (Tekuya 2018). Recently, after attaining stability and some economic progress, Ethiopia has started using the Nile and is now constructing a giant hydrologic project—the Grand Ethiopian Renaissance Dam (GERD)—on the Blue Nile. The GERD is designed to generate 5150 megawatts of electricity from thirteen turbines and will have an enormous reservoir that can hold seventy-four billion cubic meters (bcm) of water (Ezega News 2019). Solely financed by Ethiopia, the GERD is expected to provide electricity access for estimated sixty-five million Ethiopians. It will also support Ethiopia's development endeavors through energy export and serve to lift millions of people out of poverty (Yihdego 2020).

Ethiopia believes that the dam will not have a significant adverse impact on the two downstream states, Egypt and Sudan. On the contrary, Ethiopia contends the GERD confers enormous benefits to Egypt and Sudan, including a more regular flow of water, better siltation prevention, a reduction in evaporation, and cheaper electricity (Tawfik 2016). In a historic break with its past practice of moving in lockstep with Egypt, Sudan has supported the GERD since 2012 because of these benefits. Egypt, on the other hand, maintains its view that any upstream dam on the Nile River poses an existential threat (Tawfik 2015; Salman 2016). But in 2015, after painstaking negotiations, Egypt recognized the importance of the dam (Salman 2016), and the three states—Ethiopia, Sudan, and Egypt—signed the Declaration of Principles (DoP) on the GERD.

After the DoP, the three states negotiated over the filling and annual operation of GERD for about five years but failed to strike a way forward deal acceptable to all of them. In August 2019, Egypt submitted proposals on the filling and operation of the dam (Addis Standard 2019; Al Jazeera 2019a) and later effectively internationalized the GERD negotiations (Daily News Egypt 2019) by involving the US government and the World Bank as observers (US Department of the Treasury 2019). The three states have held meetings with the US Department of Treasury and the World Bank's representatives in both Africa and Washington, D.C. The Washington, D.C., talks, which were progressing well at first (Wirtschafter 2019), took a turn for the worse in January 2020, resulting in a stalemate (Widakuswara 2020; Eyssen 2020). The USA, evidently going beyond its status as an observer in the talks, proposed an agreement that adversely affects Ethiopia's national interest. Ethiopia rejected the proposal and withdrew from the last meeting held in Washington, D.C. (Meseret 2020).

The US Department of the Treasury requested that Ethiopia signs the proposed agreement and cautioned Ethiopia to refrain from testing and filling the GERD without an agreement with Egypt and Sudan (US Department of the Treasury

2020a). Ethiopia expressed its disappointment with the statement and announced that it would proceed with filling of the reservoir in parallel with the construction of the dam as agreed in the DoP (London Embassy of Ethiopia 2020). Egypt, on the other hand, signed the USA's proposal and vowed to protect its interests in the Nile River "by all available means" (Haaretz 2020).

Although disguised in talks over the GERD's filling and operation, the current tension between Ethiopia and Egypt is principally related to their long-standing disagreement over the validity of the 1902 Anglo-Ethiopian Treaty, the 1929 Anglo-Egyptian Treaty, and the 1959 Nile Treaty between Egypt and Sudan (-collectively, the "colonial Nile Water Treaties"). This disagreement—which reached an apex during the negotiation of the Cooperative Framework Agreement (CFA)—is adversely impacting the GERD negotiations.

The next part, "Background: Dispute Over the Colonial Water Treaties" briefly introduces the disputes over the colonial Nile Water Treaties. After indicating how the CFA negotiations solidified the disputes, it addresses the interplay between the colonial Nile Water Treaties and the DoP and submits that the DoP does not abrogate the colonial Nile Water Treaties. Part III analyzes the implications of the colonial Nile Water Treaties for the post-DoP talks on the filling and annual operation of the GERD. In particular, it shows the revival of the dispute over the colonial Nile Water Treaties by discussing the three states' disagreement over the baseline data for the GERD's impact studies, scrutinizing Egypt's proposal for the filling and operation of the dam, analyzing the sticking points in the Washington, D. C., process, and evaluating Egypt's arguments in the United Nations Security Council (UNSC). Part IV, "Post-DoP Negotiations and the Revival of Dispute Over Colonial Nile Water Treaties", proposal for the way forward is presented and calls upon Ethiopia, Sudan, and Egypt to address the ramifications of colonial Nile Water Treaties by limiting the scope of the current negotiations on the filling and annual operation of the GERD and leaving the long-term operation of the dam, water allocation, and dispute resolution for the CFA. Section "Conclusion" provides concluding remarks.

Background: Dispute Over the Colonial Water Treaties

The legal regime governing the Nile Basin is fragmented. The basin does not have a mutually acceptable legal framework applicable to all riparian states (McCaffrey 2019). Currently, three types of legal instruments—bilateral treaties, a multilateral agreement establishing a framework for cooperation, and the DoP—are relevant to the use and allocation of Nile waters (Tekuya 2019).

The Nile Colonial Treaties

Several bilateral treaties exist between riparian states and/or their former colonial powers concerning the flow of the Nile since the end of the nineteenth century (Mekonnen 2011; Salman 2013; Amdetsion 2007). Of these bilateral treaties, the 1902 and 1929 colonial treaties and the 1959 Nile Treaty between Sudan and Egypt are the most widely disputed (Salman 2013). They are referred to as the colonial Nile Water Treaties, although not all were entered into during colonization.

First, the 1902 Anglo-Ethiopia Treaty is a bilateral treaty concluded between Britain, on behalf of Sudan, and Ethiopia to determine the boundary between Ethiopia and Sudan. Although the agreement is about boundary delineation, it contains a provision relating to the waters of the Nile that requires Ethiopia "not to construct, or allow to be constructed, any work across the Blue Nile, Lake Tana, or the Sobat which would arrest the flow of their waters into the Nile except in agreement with His Britannic Majesty's Government and the Government of the Soudan" (1902 Anglo-Ethiopian Treaty).

The Anglo-Ethiopian Treaty is the source of a bitter dispute between Ethiopia, Sudan, and Egypt. Egypt has considered itself as a successor of this treaty and so claimed that Ethiopia must get Egypt's consent to build any project in the Nile. Ethiopia, on the other hand, rejects this, claiming, *inter alia*, that it was not ratified by Ethiopia and that the meaning of the word arrest in both the Amharic and English versions does not preclude Ethiopia from using the waters (Abdo 2004; Ibrahim 2011; Salman 2013).

Second, the 1929 Anglo-Egyptian Treaty is a bilateral treaty between Egypt and Britain, representing Sudan and its East African colonies (Kenya, Uganda, and Tanganyika). The agreement, recognizing what it called the historical and natural rights of Egypt, gave Egypt veto power over any construction projects along the Nile and its tributaries (1929 Anglo-Egyptian Treaty, art. IV, ¶ 2.). It also allocated a volumetric quantity of water to each state: forty-eight billion cubic meters (bcm) for Egypt and four bcm for Sudan. In so doing, it determined the amount of waters each state received, which the 1959 Nile Treaty then used as the "established rights" of the two states (1959 Nile Treaty, art. I).

In 1962, former British East Africa colonies Kenya, Tanzania and Uganda, adopting the Nyerere doctrine, declared that they were no longer bound by this treaty. Egypt, meanwhile, has continued to deem the treaty as valid and binding on all parties under the principle of state succession (Ibrahim 2011; Amdetsion 2007).

Third, the 1959 Nile Treaty is a bilateral treaty between Egypt and Sudan. As the 1929 Anglo-Egyptian Treaty determined the "established right[s]" of the two states, the 1959 Nile Treaty allocates only the net benefit generated from the construction of High Aswan Dam ("HAD"). Of the 32 bcm gross gain expected after the construction of HAD, the agreement deducts 10 bcm for evaporation and seepage losses and divides the remaining 22 bcm into a 2:1 ratio in favor of Sudan or 14.5 bcm for Sudan and 7.5 BCM for Egypt.

Table 3.1 1959 Nile Treaty water allocations

Mean natural river supply at Aswan	85 bcm
Less over-year storage losses	−10 bcm
Egypt's established right	−48 bcm
Sudan's established right	−4 bcm
Total net benefit	22 bcm

As indicated in Table 3.1, the 1959 Nile Treaty, adding the net benefit to the established rights, allocated the bulk of the Nile's waters, fifty-five and a half bcm, to Egypt (sixty-six percent of the total eighty-four bcm water flow), only eighteen and a half bcm (twenty-two percent) to Sudan, and left the remaining ten bcm (twelve percent) for evaporation (1959 Nile Treaty, art. II, Table 3.1). The treaty does not recognize the rights of the upstream states.

Cooperative Framework Agreement: Solidifying the Disputes

The Nile Basin Cooperative Framework Agreement (CFA) is the other important legal instrument concerning the use and allocation of the Nile watercourse. The CFA was the result of the riparian states' attempt to create a basin-wide legal and institutional framework to regulate the interstate utilization and management of the Nile River. The CFA negotiation began in the early 1990s and was formalized in the CFA Project in 1995 (McCaffrey Interview 2017).

At the time, all riparian Nile states at the time—with the exception of Eritrea—participated in the project—with financial and technical support from the United Nations Development Programme—which provided for high-level legal and political negotiations toward the conclusion of a basin-wide agreement. A separate but parallel track supported by the World Bank, the Nile Basin Initiative (NBI), focused on development in the region and involved the same nine Nile Basin states that participated in the CFA process (Ibid; Salman 2013).

During the project's negotiations, the fate of the colonial Nile Water Treaties was the subject of controversy. The upstream states believed that the purpose of the CFA Project was to produce an inclusive agreement that would replace and supersede the Nile Water Treaties. The lower riparian states—Egypt and Sudan—insisted that the new agreement must explicitly recognize the earlier treaties, referred to as "existing agreements," and would continue to be binding upon all riparian states (McCaffrey Interview 2017; Mekonnen 2010).

In an attempt to bridge this gap, the negotiators of the CFA introduced the new, non-legal concept of "water security" (Mekonnen 2010) to replace the provision governing the relationship between CFA and the existing agreements, on which no agreement could be reached (McCaffrey Interview 2017). The idea was that Egypt, concerned about water security, could be protected through a new provision and the relationship between the CFA and the "existing agreements" (whether the former supersedes the later or vice versa) and could be governed by the general rules of international law (Ibid).

However, the Nile Basin states were not able to agree on the draft provision on water security contained in Article 14 of the draft CFA (Ibid). Specifically, the lower riparian states opposed the part of Article 14 providing that the parties "agree, in a spirit of cooperation not to significantly affect the water security of any other Nile Basin state." The lower riparian states insisted that the language should obligate all Nile Basin states "not to adversely affect the water security and *current uses and rights* [emphasis added] of any other Nile Basin State" (McCaffrey Interview 2017; Mekonnen 2010).

The position of lower riparian states brought the long-standing dispute over the colonial Nile Water Treaties back to the table as a request for unequivocal recognition of their validity against upstream states. The upstream states rejected that proposal and opened the agreement for signature on May 14, 2010. The CFA has been signed by six upstream states and subsequently ratified by four states. By its terms, the CFA requires six ratifications to enter into force and, consequently, the CFA neither binds the lower riparian states nor reallocates the waters of the Nile River (CFA art. 42). The signatories are Burundi, Ethiopia, Kenya, Rwanda, Tanzania, and Uganda. The Democratic Republic of Congo, Egypt, Eritrea and Sudan did not sign.

The Declaration of Principles: A Symptomatic Treatment?

After signing the CFA, Ethiopia began construction on the GERD an estimated twenty kilometers upstream from the Sudan–Ethiopia border on the Blue Nile. Egypt and Sudan initially opposed the GERD, alleging that it would significantly and adversely affect their interests and violate the rules regulating the Nile watercourse (Salman 2011). Considering the numerous benefits conferred by the GERD, Sudan immediately changed its position to support the construction of the dam (Salman 2016; Tawfik 2016). Gradually, after painstaking negotiations, Egypt accepted the importance of the dam and the three states signed the Declarations of Principles on the GERD Project (DoP) on March 23, 2015.

The DoP is a unique addition to the legal regime governing the use of the Nile watercourse. Unlike the colonial Nile Water Treaties, the DoP expressly considers Ethiopia's interests and recognizes the significance of the Nile River for the sustainable development of its people (DoP, preamble). It also codifies some fundamental principles of international water law—including the principle of equitable and reasonable utilization, and the "no-significant-harm" rule—as the governing rubric of the Blue Nile (DoP, arts. III–IV). However, the legal effect of the DoP vis-à-vis the colonial Nile Water Treaties is unclear. There are two questions worth asking: Is the DoP binding? And, if so, does it supersede the colonial Nile Water Treaties?

First, the DoP may be a soft law instrument that does not bind Ethiopia, Sudan, and Egypt. This assertion is supported by the nomenclature of the document; declarations are not typically binding under international law (Pronto 2015). Moreover, the DoP does not specify entry into force, ratification, or deposit of the document pursuant to Article 24(4) of the Vienna Convention on the Law of Treaties (VCLT). The fact that DoP has not been ratified by the three states as required in their respective constitutions indicates the parties did not intend the DoP to be a binding instrument. If the DoP is in fact soft law, the issue of the relationship between the DoP and the colonial Nile Water Treaties disappears. But, the DoP is still important for the three states as it provides guidelines for the GERD negotiations. Ethiopia, Sudan, and Egypt may prefer the soft law nature of the DoP because it offers more flexibility in the GERD negotiations and facilitates compromise without significantly affecting their respective interests.

Second, it can be argued that the DoP is a legally binding instrument that must be enforced in good faith. While declarations are generally considered soft law, the title of a document is not wholly dispositive of its legal effect (See VCLT, art. 2). "Treaties are known by a variety of differing names, ranging from Conventions, International Agreements, Pacts, General Acts, Charters, through to Statutes, Declarations, and Covenants" (Shaw 2008). Rather, "the intent of the parties, as reflected in the language and context of the document, the circumstances of its conclusion, and the explanations given by the parties" (Nash 1994), determines its legal effect.

Further, the DoP does not necessarily need to be ratified by Ethiopia, Sudan, and Egypt in order to have legal effect, because establishing the intention to be bound by a treaty is "governed by international law" (Waldock 1962). Several aspects of the DoP suggest that it was intended to bind the three states. For instance, phrases including: the three states "respect the final outcomes of the Technical National Committee" (DoP, art. V), "the three countries have committed to..." (DoP, preamble), and "the three countries shall take..." (DoP, arts. III–IV, VII & IX) imply the DoP is binding. Hence, the agreement does not seem to be merely a "gentlemen's agreement." Subsequent state practice also appears to suggest that the three states intended to be bound by the DoP. In December 2015, Ethiopia, Sudan, and Egypt signed the 2015 Khartoum document, agreeing to implement the DoP (Salman 2016).

Even if the DoP is a hard law, the issue of the relationship between the DoP and the colonial Nile Water Treaties becomes less crucial if the two legal regimes govern different subject matters or are found compatible with each other. For the reasons explained below, the DoP cannot supersede the colonial Nile Water Treaties simply because it is binding upon the three states.

The third, and perhaps most convincing, argument would consider the DoP as a reflection of customary international law governing watercourses. The DoP binds the three states as an endorsement of existing international customs regulating transboundary watercourse. Confirming this, the International Court of Justice (ICJ) in the Nicaragua case stated that "the effect of consent to the text of such resolutions cannot be understood as merely that of reiteration or elucidation….it may be understood as an acceptance of the validity of the rule or set of rules declared by the resolutions themselves" (ICJ Judgment 1986). Even if the DoP is non-binding, the two customary principles—equitable utilization and no significant harm—are binding on Ethiopia, Sudan, and Egypt by virtue of being rules of customary international law. The three states' official endorsement of these principles in the DoP can be seen as an indication of the existence of consensus to govern the Blue Nile based on contemporary international watercourse law. This argument is further supported by the three states' letters to the UNSC in which they all invoked and subscribed to, *inter alia*, the equitable and reasonable utilization, and no-significant-harm rule as customary international law governing the Nile Basin (Egypt's Letter to UNSC; Sudan's Letter to UNSC; Ethiopia's Letter to UNSC).

The DoP is silent as to whether it supersedes the colonial Nile Water Treaties and does not address the present applicability or legal status of the colonial Nile Water Treaties. Salman Mohammed Salman considers this omission as an indirect nullification and contends that "the failure of Egypt and Sudan to refer to the 1902 Agreement, or to their existing uses and rights as per the 1959 agreement, carries with it a clear acceptance by the two countries of the new legal order established by and resulting from the DoP" (Salman 2017). Hence, this new order would replace the 1902 Treaty and the 1959 Nile Waters Agreement (Salman 2016). Although he does not provide any legal ground for this assertion, Salman appears to rely on the *lex posterior* doctrine and assumes that the DoP is binding.

The later-in-time, or *lex posterior*, principle under international law permits states to supersede past treaties by concluding a new treaty governing the same matter. However, the assertion that the DoP has replaced the colonial Nile Water Treaties is questionable. First, per Article 59 of the VCLT, the later-in-time treaty supersedes the prior treaty when the two conflict on an analogous point of international law. As indicated above, the legal status of the DoP is unclear and may not be considered as a legally binding treaty capable of abrogating a preexisting treaty. As Professor Stephen C. McCaffrey rightly described it, the DoP is a "quasi-legal document" that merely rearticulates established principles of international watercourse law (McCaffrey 2019).

Even if we assume that the DoP is a treaty, it falls short of fulfilling the requirements provided in the VCLT. The *lex posterior* doctrine applies when (1) both the former and new treaties relate to the same subject matter, and (2) the parties intended the later treaty to govern the matter, or the two treaties are "incompatible" and cannot be applied together (VCLT, art. 59). The DoP does not satisfy the first and most important test because it regulates a different matter than the 1902 Anglo-Ethiopian Treaty. While the former provides a framework for the GERD negotiations and to some extent for the Blue Nile, the latter only addresses whether Ethiopia is required to obtain prior authorization from Britain (Sudan) before undertaking projects along the Nile tributaries.

With respect to the second requirement, the DoP is arguably a special agreement governing the GERD that can be applied concurrently with the colonial Nile Water Treaties. As noted, the DoP and the 1902 Anglo-Ethiopian Treaty establish distinct regimes that do not conflict. As for the other Nile Water Treaties, the DoP is a narrow agreement that has no relevance to other Nile tributaries such as White Nile, Sobat (Baro-Akobo), and the Atbara (Tekeze and Angereb) rivers. To the extent it does not address these Nile tributaries, the DoP does not replace the colonial Nile Water Treaties.

If, however, Salman is correct in asserting that the DoP replaced the 1959 Nile Treaty, the latter treaty does not govern the Nile River as between Egypt and Sudan, who are parties to both instruments. But, as demonstrated below, Egypt and Sudan heavily relied on the 1959 Nile Treaty during the GERD negotiations. This has two potential implications. The first and least plausible is that the DoP supersedes the 1959 treaty and Egypt and Sudan's subsequent reliance was erroneous. Second, and more plausibly, the DoP did not supersede the 1959 treaty, in which case Egypt and Sudan properly relied upon it.

Post-DoP Negotiations and the Revival of Dispute Over Colonial Nile Water Treaties

The Two Studies on the GERD

As indicated above, Ethiopia, Sudan, and Egypt participated in negotiations before adopting the DoP. The three states also established an International Panel of Experts (IPoE), which consists of ten members, two from each of the three states, and four members from outside the Nile Basin (Salman 2018). The IPoE, after studying the GERD and its potential impacts on the two downstream states, found the GERD would not cause significant harm. It also recommended the three states conduct two studies: a transboundary impact study and a hydrological modeling study (Salman 2018). The three states agreed, in Article V of the DoP, to use the outcome of these studies to determine the guidelines and rules on the GERD's first filling and annual operation.

Following the adoption of the DoP, Ethiopia, Sudan and Egypt agreed that international consultants would conduct the studies under the supervision of the Technical National Committee (TNC). In September 2016, the TNC hired two French firms, BRLi Group and Artelia, to conduct the studies. However, the initiation of the studies reignited the dispute over the colonial Nile Water Treaties (Salman 2018).

Egypt insisted that the baseline data to determine the impact should be its current use of the Nile waters. In other words, Egypt claimed every drop of water that flowed into Lake Nasser, reservoir of the HAD, including the waters that Sudan failed to use. Sudan currently uses only 12 bcm from its 18.5 bcm annual "share" under the 1959 Nile Treaty while Egypt uses around 61 bcm, approximately 5 bcm of which is water from Sudan's "share" under the 1959 Nile Treaty (Ibid). Hence, Sudan demanded the 1959 Nile Treaty water allocation serves as the baseline.

Ethiopia rejected both positions, reiterating that it is not a party to the 1959 Nile Treaty and the baseline for the studies will not create rights to the three states (Ibid).

Later, Egypt attempted to include the World Bank as a mediator (Xuequan 2017) and even sought to exclude Sudan from the GERD negotiations (Tadesse 2017). Ethiopia rejected Egypt's proposals, on the grounds that "there is an opportunity for the three countries to resolve possible disputes by themselves" (Meseret 2018). They ultimately agreed to establish a new National Independent Scientific Research Group (NISRG) to study scenarios for filling and annual operation of the GERD (Oluoch 2018). However, instead of refining and agreeing on the work of the NISRG, Egypt submitted its own proposal on the filling and operation of the GERD.

Egypt's Proposal for the Filling and Operation of the GERD and Enter the Observers

In submitting the proposal in august 2019, Egypt's negotiation tactic was to open with extreme positions on the filling and operation of the GERD. According to Egypt's proposals, the GERD's first filling should "be conducted over a seven-year period, with a minimum release of forty billion meter cubic (bcm) of water annually and a guarantee to ensure that the High Aswan Dam (HAD) remained at 165 m above sea level (m.a.s.l.)". (Abraham 2019). Moreover, the proposal would require Ethiopia to obtain approval from Egypt at various stages of the filling (Solomon and Elshinnawi 2019) and to release the entire average annual flow of the Blue Nile (forty-nine bmc) once the GERD becomes operational (Lashitew 2020; Al Jazeera 2019a). Egypt also proposed that it should open an office at the GERD site (Yigzaw Interview 2020).

Ethiopia outright rejected the proposals and considered them to be an Egyptian "effort to maintain a self-claimed colonial era-based water allocation and veto power on any project in the Nile system" (Lewis 2019). Egypt, in return, claimed that the talks had reached a deadlock (Daily News Egypt 2019) and called upon the USA to mediate in order to overcome the impasse (France 24 2019). Ethiopia, in the beginning, considered Egypt's claim "an unwarranted denial of the progress" made in the negotiations and hoped that the three states could resolve their disputes without intervention (Al Jazeera 2019b). But later, Ethiopian Prime Minister, Abiy Ahmed Ali, met with the Egyptian President in Russia, and the two leaders agreed "to resume talks" on the GERD (Al Jazeera 2019c). Ethiopia accepted the USA's invitation and came to Washington, D.C., where delegates of the three states met with the US Secretary of the Treasury and the President of the World Bank. They agreed to resume talks with the USA and the World Bank serving as observers.

Egypt has long been a principal hegemon in the Nile Basin. Through a myriad of mechanisms, Egypt established an effective hydro-hegemony in the Nile Basin that prevented Ethiopia from utilizing waters of the Nile (Tekuya 2018). Egypt used its strategic position in the Middle East as leverage to block several international funds intended to help Ethiopia in developing the Nile (Swain 2002). Ethiopia lacked funds to develop irrigation projects. This resulted in "one of Africa's cruellest ironies: the land that feeds the Nile is unable to feed itself" (Thurow 2003).

By 2011, Ethiopia decided to change the game by constructing the GERD. The dam created a great sense of euphoria in Ethiopia, being the product of the blood, sweat, and tears of Ethiopians. In solely financing the dam, Ethiopians made clear that the land that feeds the Nile will finally be able to feed itself.

Soon the GERD became a *fait accompli*, and Egypt lost its leverage over Ethiopia. In this new reality, Egypt sought to convince the international community that the tripartite talks failed to deliver the expected result and that the dispute could not be resolved absent third-party involvement. In August 2019, Egypt, circumventing the NISRG, submitted proposals that proved unacceptable and then sought to internationalize the tripartite negotiation. The President of Egypt, Abdel Fattah el-Sisi, in his speech at the UN General Assembly, stated that the GERD "negotiations have not yielded the desired results" and called for international interventions to break the deadlock (Daily News Egypt 2019). Egypt further requested that the USA and World Bank join the negotiation process to resolve the impasse (France 24 2019).

Egypt, by involving the USA and the World Bank in the dispute, hoped to achieve five specific goals: (1) that Ethiopia fills the GERD slowly while releasing forty bcm of water every year; (2) that Egypt maintains its water surplus in the HAD reservoir (165 m.a.s.l.); (3) that Ethiopia obtains approval from Egypt at various stages of the filling and operation of the dam; (4) that Ethiopia releases flow of the Nile waters after the filling of the GERD favoring Egypt at the cost of the dam's power generation objectives; and (5) that Egypt's on-site office right on the upper White Nile at the Owen Falls Dam in Uganda is replicated in Ethiopia at the GERD.

To address the implications of these proposals in turn, first, it will affect Ethiopia's interest because of the delayed production of electricity. Ethiopia will not get the expected returns from the GERD and will not be able to eliminate extreme energy poverty. Second, it will make the GERD unable to meet the design power generation with marketing dependability. Egypt, for instance, can use the Nile waters unreasonably and may export waters to neighboring countries, making the water stored in the HAD below 165 m a.s.l. Ethiopia will then be forced to compensate for the loss from the GERD. Moreover, maintaining the HAD at 165 m a.s.l will be difficult in the future because of climate change. The Nile Basin is experiencing droughts that are increasing in both frequency and magnitude (Schiffler 1998). Therefore, arguably, the GERD will end up being a second reservoir for Egypt.

Agreeing to the third proposal will, in effect, impose Egypt's understanding of the 1902 Anglo-Ethiopian Treaty on Ethiopia by giving Egypt veto power over upstream projects. The fourth proposal presupposes the validity of the 1959 Nile Treaty, to which Ethiopia is not a party and thus, not bound to respect the water allocation. Even if the GERD is limited to electricity production, agreeing to release the entire flow of the Nile after the completion of the GERD will foreclose

Ethiopia's existing and future right to use the Nile waters for other essential consumptive purposes. The standard should be not the 1959 Nile Treaty, but rather, the customary rule of equitable and reasonable utilization. This requires balancing use of the Nile waters by the riparian states: a concept that Egypt is evidently finding difficult to accept. The ICJ underscored the foundational nature of this principle, declaring the principle to be recognition of a riparian state's "basic right to an equitable and reasonable sharing of the resources of an international watercourse" (ICJ Judgment 1997).

Agreeing to the last proposal would require nothing short of a surrender of Ethiopian sovereignty. Past practice shows that joint or transnational projects require shared management, as opposed to strictly national projects that are independently managed. For instance, the Manantali Dam on the Senegal River and the Kariba Dam on the Zambezi River are joint projects run together (Rangeley et al. 1994), while Chinese unilateral dams on the Mekong River are managed independently (Simone 2013). Furthermore, all the dams and reservoirs built in Egypt and Sudan are unilateral projects managed independently without the involvement of Ethiopia or other upstream states. The GERD is a unilateral project on Ethiopian territory, and therefore its operation should be left solely to Ethiopia. Ethiopia would, of course, remain obligated under general international law to use the Nile waters equitably and reasonably, exercise due diligence to avoid causing of significant harm to the lower riparian states, and regularly share data and information.

All of Egypt's proposals would have the effect of directly or indirectly imposing the colonial Nile Water Treaties on Ethiopia. Therefore, Egypt attempted to use the GERD negotiations to maintain its hydro-hegemonic status quo, an endeavor that the USA and the World Bank tried to help Egypt achieve. They did so by tabling a proposal on the filling and operation of the GERD that is biased in favor of Egypt and by subjecting Ethiopia to compulsory dispute resolution regarding the GERD.

The GERD Washington, D.C., Talks: Illuminating the Sticking Points

In complex multilateral talks, like those around GERD, it is important to scrutinize the relationship between the negotiating parties and third parties as these relationships influence the third parties' neutrality. Egypt has a close relationship with the World Bank, as its nationals populated high positions within the World Bank and Egypt benefitted from its development programs (Amdetsion 2007). The same can be said of Egypt's relationship with the USA. For example, President Al-Sisi met with President Donald Trump in April 2019 and has received high praise from President Trump (The Independent 2019). Ethiopia does not maintain as much close relationships with either the World Bank or the current United States Administration. As such, Ethiopia is warranted in its concern that the Washington, D.C., talks were not on a level playing field.

In January 2020, Ethiopia's fears were realized when the USA became a mediator as opposed to a mere observer, tabling proposals on the GERD's filling and operation. The USA then pressured Ethiopia to accept its proposals, which Ethiopia rejected (Worku 2020). The principal sticking points are related to how Ethiopia should fill and operate the GERD during drought or prolonged dry years and what mechanisms for dispute resolution would be available (Yigzaw Interview). Details of the Washington, D.C., proposal and its adverse impacts on Ethiopia are documented in Chapter 20 of this book (Abtew 2021).

Sticking Points: The Three Drought Mitigation Mechanisms

In a joint statement issued on January 15, 2020, Egypt, Sudan, and Ethiopia underscored the necessity of developing drought mitigation mechanisms for three kinds of situations: (1) drought, (2) prolonged drought, and (3) prolonged dry years during the filling and operation of the GERD (US Department of the Treasury 2020b). The USA and the World Bank tabled a proposal detailing drought mitigation mechanisms applicable in such situations.

Filling of the GERD

According to the proposals tabled by the USA and World Bank, drought is when the GERD's release is below thirty-seven bcm. If a drought coincides with the filling years, Ethiopia is expected to release the "flow" of the Nile River and supplemental water from the GERD. Prolonged drought is when the release from the GERD is below thirty-nine bcm for four years. During the filling period, a prolonged drought requires Ethiopia to release the river flow and sixty-two and a half percent of the water above 603 m.a.s.l. of the GERD for the following four years. Prolonged dry years are when the GERD's release is below forty for four consecutive years. Ethiopia must release the flow and fifty percent of the GERD storage above 603 m.a.s.l. for the next five consecutive years in the event of prolonged dry years during the dam's filling (Ethiopia Insight 2020).

The US and Egyptian proposals have two common features: (1) Ethiopia is expected to release the "flow" of the Blue Nile to the downstream states, which constitutes all of the Blue Nile's water that reaches the GERD reservoir, and (2) Ethiopia can incur certain quantities of water debt that must be paid from the GERD's reservoir. The first feature forecloses Ethiopia's right to use the Blue Nile's flow—which includes the waters of its tributaries—prohibiting Ethiopia from using the waters above the GERD equitably and reasonably.

The second feature is similarly adverse to Ethiopia's interests. Analogous international practice shows that upstream states are allowed to deliver below the minimum quantity of water during severe drought seasons and repay the water during normal seasons. For instance, the 1944 agreement between the USA and Mexico on the Rio Grande and Colorado Rivers has provisions governing possible problems resulting from drought. With respect to both the Rio Grande and Colorado Rivers, the treaty allows upstream countries to deliver below the minimum quantity of the water during severe drought seasons and repay the water during the normal seasons (the 1944 Colorado Treaty, arts. 4.B(c)–(d), 10(a)–(b)). Existing jurisprudence shows that water debt and repayment are relevant only when there is a water-sharing arrangement between riparian states. However, in the Nile Basin there is neither a water-sharing arrangement nor a minimum water quantity allocated to Egypt and Sudan. Therefore, the concept of water debt is inappropriate. International treaty law demonstrates that the riparian states must help shoulder the burden caused by drought. The proposals for the GERD's filling and operation impose the burden of drought solely on Ethiopia.

The implication of agreeing to release water from the GERD's reservoir is nothing short of recognizing the water share allocated in the 1959 Nile Treaty. In other words, if Ethiopia is required to release water from its own reservoir, it does not have any share from the Nile waters. The proposals on the long-term operation of the dam support this conclusion.

If Egypt does indeed have "established rights" in the flow of the Blue Nile, against whom are such rights enforceable? Egypt does not have a treaty right as against Ethiopia; under the 1959 Nile Treaty, Egypt only has such a right against Sudan. Therefore, any right Egypt has against Ethiopia must derive from customary international law. Ethiopia did not, however, exercise forbearance in constructing a GERD-type dam earlier out of any sense of legal obligation (*opinio juris*) (ICJ Judgment 1969). Any such forbearance was due to a combination of other factors, including threats from Egypt and a lack of funding from international financial institutions. Thus, any Egyptian right to a given flow of the Blue Nile is enforceable only against Sudan, not Ethiopia.

"Long-Term" Operation of the GERD

The definitions of drought, prolonged drought, and prolonged dry years in the long-term operation are similar to the definitions provided in the filling period. The only difference is the amount of water to be released from the GERD. During the drought period, Ethiopia must release the flow and more waters, the amount of which is to be agreed upon by the parties, from the GERD. During prolonged drought periods and dry years, Ethiopia must release the flow and all available water above 603 m.a.s.l. of the GERD within four (five in case of prolonged dry years) years (Ethiopia Insight 2020).

All the concerns raised above as to the filling of the GERD apply here. However, agreeing on the long-term operation of the GERD has three devastating consequences for Ethiopia. First, since "flow" includes the entire quantity of water that is naturally available, Ethiopia, who contributes all Blue Nile waters, would not have a share of water allocation from the Blue Nile. In sum, Ethiopia would be prohibited from using the Blue Nile's tributaries for irrigation or hydroelectric power generation. Second, the restrictions on the GERD operation will make it fail to meet its design objective of dependable hydropower generation. Third, it would recognize Egypt's claim to established rights (fifty-five and a half bcm) and, consequently, impose the 1959 Nile Treaty on Ethiopia.

In light of increased water scarcity and severe droughts in the Nile Basin due to climate change (Schiffler 1998), it will be impossible for Ethiopia to release forty bcm from the GERD, let alone get a share of the Blue Nile. Even worse, Ethiopia will be in perpetual water debt and, thus, forced to release more waters from the GERD. In effect, this will make the GERD an Egyptian storage facility. The GERD reservoir's evaporation rates will be lower than those of Lake Nasser, the HAD's reservoir. But this is an incidental benefit. It would be unrealistic to expect Egypt to plan its entire project around this goal.

The DoP provides the framework for the GERD talks, and as such, the forthcoming GERD Treaty should be in line with the DoP. The scope of the GERD Treaty envisaged in Article V of DoP is limited to the GERD. Ethiopia should, therefore, agree to release only the inflow of the GERD (i.e., the water that enters the GERD reservoir minus normal evaporation) during low flow years.

Given that the proposals did not address obligations when the GERD's release is above forty bcm, Ethiopia will likely have a water share during normal operation of the GERD when the Blue Nile's average flow is above forty-nine bcm. Egypt already requested that Ethiopia released the entire flow of the Blue Nile during the GERD's normal operations, but Ethiopia rejected the proposal outright (Yigzaw Interview 2020). Even if we assume that Egypt, Ethiopia, and Sudan share the remaining nine bcm of water, it seems that Ethiopia would not get much water from the Blue Nile. Given the increase in average global temperature due to climate change, the GERD's reservoir is projected to evaporate at an annual rate of two bcm (Salman 2016). This is much less than the ten to fifteen bcm annual evaporation in Egypt from Lake Nasser, but it still affects the water share. That means the net water the three states must share will be less than seven bcm. Considering this, one may wonder how much Ethiopia will get from the Blue Nile.

Generally, it can be said that drought is a natural phenomenon that should be addressed collectively (Tembata and Takeuchi 2018). The burden of drought mitigation is not solely Ethiopia's to bear. Egypt, Sudan, and Ethiopia should support one another in mitigating drought, without placing the entire burden on any one state.

Disagreement Over Dispute Resolution

Although the above-discussed proposals by Egypt effectively impose the colonial Nile Water Treaties on Ethiopia, Egypt sought additional means to safeguard its status quo. To that end, Egypt proposed and included a binding arbitration clause in the US draft agreement. Ethiopia rejected this provision, demanding instead that the three states resolve disputes by themselves and resort to mediation only when necessary (Yigzaw Interview 2020).

Ethiopia's approach seems appropriate for several reasons. First, the DoP does not require compulsory and binding dispute resolution. Article X of the DoP contemplates disputes arising from the GERD and indicates that they be resolved through the consultation of the national ministries tasked with water management.

If the dispute cannot be resolved, the matter is referred to the respective heads of state, who may resolve the dispute or refer it to mediation (DoP, art X).

Moreover, imposing compulsory resolution mechanisms for disputes arising from the GERD is contrary to the principle of reciprocity. In the past, Egypt and Sudan constructed several dams and reservoirs without consulting—and even over objections from—Ethiopia. These dams and reservoirs significantly harmed Ethiopia by foreclosing its future use of Nile waters (Salman 2010), and yet, Ethiopia did not have recourse to compulsory dispute resolution. The approach that subjects the GERD to compulsory dispute resolution and leaves other downstream dams and reservoirs without any analogous recourse is rather problematic. Currently, there is no structure in place to ensure accountability in the use and activities of downstream states. Ethiopia is unable to challenge Egypt's possible exportation of water from HAD in the absence of any mechanism for dispute resolution. By parity of reasoning, the GERD should also not be subject to binding dispute resolution. One possible compromise, however, might be to subject the GERD to compulsory dispute resolution with the other downstream dams and reservoirs through basin-wide legal and institutional framework as suggested below.

The Letters to the UNSC

Egypt has demonstrated its goal for a third-party mediation and even formally requested the UNSC to intervene in the GERD dispute. It submitted letter to the UNSC in June 2020 to this effect (UN Security Council 2020a). Ethiopia (US Security Council 2020b) and Sudan (US Security Council 2020c) also submitted letters stating their positions. In response, the UN Secretary-General António Guterres encouraged the three states to resolve outstanding issues and "achieve a mutually beneficial agreement" (UN Press Release 2020). The UNSC made an unprecedented move and held an open session on the GERD dispute where most of the members encouraged the three states to resolve their dispute through the African Union (AU) (Al Jazeera 2020). In its latest letter, Egypt made the flowing arguments:

(1) Egypt never tried to impose any colonial treaties on Ethiopia because "Ethiopia was never a colony."

(2) Ethiopia violates the 1902 Anglo-Egyptian Treaty and the 1993 framework for cooperation between Sudan and Egypt.

(3) Ethiopia is required to respect the current and existing uses of the Nile watercourse (Egypt's Letter to the UNSC).

To take those arguments in turn, the first fails to stand on Egypt's practice. As indicated above, Egypt tried to impose the colonial Nile Water Treaties on Ethiopia in 2019 by submitting a proposal that, *inter alia*, requires Ethiopia to obtain approval from Egypt at various stages of the filling and operation of the dam and release the entire flow of the Nile waters after the filling of the GERD. Similarly,

Egypt, with the help of the USA and World Bank, tried to impose the colonial Nile Water Treaties on Ethiopia by proposing drought mitigation mechanisms that, *inter alia*, require Ethiopia to release the flow of the Nile and additional supplemental water from the GERD. Moreover, while it is true that Ethiopia was never a colony, the 1902 Anglo-Ethiopian Treaty is in fact a colonial treaty concluded during colonization between a colonial power, Britain and Ethiopia. Therefore, Egypt is imposing a colonial treaty on Ethiopia while invoking the 1902 Anglo-Ethiopia Treaty in the UNSC.

The second argument combines a pair of claims: (a) Ethiopia violates the 1902 Anglo-Egyptian Treaty, and (b) the 1993 framework for cooperation between Sudan and Egypt. The first claim is inappropriate because Egypt cannot invoke the 1902 Anglo-Egyptian Treaty against Ethiopia. Britain concluded the treaty on behalf of Sudan, not Egypt. While Sudan may arguably inherit the right and obligation in the 1902 Anglo-Egyptian Treaty, there is no legal ground that makes the treaty opposable to Ethiopia by Egypt. Moreover, assuming, *arguendo*, that the treaty is opposable, the obligation imposed in the 1902 Anglo-Ethiopian Treaty does not prohibit Ethiopia from using the Nile waters, even without the consent of Britain (now Sudan). What is prohibited in the treaty is arresting the flow or totally blockage of the entire flow of the Nile water (1902 Anglo-Egyptian Treaty, Art III). As a hydroelectric dam, the GERD does not block the water, and it is not therefore prohibited by the 1902 Anglo-Ethiopian Treaty.

The second claim requires proper scrutiny of both Ethiopia and Egypt's practices. Indeed, the 1993 framework for cooperation requires Ethiopia and Egypt to resolve the Nile issues through technical discussion and to refrain from engaging in "any activity related to the Nile waters that may cause appreciable harm to the interests of the other party" (1993 General Framework, arts IV–V). Ethiopia has always been committed to respect the framework. In relation to the GERD, for instance, Ethiopia has taken various measures in accordance with its international obligations to prevent harm to the downstream states. Ethiopia conducted transboundary impact studies, initiated a tripartite committee, and established an International Panel of Experts (IPoE) which confirmed that the design and construction process of the dam was in line with "a number of international Standards, Codes, and Guidelines" (IPoE 2013; Qerenso 2018). Moreover, while Ethiopia can fill the GERD within two years, it extended 4–7 years filling period not to significantly affect the interest of downstream states. On the contrary, Egypt persistently violated the framework by undertaking various unilateral projects including Toshka Project and the Peace Canal that significantly affect the interest of Ethiopia, foreclosing its current and future uses (Salman 2010).

The third argument, which uses existing usage rates as a baseline, is actually a more severe imposition than a return to the terms of the colonial Nile Water Treaties because Egypt is currently using 61 bcm, preventing Sudan from using the full share to which it is entitled under the 1959 Nile Treaty (Salman 2018). Moreover, the argument is contrary to international watercourse law, which treats both existing and potential uses equally as factors for the determination of equitable use (UN Watercourse Convention, art. 6(1)(e)). It is worth noting that as the Nile watercourse is already appropriated by Egypt and Sudan, Ethiopia's future utilization will cause harm, perhaps significant, to current and existing uses of the two downstream states. Under international law, such use will be considered a violation of the no-significant-harm principle only when it exceeds Ethiopia's equitable entitlement (International Law Commission 1986; McCaffrey 2019). To the extent the harms inflicted upon Sudan and Egypt are within the limit of Ethiopia's exercise of equitable utilization, such exercise should not be regarded as infringing the rights of the downstream states.

Way Forward: Toward Taming the Elephant in the Room

The colonial Nile Water Treaties comprise the principal obstacle to the GERD negotiations; however, Egypt, Sudan, and Ethiopia are not openly discussing them. Addressing the colonial Nile Water Treaties is *a sine qua non* for the success of GERD negotiations. Therefore, the three states should address the ramifications of the colonial Nile Water Treaties by (1) limiting the scope of the forthcoming treaty on the GERD's filling and annual operation, and (2) resorting to the CFA for water allocation and long-term operation of the GERD.

GERD Treaty: First Filling and Annual Operation

Delineating the scope of a possible GERD treaty is the first, and perhaps most important, way of addressing the inequitable ramifications of the colonial Nile Water Treaties. The three states can and should address the problems associated with the colonial Nile Water Treaties by explicitly stating in the forthcoming treaty:

(1) That nothing in this treaty shall be construed as recognition of the colonial Nile Water Treaties.
(2) That nothing in this treaty shall be construed as allocation of waters between the three countries.
(3) That nothing in this treaty shall prohibit Ethiopia from equitably using the Nile waters upstream of the GERD.
(4) That this treaty is only about the GERD and therefore nothing in the treaty shall associate the GERD with the HAD.

The forthcoming treaty should further ensure equitable filling of the GERD reservoir without causing significant harm to the downstream states. The three states must consider the factors provided in Article IV of the DoP, including climatic and hydrological conditions of the river, the effect of the use on other riparian states, population, existing and potential water uses, efficient utilization of water resources, and the water contribution of each state for ensuring equitable filling of the GERD.

It is worth mentioning that the January 15, 2020, joint statement (US Department of the Treasury 2020b) did not consider all of the factors provided for in the DoP. The statement was selective (Yihdego 2020) in only considering drought mitigation, "the hydrological conditions of the Blue Nile and the potential impact of the filling on downstream reservoirs." This cherry-picking approach, which disregards equitable and reasonable utilization and focuses only on potential impact, is inappropriate and against the cardinal principles of international watercourse law. The forthcoming treaty should, therefore, consider drought and hydrological conditions "as part of the factors to determine an equitable and reasonable filling of the GERD" (Yihdego 2020).

Under the statement's proposed framework, Ethiopia would further need to fill the dam "during the wet season, generally from July to August … [while taking] appropriate mitigation measures for Egypt and Sudan during drought, prolonged drought period and prolonged dry years" (US Department of the Treasury 2020b). Based on this, the US proposal agreement included the three aforementioned drought mitigation mechanisms. As noted, the mechanisms would compel Ethiopia to release the "flow" of the Blue Nile and more water from the GERD reservoir without any *quid pro quo*. The mechanisms also presuppose the water share allocated in the colonial Nile Water Treaties and impose all drought mitigation-related burdens on Ethiopia. Hence, they should be excluded from the forthcoming treaty. Because there is no water-sharing arrangement between the three states, Ethiopia's obligation to release water from the GERD reservoir should be equitable and reasonable; Ethiopia's drought-related treaty commitment should be limited to releasing the "inflow" of the GERD.

Concerning the filling period, there is some suggestion that filling the GERD in a two- to four-year time frame will significantly affect Sudan and Egypt (Yihdego 2020). Filling the GERD slowly—over the course of ten or more years—will significantly and adversely impact Ethiopia (Yihdego 2020). Therefore, the forthcoming treaty should establish a fair and reasonable filling plan that "considers factors such as the right season, the impact of swift and prolonged filling on all parties, and the attainment of a middle ground that considers filing the dam primarily, but not exclusively, during the wet season between five to seven years" (Yihdego 2020). In so doing, the treaty "will significantly mitigate the impact of filling the dam on downstream states while granting an equitable, reasonable and timely return to Ethiopia's investment and entitlement to produce electricity" (Yihdego 2020).

Concerning the annual operation of the dam, the forthcoming treaty, as envisaged in Article V of the DoP, should establish a coordination mechanism through the three ministries responsible for water. The three states should use this mechanism to exchange data and information. Ethiopia should annually notify the other parties of its operation plan and how much water it will release from the GERD. Egypt and Sudan should also notify Ethiopia and one another of how they are using the Nile waters and how much water is stored in their respective reservoirs on an annual basis.

The treaty should also be flexible and adaptive to address the uncertainty associated with climate change and population growth in the Nile Basin (Tekuya 2019). The treaty should include an amendment provision, a review procedure, and a termination or revocation clause. These provisions will give the three states the resilience needed to revise filling and operational guidelines as hydrological and existing conditions change. Concerning the termination clause, it must be noted that revoking a treaty through an abbreviated period of notice—say six months or a year—is inappropriate in a treaty regulating a permanent structure such as the GERD (McCaffrey 2003). So, the treaty must reconcile the flexibility required for adapting to climate change with the certainty required for the proper management of dams by requiring a long period of notice, anywhere between ten to fifteen years, to withdraw from the treaty (Tekuya 2019).

The Cooperative Framework Agreement: Water Allocation and Long-Term Operation

The CFA, if accepted by Sudan and Egypt, will establish a new legal regime governing the use and allocation of the Nile waters. The CFA does not use a fixed and volumetric strategy for water allocation, which encourages flexibility. Instead, it requires equitable and reasonable utilization as its allocation strategy (CFA, art. 4, ¶ 1).

The CFA also foresees the establishment of the Nile Basin Commission (NBC) as an institutional framework for Nile Basin governance (CFA, art. 15). The NBC would possess a wide range of powers, including the ability to examine and determine optimal water use and distribution among the Nile Basin countries (CFA, art. 24, ¶ 12). It would also have a broad scope; it would be entrusted with rule-making authority and empowered to resolve disputes within the Nile Basin (CFA, art. 24). Considering the need for coordinated dam operation and integrated water resource management, the NBC is best positioned to manage the long-term operation of the GERD. But, for this to happen, Egypt and Sudan must accede to the CFA.

Given both states' previous oppositions, the question would, then, be why Egypt and Sudan accede to the CFA now? The current hydro-political context of the Nile Basin is significantly different than it was during the CFA negotiation. The GERD, which brought about a de facto change in the status quo, will affect the Nile's flow during the filling period. Ethiopia can, therefore, use the GERD as a bargaining chip to negotiate concessions from Egypt and Sudan.

As noted, Sudan is allied with Ethiopia due to the advantage it would get from the GERD. Egypt, on the other hand, is concerned about the GERD's impact on the status quo of the current Nile water allocation. It follows that Egypt will accept the CFA if it regulates the operation of the dam in a way that protects Egyptian interests. The NBC, as a basin management institution, will develop guidelines for the coordinated operation of all dams and other water control structures on the Nile

River. As the NBC makes binding decisions by consensus (CFA, art. 23, ¶ 5), Egypt's interests will be protected better.

Egypt may be further incentivized to accept the CFA by the need to avoid unilateral exploitation through cooperative use of the Nile River. Egypt, after all, cannot prevent Ethiopia from constructing the GERD. There is a risk that other riparian states would follow Ethiopia in unilaterally developing the Nile River. Therefore, the CFA would protect Egypt's interests by ensuring cooperative use of the Nile waters.

Upstream states are beginning to assert their right to use the Nile. Given Egypt's geographic and hydrologic vulnerability by virtue of being a downstream state, a legal regime that protects its interest is a necessity. Egypt (and Sudan) are therefore highly recommended to accept the CFA as it sufficiently safeguards their interests through equitable and reasonable utilization, cooperative utilization, the "no-significant-harm" principle, and binding dispute resolution mechanisms. In addition to empowering the NBC to make binding decisions (CFA, art. 23, ¶ 6), the CFA already contains terms that would allow parties to submit to binding arbitration and judicial settlement through the ICJ (CFA, art. 33, ¶ 1(a)). Since these dispute resolution mechanisms have basin-wide applications, the CFA is a good compromise to resolve the three states' disagreement over the need for including a compulsory dispute resolution for the GERD treaty.

Conclusion

Ethiopia, Sudan, and Egypt have long disagreed about the validity of the colonial Nile Water Treaties. The dispute, exacerbated by the construction of the GERD, is now threatening the peace and stability of North-Eastern Africa. Egypt, considering the dam as an existential threat, has recently requested the UNSC to intervene into the GERD dispute. The UNSC held an open session on the GERD dispute and encouraged the three states to resolve their dispute through the AU.

At the time of writing this chapter, the three states are undertaking the tripartite negotiations under the auspices of the AU. For the negotiations to move forward, the implications of the colonial Nile Water Treaties must be dispensed with. The AU should help the three states to create a positive bargaining zone by (1) restricting the scope of the ongoing talks to the filling and annual operation of the GERD and (2) encouraging the states to resort to the CFA for water allocation, dispute resolution and long-term operation of the GERD.

As for Sudan and Egypt, there are suggestions that neither state will be immediately affected by the filling as there should be sufficient water in the system to compensate for the amount Ethiopia plans to hold back (Solomon and Elshinnawi 2019; Kaba and Moges 2020). Moreover, as the GERD is a non-consumptive hydroelectric project, it will not have any adverse impact after the filling is completed. Therefore, regardless of whether AU-facilitated talks become successful or not, the real dispute between the three states will boil down to whether and to what extent Ethiopia can use the Nile waters for consumptive purposes including irrigation.

As the Nile watercourse is already appropriated by Egypt and Sudan, Ethiopia's future utilization will cause harm, perhaps significant, to current and existing uses of the two downstream states. Under international law, such use will be considered a violation of the no-significant-harm principle only when it exceeds Ethiopia's equitable entitlement. To the extent the harms inflicted upon Sudan and Egypt are within the limit of Ethiopia's exercise of equitable utilization, such exercise should not be regarded as infringing the rights of the downstream states. Therefore, Ethiopia should go ahead and start equitably utilizing the Nile watercourse for consumptive purposes, including irrigation.

Acknowledgements I am immensely grateful to my supervisor, Professor Stephen McCaffrey, for his precious comments on the earlier version of this work. I am also grateful to experts in the Ethiopian negotiating team for giving me an insider insight to the GERD dispute. I am responsible for any omissions.

Cases and Legal Documents

1. ICJ Judgment (1986) Military and Paramilitary Activities in and Against Nicaragua (Nicar. v. U.S.), Judgment, 1986 I.C.J. 14.
2. ICJ Judgment (1997) Gabčíkovo-Nagymaros Project (Hung./Slovk.), Judgment, 1997 I.C.J. 7.
3. ICJ Judgment (1969) North Sea Continental Shelf Cases (Ger. v. Den.; Ger. v. Neth.), Advisory Opinions, 1969 I.C.J. 44, 77 (Feb. 20).
4. Treaties Relative to the Frontiers Between the Soudan, Ethiopia, and Eritrea, May 15, 1902, [1902] U.K.T.S. 16 [1902 Anglo-Ethiopian Treaty].
5. Exchange of Notes between Great Britain and Northern Ireland and Egypt regarding the Use of the Waters of the River Nile for Irrigation Purposes (1929) [the 1929 Anglo-Egyptian Treaty].
6. Agreement for the Full Utilization of the Nile Waters, Egypt-Sudan, Nov. 8, 1959, 453 U.N.T.S. 51 (1959) [1959 Nile Treaty].
7. Framework for General Co-operation between the Arab Republic of Egypt and Ethiopia, 1 July 1993, UNTS, Vol. 2693, No. 47816 [1993 Framework for cooperation].
8. Agreement on the Nile River Basin Cooperative Framework, opened for signature May 14, 2010 [CFA].
9. Agreement on Declaration of Principles on the Grand Ethiopian Renaissance Dam Project, Mar. 23, 2015 [DoP].
10. Convention on the Law of the Non-Navigational Uses of International Watercourses, May 10, 1997, 2999 U.N.T.S. 52,106 [U.N. Watercourse Convention.

11. Vienna Convention on the Law of Treaties art. 24.4, May 23, 1969, 1155 U.N. T.S. 331 [VCLT].
12. Utilization of the Waters of the Colorado and Tijuana Rivers and of the Rio Grande, Mex.-U.S., arts. 4.B(c)–(d), 10(a)–(b), Feb. 3, 1944, 59 Stat. 1219 [the 1944 Colorado Treaty].

References

Abdo M (2004) The Nile question: the accords on the water of the Nile and their implications on cooperative schemes in the basin. Perceptions J Int Aff 9:47, 48

Abraham D (2019) Cooperation is imperative for an agreement on filling, and operation of GERD. In: Walta. http://www.waltainfo.com/index.php/FeaturedArticles/detail?cid=50736

Abtew W (2021) Chapter 20: The grand Ethiopian renaissance dam filling and operation negotiations, and Eastern Nile water sharing. In: Melesse MA, Abtew W, Moges S (eds) Nile and grand Ethiopian renaissance dam: past, present and future. Springer, New York

Addis Standard (2019) Ethiopia rejects Egypt's proposal on the filling, operation of grand dam. In: Addis Standard. https://addisstandard.com/news-ethiopia-outright-rejects-egypts-proposal-on-the-fillingoperation-of-grand-dam/

Al Jazeera (2019a) Ethiopia rejects Egypt proposal on the Nile dam. In: Al Jazeera. https://www.aljazeera.com/news/2019/09/ethiopia-rejects-egypt-plan-operate-nile-dam-190919073610904.html

Al Jazeera (2019b) Egypt says 'Deadlock' in Nile dam talks with Ethiopia. In: Al Jazeera. https://www.aljazeera.com/news/2019/10/egypt-deadlock-nile-dam-talks-ethiopia-191006070043565.html

Al Jazeera (2019c) Egypt, Ethiopia 'agree' to resume talks on massive Nile dam. In: Al Jazeera. https://www.aljazeera.com/news/2019/10/egypt-ethiopia-agree-resume-talks-massive-nile-dam-191024140822273.html

Al Jazeera (2020) Egypt warns of 'existential threat' from Ethiopia dam. In: Al Jazeera (June 30, 2020). https://www.aljazeera.com/news/2020/06/egypt-warns-existential-threat-ethiopia-dam-200630103858360.html, https://perma.cc/N229-RGWC

Amdetsion F (2007) Scrutinizing the scorpion problematique: arguments in favor of the continued relevance of international law and a multidisciplinary approach to resolving the Nile dispute. Tex Int LAW J 44:45

Daily News Egypt (2019) Egypt seeks international support to end GERD talks deadlock. In: Daily News Egypt. https://wwww.dailynewssegypt.com/2019/10/01/egypt-seeks-international-support-to-end-gerd-talks-deadlock/

Ethiopia Insight (2020) Why Ethiopia rejected the U.S.-drafted GERD deal. In: Ethiopia Insight. https://www.ethiopia-insight.com/2020/04/02/why-ethiopia-rejected-the-u-s-drafted-gerd-deal/

Eyssen BV (2020) Trouble as Trump dives into the dispute over Ethiopia's Nile mega-dam. In: Deutsche Welle. https://www.dw.com/en/trouble-as-trump-dives-into-the-dispute-over-ethiopias-nile-mega-dam/a-52663126

Ezega News (2019) Power generation capacity of the GERD Slashed to 5150 MW—Ethiopian Minister. In: Ezega. https://www.ezega.com/News/NewsDetails/7331/Power-Generation-Capacity-of-GERD-Slashed-to-5150MW-Ethiopian-Minister, https://perma.cc/4Z85-6E7G

France 24 (2019) Egypt calls on 'active' US role mediating Nile dam impasse with Ethiopia. In: France 24 (Sep 10, 2019). https://www.france24.com/en/20191007-egypt-ethiopia nile-river-dam-mediating, https://perma.cc/QS32-LUWY

Haaretz (2020) Egypt vows to use 'all means' to defend Nile interests after Ethiopia skips U.S. talks. In: Haaretz. https://www.haaretz.com/middle-east-news/egypt/egypt-vows-to-use-all-means-to-defend-nile-interests-after-ethiopia-skips-talks-1.8600505
Ibrahim A (2011) The Nile basin cooperative framework agreement the beginning of the end of Egyptian hydro-political hegemony. Mo Envtl L Pol'y Rev 18
ICJ Judgment (1969) North Sea continental shelf cases (Ger. v. Den.; Ger. v. Neth.), Advisory Opinions, 1969 I.C.J. 44
ICJ Judgment (1986) Military and paramilitary activities in and against Nicaragua (Nicar. v. U.S.), Judgment, 1986 I.C.J. 14
ICJ Judgment (1997) Gabčíkovo-Nagymaros project (Hung/Slovk), Judgment, 1997 I.C.J. 7
International Law Commission (1986) The law of the non-navigational uses of international watercourses, [1986] 2 Y.B. INT'L L. COMM'N 133, U.N. Doc. A/CN./SER.A/1986/Add.1
IPoE (2013) Int'l panel of experts on the grand Ethiopian renaissance dam project [IPoE]. Final report, International panel of experts on grand Ethiopian renaissance dam project. https://www.scidev.net/filemanager/root/site_assets/docs/international_panel_of_experts_for_ethiopian_renaissance_dam-_final_report.pdf
Kaba E, Moges S (2020) Why is now the right time to start filling the grand Ethiopian renaissance dam (GERD)? E. Africanist. https://eastafricanist.com/2020/06/01/why-is-now-the-right-time-to-start-the-filling-of-the-grand-ethiopian-renaissance-dam-gerd/, https://perma.cc/7QTP-EAVC
Lashitew A (2020) Why Ethiopia, Egypt, and Sudan should ditch a rushed Washington-Brokered Nile treaty. In: Brookings Inst. https://www.brookings.edu/blog/africa-in-focus/2020/02/18/why-ethiopia-egypt-and-sudan-should-ditch-a-rushed-washington-brokered-nile-treaty/
Lewis A (2019) Ethiopia Says Egypt trying to maintain "Colonial Era" grip over Nile. In: Reuters. https://www.reuters.com/article/us-ethiopia-dam/ethiopia-says-egypt-trying-to-maintain-colonial-era-grip-over-nile-idUSKBN1WN1OG, https://perma.cc/U5MW-Y3NV
London Embassy of Ethiopia (2020) Press release, Embassy of Ethiopia, London, statement of Ethiopia on the negotiations on the grand Ethiopian renaissance dam. https://www.ethioembassy.org.uk/statement-of-ethiopia-on-the-negotiations-on-the-grand-ethiopian-renaissance-dam
McCaffrey S (2003) The need for flexibility in freshwater treaty regimes. Nat Resour Forum 27(2): 156–160
McCaffrey S (2017) Interview. Legal and institutional consultant, Nile cooperative framework project, in Sacramento (Nov 2017)
McCaffrey S (2019) The law of international watercourses. Oxford University Press
Mekonnen D (2010) The Nile basin cooperative framework agreement negotiations and the adoption of a "water security" paradigm: flight into obscurity or a logical cul-de-sac? Eur J Int Law 21:421–440. https://doi.org/10.1093/ejil/chq027
Mekonnen D (2011) Between the scylla of water security and charybdis of benefit sharing: the Nile basin cooperative framework agreement—failed or just teetering on the brink? Goettingen J Int Law. https://doi.org/10.3249/1868-1581-3-1-MEKONNEN
Meseret E (2018) Ethiopia Refuses World Bank Arbitration over Nile River Dam. In ASSOCIATED PRESS https://apnews.com/42dba7a51c0b4d0c90829d0c2c4e596b
Meseret E (2020) Ethiopia won't be forced by the US on dam, Foreign Minister says. In: Associated Press. https://apnews.com/39183ccfeed1c0796ad38796d459ff3b
Nash (1994) Contemporary practice of the United States relating to international law. Am J Int L 88(1):89–104
Nile Basin Initiative (2012) The water resources of the Nile basin
Oluoch F (2018) Joint team to study likely effects of Ethiopian dam on Nile waters. In: East African. https://www.theeastafrican.co.ke/news/africa/Joint-study-effects-Ethiopian-dam-on-Nile/4552902-4570178-o5ju7nz/index.html
Pronto A (2015) Understanding the hard/soft distinction in international law. Vand J Transnat'l L 48
Qerenso W (2018) Egypt: better to unshackle the Nile. In: Reporter (May 12, 2018). https://www.thereporterethiopia.com/article/egypt-better-unshackle-nile

Rangeley R et al (1994) International river basin organizations in Sub-Saharan Africa. World Bank, Technical Paper No. 250
Salman S (2010) Downstream riparians can also harm upstream riparians: the concept of foreclosure of future uses. Water Int:350–352
Salman S (2011) Grand Ethiopian renaissance dam: challenges and opportunities. CIP Rep
Salman S (2013) The Nile Basin Cooperative Framework Agreement: a peacefully unfolding African spring? Water Int 38:17–29
Salman S (2016) The grand Ethiopian renaissance dam: the road to the declaration of principles and the Khartoum document. Water Int 41:512–527
Salman S (2017) The declaration of principles on the grand Ethiopian renaissance dam: an analytical overview. In: Yihdego Z, Desta MG, Merso F (eds) Ethiopian yearbook of international law
Salman S (2018) The GERD and the revival of the Egyptian-Sudanese dispute over the Nile waters. In: Yihdego Z et al (eds) Ethiopian yearbook of international law
Schiffler M (1998) Conflicts over the Nile or conflicts on the Nile? In: Scheumann W, Schiffler M (eds) Water in the Middle East: potential for conflicts and prospect for cooperation, pp 137–146
Simone M (2013) On the forecast: easing tension over hydraulic power along the Mekong. Project on Int'l Peace & Sec., Policy Brief No. 5.2
Shaw M (2008) International law. Cambridge University Press
Solomon S, Elshinnawi M (2019) Ethiopia condemns Egyptian proposal for Nile water usage. In: Voice of America. https://www.voanews.com/africa/ethiopia-condemns-egyptian-proposal-nile-water-usage
Swain A (2002) The Nile river basin initiative: too many cooks, too little broth. SAIS Rev 22:293–308. https://doi.org/10.1353/sais.2002.0044
Tadesse F (2017) Egypt proposes to exclude Sudan from dam talks. In: Addis Fortune. https://addisfortune.net/articles/egypt-proposes-to-exclude-sudan-from-dam-talks/
Tawfik R (2015) Revisiting hydro-hegemony from a benefit-sharing perspective: the case of the grand Ethiopian renaissance dam. Deutsches Institut für Entwicklungspolitik, Bonn
Tawfik R (2016) The grand Ethiopian renaissance dam: a benefit-sharing project in the Eastern Nile? Water Int 41:574–592. https://doi.org/10.1080/02508060.2016.1170397
Tekuya M (2018) The Egyptian hydro-hegemony in the Nile basin: the quest for changing the status quo. J Water Law 11
Tekuya M (2019) Governing the Nile under climatic uncertainty: the need for a climate-proof basin-wide treaty. Nat Res J 59:321
Tembata K, Takeuchi K (2018) Collective action and cooperation in drought response. In: Global Water Forum. https://globalwaterforum.org/2018/06/18/collective-action-and-cooperation-in-drought-response/
The Independent (2019) Where's my favorite dictator?' Trump comment on Egyptian President 'Met with Stunned Silence', report says. In: Independent. https://www.independent.co.uk/news/world/americas/us-politics/trump-egypt-president-sisi-favorite-dictator-meeting-a9104951.html
Thurow R (2003) Ravaged by famine: Ethiopia finally gets help from the Nile. Wall St J
UN Press Release (2020) Press release, Secretary-General, Secretary-General encouraged by good progress in trilateral talks to reach agreement over grand Ethiopian renaissance dam, U.N. Press release SG/SM/20088 (May 19, 2020)
UN Security Council (2020a) Letter dated 19 June 2020 from the permanent rep. of Egypt addressed to the President of the security council, U.N. Doc. S/2020/566 (Jun 19, 2020)
UN Security Council (2020b) Letter dated 22 June 2020 from the permanent rep. of Ethiopia addressed to the President of the security council, U.N. Doc. S/2020/567 (Jun 22, 2020)
UN Security Council (2020c) Letter dated 24 June 2020 from the permanent rep. of the Sudan addressed to the President of the security council, U.N. Doc. S/2020/586 (Jun 24, 2020)

US Department of the Treasury (2019) Joint statement of Egypt, Ethiopia, Sudan, the United States, and the World Bank. US Department of the Treasury. https://home.treasury.gov/news/press-releases/sm827, https://perma.cc/UE78-MDZH

US Department of the Treasury (2020a) Press release, Steven T Mnuchin, U.S. Sec'y of the Treasury, Statement on the grand Ethiopian renaissance dam. https://home.treasury.gov/news/secretary-statements-remarks/statement-by-the-secretary-of-the-treasury-on-the-grand-ethiopian-renaissance-dam

US Department of the Treasury (2020b) Press release, U.S. Dep't of the treasury, joint statement of Egypt, Ethiopia, Sudan, the United States and the World Bank (Jan 15, 2020). https://home.treasury.gov/news/press-releases/sm875

Waldock H (1962) First report on the law of treaties 2 Y.B. INT'L L. COMM'N 27, 32, U.N. Doc. A/CN. 4/144

Widakuswara P (2020) No deal from US-brokered Nile dam talks. In: VOA News. https://www.voanews.com/africa/no-deal-us-brokered-nile-dam-talks

Wirtschafter J (2019) U.S.-led talks lead to breakthrough as African rivals inch closer to war over Nile. Washington Times. https://www.washingtontimes.com/news/2019/nov/11/steven-mnuchin-mediated-talks-see-grand-ethiopian, https://perma.cc/5NP5-U5YQ

Worku K (2020) Egypt is sailing against the wind, squandering its opportunity. Diplomat Zerihun Abebe. Ethiopian Herald/Ethiopian Press (Mar 26, 2020). https://www.press.et/english/?p=20054

Xuequan M (2017) Egypt seeks world bank as technical mediator in Ethiopia's dam issue. In: Xinhua. http://www.xinhuanet.com/english/2017-12/27/c_136853702.htm

Yigzaw AY (2020) Telephone interview. Diplomat, Ministry of Foreign Affairs of Ethiopia (Apr 13, 2020)

Yihdego Z (2017) The fairness 'dilemma' in sharing the Nile waters: what lessons from the grand Ethiopian renaissance dam for international law? Brill Res Perspect Int Water Law 2:1–80. https://doi.org/10.1163/23529369-12340006

Yihdego Z (2020) International law connotations of US-'mediated' Nile dam negotiations and outcomes. 24

Chapter 4
Resolving the Grand Ethiopian Renaissance Dam Conflict Through the African Union Nexus Approach

Emmanuel Kasimbazi and Fredrick Bamwine

Abstract Egypt, Sudan and Ethiopia have been negotiating for nearly a decade to reach an agreement on key technical and legal issues related to the impact of the GERD. Some of the major milestones in the negotiation process are: the formation of the International Panel of Experts, Declaration of Principles, formation of a Joint Research Group, involvement of the USA and the World Bank to observe tripartite talks and the request by Egypt to the United Nations Security Council to intervene. The outstanding issues to be resolved include: drought mitigation, binding agreement, dam safety and dispute resolution. African Union (AU) has been approached to intervene in the dispute. The involvement of AU provides an opportunity for the continental peace architectural framework through three ways: the Assembly of Heads of state and government which is the AU's supreme policy and decision-making organ, the AU Peace and Security Council (APSC) which is the pinnacle of the AU architecture framework of conflict prevention, management and resolution and the involvement of COMESA which is the largest Regional Economic Community of the African Union aimed at promoting regional integration through trade and the development of natural and human resources. The main purpose of the chapter is to analyse how the AU peace building frameworks can be used to resolve the GERD dispute. The chapter will explain how the AU frameworks can be used to promote integrated and sustainable management outcomes and the significant steps of dispute resolution mechanisms that can be taken through the AU framework.

Keywords African Union · COMESA GERD · Dispute · Transboundary water negotiation · Egypt · Ethiopia · Sudan · Blue Nile River

---

E. Kasimbazi (✉)
Environmental Law Centre, School of Law, Makerere University, Kampala, Uganda
e-mail: ekasimbazi@law.mak.ac.ug

F. Bamwine
Elisabeth Hub School of Law, Pace University, New York, USA

A. M. Melesse et al. (eds.), *Nile and Grand Ethiopian Renaissance Dam*,
Springer Geography, https://doi.org/10.1007/978-3-030-76437-1_4

Introduction

The Grand Ethiopian Renaissance Dam formerly known as the Millennium Dam and sometimes referred to as Hidase is a gravity dam on the Blue Nile River in Ethiopia under construction since 2011 (Dunne 2020). The dam is in the Benishangul-Gumuz Region of Ethiopia, about 15 km (9 mi) east of the border with Sudan. The primary purpose of the dam is electricity production to meet Ethiopia's domestic electricity demand as well as electricity for export to neighbouring countries (Attalla 2013). With a planned installed capacity of 6.45 gigawatts, the dam will be the largest hydroelectric power plant in Africa when completed (Attalla 2015).

The eventual site for the GERD was identified by the United States Bureau of Reclamation in the course of the Blue Nile survey which was conducted between 1956 and 1964 during the reign of Emperor Haile Selassie (Asim 2018). Due to the coup d'état of 1974, however, the project failed to progress. The Ethiopian government surveyed the site in October 2009 and August 2010 and in November 2010, and a design for the dam was submitted by James Kelson. On 31 March 2011, a day after the project was made public, a US$4.8 billion contract was awarded without competitive bidding to Salini Impregilo, and the dam's foundation stone was laid on 2 April 2011 by then Prime Minister Meles Zenawi.

The potential impacts of the dam have been the source of severe regional controversy. Egypt, located over 2500 km downstream of the site, opposes the dam, which it believes will reduce the amount of water available from the Nile (Abdinor 2020). The Government of Egypt, a country which relies heavily on the waters of the Nile River, has demanded that Ethiopia cease construction on the dam as a precondition to negotiations has sought regional support for its position, and some political leaders have discussed methods to sabotage it (Soliman et al. 2016). Egypt has planned a diplomatic initiative to undermine support for the dam in the region as well as in other countries supporting the project particularly donor countries (Bayeh 2016). However, other nations in the Nile Basin Initiative have expressed support for the dam, including Sudan, the only other nation downstream of the Blue Nile (Asiedu 2018a, b).

The GERD represents another challenge on the Nile Basin that remains the only major international basin without an inclusive institutional framework for its utilization and management (Abdinor 2020). The result of the absence of an inclusive institutional framework coupled with the lack of a mutually acceptable and legally binding agreement to govern the Nile waters has continued to escalate the dispute between the two nations (Egypt and Ethiopia). Indeed, the GERD continues to raise tension between the two downstream nations over the utilization of the Blue Nile River. For instance, Ethiopia denies that the dam will have a negative impact on downstream water flows and contends that the dam will, in fact, increase water flows to Egypt by reducing evaporation on Lake Nasser. On the other hand, Egypt is demanding to increase its share of the Nile water flow from 66 to 90%.

Background and Context to the GERD Dispute

The GERD dispute is as a result of two major issues. The first one relates to water security which leads to the second which is the lack of a clear legal framework for Nile water allocation (Bulto 2008). These issues are directly linked to colonial-era Nile treaties that were concluded as a result of the Scramble for Africa. During this period, controlling the source of the Nile was a major colonial goal for the British and indeed several treaties were signed and outstanding of these treaties was the 1902 treaty that was concluded between Britain and Ethiopia. This is famously known as the Anglo-Ethiopian Treaty in which Ethiopia agreed not to arrest or totally block the flow of the Nile. Another treaty was concluded in 1929 between Britain (on behalf of its colonies, Sudan, Kenya, Tanzania and Uganda) and Egypt. The implication of this treaty is that it gave Egypt veto powers over the Nile River waters and prevented British East African colonies from using the waters of the Nile without Egypt's consent.

The other treaty that has implications for the GERD is 1959 agreement concluded between Egypt and Sudan. The role of this treaty in the current GERD dispute is that through this treaty Egypt and Sudan allocated the entire flow of the Nile to the two countries (Egypt and Sudan) without considering the interests of other upstream states. It is this act that caused other upstream states to reject the provisions of the treaty and to call for an equitable allocation of the waters of the Nile River based on a basin-wide treaty. Indeed, all the basin countries at the time negotiated and came up with the Nile Basin Cooperative Framework Agreement in 2010. However, Egypt and Sudan rejected the deal because it did not recognize their "historic right" and "veto power" over upstream projects (Shams 2013). Indeed, both Egypt and Sudan claim that the Nile Basin Cooperative Framework Agreement violates the 1959 treaty, in which Sudan and Egypt give themselves exclusive rights to all of the Nile River waters. This legacy meant that Ethiopia was left with no option but to start constructing the GERD by itself which has created tension between downstream Egypt and Sudan on the one hand and upstream Ethiopia on the other, thus culminating into the current GERD dispute.

The crucial leverage regarding Egypt's water security lies with the Blue Nile countries Ethiopia and Sudan, as the Blue Nile is the main contributor to the Nile River's flow downstream. In fact, about 85% of the overall Nile flow originates on Ethiopian territory (Swain 2011). Ethiopia's determination to build the GERD for hydropower purposes has been the flashpoint of current conflicts in the Eastern Nile Basin (Gebreluel 2014). The dam has the potential to change the status quo established in the colonial-era Nile treaties which give veto powers over the Nile water to the two downstream nations of Egypt and Sudan. At the same time, any agreement also has the potential to maintain the status quo. So, behind the ongoing talks is the struggle between changing, or maintaining, colonial legacies; however, the Nile Basin Initiative provides a framework for dialogue among all Nile riparian countries.

The Nature of Legal Disputes Raised by GERD

The GERD dispute is a transboundary water resource conflict between Egypt, Ethiopia and Sudan. The dispute is exacerbated by lack of a clear and universally binding legal agreement among Nile Basin states (Bulto 2008). From the dispute, Egypt fears a temporary reduction of water availability due to the filling of the dam and a permanent reduction because of evaporation from the reservoir (Abdinor 2020). Studies indicate that the primary factors which will govern the impacts during the reservoir filling phase include the initial reservoir elevation of the Aswan High Dam, the rainfall that occurs during the filling period and the negotiated arrangement between the three countries. These studies also show that only through close and continuous coordination, will the negative impacts be minimized or eliminated.

The reservoir volume, 74 bcm (billion cubic metres), is about 1.5 times the average annual flow (49 bcm) of the Blue Nile at the Sudanese–Egyptian border. This loss to downstream countries could be spread over several years if the countries reach an agreement. Depending on the initial storage in the Aswan High Dam and the filling schedule of the GERD, flows into Egypt could be temporarily reduced, which may affect the income of two million farmers during the period of filling the reservoir. Allegedly, it would also "affect Egypt's electricity supply by 25–40%, while the dam is being built". However, hydropower accounts for less than 12% of total electricity production in Egypt in 2010 (14 out of 121 billion kWh), so that a temporary reduction of 25% in hydropower production translates into an overall temporary reduction in Egyptian electricity production of less than 3% (Soliman et al. 2016).

The GERD could also lead to a permanent lowering of the water level in Lake Nasser if floods are stored instead in Ethiopia. This would reduce the current evaporation of more than 10 billion cubic metres per year, but it would also reduce the ability of the Aswan High Dam to produce hydropower to the tune of a 100 MW loss of generating capacity for a 3 m reduction of the water level. However, the increased storage in Ethiopia can provide a greater buffer to shortages in Sudan and Egypt during years of future drought if the countries can reach a compromise. Ethiopia further contends that the dam will retain silt. It will thus increase the useful lifetime of dams in Sudan—such as the Roseires Dam, the Sennar Dam and the Merowe Dam—and of the Aswan High Dam in Egypt.

The beneficial and harmful effects of flood control would affect the Sudanese portion of the Blue Nile, just as it would affect the Ethiopian part of the Blue Nile valley downstream of the dam. Specifically, the GERD would reduce seasonal flooding of the plains surrounding the reservoir of the Roseires Dam located at Ad-Damazin, just as the Tekeze Dam had done by retaining a reservoir in the deep gorges of the Northern Ethiopian Highlands and reduced flooding at Sudan's Khashm el-Girba Dam.

The reservoir, located in the temperate Ethiopian Highlands and up to 140 m deep, will experience considerably less evaporation than downstream reservoirs such as Lake Nasser in Egypt, which loses 12% of its water flow due to evaporation as the water sits in the lake for 10 months. Through the controlled release of water from the reservoir to downstream, this could facilitate an increase of up to 5% in Egypt's water supply and presumably that of Sudan as well.

The colonial-era Nile treaties and the subsequent agreements continue to dictate the GERD dispute negotiations. The colonial-era Nile treaties protect the interests of the two downstream nations of Egypt and Sudan in total disregard for the other upstream nations including Ethiopia whose current developments in building and filling of the GERD are source of the current dispute. On the other hand, the modern developments among the Nile Basin states such as the NBI and the CFA focus on universal allocation, use and sustainable management of the Nile River which is a departure from the interests of both Egypt and Sudan that seek to maintain their original ancient monopoly over the Nile River waters stipulated in the 1902, 1929 and 1959 treaties.

Analysis of the Nile Agreements and Their Implications for GERD Disputes

The Anglo-Ethiopian Treaty 1902

This treaty was designed to determine the borders between Ethiopia and Sudan, and the treaty was signed between the Emperor Menelik and the British agent in Ethiopia, John Lane Harrington. Article III of the treaty "achieved a long-standing British aim to safeguard the unimpeded flow of the waters from the Blue Nile and Lake Tana". Text from the article reads:

> His majesty the emperor MENELEK II, King of Kings of Ethiopia, engages himself toward the government of his Britannic Majesty not to construct or allow to be constructed any work across the Blue Nile, Lake Tsana or the Sobat, which would arrest the flow of their waters into the Nile, except in agreement with his Britannic Majesty's Government and the Government of the Sudan.

Although this agreement was to regulate the frontiers between Ethiopia and colonial Britain (Anglo-Egyptian Sudan), it contained a peculiar stipulation on the use of Nile waters, which the Ethiopian emperor Menelik apparently had signed on to due to a mistranslation between the English and Amharic versions. Ethiopia has, moreover, ever since renounced this agreement—calling it illegitimate—invoking the Egyptian and Sudanese practice of denouncing "unequal" colonial-era treaties when these are not in their interest. Ethiopia has thus not considered the purported obligation to obtain Egyptian and Sudanese consent binding (Tadesse 2008).

The Anglo-Egyptian Nile Waters Agreement 1929

The agreement was signed between Egypt and Great Britain, which at the time represented Uganda, Kenya, Tanganyika (now Tanzania) and Sudan. It granted Egypt the right to veto projects higher up the Nile that would affect its water share. Britain acknowledged these natural and historical rights as "acquired rights" by allocating the entire utilizable annual discharge to Egypt and Sudan. In effect, this put severe restrictions on upstream water use. None of the upstream riparian states were part of this agreement, and thus, after independence, all upper riparian states have rejected this treaty. Egypt and Sudan still consider it binding with reference to the principle of universal state succession.

However, the upstream states refute this principle and the agreement under the auspices of the "Nyerere Doctrine" of selective succession to treaties, arguing that international agreements dating from colonial times should be renegotiated when a state becomes independent as the nation should not be bound by something it was not in a sovereign position to agree to at that time (UNEP). Nevertheless, the wordings of acquired, natural and historical rights constitute the backbone of Egypt's subsequent approach to Nile issues, as reproduced in the superseding 1959 agreement (Lumumba 2007).

Nile River Bilateral Agreement Between Egypt and Sudan, 1959

The agreement gave Egypt the right to 55.5 bcm of Nile water a year and Sudan 18.5 bcm per year (Bayeh 2016). The agreement allocated all Nile waters to the two countries, reinforced the downstream claim of "natural and historical rights" to Nile waters and became both Sudan and Egypt's "redline" for future negotiations in the basin. The agreement has been described as patently anomalous as "while it is purely bilateral, it seeks to apportion the entire flow of the Nile to Egypt and Sudan, excluding the interests of any other riparian, notably Ethiopia" (Bayeh 2016).

The treaty has been lauded as the first ever agreement between two independent riparians signalling a new era in the management of the Nile Basin, but has also been criticized for replicating colonial-style treaties as its main thrust is to sanction a monopoly on the waters of the Nile by Egypt and the Sudan (Tafesse 2001). The upper riparian states have rejected the legal foundation of the agreement and its binding force under the auspices of the principle that a treaty binds the parties and only the parties—bilateral treaties do not impose any obligations, nor confer any rights, on third states.

The agreements so far made in regard to the Nile water are of limited scope in their application as none have managed to involve all riparian states, and thus they have been disputed by those not part of, but affected by, the agreements (Abdo 2004). Owing to their bilateral nature, previous treaties undermine the emergence of a basin-wide shared understanding and a communal identity as riparian societies, thus also preventing regional integration and trust building. As with the case of upstream Ethiopia and downstream Egypt and Sudan there is no mutually agreed or acceptable framework.

Indeed, the lack of a comprehensive and binding agreement has precluded management and development of Nile water. The downstream states' selective insistence on existing agreements has also become an obstacle for regional cooperation and comprehensive treaties.

Framework Agreement for General Cooperation Between Egypt and Ethiopia, 1993

This agreement was signed between former Egyptian President Hosni Mubarak and Meles Zenawi, the president of the transitional government of Ethiopia in July 1993. The two countries agreed to settle their Nile water disputes under the framework of international law and based on expert discussions. They also agreed that neither country would engage in any activity deemed harmful to the other's interests.

The Nile Basin Initiative (NBI) Partnership, 1999

This is an intergovernmental partnership of ten (10) Nile Basin countries, namely: Burundi, DR Congo, Egypt, Ethiopia, Kenya, Rwanda, Sudan, Tanzania and Uganda. Eritrea is included but participates as an observer. On 22 February 1999, NBI was established as a forum for consultation and coordination among the eleven (10) countries for mutually beneficial sustainable management and development of the shared Nile Basin water and related resources. The countries participating in the NBI are now 11 because the Republic of South Sudan was admitted to the NBI by the Nile Council of Ministers during their 20 regular meeting held on 5 July 2012 in Kigali, Rwanda. One of the key reasons for South Sudan joining the NBI is that geographically, the country falls wholly (98%) within the River Nile Basin and therefore its growth and prosperity are undoubtedly linked to the developments within the River Nile Basin.

The NBI constitutes a departure from the past trend of unilateral water use and management, particularly in emphasizing the issues of fair water allocation, joint management and development of the resources. This signals a fundamental shift in the status quo in upstream–downstream relations, particularly in the eastern Nile Basin and in the relationship between Egypt, Sudan and Ethiopia.

Although the NBI has been successful in bringing all the riparian states together, there are challenges ahead. Consensus over the shared vision and the action programmes was not achieved without difficulties, but a mutually acceptable legal and institutional framework has been still more difficult to reach (Arsano 2010). The NBI has also been the most inclusive technical endeavour on Nile water to date. However, the participation of several countries has varied over the years. Eritrea only participates as an observer. When negotiations over the CFA ended in disagreement in 2010, Egypt and Sudan suspended their participation in the NBI, with Sudan resuming participation in 2014. Non-participation has affected implementation of some programmes within the NBI, in particular in the eastern Nile which includes the development of GERD.

The Cooperative Framework Agreement (CFA), 2010

This was developed over for more than a decade, and the CFA has yet to be ratified by Egypt and Sudan due to their reservations over Article 14b. The article reads as follows: "not to significantly affect the water security of any other Nile Basin States". All countries (Burundi, DR Congo, Ethiopia, Kenya, Rwanda, Tanzania and Uganda) agreed to this text except Egypt and Sudan. The two countries propose the following wording: "not to adversely affect the water security and current uses and rights of any other Nile Basin State".

The CFA intends to establish a framework to "promote integrated management, sustainable development, and harmonious utilization of the water resources of the Basin, as well as their conservation and protection for the benefit of present and future generations". It aims for the establishment of a permanent commission, the Nile River Basin Commission (NRBC) that would serve to promote and facilitate the implementation of the agreement. After Egypt and Sudan failed to agree on the CFA in 2010, Ethiopia embarked on the dam project.

The Nile Cooperative Framework Agreement (CFA) aims to provide an agreement on legal principles which will determine a reasonable and equitable solution for sharing Nile waters among the basin states. Upstream countries have insisted this new framework must disregard all previous agreements to which they were not part. Conversely, downstream countries seek a new framework to incorporate such earlier agreements. Further, the NBI still battles with establishing a legal framework involving all stakeholders, which is within the purview of the NBI's shared vision programme. The Nile River Basin Cooperative Framework signed and ratified by Ethiopia is based on principles of international water law. The agreement did not answer the main question of States' water share or distribution of water among them.

The CFA has thus far been ratified by Ethiopia, Tanzania, Rwanda and Uganda. It needs a total of six instruments of ratification/accession to enter into force. Egypt and Sudan continue to vehemently reject the CFA.

Issues Dominating Negotiations

As observed, the issues dominating negotiations ideally emanate from the colonial treaties pertaining the Nile which have hampered any effort to establish a comprehensive and binding treaty. Through the 1959 agreement, the Nile water is divided between the Sudan and Egypt to the total exclusion of the upstream states. It explicitly created rights enjoyed by both Egypt and Sudan and was silent about the rights of the other upstream states (Shams 2013).

According to Ethiopia, the GERD confers enormous benefits to Egypt and Sudan. These include ensuring a regular flow of water, preventing siltation, reducing evaporation and providing cheaper electricity. Sudan has supported the project since 2012 because of these benefits. But Egypt maintains that any upstream dam on the Nile River threatens the flow of the Nile, thus raising the following issues which have dominated the negotiations.

The main issues that are dominating the negotiations are as follows.

Drought Mitigation

Egypt is mainly concerned about the management of drought during the years of filling and operating the dam, as well as in the years that follow a natural drought. Egypt says if flood levels are high during the first filling, Egypt will not encounter any major shortage of water (Middle East Eye 2020). The problem will occur if there is a natural drought either during the filling or initial operation years of the dam. A drought can be either man-made or natural. A man-made drought in the case of the GERD will occur because any water entering the dam's reservoir will reduce the amount of water going downstream to Egypt and Sudan. In the event of a natural drought, with a dearth of rainfall upstream, Ethiopia may be called on to release water to Egypt and Sudan from the GERD, thus depleting the dam's reservoir. It will then need to refill it, causing additional shortages downstream.

The Dam Filling

Egypt proposes that in case of drought during the filling, Ethiopia will generate a maximum of 85% of its electricity needs, slow down the filling process and allow water to reach Egypt to make up for the natural drought. Egypt needs a guarantee that the filling and operation of the Renaissance Dam will not affect this arrangement and what it calls "existing use and rights". In other words, Egypt needs a guarantee that Ethiopia will not use the Nile waters for consumption purposes, including irrigation, in future. Cairo wants an agreement that the dam's reservoirs be filled over a long period, lasting about twenty (20) years. And it wants veto power. Apart from some safety-related concerns, Sudan supports the project. The parties have agreed on how Ethiopia should fill the dam and how much water it would release when there is sufficient rainfall. Ethiopia has also agreed to release predetermined amounts of water—depending on the dam level and Blue Nile inflow into the reservoir, even in drought years.

The Dam Operation

The Aswan Dam reservoirs will be diminished after the GERD filling. Therefore, Egypt further insists that if drought occurs at that time, the GERD reservoir should be used to provide Egypt with water to make up for the reduction in the Aswan Dam. Ethiopia initially agreed on the above proposals in Washington in February and then later retracted its decision. However, Addis Ababa insists that committing a specified volume of water during periods of drought will ultimately drain the reservoir, thereby impeding its ability to generate the electricity it badly needs, according to Daniel C Stoll, Associate Dean for global affairs at St. Norbert College in the USA. It also believes that Egypt is trying to perpetuate what it regards as Egypt's unfair claim to substantial amounts of the Nile's waters. Ethiopia argues that any drought should also be mitigated by water stored in Sudan's and Egypt's dams, not just by the GERD.

Binding Agreement

Egypt and Sudan have demanded a written, binding agreement between the parties regarding Ethiopia's commitments to prevent harm to the downstream countries. Ethiopia has so far only made verbal pledges and refused to sign a binding agreement. Addis Ababa believes that the 2015 Declaration of Principles (DoP) is sufficient to demonstrate its respect for the no-harm principle but Egypt and Sudan argue that they cannot rely on its goodwill. The three countries signed the Declaration of Principles in 2015. This provides the framework for the talks about the first filling and annual operation of the dam. But Egypt's concerns seem to have changed towards ensuring it gets its "historic water share" as stated under the 1959 treaty. That would be 55.5 bcm, 66% of the river's total flow. The treaty also gave Sudan 22% and left the rest (12%) for evaporation. It did not recognize the rights of nine upstream countries, including Ethiopia, whose territory contributes more than 85% of the Nile.

Dispute Resolution

Whereas Ethiopia would like to settle future disputes through negotiations, Egypt and Sudan have been in favour of binding international arbitration. Ethiopia opposes arbitration due to the absence of a Nile Basin agreement that could be used by international arbitrators to settle water allocation disputes. Moreover, according to Stoll, Ethiopia believes that it enjoys a privileged position, given that it is the source of the Blue Nile and it has the ability to control the dam and what the flow will look like downstream. It fears that arbitration will be a limit on its power and influence, and therefore it prefers direct negotiations.

Dam Safety

The Sudan and Egypt have concerns about the safety measures implemented during the construction of the dam and the potential consequences of any faults in the dam for their countries. For example, Sudan, which lies only 20 km from the dam, is concerned that releases of water from the GERD have the potential to flood its Roseires Dam, if not coordinated properly. It is demanding that Ethiopia provides more assurances on the management of the GERD's reservoir and its safety standards. In the event of the collapse of the dam due to faults in its construction, all of Sudan would be at risk.

Chronological Developments of the GERD Negotiation Process

International Panel of Experts (IPoE), 2012

The International Panel of Experts (IPoE) was founded on 15 May 2012, comprising experts from Egypt, Sudan, Ethiopia as well as international specialists. The panel consisting of ten (10) members; six (6) from the three countries and four (4) international in the fields of water resources and hydrologic modelling, dam engineering, socioeconomic, and environmental fields held its fourth meeting in Addis Ababa in November 2012. It reviewed documents about the environmental impact of the dam and visited the dam site. The panel submitted its preliminary report to the respective governments at the end of May 2013. The Ethiopian government stated that, according to the report, "the design of the dam is based on international standards and principles" without naming those standards and principles. It also said that the dam "offers high benefit for all the three countries and would not cause significant harm on both the lower riparian countries". According to Egyptian government, however, the report "recommended changing and amending the dimensions and the size of the dam".

Declaration of Principles (DoP), 2015

This was signed by Egypt, Sudan and Ethiopia in Khartoum in March 2015; the DoP outlined a set of principles that should guide the establishment of the GERD, including mutually beneficial coordination; sustainable energy supply; prevention of significant harm to the three countries; equitable and reasonable utilization of water resources; agreement on the first filling of the dam in accordance with expert recommendations; and the peaceful settlement of disputes arising from the dam construction.

Egypt, Sudan and Ethiopia Form Joint Research Group, 2018

The three countries formed the National Independent Scientific Research Study Group in May 2018 to assess the project's impact, filling and operation. Further, the fifteen (15)-member group was to discuss means of enhancing the levels of understanding and cooperation among the three countries with regard to the GERD and would mainly address "equitable and reasonable utilization of shared water resources while taking all appropriate measures to prevent the causing of significant harm".

US Treasury and the World Bank Observe Tripartite Talks, 2019

The 2018 study group failed to reach an agreement, giving way to further talks by the three countries in November 2019, this time in the presence of the US Treasury and the World Bank.

Negotiations, 2020

In 29 February 2020, Ethiopia refused a drought mitigation agreement (Middle East Eye 2020) According to Ethiopia, the negotiation for long-term operation became about water sharing, which should not have been part of the agenda. Rather, Ethiopia preferred discussions only on the reservoir operation that is restricted to the dam's inflows and outflows. To Ethiopia, Egypt and USA proposed an unamendable plan for permanent operation, which amounted to a "water allocation" arrangement that effectively protects Egypt's claimed 55.5 bcm annual share of Nile waters (Xinhua 2020). However, in 10 April 2020, Ethiopia proposed an agreement on the first two years of filling. Ethiopian Prime Minister Abiy Ahmed proposed an interim agreement to cover the first two years of filling the dam reservoir, but Cairo rejected it, accusing Addis Ababa of trying to delay a comprehensive deal while going ahead with the filling process. The parties resumed negotiations on 9 June 2020.

On June 19, Egypt asked UN Security Council to intervene. Egypt formally requested the UN Security Council to intervene and call Ethiopia back into the talks. In response, Ethiopia sent a letter to the council accusing Egypt of trying to exert pressure on it through the international venue. Despite resumption of negotiations, on June 15, technical teams failed to reach a deal. The technical teams of Egypt, Sudan and Ethiopia convened in mid-June to reach a deal, but they failed to reach agreement on the key issues of drought mitigation protocols and a dispute resolution mechanism (Middle East Eye 2020). On June 26, Egypt, Ethiopia and Sudan leaders agreed on African Union-led talks, and on June 29, UN Security Council encourages talks between the three countries. Under-Secretary-General Rosemary DiCarlo briefed the UN Security Council on the latest GERD crisis and hailed a recent agreement between the three countries to join an AU-led process of talks on outstanding issues.

African Union Architectural Approach to the GERD Negotiation and Dispute Resolution

There are three ways through which the GERD dispute can be resolved through the AU architectural framework.

The Role of Assembly of Heads of State and Government of AU

This is the AU's supreme policy and decision-making organ. It comprises all Member State Heads of state and governments and the Regional Economic Communities (RECs) which are regional groupings of African states (Williams 2011). It determines the AU's policies, establishes its priorities, adopts its annual programme and monitors the implementation of its policies and decisions.

The assembly is mandated to accelerate the political and socio-economic integration of the African continent. It may give directives to the Executive Council and the Peace and Security Council on the management of conflicts, emergency situations and the restoration of peace. Tapping into the mandate of the Assembly of Heads of state and government of AU to champion negotiations gives the whole process of GERD dispute negotiation a sense of ownership at the same time serving the interests of the African continent since the impact of the dispute is beyond the three rival states of Egypt, Ethiopia and Sudan but covers all the Nile Basin states.

AU Peace and Security Council (APSC)

This is the pinnacle of the AU architecture framework because it is the standing decision-making organ of the AU for the prevention, management and resolution of conflicts. It is a collective security and early warning arrangement intended to facilitate timely and efficient responses to conflict and crisis situations in Africa. It is mandated to promote harmonization and coordination of efforts between the regional mechanisms and the AU in the promotion of peace, security and stability in Africa.

The pillars of APSA include the Panel of the Wise (PoW) which is comprised of a five-person panel of "highly respected African personalities from various segments of society and support the APSC and the Chairperson of the AUC in the promotion and maintenance of peace, security and stability in Africa, particularly in the areas of preventive diplomacy and mediation" (Williams 2011).

The Common Market for Eastern and Southern Africa (COMESA)

This is the largest Regional Economic Community (REC) of the African Union formed with the aim of promoting regional integration through trade and the development of natural and human resources for the mutual benefit of all people in the region. COMESA has twenty-one (21) member states which include Egypt, Ethiopia and Sudan hence providing another opportunity through which the three (3) states can negotiate through the AU framework.

The role of COMESA in the negotiations of the GERD dispute can be played through the COMESA Committee of Elders established during the 11th Summit of COMESA, held in Djibouti in 2006, to address conflict trends among and within its member states. The committee's mandate is to steer the course of preventive diplomacy for conflict resolution and to augment COMESA's peace-building and peace-making processes, many of which require the participation of multiple actors and well-coordinated interventions in order to effectively address the region's complex challenges.

Lessons from Other Transboundary Water Disputes

Euphrates Catchment Disputes

Since the dawn of civilization, a number of cities were built within the Euphrates River basin. This river is the longest river in southwest Asia, and now there are about 23 million people who rely on the water of this river to maintain their living.

Before 1974, hydraulic structures and irrigation projects were not as many or extensive as to affect the discharge of the river so much. After that and due to increased population growth rate and development, the riparian countries started to utilize its water extensively. Turkey started an ambitious plan to develop the GAP project in this context.

In addition, the Middle East is experiencing drought periods due to climate change, which made water shortage problems more severe. As a consequence, the lower riparian countries (Syria and Iraq) are experiencing water shortage problems and deterioration of water quality. Large agricultural areas are expected to turn into desert in the near future. This has caused friction among riparian countries in the past and could lead to even greater conflicts in future. International, regional and national actions are mostly needed, and stronger cooperation is required to overcome this problem and to avoid its negative consequences on the people and the environment within the river's basin.

A key lesson from the Euphrates catchment dispute is that addressing trans-boundary water resource disputes requires the disputing parties to look beyond individual national interests and come up with a common goal of protecting the resource. In fact, despite the long-standing dispute in the Euphrates catchment, it is evident that the states have always come to a compromise to address key issues such as food security. Unlike in the GERD dispute where all parties are struggling to assert their individual national interests, the joint efforts of Iraq, Syria and Turkey in working towards an amicable solution have made some progress. For example, Turkey has proposed a three-step plan: (1) establish a database for available water resources assessment; (2) create a database of land resources; and (3) create a database of water resources (Schneider 2017). The Euphrates catchment dispute therefore should guide the current GERD dispute negotiations to search for a common ground in the form of proposals for protecting the Nile water resource.

Indus River Dispute

The roots of the conflict between India and Pakistan can be traced to the bitter and bloody circumstances under which the two South Asian nations emerged onto the global stage in 1947. The intertwined nature of the Kashmir and Indus disputes has direct linkage to the Radcliffe boundary award, according to which the British Punjab was divided between India and Pakistan at the time of Partition of the Subcontinent, and under which India gained control of the head-works of two rivers providing irrigation in West Punjab (Pakistan) and the only land link (from Indian territory) to the princely state of Kashmir, through a road over Madhopur head-works. Consequently, by capturing parts of Kashmir, India gained access to the catchment areas of the whole of the Indus river system, where its five tributaries—the Jhelum, Chenab, Ravi, Sutlej and Beas—originate. Kashmir has continued to be the bone of contention in their relations.

Maharaja Hari Singh, the ruler of the "princely state" of Kashmir, sought the continuation of independent status and offered a "standstill agreement" to both India and Pakistan. The offer was accepted by the latter but rejected by the former. The Muslims of Kashmir revolted against the Maharaja, allegedly demanding accession of the state to Pakistan. India launched a military offensive on 26 October 1947, claiming that the Maharaja had signed an instrument of accession with its leaders. On 1 April 1948, India cut off the irrigation water from the rivers flowing into Pakistan. Then, in May 1948, Pakistan also mobilized its troops. Both sides captured parts of Kashmir territory.

Posturing for a peaceful resolution, India referred the issue to the United Nations Security Council (UNSC) and both countries accepted the UN supervised ceasefire, agreeing to its resolution of instituting a plebiscite under its supervision, which has not been implemented so far. Since then, the only projected ongoing cause of the Kashmir conflict centres around the idea of conflicting ideologies: on the one hand, India is seeking to maintain its "secular outlook" and negate the very rationale behind the creation of Pakistan, the "two-nation theory", by retaining control over a Muslim majority state, Jammu and Kashmir, while on the other hand Pakistan is struggling for the region's "liberation" from the Indian "yoke", aiming for its integration with it. To resolve the dispute, the Indus Waters Treaty 1960 was concluded between India and Pakistan. The Indus water treaty is perhaps quoted as the most successful water-sharing mechanism in the recent times (Qamar et al. 2019). The treaty did not divide the waters of the Indus River but the rivers, and the treaty also created a commission particularly to exchange information and provide notice on the parties' developments on the river. This arrangement has always been the safeguard against disputes and conflict over the Indus River from which the GERD negotiations can borrow a leaf to come up with an amicable agreement to regulate the activities of Egypt and Ethiopia on the Blue Nile.

Conclusion and Recommendations

The GERD dispute negotiation between Ethiopia, Egypt and Sudan remains deadlocked, and if not resolved soon, it creates risks for three countries and other Nile Basin countries to be drawn into conflict because the stakes are so high. Since the GERD dispute is an African-based dispute, using approaches that are based on the African continent to resolve the dispute may be realistic options. The African continental peace architecture can provide an institutional framework for implementing the concept of a comprehensive peace that encompasses conflict prevention, peace-making, peacekeeping, post-conflict reconstruction and peace building. The African Union approach can adopt a three-step approach: first, through the Assembly of Heads of state and government; Heads of states and governments should work towards reducing mutual suspicion by taking a number of confidence-building measures. For example, they can facilitate joint study tour to the GERD construction site of political leaders and experts from Egypt, Ethiopia and Sudan as well neighbouring states such as South Sudan, Kenya and Uganda as the Chairing Head of State and African Union. Secondly, the three countries can negotiate a new, transboundary framework for resource sharing to avert future conflicts facilitated by AU Peace and Security Council (APSC) and the AU can facilitate the negotiation and oversee the enforcement of mutually agreed and legally binding agreement concluded by the parties. Thirdly, through COMESA joint feasibility studies that cover environmental, economic and legal aspects can be conducted to feed into the approaches above.

References

Abdinor HD (2020) Hydro-politics: the Grand Ethiopian Renaissance Dam and the future of the Nile River Basin. University of Oxford

Abdo M (2004) The Nile question: the accords on the water of the Nile and their implications on cooperative schemes in the Basin. https://www.researchgate.net/publication/294124784_The_Nile_Question_The_Accords_on_the_Water_of_the_Nile_and_Their_Implications_on. Accessed 28 Dec 2020

Arsano Y (2010) Institutional development and water management in the Ethiopian Nile Basin. In: Tvedt T (ed) The River Nile in the post-colonial age. I.B. Tauris, London, pp 161–178

Asiedu M (2018a) The construction of the Grand Ethiopian Renaissance Dam (GERD) and geopolitical tension between Egypt and Ethiopia with Sudan in the mix. Policy Brief Global Political Trends Center (GPoT). http://www.gpotcenter.org/dosyalar/PB_50.pdf. Accessed 28 Dec 2020

Asiedu M (2018b) The Grand Ethiopian Renaissance Dam (GERD): Ethiopia, Egypt and Sudan push for cooperation. GPoT PU No. 15. http://www.gpotcenter.org/dosyalar/PU.pdf. Accessed 28 Dec 2020

Asim IM (2018) On the Grand Ethiopian Renaissance Dam (Gerd). Int J Environ Sci Nat Resour 11(2):40–44

Attalla RA (2013) The Grand Ethiopian Renaissance Dam. An interactive qualifying project report. Worcester Polytechnic Institute, 2015

Attalla RA (2015) Grand Ethiopian Renaissance Dam (GERD). An interactive qualifying project report submitted to the faculty of worcester polytechnic institute, October 15, 2015

Bayeh E (2016) Agreement on declaration of principles on the Grand Ethiopian Renaissance Dam project: a reaffirmation of the 1929 and 1959 agreements? ResearchGate. https://doi.org/10.4172/2151-6200.10000170

Bulto TS (2008) Between ambivalence and necessity: Occlusions on the path toward a basin-wide treaty in the Nile basin. 20 Colo J Int'l Envtl L & Pol'y 291, HEINONLINE

Dunne CW (2020) The Grand Ethiopian Renaissance Dam and Egypt's military options. http://arabcenterdc.org/policy_analyses/the-grand-ethiopian-renaissance-dam-and-egypts-military-options/. Accessed 28 Dec 2020

Gebreluel G (2014) Ethiopia's Grand Renaissance Dam: ending Africa's oldest geopolitical rivalry? The Washington Quarterly 37(2):25–37

Lumumba LO (2007) The interpretation of the 1929 treaty and its legal relevance and implications for the stability of the region. Afr Sociol Rev 11(1):10–24

Middle East Eye (2020) The Grand Ethiopian Renaissance Dam (GERD): Ethiopia, Egypt and Sudan push for cooperation. Middle East Eye, July 2020, explained. The Grand Ethiopian Renaissance Dam. https://www.middleeasteye.net/news/grand-ethiopia-renaissance-dam-explained-egypt-sudan. Accessed 28 Dec 2020

Qamar MU, Azmat M, Claps P (2019) Pitfallsin transboundary Indus Water Treaty: a perspective to prevent unattended threats to the global security. JO-npj clean water, vol 2. https://doi.org/10.1038/s41545-019-0046-x. Accessed 28 Dec 2020

Soliman G, Soussa H, El Sayed S (2016) Assessment of Grand Ethiopian Renaissance Dam impacts using decision support system. J Comput Eng 18(5):8–18

Tadesse D (2008) The Nile: is it a curse or blessing? Pretoria and Addis Ababa: institute for security explained: The Grand Ethiopian Renaissance Dam, 10 July 2020. www.middleasteye.net/news. Accessed 5 Sep 2020

Tafesse T (2001) The Nile question: hydropolitics, legal wrangling, modus vivendi and perspectives: Lit Verlag. Munster, Germany

Schneider P (2017) Nexus approach as tool for transboundary water conflict resolution. Lecture Nexus Seminar series, UNU-FLORES and TU Dresden. https://flores.unu.edu/en/news/news/resolving-transboundary-water-conflicts-through-a-nexus-approach.html. Accessed 31 Dec 2020

Shams DH (2013) The long empty canyon: a study of the old/new legal problems of the Nile Basin. J Water Resour Ocean Sci 2(5):141

Swain S (2011) Challenges for water sharing in the Nile basin: changing geo-politics and changing climate, Hydrolog Sci J 56(4):687–702. https://doi.org/10.1080/02626667.2011.577037

Williams PD (2011) The African Union's conflict management capabilities. Council on Foreign Relations Inc., USA

Xinhua (2020) Egypt, Sudan, Ethiopia agree to form committee to finalize deal on controversial Nile dam. http://www.xinhuanet.com/english/2020-06/27/c_139170644.htm. Accessed 31 Dec 2020

Chapter 5
Conceptualization of Equitable and Reasonable Water Sharing in the Nile Basin with Quantification of International Transboundary Water-Sharing Principles

Mekdelawit Messay Deribe and Belete Berhanu

Abstract Transboundary waters account for a significant portion of the global freshwater resource. As such, effective water resource management should ideally have transboundary water management at its core. Despite this, however, two-thirds of the world's transboundary rivers do not have a cooperative management framework. Nonetheless, numerous international treaties, laws, and principles have existed throughout history to govern transboundary water sharing. The principle of limited territorial sovereignty forms the basis for modern customary water laws. There are notable widely accepted transboundary water management rules and principles based on the prinicple of limited territorial soverginity such as the Helsinki rules, the U.N. 1997 convention, and the Cooperative Framework Agreement (CFA) in the case of the Nile basin. However, the contextualization and quantifications of such rule to useable frameworks are still mostly lacking. This study outlines the evolution of transboundary water-sharing rules and principles and the history of water sharing in the Nile basin. It then presents factors suggested by international water-sharing principles to determine equitable use. The study contextualizes, quantifies, and weighs these factors for the Nile basin to evaluate different scenarios, which can be a base for fair and equitable water sharing. Finally, the authors forward possible recommendations for equitable and sustainable water use in the basin to move forward collectively.

Keywords CFA · Egypt · Equitable use · Ethiopia · Helsinki rules · International water-sharing principles · Nile · Sudan · Sustainable water use

---

M. M. Deribe (✉) · B. Berhanu
Addis Ababa Institute of Technology, Addis Ababa, Ethiopia
e-mail: mderi017@fiu.edu

B. Berhanu
e-mail: Belete.Berhanu@aait.edu.et

M. M. Deribe
Florida International University, Miami, FL, USA

A. M. Melesse et al. (eds.), *Nile and Grand Ethiopian Renaissance Dam*,
Springer Geography, https://doi.org/10.1007/978-3-030-76437-1_5

Introduction

The world's population has increased from around 1 Billion in 1800 to more than 7.7 billion today (Roser et al. 2013). Water demand has also been steadily rising, especially in the last century, growing at twice the population rate (McKinney 2011; FAO 2007). However, the global freshwater reserve has stayed constant if not diminished because of pollution and overuse over the centuries (Boretti and Rosa 2019). Moreover, climate change is altering weather patterns across the world and reducing the available water per capita in many areas (Lakshmi et al. 2018). Water is a renewable resource. However, the global freshwater reserve cannot indefinitely keep up with the ever-growing demands of the world. The steeply rising demand for water driven by economic growth and population increase, coupled with the effects of climate change, reduction, and pollution of water resources, put water scarcity front and center in the global agenda.

Currently, around 20% of the global population, more than 1.5 billion people, suffer from physical water scarcity, and a similar number suffers from economic water scarcity (Petruzello 2019). The U.N. estimates that 6 billion people will suffer from clean water scarcity by 2050 (WWAP 2018). Without adequate water resource management, freshwater availability will fail to keep up with global water demand as early as 2040 (Global Water Security 2012). Hence, sustainable management of global water reserves is key to ensuring the continuous supply of adequate water quantity and quality to the global population.

The world has a limited freshwater supply, accounting for less than 3% of the total water available on earth, and transboundary waters across the globe account for a significant portion of this global freshwater supply. With 276 transboundary rivers and lakes worldwide, transboundary waters account for 60% of the global freshwater supply (USAID 2019; Wolf and Newton 1999). These waters cover almost half of the earth's surface and cater to more than 40% of the global population's water needs (Munia et al. 2016). As such, any endeavor on global water management should ideally have management of transboundary water resources at its heart. That is why the U.N. Sustainable Development Goals (SDGs) have put transboundary water management in its agenda under SDG 6 (SDG 6.5.1 Integrated water resources management and 6.5.2 Transboundary basin area with water cooperation). Regardless of this, however, two-thirds of all transboundary rivers still do not have a Cooperative Framework (U.N. Water 2018).

Management of transboundary watercourses is often a complicated task. For one, riparians of a transboundary watercourse often have conflicting interests when using shared water (Arjoon et al. 2016; Brochmann and Gleditsch 2012). Moreover, countries' demands are often "more than their endowments" (Ansink and Weikard 2009). These factors have made transboundary water management complex and laced with tension. Research by the pacific institute (http://www.worldwater.org/conflict/list/), documenting water conflicts chronologically, also shows that there have been 926 conflicts related to water from 3000 BC to 2019, while the U.N. has documented 37 "acute" disputes over water in the last 50 years (U.N. 2014). The U.

N. has warned that water scarcity in these areas, coupled with climate change stress, can cause serious conflicts worldwide two decades ago (WWAP 2009). Nevertheless, studies indicate that two-thirds of the time, engagements over transboundary water, often lead to cooperation than conflict (UNESCO 2015; U.N. 2014; De Stefano et al. 2010; Zeitoun and Mirumachi 2008; Wolf 2007).

There have been serious attempts at governing shared resources since ancient times. According to the U.N., the first legal agreement of any kind in the world was a transboundary water-sharing agreement (U.N. 2014). It was signed between the city-states of Lagesh and Umma during the Sumerian civilization in 2550 B.C. This treaty between these two city-states incorporated crop sharing and land tenancy schemes as part of the water-sharing arrangement they had over the Tigris river (Sand 2018). The world has come a long way in terms of transboundary water sharing since ancient times. More than 3600 treaties relating to water have been signed since 805 A.D., with more than 150 treaties just in the last 50 years (U.N. 2014).

Over the centuries, customary Roman and English laws have dictated the mutual use of shared resources, mostly dealing with border demarcation and navigation (Bannon 2017). During the industrial revolution, transboundary waters were important international highways transporting cargo, and hence, laws on the navigational use of these shared watercourses were dominant (Rahman 2009; Salman 2007). However, the industrial revolution itself came up with more efficient transport methods, and the relevance of transboundary waters for navigation started to decline. During the nineteenth and early twentieth centuries, population growth and the rising water demand necessitated using transboundary waters for irrigation, hydropower production, and industrial uses (McCaffrey 1996). As such, numerous principles and doctrines have risen to govern such non-navigational uses of transboundary waters.

The doctrine of absolute territorial sovereignty also called the Harmon doctrine grants upstream riparian countries absolute and unrestrained use of a shared resource in their territory regardless of the effect downstream (McCaffrey 1996). Contrary to this, the doctrine of absolute territorial integrity grants downstream riparians uninterrupted flow from upstream countries, essentially stripping upstream countries of any significant water use (McCaffrey 1996). The doctrine of prior appropriation, on the other hand, guarantees the first users of a shared resource the full amount of water they priorly utilized, regardless of subsequent users or changes in the basin (Beaumont 2000). The above doctrines are some of the principal ideals in the history of water management. However, because of the inequitable, short-sighted, and biased nature of the above doctrines, they have failed to have a significant stronghold on modern transboundary water laws.

Modern transboundary water laws hinge on the doctrine of limited territorial sovereignty, which asserts that all riparian countries have equal rights to a fair share of the shared resource (Beaumont 2000). More progressive concepts of transboundary water use such as the concept of a community of interest, which considers the entire river basin as a single economic unit and gives the right to govern the water to a community of interest from all basin states, have also been floating in transboundary water governance discourse (Salman 2007). However, modern transboundary water sharing hinges on three principles, also called the three prongs of International Water Law: the principle of reasonable and equitable use, the principle of causing no significant harm, and the principle to cooperate. These three tenants have been the guiding principles behind influential modern-day transboundary water-sharing rules such as the Helsinki and Berlin Rules developed by the International Law Association in 1966 and 2004 and the U.N. Convention on the Law of Non-Navigational Uses of International Watercourses in 1997.

These rules and principles establish the principles of equitable and reasonable use and the causing of no significant harm as foundations and guiding principles on transboundary water sharing. However, because of their law-based nature, they only provide loose guidance to formulate equitable water-sharing frameworks, and these principles leave the contextualization of what "equitable and reasonable use" means and what constitutes "significant harm" to each transboundary basin (Salman 2007).

The Helsinki rules were the first to outline the factors to be considered to determine equitable and reasonable use. The U.N. 1997 Convention adopted and modified the Helsinki factors afterward. These factors offer a starting point for evaluating equitable and reasonable water use in a transboundary watercourse. However, as mentioned earlier, contextualizing these factors to the basin in question is left to the basins themselves. It is up to the basin states to define and contextualize what these principles practically mean to their basin and translate the laws into operational and usable water-sharing frameworks.

The rationale behind this chapter is to quantify the factors outlined by these international rules and principles in the context of the Nile basin and develop a basic framework for water sharing to evaluate equitable and reasonable shares of countries in the Nile basin.

The Nile Basin

Arguably the most famous river in world history, the Nile River is the longest river in the world. This river crosses 11 countries along its trajectory, flowing 6695 km from its most distant source in the equatorial lakes region until it empties into the Mediterranean Sea (NBI Atlas 2016). The Nile River has two major tributaries, the White Nile and the Blue Nile. The longest tributary, the White Nile, starts around the equatorial lakes region. Rivers from Burundi, the Democratic Republic of Congo, Rwanda, Tanzania, Uganda, and Kenya flow into Lake Victoria, giving rise to the White Nile as it exits the lake. The White Nile then flows through Uganda and South Sudan, collecting water from other tributaries such as the Baro-Akobo River from Ethiopia and others. The White Nile loses a considerable amount of its water in the Sudd wetlands in South Sudan but emerges and flows to meet the other major tributary, the Blue Nile.

The shorter tributary, which contributes the lion's share of the Nile waters, is called the Blue Nile and originates from the highlands of Ethiopia. It flows to Sudan to meet the White Nile in the capital city of Sudan, Khartoum, later being joined by the Tekeze-Atbara River, originating from Ethiopia to flow as one Nile through Sudan and Egypt emptying into the Mediterranean. The Blue Nile or Abay, together with the Baro-Akobo-Sobat, Mereb river, and the Tekeze-Atabra River, contributes more than 86 billion cubic meters (BCM) of water to the annual Nile flow (Berhanu et al. 2014).

Even though the Nile crosses 11 countries, Egypt and, to a lesser extent, Sudan dominate the water use (Swain 2011). The dissonance in endowments versus the use of a shared river is nowhere more evident than in the Nile basin. The upper riparian countries, which are the water sources, use minimal amount of the water, while the lower riparian countries with minimal to zero contribution to the Nile waters use all of the water (Obeng 2016).

Uninclusive historical agreements have resulted in the inequitable water use arrangement and skewed status quo in the region. Colonial powers in the Nile basin region, in particular, the U.K. had vested interests in ensuring the amount of the Nile water flowing to Egypt and later on to Sudan would not decrease, as the textile industries in the U.K. were fed by cotton plantations in these two downstream countries (Conniff et al. 2012). Since 1891, there have been a series of treaties and agreements to ensure the full flow of the Nile reaches downstream countries (Ferede and Abebe 2014). Table 5.1 shows summary water use treaties in the basin.

For quite some time, these agreements have been in place and have given rise to claims of "historical rights" by downstream countries. However, these agreements are uninclusive, inequitable, and unsustainable, not to mention they infringe on the national sovereignty of the basin states. These agreements are also not in line with the current trend in modern transboundary water-sharing principles that advocate for equitable use of all basin states. The need for equitable and reasonable water use in the Nile basin cannot be overstated.

The Nile basin already supports 40% of the African population, with more than 260 million people directly living in the basin and almost half a billion people living in the basin countries (NBI 2016). The increasing population growth and impressive economic development observed in the basin, especially in upstream riparian countries in the past two decades, have increased the water demand in all the basin states. Population growth, coupled with the stress from climate change and water scarcity in the basin, forces nations to utilize their resources more aggressively, resulting in riparians demanding for their fair share from this shared resource (Swain 2011). However, the Nile basin is not equipped to entertain such demands. The whole of the Nile waters is already allocated for use between the two downstream countries of Egypt and Sudan, stripping the other nine riparian countries of their fair share of this shared resource. Such an arrangement is the root cause of the tension and disagreement in the Nile basin and underlines the need for an equitable water-sharing framework.

Even though there is a global consensus on the equitable and reasonable sharing of transboundary water resources, the Nile River basin has not progressed with the

Table 5.1 Summary of water use treaties in the basin

Year	Parties to the treaty	Objective	Article
1891	Italy and U.K.	The Italian government shall undertake not to initiate any irrigation works on the Atbara which may alter the rate of flow of the Nile water	Article 3
1902	Ethiopia and U.K.	His Majesty the Emperor Menelek II King of Kings of Ethiopia engages himself toward the Government of his Britannic Majesty not to construct, or allow to be constructed, any works across the Blue Nile, Lake Tsana or the Sobat which would arrest the flow of their waters into the Nile except in the agreement with his Britannic government and the government of the Sudan	Article 3
1906	France, Italy, and U.K.	…ensure the interest of Great Britain and Egypt in the Nile basin, more especially as regards the regulating of the waters of that river and tributaries (due consideration being paid to local interests) without prejudice to Italian interest…	Article 4A
1906	Congo and U.K.	The government of the independent state of the Congo undertake not to construct, or allow to be constructed, any works on or near the Similki or Isango River, which would diminish the volume of water entering Lake Albert, except in agreement with the Sudanese government	Article 3
1925	Italy and U.K.	The Italian government recognizing the prior hydraulic rights of Egypt and the Sudan and engage not construct on the headwater of the Blue or White Nile or their tributaries or effluents any works which might sensibly modify their flow into the main river	
1929	Egypt and U.K.	Allocates 48 billion cubic meters (BCM) to Egypt and 4 BCM to Sudan	
1934	Belgium and U.K.	Required Belgium to return waters from streams originating in areas under its control without substantial reduction of the natural bed before forming a common boundary with British-controlled regions	
1949	Egypt and U.K.	Uganda would regulate the discharges to be passed through the dam on the instruction of the Egyptian resident engineer be stationed with his staff at the dam	Article 4
1959	Egypt and Sudan	Allocates 55.5 BCM to Egypt, 18.5 BCM to Sudan, 10 BCM for evaporation, allocating zero shares for the other riparian countries and gives veto power to Egypt over upstream projects	

global pace. It is still clinging to residual colonial era treaties that do not embody the cardinal principles of equitable and reasonable use. Even the framework for general cooperation between Ethiopia and Egypt signed in 1993, which highlighted cooperation and the causing of no significant harm, did not address equitable use in the basin. The sole exception to this trend is the Cooperative Framework Agreement on the use of the Nile waters (CFA).

The CFA is the only framework to be negotiated and discussed by all the riparian countries for the first time in the basin's history. It is a framework rooted in the principles of equitable use and causing no significant harm, negotiated by the countries for thirteen years since 1997 and finally became ready for signing in 2010. To date, six of the upper riparian countries have signed it while four have ratified it, although the current dominant users of the Nile, Egypt and Sudan, have neither signed nor ratified it even though they have been willing participants of the negotiation for 13 years.

The CFA primarily adopted the factors to be used in the determination of equitable use from the U.N. 1997 convention and added two more factors deemed relevant to the basin. As such, it provides a contextual and relevant starting point in the development of equitable and reasonable water sharing in the Nile basin.

In this chapter, we quantitatively evaluate equitable and reasonable water sharing in the Nile basin, taking a mix of factors outlined by the Helsinki rules, the U.N. 1997 Convention, and the CFA. These factors are not exhautive lists and other factors can be added upon consensus of the basin states. Table 5.2 summarizes the factors to be considered in the determination of equitable and reasonable use according to these three legal frameworks.

Methodology

Although many of the factors suggested for use in the determination of equitable and reasonable use are similar, simplification and limitation of the factors' scope were necessary for this study's purposes. As such, the factors used in this study, their definitions, and quantifications are described in Section "Definition and Quantification of Factors Used in the Determination of Equitable Use". Then a multi-criteria analysis is employed to evaluate equitable shares in the basin. The factors considered in the determination of equitable and reasonable use are given different weights, and the relative shares of countries are calculated based on these weights.

Definition and Quantification of Factors Used in the Determination of Equitable Use

Note that most of the factors are rather broad and can have multiple interpretations and, consequently, various indicators. The authors fully acknowledge that other nuanced indicators can be developed to quantify the factors. The indicators used in this study are a simplified interpretation of the factors with their scope limited to fit the study's purpose. When there was a need to limit the extent of the factors for lack of data or resource, simplifications following phrasings in the Helsinki rules were followed as the Helsinki rules provided indicators in some instances.

Table 5.2 Summary of factors used in the determination of equitable and reasonable use

Factors	Helsinki 1966	UN 1997	CFA 2010
F1	Geography of the basin including, in particular, the extent of the drainage area in the territory of each basin state	Geographic, hydrographic, hydrological, climatic, ecological, and other factors of a natural character	Geographic, hydrographic, hydrological, climatic, ecological, and other factors of a natural character
F2	Hydrology of the basin, including in particular the contribution of water by each basin state	The social and economic needs of the watercourse states concerned	The social and economic needs of the watercourse states concerned
F3	Climate affecting the basin	The population dependent on the watercourse in each watercourse state	The population dependent on the watercourse in each watercourse state
F4	Past utilization of the waters of the basin, including in particular existing utilization	The effects of the use or uses of the watercourses in one watercourse state on other watercourse states	The effects of the use or uses of the watercourses in one watercourse state on other watercourse states
F5	Economic and social needs of each basin state	Existing and potential uses of the watercourse	Existing and potential uses of the watercourse
F6	The population dependent on the waters of the basin in each basin state	Conservation, protection, development, and economy of use of the water resources of the watercourse and the costs of measures taken to that effect	Conservation, protection, development, and economy of use of the water resources of the watercourse and the costs of measures taken to that effect
F7	Comparative costs of alternative means of satisfying the economic and social needs of each basin state	The availability of alternatives, of comparable value, to a particular planned or existing use	The availability of alternatives, of comparable value, to a particular planned or existing use
F8	Availability of other resources in the basin states		The contribution of each basin state to the waters of the Nile River system
F9	Avoidance of unnecessary waste in the utilization of waters of the basin		The extent and proportion of the drainage area in the territory of each basin state
F10	The practicability of compensation to one or more of the co-basin states as a means of adjusting conflicts among uses		
F11	The degree to which a basin state's needs may be satisfied without causing substantial injury to a co-basin state		

Geography (F1)

In this study, geography refers to the extent of a basin's drainage area in each basin state. This factor is characterized by the percentage of the basin present in each riparian state (as described in the Helsinki rules), Eq. 5.1.

$$\begin{aligned}&\%\text{ of country in the basin}\\&=\left(\frac{\text{Area of country in the basin}}{\text{Total area of the basin}}\right)*100\%\end{aligned}\tag{5.1}$$

Geography (F2)

Hydrology refers to the contribution of water by each basin state to the transboundary water (as described in the Helsinki rules). The significant water loss in the Sudd Swamp is deducted as a natural loss from the basin as a whole, and the percentage country contribution is evaluated afterward (Eq. 5.2).

$$\begin{aligned}&\%\text{ contribution to the basin}\\&=\left(\frac{\text{Water contribution by country}}{\text{Total water in the basin} - \text{Loss in the Sudd swamps}}\right)*100\%\end{aligned}\tag{5.2}$$

Climate Affecting the Basin (F3)

Climate affecting the basin is limited to the rainfall and evapotranspiration of basin states as these two factors are the most significant elements of climate affecting water balance in the Nile basin. The Blue Nile region is highly affected by rainfall, while the equatorial lakes are very susceptible to evapotranspiration. Hence, for this study's purpose, the climate affecting the basin is limited to these two major climate elements. Equation 5.3 shows rainfall and evapotranspiration contribution by country.

$$\%\text{RF or ET contribution of a country} = \left(\frac{\text{RF or ET of country}}{\text{Total RF or ET in the basin}}\right)*100\%\tag{5.3}$$

where ET is evapotranspiration, total RF/ET is total rainfall and evapotranspiration ratio in the basin and by each country.

Population Dependent on the Water Resource (F4)

The population dependent on the water resource is classified into two, the population which is directly dependent on the river, i.e., living in the basin, and the population indirectly dependent on the water resource, which will be the total population of countries minus population directly dependent/living on the water as shown in Eq. 5.4.

$$\begin{aligned}&\%\text{ of population by country (in)directly dependent on the basin}\\&=\left(\frac{\text{Population of a country (in)directly dependent in the basin}}{\text{Total number of people (in)directly dependent in the basin}}\right)*100\%\end{aligned} \tag{5.4}$$

where total population directly dependent on the basin is the sum of population directly living in the basin in each basin state; total population indirectly dependent in the basin is sum of population of countries living outside of the basin in each basin state.

Existing and Potential Uses of the Water Resources (F5)

Existing and potential water use of the resources refers to the current and future uses of major consumptive sectors in the basin states, namely irrigation, domestic/municipal use, and industrial use. Equation 5.5 is evaluation of existing and potential water demand of basin states.

$$\begin{aligned}&\%\text{ of water demand by country}\\&=\frac{\text{Amount of current or projected use of the country}}{\text{Current or projected demand in the basin}}*100\%\end{aligned} \tag{5.5}$$

Social and Economic Needs (F6)

In this study, the social and economic needs are reflected through countries' expected water needs in the foreseeable future. Quantifying a country's socio-economic needs through its projected water demand was also used in a similar study for the Jordan river by Mimi and Sawalhi (2003). Hence, the high projection scenarios developed by NBI for municipal, industrial, and irrigation demands of countries for 2050 were used in this study as an indicator for countries' social and economic needs. See the factor above for calculation.

The Availability of Alternatives, of Comparable Value, to a Particular Planned or Existing Use (F7)

Groundwater resources and desalination potential are considered as indicators for alternative sources in this study. Groundwater resources (both renewable and non-renewable) and desalination potential expressed in coastlines of countries in km are considered indicators for alternative sources. Equation 5.6 shows evaluation of alternative resource contribution.

$$\begin{aligned}&\%\ \text{share of alternative resource of a country}\\&=\frac{\text{Amount of ground water or Shore line of a country}}{\text{Total amount of ground water or shore line in the basin}}*100\%\end{aligned}\tag{5.6}$$

The Effects of the Use or Uses of the Water Resources in One Basin State on Other Basin States (F8)

The effect of use by one basin state on others is evaluated in terms of the extent of countries' consumptive water use. The majority of the water in the basin goes to irrigation use. Therefore, the potential of harm to other basin states is evaluated through each basin state's expected irrigation water use. Equation 5.7 shows evaluation of the effect of use on others.

$$\begin{aligned}&\%\ \text{Share of irrigation demand}\\&=\frac{\text{Irrgation demand of a country for 2050}}{\text{Total irrigation demand in the basin}}*100\%\end{aligned}\tag{5.7}$$

Conservation, Protection, Development, and Economy of Use of the Water Resources and the Costs of Measures Taken to that Effect; (Including the Avoidance of Unnecessary Waste in the Utilization of Waters of the Basin) (F9)

This factor is rather broad, but because of the lack of data and for this study's scope and purpose, the potential water savings possible for each country were considered an indicator of unnecessary waste avoidance. The NBI has evaluated the irrigation savings that are possible in the basin by implementing various measures. Equation 5.8 shows evaluation of the effect of use on others.

$$\begin{aligned}&\%\ \text{Share in basin water savings}\\&=\frac{\text{Potential water saving by country}}{\text{Total water saving in the basin}}*100\%\end{aligned}\tag{5.8}$$

Analytical Framework

After the factors are quantified, they are assigned different weights to calculate countries' relative shares. In calculating water shares based on outlined factors, some factors positively affect water shares while others take away from a country's share. For instance, the more water a country contributes to the basin, the more share it is entitled to; however, the more significant alternative resources a country has, the less share it is entitled to. As such, "negative" factors multiplied by their respective weights deducted from the product of "positive" factors with their respective weights frame the countries' water share of countries. This is expressed by Eq. 5.9.

$$\text{Water Share of a country } (Wsj) = \frac{\text{Positive } (Wij * Fij) - \text{Negative } (Wij * Fij)}{\sum_{j=1}^{11} \text{Positive WiFi} - \sum_{j=1}^{11} \text{Negative WiFi}} \quad (5.9)$$

where WSj is water share of countries; j is countries; i is factor indicators, and w is weights for indicators in %.

Factors that are represented by more than one indicator are assigned different relative weights to the corresponding indicators to reflect their relative importance. As such, rainfall and evapotranspiration, as the two dominant climate factors affecting the basin water balance, are given equal weight balance under the factor climate. Groundwater and desalination potential of countries under alternative resources are also given equal weights. A country with a larger population directly living in the basin has more dependency on the basin than the population indirectly living in the basin. A factor of 0.7 is assigned to population directly dependent on the basin and 0.3 for population indirectly dependent on the basin to account for this rationale. These weights are only indicators of the relative importance of the factors and can be changed as part of the allocation model's flexible nature. The summary of the factors and relative weights are given, and their effect on allocation (positive vs. negative effect on shares) is summarized in Table 5.3.

Four scenarios were then developed by varying the weights given to the factors to show the range of possible water allocation and the resulting water shares of basin states. The first scenario gave equal weights for all the factors and calculated the resulting shares. Each factor was given an equal weight of 1/9, and the resulting shares of each country are evaluated.

The second scenario maximized the allocation share for individual countries by allowing the weights to be optimized to benefit individual countries. This exercise is done for the eleven countries, and the impact of maximizing the share of one country on others is evaluated. Equation 5.10 is optimization equation in Scenario 2.

Table 5.3 Summary of factor influences

Factors	Effect		Note
Geography	Positive	$Fi * wi$	
Hydrology	Positive	$Fi * wi$	
Climate: rainfall	Positive	$0.5 * Fi * wi$	Equal weights of 0.5 assigned to rainfall and evapotranspiration under the factor of climate
Climate: evapotranspiration	Negative	$-0.5 * Fi * wi$	
Population: directly dependent and indirectly dependent	Positive	0.7 (*Fi* direct * *wi*) + 0.3 (*Fi* indirect * *wi*)	A weight of 0.7 is given to directly dependent population and 0.3 to indirectly dependent population
Existing use	Positive	$Fi * wi$	
Socio-economic demands	Positive	$Fi * wi$	
Effect of use on others	Negative	$-Fi * wi$	
Alternative resources: desalination and groundwater	Negative	− 0.5 (*Fi* groundwater * *wi*) − 0.5 (*Fi* desalination * *wi*)	Equal weights of 0.5 assigned to desalination potential and groundwater potential
Avoidance of unnecessary waste	Positive	$Fi * wi$	

$$\text{Maximize water share of a country } (WSj) = \frac{\text{Positive } (Wij * Fij) - \text{Negative } (Wij * Fij)}{\sum_{j=1}^{11} \text{Positive WiFi} - \sum_{j=1}^{11} \text{Negative WiFi}} \quad (5.10)$$

By changing variables *Wi* for country *j*.
Constraints: $0 < Wij < 1$; $\sum Wij = 1$.

where *WSj* is water share of countries; *j* is countries; *i* is factor indicators; *w* is weights for factor indicators in %; and $\sum Wij$ is sum of the weights of all the factors.

The third scenario maximizes countries' shares by putting limiting constraints on the weights given to the factors so no one factor is overly dominant. Equation 5.11 is optimization equation in Scenario 3.

$$\text{Maximize water share of a country } (WSj) = \frac{\text{Positive } (Wij * Fij) - \text{Negative } (Wij * Fij)}{\sum_{j=1}^{11} \text{Positive WiFi} - \sum_{j=1}^{11} \text{Negative WiFi}} \quad (5.11)$$

By changing variables Wi for country j.
Constraints: $0.05 < Wij < 0.5$; $\sum Wij = 1$.

where WSj is water share of countries; j is countries; i is factor indicators; w is weights for indicators in %; and $\sum Wij$ is sum of the weights of all the factors.

The fourth scenario co-maximizes all the countries' shares in the basin by optimizing the factors listed to maximize the sum of shares in the basin. Equations 5.12a and 5.12b are optimization equation in Scenario 4.

$$\text{Maximize} \sum WSj = \sum_{1}^{j} \text{Positive}\,(Wij * Fij) - \text{Negative}\,(Wij * Fij) \tag{5.12a}$$

By changing variables Wij for all countries with constraints: $0 < Wij < 1$; after the optimization, the shares of each country are evaluated as:

$$WSj = \sum_{1}^{j} \text{Positive}\,(Wij * Fij) - \text{Negative}\,(Wij * Fij) * 100\% \tag{5.12b}$$

where WSj is water share of countries; j is countries; i is factor indicators; w is weights for indicators in %; and $\sum WSj$ is sum of the shares of all the factors.

The optimization analysis was done in Excel using the Solver function and the Generalized Reduced Gradient (GRG) nonlinear method.

Assumptions and Limitations

The (un)availability of data necessitated some informed assumptions and approximations in the quantification of the factors. Data for Congo and Eritrea was mostly lacking from the NBI system and was supplemented from other sources where possible. A study by Macdonald et al. (2012) was used as a comprehensive data source for groundwater resources. However, since the study was done right after the independence of South Sudan, data for South Sudan alone could not be found. Sudan's groundwater potential was divided equally between the two countries, after the same proportion as observed in their surface water shares. The final factor, i.e., conservation, protection, development, and economy of use of the water resources and the costs of measures taken to that effect (including the avoidance of unnecessary waste in the utilization of waters of the basin), is extensive. Data to quantify all aspects of the factor was not available; hence, the indicator was limited to the avoidance of unnecessary waste in the utilization of waters following the Helsinki rules.

In the factor concerning alternative resources, the factor was limited to groundwater and countries' desalination potential. However, the availability of "other" surface water resources available in the basin states, besides the Nile, can be considered as alternatives. There is no distinction made between groundwater systems that feed the Nile system versus separate ones. This distinction, however, is necessary to avoid double counting of water resources contributing to the basin. The environmental water requirement is also not conisdered in water allocation in this study. These limitations and nuances were beyond the study's current scope because of time and resource constraints but need to be addressed in future studies.

Results

The summary of the quantified factors evaluated according to the definitions given in the methodology section for all eleven countries is summarized in Table 5.4. Based on this quantification, the scenarios are evaluated, the results of which are presented below.

Scenario 1: Equal Weights for All Factors

When equal weight is assigned to all the factors, i.e., a weight of 1/9 assigned to each factor, the resulting water shares in the basin are given in Fig. 5.1. Ethiopia would get the highest share, around 32%, with Sudan and Egypt following, with approximately 20% shares each. South Sudan will take the fourth largest share, with almost 10%, while the other upper riparian shares range between 2 and 6%.

Scenario 2: Country-Based Optimization (Identification of Dominant Factor for Each Country)

Scenario 1 gave equal weights to all the factors used in the determination of equitable shares. However, it is evident that no country in the basin would realistically assign equal weights to all the factors but instead try and optimize the factors to maximize its share. The second scenario simulates the share of each country if the factors were to be optimized to maximize individual country shares. The summary of the resulting shares from such optimization is given in Table 5.5.

As shown in the table, countries get extremely high shares when the system is optimized for their benefit. For example, Ethiopia would be entitled to 71% of the Nile waters when the system is optimized for Ethiopia, while Egypt would get around 67%. To maximize the shares of individual countries, the model assigns higher values for the factors that benefit a certain country while giving lower weights for the factors that take away from the country's share. In this scenario, the model gave a factor of 1 (100%) for the factors beneficial for individual countries

Table 5.4 Summary of factors quantified

Country	F1 (%)	F2 (%)	F 3.1 (%)	F 3.2 (%)	F 4.1 (%)	F 4.2 (%)	F5 (%)	F6 (%)	F 7.1 (%)	F 7.2 (%)	F8 (%)	F9 (%)
Burundi	0.44	3.17	13.36	6.87	2.30	2.60	0.05	0.05	0.03	0	0.02	0.1
DRC	0.69	3.37	13.01	7.1	1.17	32.07	0.01	0.02	20.79	0.49	0	0
Egypt	9.52	0	0.45	12.42	34.55	2.69	79.07	57.9	29.96	32.52	58.47	11.95
Eritrea	0.81	0.59	3.41	11.09	0.89	1.42	0	0	0.18	29.65	0	0
Ethiopia	11.5	64.92	11.49	8.65	15.14	29.21	1.83	15.57	6.89	0	16.93	31.32
Kenya	1.62	5.5	8.1	9.98	6.93	12.19	1.67	3.35	4.8	7.11	1.15	1.37
Rwanda	0.65	3.87	11.32	5.99	3.50	0.85	0.28	0.41	0.03	0	0.07	0.14
South Sudan	19.54	6.76	11.21	9.09	4.79	0.05	0.11	3.27	17.15	0	3.38	18.5
Sudan	43.95	4.58	3.01	12.64	12.65	2.22	16.35	17.92	17.15	11.32	18.96	35.25
Tanzania	3.73	3.66	10.82	7.76	4.55	15.88	0.09	0.75	2.85	18.9	0.79	1.05
Uganda	7.56	3.58	13.84	8.43	13.53	0.19	0.54	0.75	0.18	0	0.24	0.33

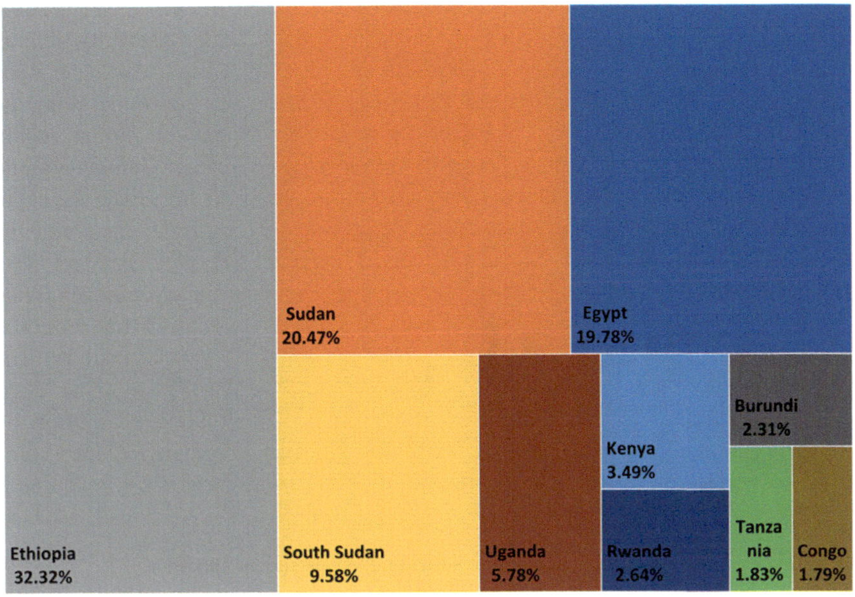

Fig. 5.1 Water shares according to Scenario 1

Table 5.5 Country water shares and corresponding dominant factors according to Scenario 2

Country	Country optimized (%)	Dominant factor
Burundi	8	Hydrology
Congo	18	Population
Egypt	67	Existing use
Eritrea	2	Population
Ethiopia	71	Hydrology
Kenya	15	Population
Rwanda	9	Hydrology
South Sudan	30	Geography
Sudan	53	Geography
Tanzania	14	Population
Uganda	17	Population

(the dominant factors) while assigning zero weights for all the other factors resulting in the ultimate maximum share a country can get under the prescribed factors.

It is noteworthy to observe that the maximized shares of individual countries come at the cost of other countries' shares, as shown in Fig. 5.2. The figure shows other countries' shares when the system is optimized for Egypt, Ethiopia, and Sudan. For example, when the system was optimized for Egypt, Egypt's share was 67%, while Ethiopia and Sudan's shares were 13 and 9%. Similarly, when the

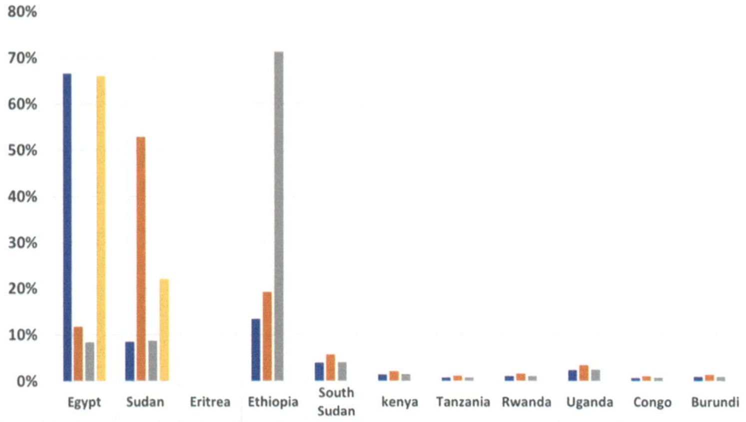

	Egypt	Sudan	Eritrea	Ethiopia	South Sudan	kenya	Tanzania	Rwanda	Uganda	Congo	Burundi
Egypt Optimized	67%	9%	0%	13%	4%	1%	1%	1%	2%	1%	1%
Sudan Optimized	12%	53%	0%	19%	6%	2%	1%	2%	3%	1%	1%
Ethiopia Optimized	8%	9%	0%	71%	4%	1%	1%	1%	2%	1%	1%
Current allocation	66%	22%	0%	0%	0%	0%	0%	0%	0%	0%	0%

Fig. 5.2 Country shares comparison under Scenario 2 versus current allocation

system was optimized for Ethiopia, Ethiopia's share was 71% while Egypt and Sudan's shares were 8% and 9%, respectively. The shares of the upper riparian countries did not fluctuate much from their shares in Scenario 1 when the system was optimized for the three countries as the initial shares were low in the first place.

Another interesting thing to note is the comparison of this scenario to the current allocation scenario, which gives Egypt a share of 66%, Sudan 22%, 12% for evaporation losses and zero shares for the rest of the countries. When the system is optimized for Egypt, the resulting share of Egypt (67%) is very close to the country's current claimed share (66%). This indicates that the current allocation framework is essentially optimized for Egypt. However, even the model optimized scenario for Egypt allocates some water for the other countries while the current allocation does not. Even if the water left for evaporation, i.e., 12% left from Egypt and Sudan's share, was distributed between the other nine countries, it would result in much lower shares than what the model allocates for all countries while maximizing Egypt's share. This shows the extent of inequity currently prevalent in the basin.

Scenario 3: Country Optimization Under Constraints

As explained above, the way the model maximized shares of individual countries was by assigning 100% weight to one dominant factor beneficial for a country while giving zero weight for the other factors. To ensure that all factors were fully included in the allocation framework, and no one factor has dominance over others, constraints on the weights given to each factor were fixed in Scenario 3. In this scenario, no one factor could be given a weight of more than 50% or less than 5%. This way, the preferential choice of factors can be established while ensuring the relative importance of all factors. The shares of countries, according to Scenario 3, are given in Table 5.6.

In this scenario, the model still optimized the weights given to the factors by allocating a share of 50% for the dominant factors for each country, without disregarding the other factors, assigning minimal weights. The resulting optimized shares for individual countries are less than those in Scenario 2, while other countries' shares have slightly increased compared to Scenario 2. Figure 5.3 shows a comparison of scenarios 2 and 3 for the four largest water users.

Scenario 4: Basin Sum Maximization

In scenarios 2 and 3, the maximized shares of one country came at the cost of the other countries' shares. In Scenario 4, the aim was to maximize all countries' shares concurrently by maximizing the basin sum. In this scenario, as shown in Fig. 5.4, the shares of all the basin countries are optimized without harsh trade-offs between countries. A closer look at the weights given to the factors under this scenario shows that the model assigned a weight of 100% for all positive factors and zero weights for all negative factors for each country. This way, each country would have maximized shares, and hence the basin sum will also be maximized (more than 100%), which was then reproportioned to 100%.

Comparing all of the above scenarios with the current allocation scenario, it can be seen that with the sole exception of Egypt and Sudan, all other countries are better off in all other scenarios compared to the current allocation scenario. The current allocation scenario mimics the scenario where the model was optimized for Egypt. Still, while this scenario allocates some amount of water, however minimal,

Table 5.6 Country shares under Scenario 3

Countries	Shares (%)
Egypt	54
Sudan	42
Ethiopia	59
South Sudan	22
Kenya	10
Tanzania	8
Rwanda	6
Uganda	12
Burundi	5
Congo	10
Eritrea	0

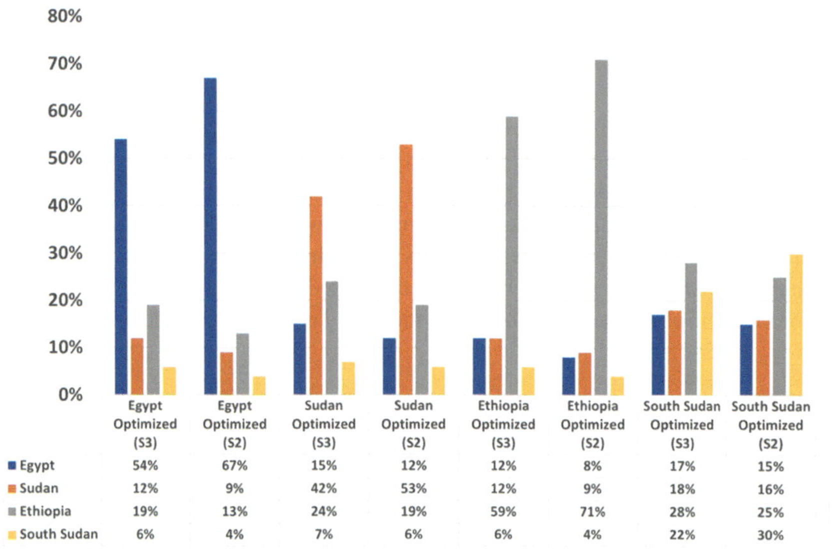

Fig. 5.3 Comparison of Scenario 2 versus 3

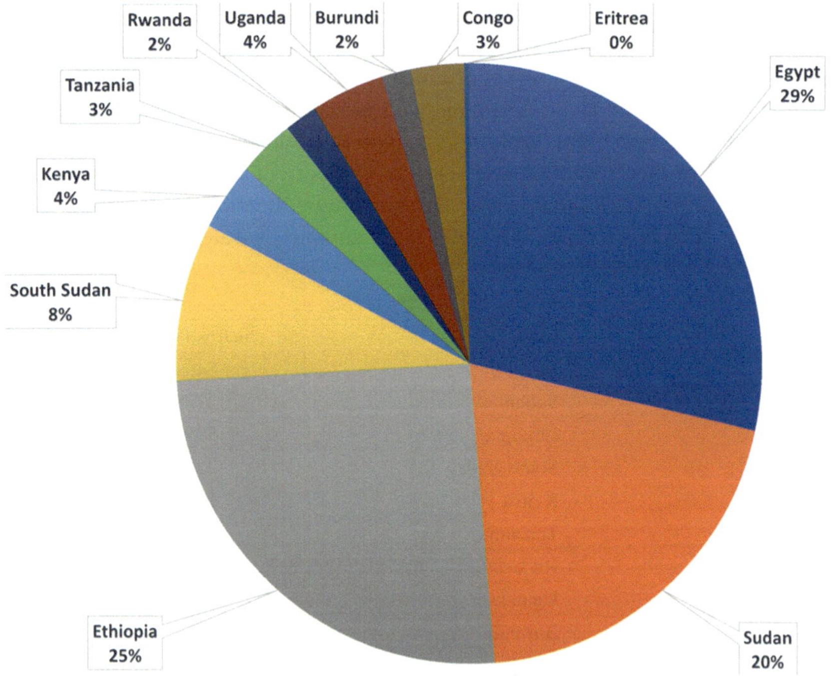

Fig. 5.4 Country shares under Scenario 4

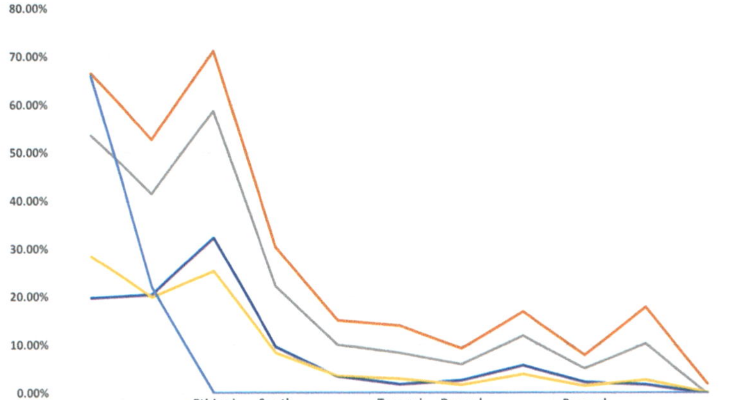

	Egypt	Sudan	Ethiopia	South Sudan	Kenya	Tanzania	Rwanda	Uganda	Burundi	Congo	Eritrea
Scenario 1	19.78%	20.47%	32.32%	10%	3.49%	1.83%	2.64%	6%	2.31%	1.79%	0%
Scenario 2	67%	53%	71%	30%	15%	14%	9%	17%	8%	18%	2%
Scenario 3	54%	42%	59%	22%	10%	8%	6%	12%	5%	10%	0%
Scenario 4	29%	20%	25%	8%	4%	3%	2%	4%	2%	3%	0%
Current Allocation	66%	22%	-	-	-	-	-	-	-	-	-

Fig. 5.5 Scenario envelope

to other basin states, the current allocation scenario allocates zero water for the other basin states. The current allocation scheme has no basis when evaluated against equitable and reasonable values, and no conceivable equitable scheme can result in shares resembling the current allocation.

It is also interesting to note that Sudan's shares under scenarios 1 and 4 are very close to its current claimed share. Ethiopia, Sudan, Egypt, and South Sudan constitute the top four largest shares under all the scenarios, while the Equatorial countries' shares are broadly similar, with Uganda taking the largest share among them. The scenario envelope in Fig. 5.5 shows the resulting shares under all evaluated scnearios.

Conclusion and Recommendation

In this study, factors outlined by international transboundary water-sharing principles were quantified and used to evaluate equitable use scenarios in the Nile basin. This study clearly showed that the current allocation scheme in the basin has no basis under the principles of equitable and reasonable use. The scenarios developed gave broad thresholds for the rough shares to be expected for countries under an equitable water use framework. However, under no conceivable equitable water sharing can the current allocation framework stand to be justified.

If countries were to re-negotiate the Nile waters' redistribution, the allocation would probably resemble a pattern somewhere between scenarios 1 and 4 as lower thresholds and Scenario 3 as an upper threshold. All the scenarios allocated more water to the upper riparian countries, which currently have zero shares under the existing water-sharing scheme. In contrast, the current dominant user of the Nile water, Egypt, would not get the amount of water it is accustomed to using under all scenarios except in Scenario 2, where the system was optimized to benefit Egypt. This does not mean, however, that Egypt can/will suddenly start using less water. The Nile waters are already accounted for between Egypt and Sudan for practical uses. Millions of people, homes, industries, and cities depend on the water in these two countries.

Moreover, Egypt's expected water demands and all the Nile basin countries are expected to increase drastically in the near future. This highlights the need to go beyond numerical water allocation and find innovative and sustainable ways of satisfying all basin states' water demands while ensuring equitable shares for all basin states. To this end, eight recommendations are forwarded for the sustainable and equitable utilization of the shared resource.

Physical Water Trade

Assuming equitable and reasonable redistribution of water between basin states exists, it is improbable that all countries will immediately start utilizing their fair shares. By pricing water on a regional basis, countries which require more water than their fair shares can buy surplus water from countries that are not fully utilizing their fair shares. This is especially doable in the short term, as countries would need time to develop infrastructures to utilize their resources fully.

Physical water trade would also incentivize basin-wide water-saving practices. Upstream countries would save water to export while downstream water countries could reduce their imports through water-saving practices, resulting in a basin-wide positive saving.

The European Emission Trading System (ETS) can be an excellent international example of such a scheme. Under this system, the E.U. set a "cap and trade" system where countries can trade carbon credits within the E.U., incentivizing the reduction of greenhouse gas emissions across the E.U. By implementing this system, the E.U. was able to surpass its goal of reducing emissions by 21% from 2005 levels by 2020 and is on track to reduce emissions by 43% by 2030 (https://ec.europa.eu/clima/policies/ets/markets_en). The same kind of water savings can be expected in the Nile basin if the mechanism for water trade can be established while satisfying the demands of all basin countries.

Virtual Water Trade

Virtual water trade, i.e., trading water embedded in commodities, is another way of overcoming global and regional water scarcity (Allan 2011). Water-rich and climatically favorable countries can export water virtually through the export of energy and food, capitalizing on location efficiency, and judiciously using water. Horlemann and Neubert (2007) reported that the world used 8% less water between 1997 and 2001 as a result of virtual water trade through agricultural products'. Mechanisms for virtual water trade through regional trading of products can help alleviate water scarcity in the Nile basin. The trade need not be limited to agricultural produces. There is a vast opportunity to trade hydropower as well in the basin. Ethiopia alone has the potential to produce an estimated 45,000 MW (https://www.hydropower.org/country-profiles/ethiopia) of clean and affordable hydropower energy, which can help feed the demands of the basin. Initiatives such as the Eastern Nile Power Export and Equatorial lakes power export schemes can be good starting points for expanding the region's virtual water trade.

Increased Water Use Efficiency

Current water use efficiency in the Nile basin is extremely low. Overall irrigation efficiency in Egypt stands between 40 and 60% (Mahmoud and El Bably 2017). The Gezira scheme, which accounts for more than 50% of the irrigation in Sudan, also stands at 22% efficiency (Mohamed et al. 2011). Moreover, the amount of water lost to seepage and evaporation from open canals and dams in arid areas is immense. A study by the NBI and IWMI (2019) estimated potential water saving as much as 38 BCM per annum is possible by implementing water-saving measures such as deficit irrigation, regional cropping, increasing the efficiency of rain-fed agriculture, and increasing water use efficiency across the basin. This marks the urgent need to enhance water use efficiency in the basin.

Water-Smart Investments

Often myopic and wasteful water projects are commissioned in the basin to maximize national interests because basin-wide collaboration is not in place. However, the deficit between water demand in the basin and the available water can only be bridged if water-smart investments replace wasteful projects. Integrated basin-wide water resource management, such as storing water in low evaporation areas, capitalizing on location efficiency by utilizing suitable irrigable land and hydropower topography, investing in water-storage enhancement and conservation practices upstream on the Nile's headwaters are critical for ensuring the Nile is a sustainable resource. This requires looking at the basin as an economic unit and making water-smart decisions. For such a system to materialize, serious political will, trust, strong economic ties, alliance, and political relationships among the basin states are necessary. It is imperative to look at the basin as a unit beyond national borders to ensure sustainable and equitable use of the Nile waters.

Regional Integration for Synchronized Planning and Operation

A scenario study by the NBI and IWMI (2019) found out that if all basin countries were to utilize their irrigation potentials in the basin fully, the projected water demand would be around 137 BCM by 2050. This is not including other needs such as domestic, industrial, and municipal demands in the region. Unless integrated planning and use of the Nile water is in place, there is no way the Nile River's current water supply can sustain the basin states' existing and future demands. A study done by the NBI and the Eastern Nile division (Eastern Nile Technical Regional Office, ENTRO) underscores that against the backdrop of climate change, increased population and demand, the Nile cannot sustain its population and stressed the need for integrated planning. The study recommends a regional, basin-wide perspective in investments, regional planning and operation of irrigation and hydropower plants in the basin, multi-sector coordination, and a regional approach to agriculture, food, and energy security in the basin (ENSAP 2017).

Benefit-Sharing Schemes and Co-owned Infrastructures

Experiences from other international transboundary river basins such as the Senegal river basin in West Africa and the Parana River basin in South America are shining examples of how co-owned infrastructures and benefit-sharing schemes can be efficient ways of sharing a scarce resource and satisfying contested demands.

The Manantali dam, located in Mali but co-owned by three of the four riparians of the Senegal river basin, i.e., Senegal, Mali, and Mauritania, was able to satisfy the navigation demand of Mali as well as the hydropower power and irrigation demand of the other two countries (Tignino 2016; WWAP 2003). The same story goes for the Itaipu Dam, the second-largest operational energy-producing dam globally, located on Brazil and Paraguay's border (https://www.internationalwaterlaw.org/documents/regionaldocs/parana2.html). This dam is co-owned and co-operated by these two countries since 1984, fostering cooperation and mutual benefit between these countries and others in the region, most notably Argentina. These are just a few examples. Other notable examples in North America, Europe, and Asia are present to take note of for the Nile basin.

Exploring Alternative Water Resources

The growing demands for water in the Nile basin are sure to surpass the river's limited supply of water. Therefore, it is not an option, rather a necessity to look into alternative water resources and reduce dependency on the Nile. Rainwater harvesting, sustainable groundwater use, and recharge (where possible), as well as new age alternative resources such as wastewater reuse and desalination, are worthy avenues to pursue.

Environmental Conservation and Creating Improved Water Consciousness

Concerted environmental conservation endeavors by all riparian countries—to increase the water stored in the basin and reduce pollution and enhance the ecosystems—are necessary for the Nile River's sustainable use. Environmental conservation activities should be synchronized and implemented with the collaboration of all basin states. Environmental conservation and protection upstream in the basin benefit the whole basin and should not be the responsibility of upstream states alone. Similarly, pollution of the river downstream will have implications on the water quality and extent of use upstream. Hence, a coordinated environmental reclamation, conservation, and protection mechanism must be in place by all basin states.

In conjunction, creating a water-conscious population, one that knows the value of water and uses it judiciously, by starting with children, enforcing water-conservation measures, and promoting water-saving practices is necessary for the basin. The sustainable use of the Nile water largely hinges on the population's awareness and water consciousness.

Legal Agreements (Chronologically)

1. Protocols between Great Britain and Italy on the demarcation of their respective spheres of influence in East Africa: Rome, Italy, 15th April 1891.
2. Treaty between Ethiopia and the Great Britain on the Delineation of the Frontier between Ethiopia and Sudan: Addis Ababa. Ethiopia, 15th May 1902.
3. Agreement between Great Britain and the independent State of the Congo: London, UK, 9th May 1906.
4. Agreement between the Great Britain, France and Italy respecting Abyssinia/Ethiopia: London, UK, 13th December 1906.
5. Exchange of Notes between Great Britain and Italy respecting concessions for a barrage at Lake Tsana and a railway across Abyssinia from Eritrea to Italian Somali land: Rome Italy. 20th December 1925.

6. Exchange of Notes between her Majesty's Government in the United Kingdom and the Egyptian Government on the use of the waters of the Nile for Irrigation: Cairo. Egypt, 7th May 1929.
7. Exchange of notes constituting an agreement between the government of the United Kingdom of Great Britain and Northern Ireland and the Government of Egypt regarding the construction of the Owen Falls Dam, Uganda: Cairo, Egypt, 30 and 31st 1949.
8. Agreement between the republic of Sudan and the United Arab republic on the full utilization of the waters of the Nile: Cairo. Egypt, 8th November 1959.
9. ILA (International Law Association) (1966) Helsinki Rules on the Uses of the Waters of International Rivers Report of the Fifty-Second Conference, Helsinki, pp. 447–533 (London: ILA).
10. Framework for General Cooperation between Ethiopia and the Arab Republic of Egypt: Cairo, 1st July 1993
11. Agreement on the Nile River basin cooperative Framework (CFA). Entebbe, Uganda, 14th May 2010
12. ILA (2004) The Berlin rules on water resources Report of the Seventy-First Conference, Berlin (London: International Law Association).

References

Allan T (2011) Virtual water: tackling the threat to our planet's most precious resource, vol 36, issue 7, pp 948–950. November 2011 Water International. https://doi.org/10.1080/02508060.2011.628575

Ansink E, Weikard H-P (2009) Contested water rights. Eur J Polit Econ 25(2):247–260

Arjoon D, Tilmant A, Herrmann M (2016) Sharing water and benefits in transboundary river basins. Hydrol Earth Syst Sci 20(6):2135–2150

Bannon C (2017) Fresh water in roman law: rights and policy. J Roman Stud 107:1–30. https://doi.org/10.1017/S007543581700079X

Beaumont P (2000) The 1997 U.N. convention on the law of non-navigational uses of international watercourses: its strengths and weaknesses from a water management perspective and the need for new workable guidelines. Water Resour Dev 16(4):475–495

Boretti A, Rosa L (2019) Reassessing the projections of the World Water Development Report. npj Clean Water 2:15. https://doi.org/10.1038/s41545-019-0039-9

Brochmann M, Gleditsch NP (2012) Shared rivers and conflict—a reconsideration. Peace Research Institute, Centre for the Study of Civil War, Political Geography, vol 31

Conniff K, Molden D, Peden D, Awulachew SB (2012) Nile water and agriculture: past present and future. In: Awulachew SB, Smakhtin V, Molden D, Peden D (eds) The Nile River Basin: water, agriculture, governance and livelihoods. Routledge–Earthscan, Abingdon, U.K., pp 5–29

De Stefano L, Edwards P, de Silva L, Wolf AT (2010) Tracking cooperation and conflict in international basins: historic and recent trends. Water Policy 12(6):871–884

ENSAP (2017) Eastern Nile subsidiary action program. Eastern Nile multi-sector investment opportunity analysis. Summary report

FAO (2007) Coping with water scarcity: challenge of the twenty-first century. U.N. Water 24 (1):28–29

Ferede W, Abebe S (2014) The efficacy of water treaties in the Eastern Nile Basin. Africa Spectr 49(1):55–67. ISSN: 1868–6869 (online), ISSN: 0002–0397 (print)
Global Water Security (2012) Intelligence community assessment ICA 2012–08, 2 February
Horlemann L, Neubert S (2007) Virtual water trade: a realistic concept for resolving the water crisis. Dt. Institut für Entwicklungspolitik, Bonn. ISBN: 978–3–88985–327–1
Berhanu B, Seleshi Y, Melesse A (2014) Surface water and groundwater resources of Ethiopia: potentials and challenges of water resources development. In: Melesse A, Abtew W, Setegn S (eds) Nile River basin. Springer, Cham. https://doi.org/10.1007/978-3-319-02720-3_6
Lakshmi V, Fayne J, Bolten J (2018) A comparative study of available water in the major river basins of the world. J Hydrol 567:510–532. https://doi.org/10.1016/j.jhydrol.2018.10.038
MacDonald AM, Bonsor HC, Dochartaigh BÉÓ, Taylor RG (2012) Quantitative maps of groundwater resources in Africa (Published 19 April 2012, 2012 IOP Publishing Ltd.) Environ Res Lett 7(2)
Mahmoud MA, El-Bably AZ (2017) Crop water requirements and irrigation efficiencies in Egypt. In: Negm AM (ed) Conventional water resources and agriculture in Egypt. The handbook of environmental chemistry, vol 74. Springer, Cham
McCaffrey SC (1996) The harmon doctrine one hundred years later: buried, not praised the harmon doctrine one hundred years later: Buried Not Praised 36(4)
McKinney DC (2011) Transboundary water challenges: case studies, 120 Boretti and Rosa, 2019
Mimi ZA, Sawalhi BI (2003) A decision tool for allocating the waters of the Jordan River Basin between all Riparian parties, pp 447–461
Mohamed Y, Thiruvarudchelvan T, Mamad N, Mul M, Van der Zaag P (2011) Performance assessment of large irrigation systems using satellite data: the case of the Gezira scheme, Sudan. In: 21st congress on international commission on irrigation and drainage proceeding (ICID), Tehran, Iran
Munia H, Guillaume JHA, Mirumachi N, Porkka M, Wada Y, Kummu M (2016) Water stress in global transboundary river basins: significance of upstream water use on downstream stress. Environ Res Lett 11(1)
NBI and IWMI (2019) Benchmarking irrigation performance and projection of irrigation water demand in the Nile Basin component III—scenarios for water saving in irrigated agriculture funded by EU-BMZ Project "Support to trans-boundary water cooperation in the Nile Basin" Prepared and to be Implemented by IWMI and NBI
Nile Basin Water Resources Atlas (2016) Nile basin initiative, July 2016. ISBN 978–9970–444–02–01
Obengo JO (2016) Hydropolitics of the Nile: the case of Ethiopia and Egypt. African Secur Rev 25 (1):95–103
Petruzello M (2019) Water scarcity. Encyclopedia Britannica, December 17, 2019. https://www.britannica.com/topic/water-scarcity
Rahaman MM (2009) Principles of international water law: creating effective transboundary water resources management, vol 1, issue 3, pp 207–223
Roser M, Ritchie H, Ortiz-Ospina E (2013) World population growth. Published online at https://ourworldindata.org/
Salman MAS (2007) The Helsinki rules, the U.N. watercourses convention and the Berlin rules: perspectives on international water law. Water Res Dev 23(4):625–640
Sand PH (2018) Mesopotamia 2550 B.C.: the earliest boundary water treaty. Glob J Arch Anthropol 5(4):555667. https://doi.org/10.19080/GJAA.2018.05.555669
Swain A (2011) Challenges for water sharing in the Nile basin: changing geo-politics and changing climate. Hydrol Sci J 56(4):687–702
Tignino M (2016) Joint infrastructure and the sharing of benefits in the Senegal and Niger watersheds. Water Int 41(6):835–850. https://doi.org/10.1080/02508060.2016.1214894
UN (1997) Convention on the law of the non-navigational uses of international watercourses. General Assembly resolution 51/229, annex, Official Records of the General Assembly, Fifty-first Session, Supplement No. 49 (A/51/49)
U.N. (2014) Transboundary waters | international decade for action 'Water for Life' 2005–2015

UNESCO (2015) UN-water transboundary waters thematic priority area. Good practices in transboundary water cooperation. http://www.ais.unwater.org/ais/TPA_Transboundary/map/
UN-Water (2018) Transboundary waters. http://www.unwater.org/water-facts/transboundary-waters
USAID (2019) Water currents: transboundary waters. www.Globalwaters.org
Wolf AT (2007) Shared waters: conflict and cooperation. Annu Rev Environ Resour 32:3.1–3.29. https://doi.org/10.1146/annurev.energy.32.041006.101434
Wolf AAT, Newton JT (1999) Case study of transboundary dispute resolution: the Nile waters agreement, pp 1–8
World Water Assessment Programme (Nations Unies) (2018) The United Nations World Water Development Report 2018 (United Nations Educational, Scientific and Cultural Organization, New York, United States). www.unwater.org/publications/world-water-development-report-2018/
World Water Assessment Programme (2003) U.N. World Water Development Report 1: water for people, water for life; Paris, UNESCO and New York, Berghahn Books, Chapter 20: Senegal River Basin, Guinea, Mali, Mauritania, Senegal, pp 447–461
World Water Assessment Programme (2009) The United Nations World Water Development Report 3: water in a changing world. UNESCO, Earthscan, Paris, London
Zeitoun M, Mirumachi N (2008) Transboundary water interaction I: reconsidering conflict and cooperation. Int Environ Agreements Polit Law Econ 8(4):297–316

Chapter 6
Canada-US Columbia River Treaty: A Review

Getnet Y. Muluye

Abstract The Columbia River Treaty is an international agreement between the governments of Canada and the USA for the cooperative development and operation of the water resources of the Columbia River Basin for the benefit of flood control and hydropower generation. The Treaty is known throughout the world as one of the most successful models of a transboundary water Treaty and considered as a model of "cooperative development" on an international river system. This article reviews the circumstances that led to the Treaty and key Treaty provisions, including arrangements for Treaty implementation, cooperative Treaty projects and benefits, benefit-sharing arrangements, settlement of differences, and period of Treaty. The intent of this article is to demonstrate the practicality of cooperative development and benefit-sharing mechanism to govern transboundary water resources—as opposed to water sharing or water right—through a successful, 56-year-old Treaty.

Keywords The columbia river treaty · Benefit-sharing · Cooperative development · Transboundary rivers

Introduction

The Columbia River is the predominant river in the Pacific Northwest and the fourth largest river in North America. The river emerges in the Rocky Mountains in British Columbia (Canada) and empties into the Pacific Ocean near Astoria, Oregon (United States). In its 1,200 miles course, the river drains about 258,000 square

---

Disclaimer: All the views and opinions presented here are my own and do not represent the views and opinions of any entity whatsoever with which I have been, am now, or will be affiliated.

G. Y. Muluye (✉)
Manitoba Hydro, Winnipeg, Canada

A. M. Melesse et al. (eds.), *Nile and Grand Ethiopian Renaissance Dam*,
Springer Geography, https://doi.org/10.1007/978-3-030-76437-1_6

miles in the northwestern USA and southwestern Canada, including parts of British Columbia (Canada) and parts of seven U.S. states (Oregon, Washington, Montana, Idaho, Wyoming, Utah and Nevada). The Canadian portion of the Columbia River Basin covers only about 15% of the total area of the basin, but contributes, on average, 38% of the total runoff. The river experiences significant seasonal variations in flow and about 60% of the natural runoff occurs during May, June, and July (Bonneville Power Administration 2001). The Columbia River is the largest hydropower-producing river system in the region and provides about half the region's supply of electricity, in addition to other benefits (Stern 2019). Figure 6.1 depicts the Columbian River Basin and major dams.

Fig. 6.1 Columbia River Basin and major dams (2011 U.S. Army Corps of Engineers)

The Columbia River Treaty

The Columbia River Treaty (the Treaty), signed in 1961 and ratified in 1964, is an international agreement between the governments of Canada and the USA for the cooperative development and operation of the water resources of the Columbia River Basin for the benefit of flood control and hydropower generation (B.C. Government, n.d.). The Treaty is known throughout the world as one of the most successful examples of a transboundary water Treaty and characterized as a model of "cooperative development" on an international river system (Environment and Climate Change Canada, n.d.). The Treaty took a lengthy, twenty years negotiations.

The Treaty is an agreement in perpetuity and has no specific end date, with a minimum length of 60 years. Most of its provisions, except those related to flood control operations, would continue indefinitely without action by the USA or Canada. Currently, either nation can terminate the Treaty, any time after 2024 (60 years after its ratification), with a minimum of 10 years' written notice.

The Treaty is implemented by entities from both countries: BC Hydro is the designated Canadian entity and the Bonneville Power Administration and the U.S. Army Corps of Engineers are the designated U.S. entities (Stern 2019). An independent board ("Permanent Engineering Board"), established under the provisions of the Treaty, periodically reviews and reports to the entities on the operation of Treaty projects and investigates and reports on any other matter coming within the scope of the Treaty at the request of either entity.

Treaty Drivers

The negotiation and ratification of the Treaty were prompted by two major events in the basin: periodic and sometimes devastating flooding, most notably, the Vanport flood in 1948; and growing power demand in the Pacific Northwest due to an upswing in the economy after WWII (Stern 2019).

A number of Joint Canada-US Engineering studies were initiated in 1944 through formation of the International Columbia River Engineering Board under the International Joint Commission (International Joint Commission, n.d.-a). The IJC is a joint international body, formed by the Boundary Waters Treaty of 1909, with established Rules and Principles to help resolve disputes concerning water quantity and quality along the U.S.-Canada boundary (International Joint Commission, n.d.-b). Based on several technical studies, the IJC recommended development of storage dams in Canada to help regulate flows on the Columbia River for the purpose of flood control and hydropower production on both sides of the border (U.S. Army Corps of Engineers and Bonneville Power Administration 2011).

Key Treaty Provisions

Under the terms of the Treaty, Canada would:

(i) provide 15.5 million acre-feet of additional storage on the Canadian side of the basin through the construction of three dams: Duncan (completed in 1968), Hugh Keenleyside, or Arrow (completed in 1969), and Mica (completed in 1973).
(ii) operate the storage for optimal power generation and flood control on both sides of the border.

And the USA would:

(i) pay Canada 50% of future flood control benefits resulting from operation of Canadian storage.
(ii) pay Canada 50% of downstream power benefits (known as the "Canadian Entitlement"), which is the forecast additional electricity generated on the U.S. side as a result of operation of Canadian storage.

Other provisions include:

(i) the USA would have the option to build a dam on the Kootenai River near Libby, Montana (Libby Dam completed in 1973), and Canada agreed to the flooding of its territory (42 miles into Canada) that would come with the construction of the Libby Dam in Montana.
(ii) Canada would have the right, after the expiration of 20 years from the ratification date, to divert Kootenay River at Canal Flats into the headwaters of the Columbia River (Kootenay Canal Generating Station completed in 1976).

Treaty Benefits

The Treaty is considered successful in terms of meeting its primary objectives: flood control and power generation on both sides of the border (Stern 2019).

The main benefits of the Treaty to the Province of British Columbia (Canada) include (U.S. Army Corps of Engineers and Bonneville Power Administration 2011; Columbia Basin Trust, n.d.):

(i) A lump sum of US$254 million in Canadian Entitlement (a "pre-sale" of the first 30 years) and a 60-year pre-payment of assured annual flood control worth US$64.4 million. These payments covered a large portion of the construction of Mica, Keenleyside, and Duncan dams;
(ii) A full Canadian Entitlement benefits worth approximately US$120 million each year since 2003;

(iii) On-site hydropower generation at Treaty dams (Mica and Keenleyside);
(iv) Increased water storage and flow regulation led to the development of additional generating facilities, including Kootenay Canal Generating Station (1976), Revelstoke Dam (1984), and Brilliant Expansion project (2007).
(v) Increased employment during construction, ongoing operations, and periodic upgrades;
(vi) Flood damage reduction;
(vii) Low-cost hydroelectricity to industry and communities across the province;
(viii) Increased economic activity.

The main benefits of the Treaty to the USA include (U.S. Army Corps of Engineers and Bonneville Power Administration 2011; Columbia Basin Trust, n.d.):

(i) Doubled the available flood storage in the Columbia River Basin;
(ii) Greatly reduced risk of devastating floods. No major flood damage occurred since the Treaty storage became operational, eliminating $2 billion in potential damage every year;
(iii) Significantly improved hydropower generation;
(iv) On-site hydropower generation at Treaty dams (Libby);
(v) Increased dependable capacity;
(vi) Increased firm energy and usable non-firm energy. Predictable and reliable hydro-electricity allow utilities to meet their customer load obligations;
(vii) Increased agricultural prosperity to vast arid areas of the region (about 2.9 million hectares);
(viii) Greatly improved conditions for navigation and allowed river traffic to extend and expand;
(ix) Facilitated municipal water use, industrial use, recreational and ecosystem augmentation;
(x) Increased economic activity in the region.

Settlement of Differences

Differences arising under the Treaty which both countries cannot resolve may be referred by either to the IJC or an arbitration tribunal for decision. Figure 6.2 illustrates the process involved in settlement of differences.

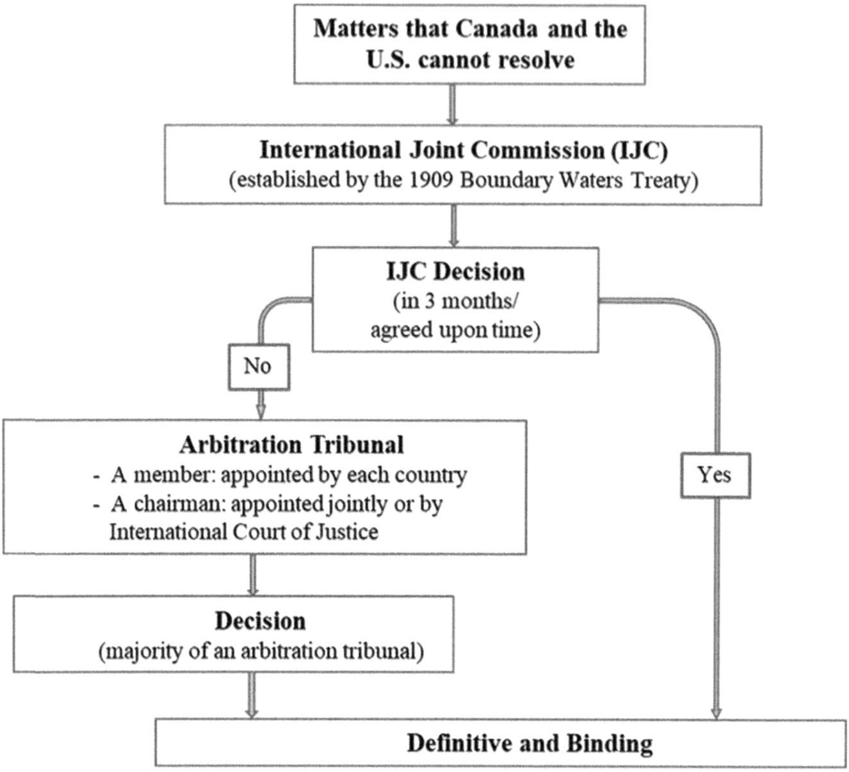

Fig. 6.2 Columbia River Treaty settlement of differences

The Future of the Treaty

The Treaty, which is seen as a successful and a benchmark on international river cooperation and benefit sharing, is not without shortcomings. It exclusively focuses on flood control and power generation, ignoring the ecological function. After 56 years in operation, the Treaty is now up for renegotiation and is expected to right historic wrongs through the inclusion of ecosystem-based function as a third primary objective of the Treaty; indigenous participation in decision making as sovereigns; and meaningful public participation.

Currently, either country has the right to unilaterally terminate most provisions of the Treaty as early as September 2024 with a minimum of 10 years' written notice. This prompted both countries to launch separate reviews of the Treaty and conduct a series of studies to evaluate post-2024 decision options, including possible continuation, amendment or termination of the Treaty, in advance of the initial date for announcement of intent to terminate (2014).

If the Treaty is terminated without renegotiation, the Canadian Entitlement would end. Canada would operate the Treaty dams for Canadian power benefits and interests only, except for called-upon flood control. The Boundary Waters Treaty of 1909 would govern the Columbia River, and each country would maintain exclusive control and use of the river on their side of the border (International Joint Commission, n.d.-b).

Perspectives on the Treaty vary considerably on both sides of the border. On the one hand, some in the USA believe that the Canadian Entitlement should be adjusted, or even be eliminated entirely; on the other hand, the Canadian entity—as outlined in its decision and principles to guide future negotiations (Government of British Columbia, n.d.)—noted that "all downstream U.S. benefits, such as flood risk management, hydropower, ecosystems, water supply (including municipal, industrial and agricultural uses), recreation, navigation and any other relevant benefits … should be accounted for and such value created should be shared equitably between the two countries" (Principles #3). Canada's position is that without the Canadian Entitlement (or considerations that would reduce its entitlement), it sees no reason for the Treaty to continue or renegotiate.

In its Regional Recommendation to the Department of State on December 13, 2013, the U.S. entity recommended continuing the Treaty with certain modifications to "modernize" the Treaty "that ensures a more resilient and healthy ecosystem-based function throughout the Columbia River Basin while maintaining an acceptable level of flood risk and preserving reliable and economic hydropower benefits" (U.S. Entity 2013, p. 2). The regional recommendation identified nine "general principles" to guide future negotiations. Likewise, on March 13, 2014, the Canadian entity, through the province of British Columbia, issued its decision to "continue the Columbia River Treaty and seek improvements within the existing Treaty framework" and identified fourteen principles to guide future negotiations (Government of British Columbia, n.d.). Canada would like to see "Treaty provisions post-2024 be fixed for a sufficient duration to provide planning and operational certainty while allowing for adaptive mechanisms to address significant changes to key components and interests" (Government of British Columbia, n.d., Principles #4) in a modernized Treaty.

On May 29, 2018, Global Affairs Canada initiated a formal negotiation process with the USA to modernize the Treaty regime (Global Affairs Canada 2018, May 22). Treaty negotiation is ongoing at present, and the tenth round of negotiations was conducted by web conference on June 29 and 30, 2020.

Closing Remarks

The Columbia River Treaty is known throughout the world as one of the most successful models of transboundary water treaties on international river system. The model is based on cooperative development and equitable benefit sharing, as opposed to water sharing or water rights. The Treaty has successfully met the primary Treaty objectives and brought enormous socio-economic benefits and prosperity to the region on both sides of the border. After nearly 60 years in operation, both countries have agreed to continue and modernize the Treaty even if either country has the option to unilaterally terminate the Treaty. This is a testament to the practicality of the benefit-sharing model to govern transboundary rivers and no wonder why the Treaty is considered an "international model of international cooperation," and many countries wish to emulate a similar water governance model.

Some key Treaty takeaways include:

(i) A cooperative development and benefit-sharing model can serve as an effective mechanism to manage shared transboundary water resources.
(ii) For a cooperative development project to be successful and effective, the planning and development aspects of the water management project would need to consider the entire river basin as one unit irrespective of country borders to realize the full potential. This form of arrangement reduces the risk of disputes and conflicts among countries and promotes regional integration, transboundary cooperation, peace, security, and sustainable development.
(iii) Coordinated operations of upstream and downstream water management facilities provide optimal power generation, reduced flood risk, and facilitates navigation, agriculture, municipal water use, industrial use, recreation, and ecosystem augmentation.
(iv) Negotiating benefits with a win-win approach is vital and key to the success and continuation of a Treaty or a water governance structure.
(v) Treaty provisions should be of a fixed duration that consider planning and operational certainty, as well as adaptable to respond to key components, interests, new information, and changing conditions.

Acknowledgements The author acknowledges the organizing committee and the reviewers of this article.

References

B.C. Government. (n.d.) The Columbia River Treaty. https://engage.gov.bc.ca/app/uploads/sites/6/2012/04/Columbia-River-Treaty-Protocol-and-Documents.pdf. Accessed 16 Nov 2020

Bonneville Power Administration (2001) The Columbia River System inside story, 2nd edn. Bonneville Power Administration, Portland, Oregon. https://www.bpa.gov/p/Generation/Hydro/hydro/columbia_river_inside_story.pdf. Accessed 13 Dec 2020

Columbia Basin Trust (n.d.) Columbia River Treaty. https://thebasin.ourtrust.org/columbia-river-treaty/. Accessed 13 Dec 2020

Columbia River Treaty Negotiations Status (2020) Global affairs Canada. https://engage.gov.bc.ca/columbiarivertreaty. Accessed 13 Dec 2020

Environment and Climate Change Canada (n.d.) Compendium of Canada's engagement. In: International environmental agreements and instruments. https://www.canada.ca/content/dam/

eccc/documents/pdf/international-affairs/compendium/2020/batch-6/canada-united-states-columbia-river.pdf. Accessed 6 Dec 2020

Global Affairs Canada (2018) Canada and United States launch negotiations to renew Columbia River Treaty. https://www.canada.ca/en/global-affairs/news/2018/05/canada-and-united-states-launch-negotiations-to-renew-columbia-river-treaty.html. Accessed 13 Dec 2020

Government of British Columbia (n.d.) BC decision. Columbia River Treaty. Retrieved November 16, 2020, from https://engage.gov.bc.ca/columbiarivertreaty/bc-decision/. Accessed 13 Dec 2020

International Joint Commission (n.d.-b) Boundary Waters Treaty. (n.d.). https://ijc.org/en/who/mission/bwt. Accessed 16 Nov 2020

International Joint Commission (n.d.-a) https://ijc.org/en. Accessed 16 Nov 2020

Stern CV (2019) Columbia River Treaty review. Library of congress. Congressional Research Service, Washington D.C.

U.S. Army Corps of Engineers and Bonneville Power Administration (2011) Columbia River Treaty: history and 2014/2024 review, Portland, Oregon. https://www.bpa.gov/news/pubs/GeneralPublications/crt-Columbia-River-Treaty-History-and-2014-2024-Review.pdf. Accessed 13 Dec 2020

U.S. Army Corps of Engineers, Northwestern Division (2011) Columbia River post-2024 flood risk management procedure. USACE, Northwestern Division, Portland, OR. https://www.critfc.org/wp-content/uploads/2019/07/Post-2024-White-Paper-09-11_FINAL.pdf. Accessed 15 Dec 2020

U.S. Entity (2013) U.S. Entity regional recommendation for the future of the Columbia River Treaty after 2024. Bonneville Power Administration. https://www.bpa.gov/Projects/Initiatives/crt/CRT-Regional-Recommendation-eFINAL.pdf. Accessed 13 Dec 2020

Chapter 7
Geographic Aspects of Hydropolitics: The Case of Nile and Mekong River Basins

Daniel K. Waktola

Abstract The filling of the Grand Ethiopian Renaissance Dam in 2020 raised the age-old confrontation between upstream Ethiopia and downstream Egypt to a new level. In this study, a geographic approach is used to ascertain whether Ethiopia could survive without touching the Blue Nile water by investigating the balance between quality of land resources and demographic pressures against the backdrop of the geohyropolitics in the Mekong River. We used a set of geospatial databases acquired from measured and modeled scientific data sources. DEM, slope, and contours were generated from the terrain data. Rasterized population data were used to discern the temporal trend and plot population density across topographic gradients. Global Climate Model (GCM) generated temperature, and precipitation data, averaged for 2020–2040, 2040–2060, 2060–2080, and 2080–2100, were extracted from five sample locations in each river basin and analyzed for December and June, representing Winter and Summer seasons, respectively. To explore alternative water sources for Egypt, multiple buffer zones were created from the Red Sea and Mediterranean Seas. Results revealed (1) In light of intensifying demographic pressure, dwindling carrying capacity of land resources, warming atmospheric temperature, and growing uncertainty of rainfall patterns, Ethiopia's vulnerability to food and economic insecurity is bound to worsen in the foreseeable future, which ultimately calls for equitable and reasonable utilization of the Blue Nile water. (2) Over 75% of settlements and over 97% of Egypt's population are seated within a 300-km distance from the two seas where desalinated water could be harvested for domestic and agricultural uses. The study concludes that Egypt should come to terms with the biophysical and demographic dynamics of the Nile Basin and diversify its water sources rather than hoping to cling to obsolete treaties of the Nile, which the upper riparian countries were not a party to.

Keywords Nile River Basin · Mekong River Basin · GERD · Geohydropolitics · Egypt · Ethiopia

D. K. Waktola (✉)
Austin Community College, 11928 Stonehollow Drive, Austin, TX 78758, USA
e-mail: daniel.waktola@austincc.edu

A. M. Melesse et al. (eds.), *Nile and Grand Ethiopian Renaissance Dam*,
Springer Geography, https://doi.org/10.1007/978-3-030-76437-1_7

Introduction

Given the growing scarcity of freshwater and the anticipated compounding impacts of climate change, World War III is believed to be fought over the scarcity of water. Some scholars assert that in a geopolitical sense, water is likely to become the "oil of the next century" (Elhance 1999). One of the presumed potential conflicts is between Egypt and Ethiopia over the contested use of the Nile River. While the tension between the two countries dates back for centuries, the conflict rose to the epicenter of the mediasphere after Ethiopia started filling the Grand Ethiopian Renaissance Dam (GERD) reservoir. As the Nile cast a long shadow over Egypt, because of the country's near-complete dependence on the river (AlSayyad 2019), and the fact that the legacy of Egypt's civilization is nursed in the womb of the Nile (Hassan 1997), the construction of GERD, which is going to be the largest dam in Africa, the stakes have never been higher. It is described as "the most fascinating riparian power struggles of modern time unfolding in East Africa" (Smith 2020). Ibrahim (2011) equated GERD to "the beginning of the end of Egyptian hydropolitical hegemony."

While Ethiopia has been pursuing its natural rights and actively championing diplomatic solutions, Egypt has not only reiterated its arguments of primary need, prior use, and acquired water rights enshrined in the obsolete treaties of 1929 and 1959, but also introduced three newer approaches to deny Ethiopia's access to the Nile: first, by proposing a newer equation for sharing the Nile's water by adding "green-water" as a variable. Unlike "blue-water," which includes the surface and groundwater in the basin, some Egyptian scholars (e.g., Abu-Zeid 2008) advanced the concept of "green-water," which includes all sources of water (atmospheric water and soil moisture in the unsaturated zone) taken up by local vegetation in the Nile Basin catchment[1], second, by spearheading an aggressive diplomatic style—like escalating the case to the United Nations Security Council to intervene or lobbying the US government to withhold security and developmental assistance to Ethiopia, third, by concocting coercive measures, viz. plotting to take military actions against Ethiopia, which was unmasked by WikiLeaks (Smith 2020) and making a discussion on aiding anti-government rebels in Ethiopia to sabotage the dam project, which was unintentionally broadcasted through a hot mic (Smith 2020). As a result, the current relationship between the two countries is in limbo, and it is difficult to predict whether they will foster an equitable and reasonable allocation of the Nile's water or resort to war.

Globally, 260 river basins are shared by two or more sovereign states (Smith 2020). Inherent power imbalances exist between states in those international rivers,

[1]If this sinister equation is implemented, vegetations, only present in the upper parts of the Nile River, would enlarge the total volume of water for sharing. This means, the volume of water to be allocated for Egypt, contributing neither blue-water nor green-water to the equation, would be derived only from the blue-water, i.e., River Nile. On the contrary, the upper riparian states would get their share in the form of green-water. In short, Egypt would have an exclusive use of the Nile water.

vis-a-vis topographic positions. The Nile River exemplifies a group of transnational rivers where a weaker riparian occupies the upstream position, whereas the powerful riparian is situated downstream. There is also a group of river basins where a powerful riparian occupies the upstream position, and the weaker one occupies the downstream position, as manifested in the Mekong River.

As Amery and Wolf (2000) put it, all things being equal, geographical advantage confers upon the upstream riparian power to alter the quality and quantity of the water by such tactics as diversion and contamination of the water flowing downstream, thus affecting the other riparian states. According to Dinar et al. (2013), the state's physical position along the river may provide an otherwise weaker riparian state with the means to challenge the status quo and the weaker states in the upstream position to assert their power when triggered by economic or political geographies. The question is, to what extent this relationship plays out in countries of the Nile River Basin (NRB) and Mekong River Basin (MRB)?

Research conducted on transboundary rivers could be encapsulated in two domains: the environmental and hydropolitical (e.g., Varies et al. 2008; Melesse et al. 2014; Negm 2017; Abtew and Dessu 2019). In the environmental category, hydrological, geological, land use/land cover, climate change models, etc., are cited. On the hydropolitical side, hydro hegemony, treaties, diplomatic challenges, water rights, military and economic powers are addressed. Studies anchoring the physical geographic factors to the environmental or hydropolitical domains of studies are meager. As both environmental and hydropolitical aspects of international rivers are anchored onto a geographic space, a study incorporating a geographical approach would fill the void surfaced in the hitherto studies.

Against the backdrop of the forgoing gaps, this study attempts to (1) investigate the interplay of topographic positions, land resource vulnerabilities, population distribution, and power balance between the upper and downstream countries in the NRB and MRB, (2) identify the balances between the predicted supply side (land and climate resources) and demand side (demographic pressure) in the NRB and MRB, and (3) explore the accessibility of alternative water resources for Egypt. Figure 7.1 shows the NRB and MRB basins with the flow direction.

Methodology

A two-pronged approach is adopted: the biophysical appraisal, which consists of topography, climate, land resources, and the analyses of the hydrogeopolitical landscapes, using the MRB as a comparative tool for the Nile Basin study.

While there are conspicuous contrasts in terms of the biophysical, hydropolitical, and cultural landscapes, the two river basins also share resemblances that could shed light on the analysis of the circumstances in the Nile River. The two river basins are home to overwhelmingly poor and rural communities. Recently, both the Nile and Mekong Rivers are immersed in hydropolitical wrangling due to asymmetric power balances between each basin's upper and lower riparian countries.

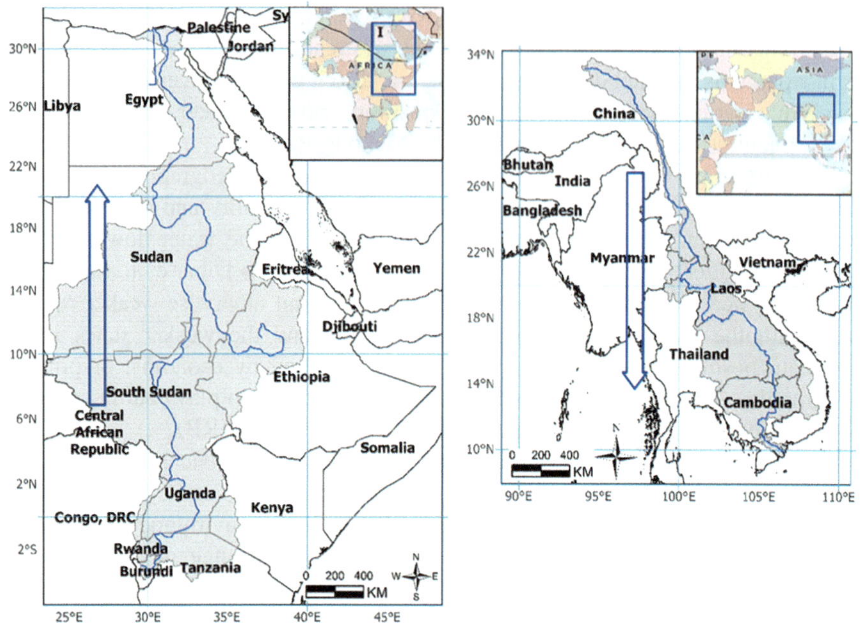

Fig. 7.1 Locations of the Nile and Mekong River Basins along with the flow direction. *Note* The maps allow comparison of the area and latitudinal positions

While the Nile River's downstream is occupied by the hydro hegemon Egypt, the Mekong River's upstream is occupied by the hydro hegemon China. Therefore, the comparison allows to deriving generalizable conclusions from the findings.

A complete list of data sets used in this study is provided in Table 7.1. Of the available pathways developed by the Meteorological Research Institute Earth System Model version 2.0 (MRI-ESM2.0) released in 2019 (Yukimoto et al. 2019), we selected the middle scenario's socioeconomic pathways, which are monthly averaged for the years 2021–2040, 2041–2060, 2061–2080, and 2081–2100. We selected five sample locations in each river basin to extract digital data from the continuous layers of climate prediction models (Fig. 7.2). Those sample locations run in a north–south direction. For each sample location, we derived the projected mean minimum temperature (Tmin), mean maximum temperature (Tmax), and monthly precipitation (Prec) from the Global Climate Model (GCM) , MRI-ESM2-0. The spatial resolution is 0.5-min. The comparative assessment relied on two monthly data: June, representing Summer, and December, representing Winter seasons.

Rasterized population data were used to discern the temporal pattern and plot population density across topographic gradients. To explore alternative water sources for Egypt, multiple buffer zones were generated from the Red Sea and the Mediterranean Sea.

Table 7.1 Types, sources, and characteristics of data used in the study

Data	Source	Description
Shape files	http://geoportal.icpac.net/documents/47; https://data.opendevelopmentmekong.net/	Basin and administrative boundaries
DEM	The Shuttle Radar Topography Mission (SRTM)	30 s (elv 30 s)
Population	https://www.unocha.org/; https://sedac.ciesin.columbia.edu	Gridded Population of the World, Version 4 (GPWv4) for 2000, 2005, 2010, 2015, and 2020.
Bioclimatic variables (Historical and Projected)	https://www.worldclim.org/	Rasterized Global Climate Models (GCMs) data- only MRI-ESM2-0- (2021–2040, 2041–2060, 2061–2080, 2081–2100). 0.5-min spatial resolution
Soil	ISRC (https://www.isric.org/)	250 m resolution: soil depth, OM, CEC,…
Shape files	http://geoportal.icpac.net/documents/47; https://data.opendevelopmentmekong.net/	Basin and administrative boundaries

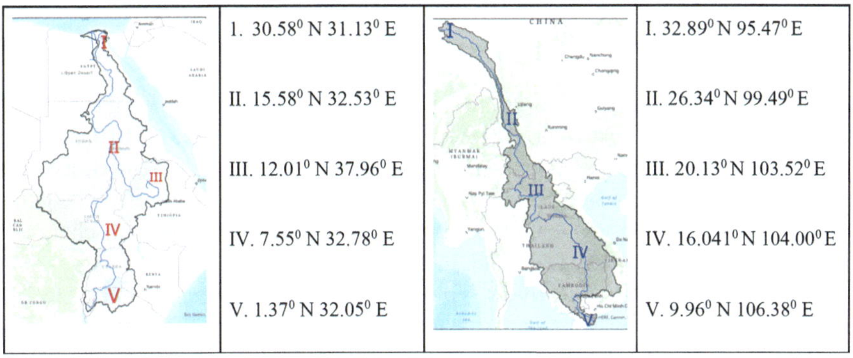

Fig. 7.2 Location of sample points in the Nile and Mekong River Basins

Results and Discussion

Location and Hydrogeologic/Physiographic Characteristics

Both the NRB and MRB are characterized by complex physiological, climatic, and geopolitical environments. The Nile River traverses a larger geographic area and home to many countries than the MRB (Fig. 7.1). The Nile River flows from 4°S to 31°N, making a latitudinal extent of 35 degrees. The Mekong River, in contrast, flows from 33°N to 10°N, making a latitudinal extent of 22 degrees. The NRB drains eleven countries, vis. Tanzania, Uganda, Rwanda, Burundi, the Democratic Republic of Congo, Kenya, Ethiopia, Eritrea, South Sudan, Republic of Sudan, and Egypt. The MRB, on the other hand, drains six countries, viz. Vietnam, Thailand, China, Laos, Cambodia, and Myanmar (Burma).

Three principal streams from the NRB: The Blue Nile, which flows out of Lake Tana in the northwestern highlands of Ethiopia with several tributaries joining along the way, and the White Nile, which flows from Lakes Victoria and Albert and joined by the Sobat from Southwestern Ethiopia (Baro and Akobo Rivers). At Khartoum (Sudan), the White Nile meets the Blue Nile. From there, it runs a 2500 km journey across the world's largest desert, the Sahara, only interrupted by the confluence of the wild Tekeze-Atbara River from Ethiopia and Eritrea.

Originating from the Tibetan Plateau, Mekong flows through Yunnan Province and then to the Three Parallel Rivers Area in the Hengduan Mountains. The river flows through the China–Myanmar border until it reaches the tripoint of China, Myanmar, and Laos. From there, it flows southwest until it arrives at the tripoint of Myanmar, Laos, and Thailand, called the Golden Triangle. Then, Mekong turns southeast to form the border of Laos with Thailand for some 850 km. Thereafter, it crosses Cambodia and Vietnam before emptying into the South China Sea. The arc-shaped Nile Delta covers over 20 million square km of fertile soil and homes to 39 million Egyptians. The Mekong River Delta, dubbed the "biological treasure trove," has 15,600 sq km and is inhabited by 17 million Vietnamese.

Both the length and drainage areas of the Nile River exceed the Mekong River. The Nile River is 6650 km long, and its basin area is 3,400,000 km^2, making it the first and the third in the world, respectively. On the other hand, the Mekong River is the 12th longest river (4350 km) with a catchment area of 810,000 km^2. While the Nile River's catchment area is over four times that of the MRB, its water discharge is disproportionately low. The Mekong's discharge volume at the China Sea is 16,000 m^3/sec, which exceeds five times the Nile River discharge of 2830 m^3/sec at the Aswan Dam and 1400 m^3/sec at Cairo.

Like most of the African rivers, viz. Niger, Congo, Orange, Senegal, etc., the White Nile receives water from its equatorial belt. According to Fielding et al. (2017), the modern Nile's drainage is controlled predominantly by local geology and geomorphology changes. On the other hand, the tributary networks of the Mekong are complex, with different sub-basins often exhibiting distinct drainage patterns (Tandon and Sinha 2007), controlled by heterogeneous underlying geological structures.

The flow direction of the Nile and Mekong Rivers is in opposite directions (Fig. 7.1). Descending from Lake Victoria (of White Nile) and Lake Tana (of Blue Nile), the Nile River flows northward through contrasting climatic belts. In contrast, the Mekong River descends from the glaciated Tibetan region of China, flowing southward through rapids and waterfalls to empty into the China Sea. Abdelkareem and El-Baz (2015) classified the Nile Basin into three zones: (1) arid to hyper-arid zone, which contributes neither sediments nor runoff in most seasons and occupies the northern part, (2) semiarid zone, which contributes sediments and runoff during the summer season and (3) tropical zone, nearly perennial, including equatorial lakes, and representing the main source of the White Nile. The Mekong Basin is commonly divided into two parts: The upper basin in China (where the river is called Lancang) and the Lower Mekong Basin from Yunnan (China) downstream to the South China Sea. According to MRC (2010), the upper basin makes up a quarter of the total area and contributes 15–20% of the water that flows into the Mekong River. The melt from glaciers and the Tibetan area feeds the Mekong River's dry-season flow, about 17%. The rest of the water is contributed by the monsoon rain in the Lower Mekong area. On the other hand, Ethiopia contributes 86%, while Egypt contributes nothing to the Nile water.

Analysis of Topographic Positions in Relation to the River Basins

The topographic positions occupied by powerful nations in the NRB and MRB are discordant. Egypt occupies the downstream position and suffers from dry climatic conditions. And yet, Egypt is a hydro hegemonic actor in the NRB. China, on the contrary, is situated in an upstream position, receives very little rainfall, asserts its power over the downstream countries of the MRB.

By virtue of being situated in an upstream position, Ethiopia is considered advantageous while, Egypt, on the other hand, is presumed to be a victim to its geographic position. Further analysis of the topographic position proves otherwise. This flawed perception emanates from the unnoticed differences between the physical, economic, and political geographies. Physically, Ethiopia is the source of the Blue Nile, situated in a mountainous position and primed to enjoy a myriad of advantages, among which defense is the one. As mountains act as natural borders, cities flourished in mountainous places are among outstanding defenders of their territories from invaders. The resounding success of Ethiopians in defending numerous foreign enemies is partly attributed by some scholars (e.g., Berkeley 1902) to the role played by the topographic position. For instance, between the Battle of Mekdela in 1868 and the Battle of Adwa in 1896, Ethiopia had entertained twelve battles (Pankhurst 1963). And yet, Ethiopia is one of the two African nations to have never been colonized. It is to be recalled that Ethiopia unequivocally defeated Egypt (1874–1876) both at the Battle of Gundet and the Battle of Gura in Northern Ethiopia.

Especially in tropical regions, mountains provide cooler temperature conditions and thereby attract denser populations. Mountains in the tropical area are also preferred places to evade tropical diseases like malaria. However, in the middle and high latitude latitudes, lowlands are already cooler. In such environments, mountainous areas make the temperature brutally cold to the detriment of human settlement.

In terms of sharing waters of the transnational rivers, countries positioned upstream could easily assert their dominance over the downstream countries.

Countries such as the USA, Turkey, China, Spain, and India are good examples of harnessing the transboundary water resources. While the physical geography of Ethiopia sets the stage for those benefits, economic and political geography factors have remained a deterrent. Elhance (1999) explained the diverse and cardinal role of river basins' physical, economic, and political geographies, based on case studies conducted in transnational river basins, including the Nile and Mekong.

Contrary to the above advantages, mountain communities are increasingly marginalized due to the downhill flow of natural assets at unsustainable rates (Pratt and Preston 1998), where Ethiopia can be cited as a perfect example. This is partly due to the ruggedness of the landscapes, making them susceptible to natural hazards (e.g., landslides), a high rate of soil erosion, which leads to shallower soils, and reduced agricultural productivity. A higher magnitude of steepness offers little or no potential for practicing mechanized farming. Rather, it is limiting the opportunity for lowering the cost of production and farm efficiency. Efficient cultivation is impossible on gradients of more than 11°, and arable cultivation is impossible on 18° slopes (Hammond 1985). The upper parts of high mountains such as the Andes, Rockies, Himalayas, Alps, Atlas, and Kilimanjaro are ruled out for any meaningful agriculture (Nkemasong and Neba 2009). A study conducted in Central Ethiopia (Saguye 2017) shows how the shortage of farmland pushed farmers to cultivate steep valley slopes of up to 80 degrees.

Moreover, rivers in their mountainous positions create deep and steep gorges, making irrigation structures expensive. Infrastructure, too, is expensive in mountainous areas. We argue that countries dominated by the mountainous positions are highly correlated to poorer countries of the world, viz. Tibet, Nepal, Peru, and Ethiopia. On the contrary, most of the developed areas have emerged in the low-lying and flattish areas. Studies (such as Cohen and Small 1998; Small and Nicholls 2003) show that areas with elevation less than 100 m and closer than 100 km to the coast house are around one-quarter of the global population.

The slope pattern (Fig. 7.3), generated from the DEM, shows how the share of the steep gradient is reversed in two river basins. While the Nile River basin is overwhelmingly dominated by flattish lands, except for Ethiopia and the Great Lakes area, the overwhelming areas of the Mekong River basin are dominated by steeper slopes. Though flat areas have the potential for mechanized farming, the flatter parts of the Nile River Basin are unsuitable for agricultural development as they lie in a barren rocky desert.

The Nile, which excludes the White Nile River segment, and Mekong River's vertical profiles demonstrate a different slope pattern (Fig. 7.4). Mekong River shows a precipitous decline of elevation. Within the first 600 km from its head, the river cascades through a higher degree of inclination. The Blue Nile, descending from the Ethiopian plateau, is flowing through a moderate slope angle.

Topography influenced the population distribution of the two river systems in a reverse pattern. The upstream Mekong is scarcely populated, while Lower Mekong River is overwhelmingly populated. It appears that the impact of

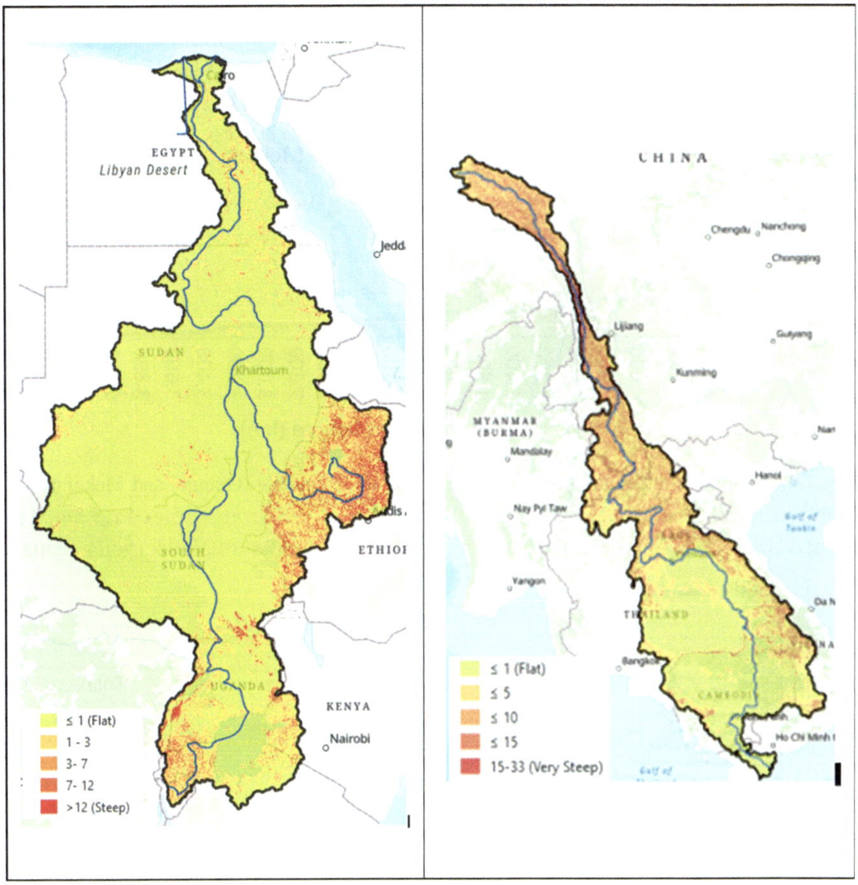

Fig. 7.3 Slope patterns of the Nile and Mekong Rivers

population on the delicate environment of the steep slopes of the Mekong basin is negligible, which ultimately contributes to a limited negative impact on the downstream area. On the other hand, population distribution in the NRB follows a bimodal pattern. Both the source and mouth of the NRB are densely populated, while the middle course is sparsely populated. Since subsistence agriculture is the mainstay of the upstream Nile, the current use of agrochemicals is negligible, which would likely change soon.

The stark difference in the population–slope relationship is shown in Fig. 7.5. Putting aside the contrast in the total number of populations residing in the two basins, most of the population in the MRB is clustered within 0–2 degrees of slope, which is situated in the delta area. Over the NRB, however, the population densely concentrated in the lower slopes around the Nile River Delta and encroached onto

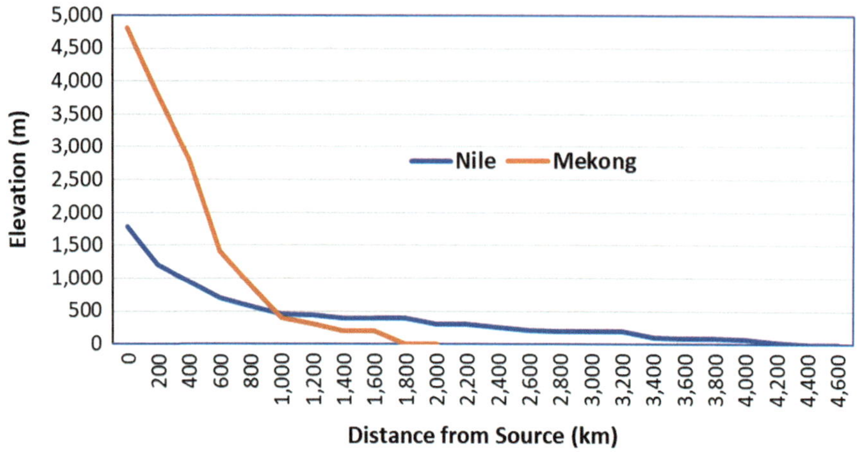

Fig. 7.4 Population distribution across elevations in the Blue Nile (Orange) and Mekong River (Blue)

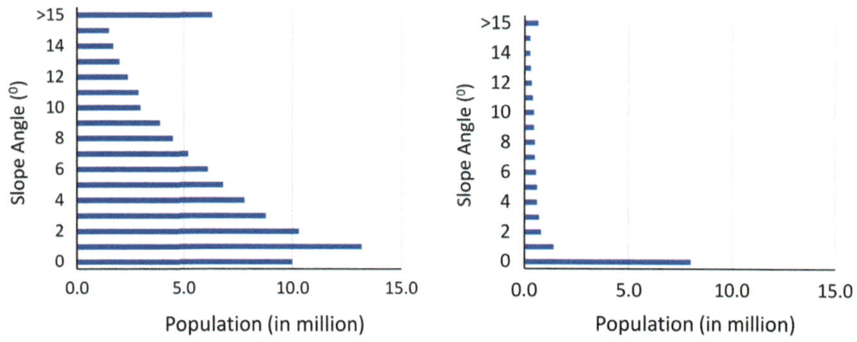

Fig. 7.5 Population distribution across slope gradient in the Nile and Mekong River Basins

the critical hillslopes of the basin. The wide gap in the proportion of the population living above 15 degrees of slope in the two basins demonstrates how the steeper terrains of the NRB are subjected to the ramifications of population pressure on the fragile land resources, where the upper Blue Nile is a part of. The pattern observed in the two river systems goes in line with the finding by Li et al. (2011), who explored geographical differences in the spatial patterns of human settlements in the Three Gorges Reservoir Area in central China.

Land Productivity-Population Balance

The two river basins differ in the magnitude of human dependence on land resources, especially for generating energy. The global database of dams[2] reveals the huge mismatch between the number of dams and the population residing in the two basins. In the NRB, only 40 dams are developed, while that number is 440 for the MRB. Over 250 million people reside in the NRB, yet the number of dams in the MRB is more than 11-fold theirs while hosting only 60 million people, a quarter of the Nile Basin population. It implies that NRB is under a more intense demand for energy than the MRB. As successful irrigation agriculture relies on river dams, agriculture in the upper Nile Basin would continue to be rainfed, which is highly vulnerable to short-term dry spells and long-term droughts.

The future of population well-being in the two river basins hinges on the balance between demand and supply. The demand factor is associated with the need for natural resources generated by a rapidly growing population, whereas the supply is associated with land productivity, which is a function of rainfall, temperature, and soil nutrients.

Demand-Side Analysis: Demographic Pressure

As developing countries inhabit both river basins, population growth in those basins is much faster than the global average. However, discernible differences could be gleaned between the two river basins (Fig. 7.6). In the NRB, the dynamics of the population show a consistent and faster trend of growth through time than MRB.

"Doubling time" is a common term used when studying population growth. It is the projected amount of time that it will take for a given population to double. Doubling time is an easy yardstick for comparing population growth among countries. The higher the population growth, the shorter the doubling time, and vice versa. It is based on the annual growth rate and is calculated by what is known as "The Rule of 70." To find the doubling rate, the growth rate is divided as a percentage into 70. As the population growth of Ethiopia, Egypt, Vietnam, and China are 2.56%, 1.94%, 0.91%, and 0.39% in 2020, the doubling time for the population of the four countries is 27, 36, 76, and 179 years, respectively. Thus, the NRB is expected to double the magnitude of demand for agricultural lands and energy within the next three decades than the MRB—taking a much longer time to do so.

In the NRB, a 52% population increase is estimated between 2010 and 2030 (NBI 2012) and a staggering tenfold increase by the end of the twenty-first century (Brown et al. 2003). Such rapid population growth leads to the rapid conversion of marginal and vulnerable lands for crop cultivation, settlement, and other service

[2]http://geodata.policysupport.org/dams

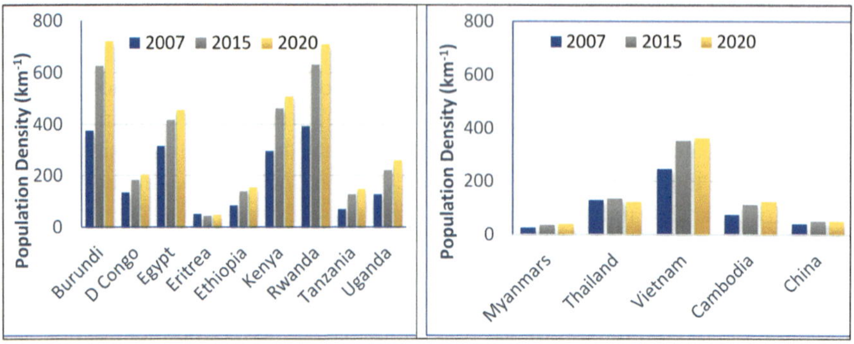

Fig. 7.6 Patterns of population growth for select countries of the Nile and Mekong River Basins

expansions. Milas (2013) warns that the Nile Basin countries would face a demographic crisis, in which constraints of land and water may severely limit rural futures in the context of rapid population increase. To cope with the ever-growing demand by the agriculture-dependent societies, the supply should grow at a matching rate, which is a tall order in the context of the two river basins.

Supply-Side Analysis

One of the most important land resources that affect food security is the quality of land, which is mainly dependent on soil quality. To sustain crop production, agricultural soils should be deep enough to store the necessary soil nutrients and soil moisture. Soil depth can be reduced by erosion due to the absence of vegetation cover and lack of sound land management practices. Already, soil depth in the Northcentral Ethiopian highlands is shallow, partly due to its position in the mountainous upstream areas of the Nile. For instance, large areas of Wello and Gonder, which are located within the upper part of the Blue Nile Basin, and Tigray, which is in the upper part of the Tekeze (Atbara) basin, are identified as "land degradation hotspots" (Sonneveld and Keyzer 2003). According to Mitiku et al. (2006), soil fertility depletion is more serious in the highlands where most of the human and livestock population are clustered, leading to complete removal of crop residue from farmlands for household energy and livestock feed, instead of using it back for soil fertility maintenance.

Shallow soils hardly buffer frequent episodes of dry spells within the growing seasons, responsible for recurrent droughts and food insecurities in Ethiopia. Unless the agricultural lands of the Blue Nile Basin are supplied with additional water sources in the form of irrigation, it is difficult to feed the existing population, let alone the projected doubled population size in the coming few decades. As Teketay (2001) succinctly put it, the availability of lands suitable for agriculture in Ethiopia is shrinking fast, while the amount of land required to feed the growing population is steadily increasing. Due to rapid population growth, farmers in the highlands of Ethiopia plow their lands with a little-to-no fallow period. Continuous cultivation in the absence of intensive agricultural input leads to reduced soil's organic matter, ultimately reducing the soil fertility. For instance, a study conducted in Northern Ethiopia by Corral Nuñez et al. (2014) noted that soil organic matter in farmland soils would decline given the current agricultural practices, including removal of crop residues and the use of manure as an energy source for cooking.

The Ethiopian highlands have one of the highest soil erosion rates in the world (Blaikie 1985; Blaikie and Brookfield 1987). The average estimated soil loss in the Ethiopian highlands ranges from 3–7 t ha^{-1} yr^{-1} (Gebremedhin and Swinton 2003) to 42 t ha^{-1} yr^{-1} (Hurni 1993) to 100 t ha^{-1} yr^{-1} (FAO 1986). In the MRB, soil erosion has been a major problem and approximately 50% of the sediment in the Mekong river comes from the upper basin. In Vietnam, Van De et al. (2008) estimated a mean loss of 85.2 t ha^{-1} y^{-1} under cultivated lands and 43.3 t ha^{-1} y^{-1} on uncultivated lands. Unless the type of green revolution which was implemented in India in the 1960s is replicated in the Ethiopian agricultural system, it is very difficult to keep up with the growing demands of its population using the current land management practices.

Climate is another core variable that affects the quality of land resources. Projections of temperature and precipitation for the years 2021–2040, 2041–2060, 2061–2080, and 2081–2100 show that the prospect of future climate is bleak both for the NRB and MRB. As the livelihood threat posed by soil erosion is expected to be compounded by climate change (Bewket 2011), analyzing the future temperature and precipitation at the sample locations illuminates the expected magnitude of challenges in the two river basins.

The projected mean minimum temperature would be increased, both in June and December, at all sample locations in the NRB (Fig. 7.7) and MRB (Fig. 7.8). The projected warmer winter will be critical for the societies residing at the upper course of the Mekong River. This is because the Mekong River is partly furnished by the melting snow from the Tibetan Plateau, which plays an important role in flow regime change and the low-flow hydrology of the lower mainstream, especially in the dry season. A snowpack could be affected by the warming of temperature both in the winter and summertime. In winter, the impact is mainly due to the reduced accumulation of snowpack. As noted by Hoanh et al. (2010), climate change and the effects on snowmelt in the upper Mekong Basin could result in changes in the flow regime of the Mekong River.

While there are no significant differences in minimum temperature increases among the five sample locations in the MRB, Hoanh et al. (2010) reported that the highest temperature increase would occur in the uppermost parts of the basin in China. The magnitude of warming will be less in the LMB, where the highest temperature changes would occur in southern Cambodia and the Mekong Delta.

The results also revealed an increased mean maximum temperature in June and December at all sample locations in the NRB (Fig. 7.9) and MRB (Fig. 7.10). This result goes in line with the predictions calculated for river basins elsewhere in the world, viz. Xiangjiang River Basin in China (Ma et al. 2016), Colorado River Basin

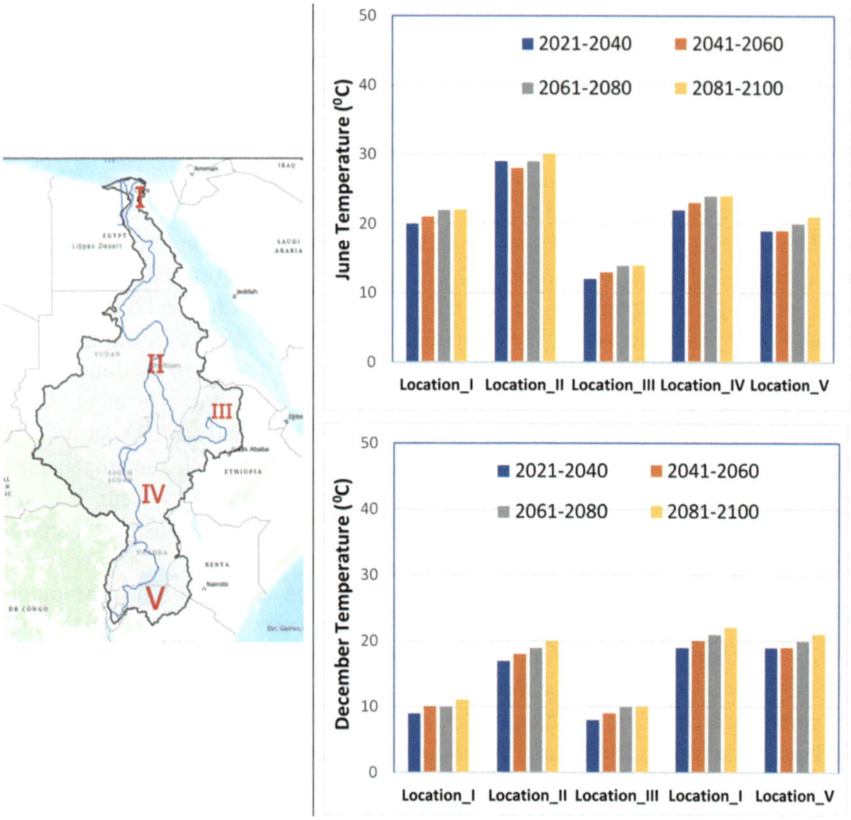

Fig. 7.7 Patterns of projected mean minimum temperature in the sample locations of the Nile Basin

in the USA (Ficklin et al. 2013), and Upper Nan River Basin in Thailand (Gunathilake et al. 2020). Such a warming trend has implications for higher evaporative demand, increased water shortages, heat waves, etc. The NRB tends to have lower resilience to climate change, as its hydrological stability depends on long-distance moisture delivery processes that are vulnerable to the influence of climate change (Pacini and Harper 2016). Mohamed et al. (2005) reported that some 89% of the NRB's precipitation originates from outside the basin's physical boundary.

In light of Hoanh et al.'s (2010) findings, where an increased temperature during the summer season melts the snow faster and increases the risk of flooding in the wet season in the upper MRB, a greater flooding risk is anticipated in the low-lying areas of the Mekong Delta.

The patterns of predicted precipitation are hardly discernible like that of the predicted temperatures in both river basins. In NRB (Fig. 7.11), the winter precipitation is nil in sample Location I, which lies in the delta area around Cairo.

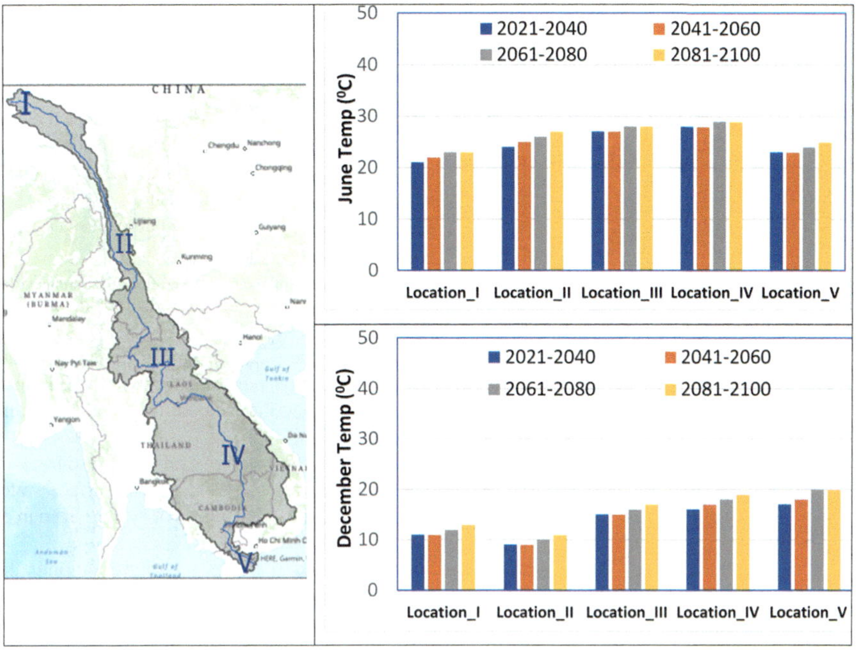

Fig. 7.8 Patterns of projected mean minimum temperature in the sample locations of the Mekong Basin

However, a slightly declining summer precipitation pattern is predicted to happen in the Nile River Delta, where the population is densely concentrated.

In the MRB, rainfall mainly occurs in summer and shows little or no increase in precipitation at sample Locations II, III, IV, and V (Fig. 7.12). However, Location I, situated in the upper part of MRB, shows two unique features: (1) exceptionally receiving precipitation in the winter season, which is only second to the Mekong delta's precipitation levels, and (2) a decreased pattern of projected precipitation across time. As the upper Mekong River contributes 15–20% of MRB annual flows (MRC 2005, Adamson et al. 2009), the projected reduction of precipitation would likely impact the volume of Mekong's water, mainly due to reduced production of ice packs. On the contrary, increased precipitation levels in the lower Mekong could trigger frequent and intense flooding, increased soil erosion, and shortened lifespan of dams and reservoirs through rapid sedimentation.

In light Hausfather's (2018) assessment, where the entire world is expected to see an increase in extreme precipitations, the total precipitation levels presented in our result may not show the real challenge. Even though the mean annual precipitation is predicted to remain constant, rainfed agriculture could still be negatively impacted due to the increasing variability of precipitation. Fischer et al. (2014) reported that the distribution, rather than the total annual precipitation, is predicted to show robust deviation due to climate change. Rosenzweig et al. (2009)

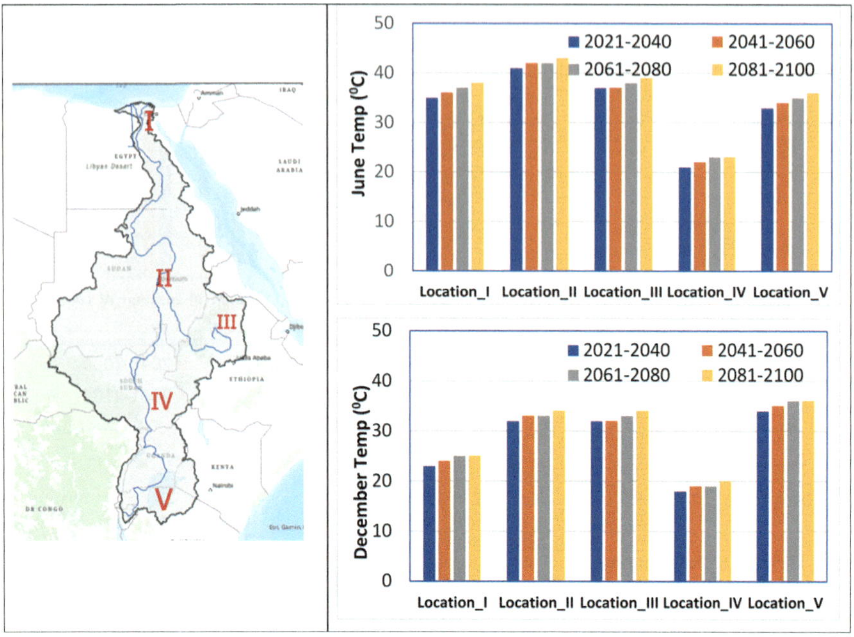

Fig. 7.9 Patterns of projected mean maximum temperature in the sample locations of the Nile Basin

suggest that due consideration should be given to changes in the frequency of extreme events associated with, and not captured by, small changes in long-term mean temperatures. Precipitation deviations from the mean could lead to a late start and early ending of the rainfall period, hugely affecting the agricultural calendar and possibly leading to partial or total failure of crop production. As Dubale (2001) described it, 15 drought incidences were observed from 1913 to 1992 in Ethiopia resulting from the failure of the short rain or its lateness.

Toward Counter Hydro Hegemony

Egypt and China are the two hegemon riparian countries, coercing their weaker riparian neighbors over the use of the Nile and Mekong River resources, respectively. To make matters worse, these two countries are growing in their dependence on those contested rivers, minimizing their flexibility and willingness to engage in principled water sharing and governance with their riparian counterparts. On the contrary, many of the weaker riparian countries employ bargaining power in reactive or active diplomacy, strategic cooperation, or the mobilization of funding to

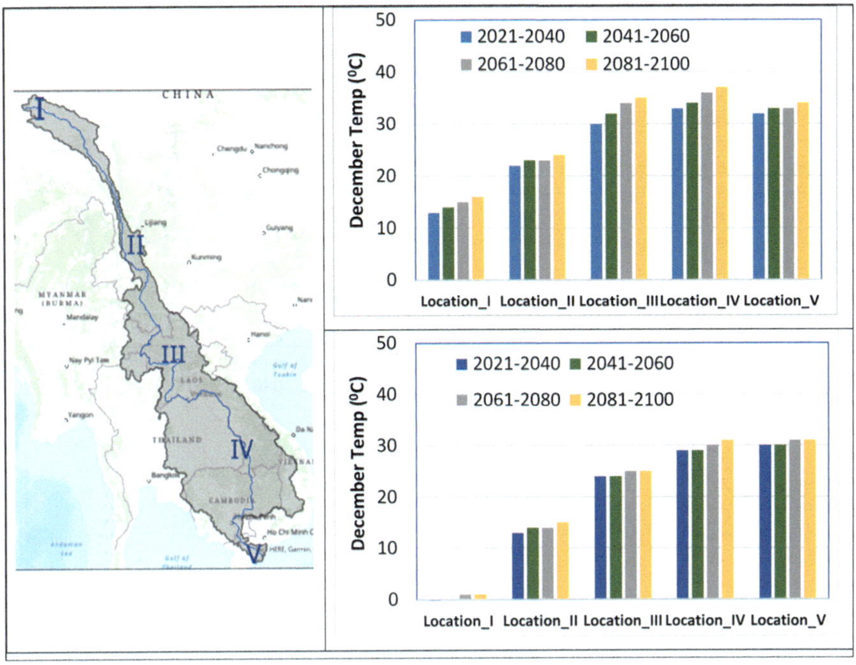

Fig. 7.10 Patterns of projected mean maximum temperature in the sample locations of the Mekong Basin

widen their options more than ever before (Cascão 2008). This section details how the two countries have intensified their dependency on those rivers and explore ways to overcome the gridlock.

Egypt

Since Herodotus, Egypt's politicians, scholars, and media have been arguing that "*No Nile means no food, no state, no historical monument, no pyramid. In short, no Egypt.*" To impede Ethiopia from using the Nile River, Egypt often cites the triple doctrines: primary need, prior use, and acquired water rights (Elhance 1999). Historically, Egypt has been very effective in blocking any assistance or cooperation toward implementing projects in the Blue Nile. Egypt's success is due to Egypt's hydropolitical advantages inherited from the earlier colonizers, its overwhelming military superiority in the Nile Basin, its diplomatic clout in the international system, and continued political instability in the upstream riparian states (Smith 2020). In short, Egypt has been doing everything to maintain the status quo while turning a blind eye to the ongoing biophysical and demographic dynamics in the basin at large.

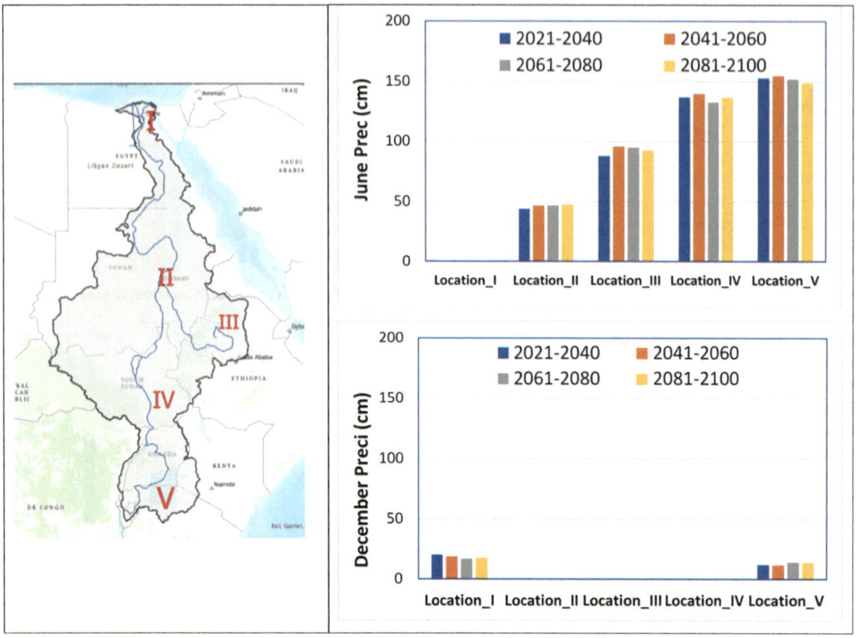

Fig. 7.11 Patterns of projected precipitation in the sample locations of the Nile River Basin

However, Egypt's demand for Nile's water increased due to population growth, but ambitious Nile-based project plans have also balooned it. Other than irrigating the adjacent floodplains of the river Nile, Egypt has taken unilateral action to launch large-scale agricultural development projects and establishment of new cities in its arid portions such as the western delta, the North Sinai, and Toshka (Swain 2011; Cascão 2009; Elhance 1999) that are hundreds of kilometers away from the main course. In addition, Egypt is ambitious to generate hard currency by exporting Nile water to Israel without seeking any agreement from its riparian counterparts. As these new projects rely on the Nile, any attempt by the Ethiopian government to use the Nile's water would likely spark diplomatic and military confrontations.

Egypt has not only stalled Nile-based initiatives and cooperative efforts by the 10 Nile Basin riparian states for equitable benefit-sharing but also vetoed any loans or funding opportunities directed toward Ethiopia's dam projects from national, regional, and international financial institutions, and openly expressed a desire to sabotage Ethiopia's effort to build the dam. Ethiopia announced the construction of the GERD in February 2011, with a total estimated cost of around US$5 billion, set to be one of the largest dams ever constructed in Africa. Ethiopia's attempt to mobilize the Nile Basin Initiative and the unilateral decision to fill up the dam have labeled Ethiopia as a counter-hydro hegemon.

Will Egypt cease to exist if the Nile's water is shared equitably with the upper riparian countries? Furthermore, is the Nile the only water source for Egyptians?

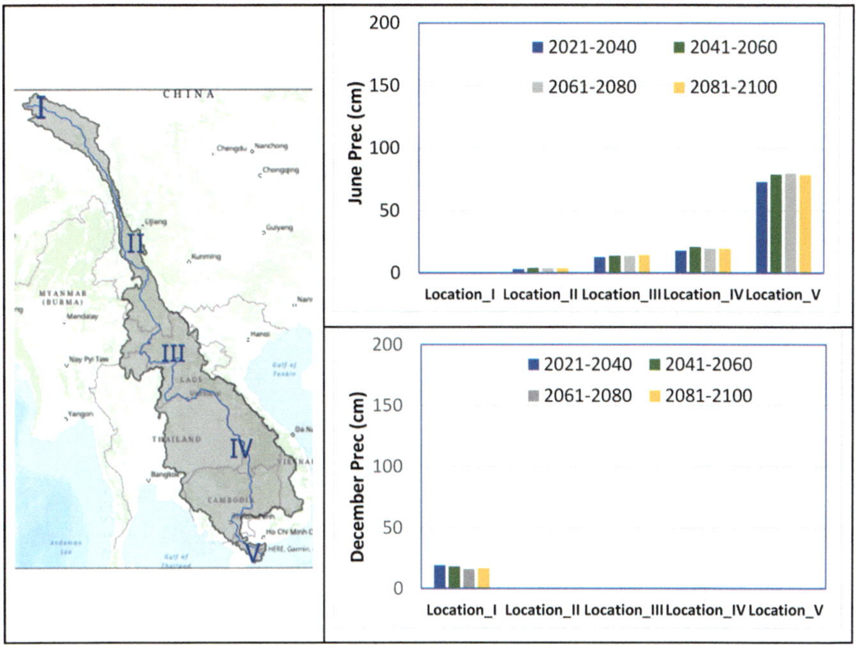

Fig. 7.12 Patterns of projected precipitation in the sample locations of the Mekong River Basin

These questions invite one to carry out a geographical analysis of water resources in Egypt. Results of the site and situation analysis of Egypt's demographic and natural resources revealed two non-conventional sources of water other than the Nile River. The first is the use of seawater through desalination, and the second is tapping water from underground reservoirs.

Over 96% of Egypt's population lives within 20 km of the Nile River and its delta in just 3.03% of Egypt's total land area (Milas 2013). Results of multiple buffer analyses (Fig. 7.13) show that over 75% of settlements and over 97% of Egypt's population fall within the 300-km buffer zone. Unfortunately, desalinated seawater in Egypt comprises only 0.08% (Salim 2012).

Arguably, compared to the attention given to its military supremacy, which allocated 11.1 billion USD in 2019, the attention given to the development of alternative sources of water is negligible. For a country where its survival is overwhelmingly dependent on the Nile, diversifying water sources should have prioritized its national security. Given its economic might and diplomatic leverage over the western countries for financial assistance, Egypt could have exerted all its resources on harnessing its natural resources: the desalination of the Red Sea and the Mediterranean Sea. Therefore, the age-old argument "*No Nile means, no food, no state, no historical monument, no pyramid. In short, no Egypt*" makes no sense.

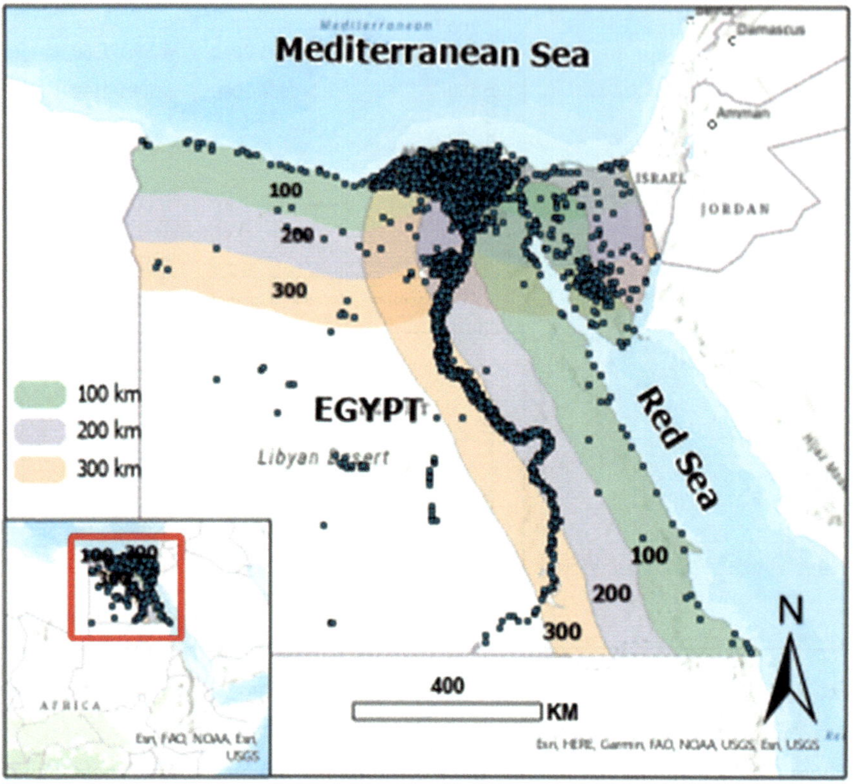

Fig. 7.13 Distribution of major populated places in Egypt across 100, 200, and 300 km buffer zones from the Red Sea and the Mediterranean Sea. The graph shows the percentages of populated places falling within each buffer zone

Desalination is a technology that is used to obtain fresh drinking water from the sea and other sources and a competitive choice of water source, especially for coastal cities. Currently, about 21,000 desalination plants are in operation in more than 120 countries (Latteman 2010). The International Desalination Association claims that 300 million people get water from desalination. The biggest ones are the United Arab Emirates, Saudi Arabia, and Israel. Singapore desalinates water for US $0.49 per cubic meter. Israel desalinates water for <US$0.40 per cubic meter. Perth and Sydney operate reverse osmosis seawater desalination plants, and as a joint venture between Israel and Jordan, a large desalination plant was piped 100 miles north through Jordan to replenish the Dead Sea.

The second source of water is drawing it from Egypt's underground reservoirs. Salim (2012) confirmed that Egypt possesses a high potential of groundwater available from different aquifers at 1.65 billion m^3/year (Abdel-Shafy and Kamel 2016). While the exploration and utilization of groundwater in Egypt date back to some centuries before Christ (Shahin 1987), the investment in this water source has been negligible in the recent history of Egypt. This failure is despite the advice of Shahin (1987), who suggested developing more groundwater resources to cover some of the needs of the ever-increasing population.

China

China is blessed with rivers and is considered the world's biggest "exporter" of transboundary water. Being positioned in the upstream and a powerful nation, China asserted its hydro-supremacy by building mega-dams. China has a long history of dams, dating back to 256 BC (Zhang 2000). Of the ten dams in the Lancang River, the oldest one is Manwan Dam, which was completed in 1995. The rest of them were completed after 2002. Of the world's total large dams, China is a leader in constructiing large dams, accounting for 20% of them (Narasaiah 2007).

Beijing insists the dams will benefit downstream countries by "evening out" the river's flow, reducing it in the rainy season, and boosting it in the dry season. However, a study conducted by Räsänen et al. (2017) found that the construction of large dams on the upper reaches of the Mekong has been fluctuating river flow, altering fish habitats, and affecting the rich aquatic and terrestrial diversity in the region. By turning the river flow into a series of reservoirs, half of the sediment load transported from the upper basin to the lower basin is trapped. As a result, downstream communities may suffer particularly in drought conditions, Vietnam and Cambodia being the most at risk.

China and Myanmar have abstained from regional initiatives, claiming a lack of deliverable gain through the commission. For a long time, China has focused too much on creating economic benefits from and beyond the river while neglecting ecological benefits to the river (Biba 2018). China influenced Laos into dynamiting the river's many rocky rapids on the Laotian side to create a deepwater channel, so large cargo and passenger ships could travel from Yunnan to the sea. Most Chinese-aided dam projects in Laos, Cambodia, and Myanmar are designed to pump electricity into China's southern electricity grid, with the lower riparians bearing the environmental and social costs.

In rejecting the 1997 United Nations convention rules on shared water resources, Beijing asserted its claim that an upstream power has the right to enforce absolute territorial sovereignty over the waters on its side of the international boundary—or the right to divert as much water as it wishes for its needs, irrespective of the effects on a downriver state (Buckley 2014).

China exerts maximum pressure on the lower MRB region—not due to the demands for energy generated in and around the Lancang/Mekong area, but mainly to satisfy the aggressively growing economic activity in its eastern parts, which is located further away from the Mekong Basin. Otherwise, the upper stream areas around Lancang/Mekong are sparsely populated due to climatic and topographic impediments. While blessed with many river basins in the country and alternative energy sources, overexploiting the MRB at the expense of endangering people of the lower MRB symbolizes the absence of compassion to its neighboring countries.

China should not necessarily be engaged in the dam's race to remain the world's leader in hydropower; this approach would jeopardize its diverse economic, political, and cultural relationship with its neighboring countries.

Conclusion

While the biophysical and political aspects of the transboundary rivers have attained a center stage in the academic, media, and political discourse, aspects of their physical geography—like the role of topographic positions—have not been adequately assessed. Taking into consideration, the role played by topographic factors; this study attempted to meet two goals: to examine the balance between the projected supply side (land and climate resources) and demand-side (demographic pressure) in the Nile River Basin, using the Mekong River Basin as a backdrop, and to identify alternative and accessible water resources for Egypt, which, in effect, relieve Egypt's ever-growing dependence on the Nile River and thereby engage in a genuine discussion to bring about an equitable share of the Nile River.

Results disclosed the mismatch between the demand and supply factors within the two river basins. On the one hand, the demand for Nile and Mekong land resources is growing due to rapid demographic changes. On the contrary, the supply factor within the river basins is projected to diminish. Given the current and future dynamics of demographic and biophysical variables, riparian countries of the NRB and MRB can hardly satisfy their growing demand without efficient and equitable use of the Nile and Mekong Rivers. Specifically, it is safe to speculate Ethiopia's inability to survive as a nation without fair use of the Blue Nile water. On the other hand, whether Ethiopia uses the Nile water or not, the volume of the Nile's discharge will decline in the future given the changing biophysical environment in the basin. In effect, Egypt can hardly continue with business as usual, anticipating enjoying 55.5 million cubic meters of water from the Nile.

Be it in the upstream or downstream positions; it appears that both Ethiopia and Vietnam are prisoners of geography. While Vietnam, the non-hegemon riparian nation in the MRB, occupying a downstream position, earn some level of sympathy and allegiance from concerned non-governmental actors, Ethiopia, the weaker riparian nation in the NRB, and yet occupying an upstream position, neither allowed to benefit from its geographic position nor receive compassion from non-governmental actors. In short, the challenges posed in Ethiopia are much more than in Vietnam.

Analyzing the NRB's hydrogeopolitical landscape with a backdrop of the MRB helped highlight the dynamics of the upstream–downstream power balance and expose the hypocrisy of many international organizations in recognizing the underlying challenges faced by weaker upstream riparian when they attempt to harness their geographical right. As neither all upstream riparians countries are culprits nor are all downstream countries angels, the power balance between riparian states should be judged by their physical geography and coupled with economic and political geographies.

To counterbalance the projected water deficit, Egypt can harness untapped water resources: seawater and groundwater. Egypt's demographic and geographic realities make desalinization more feasible than some countries currently desalinizing. Had Egypt resorted to allocating budget for exploring alternative water sources, its ultra-dependence on the Nile water would have lowered substantially. By implication, Egypt would have liberated from the yolk of its age-old bondage to the obsolete colonial treaties and engaged in a genuine negotiation in the basin-wide forums.

Obviously, there is no shortcut or magic bullet for solving problems the two river basins face, except engaging in a genuine basin-wide dialog driven by the facts on the ground. Regrettably, the burden of presenting the facts on the ground lies heavily on the shoulders of Ethiopia. The study recommends that while shoring up its ongoing diplomatic ventures, Ethiopia should demystify the "No Nile-No Egypt argument" by exposing the hidden treasures of Egypt's seawater and groundwater, especially to non-state actors, viz. regional and international NGOs.

Acknowledgements I am grateful for the support provided by Omi Kassahun, organizers of the International Conference on the Nile and Grand Ethiopian Renaissance Dam, and reviewers of the draft manuscript.

References

Abdelkareem M, El-Baz F (2015) Regional view of a trans-African drainage system. J Adv Res 6 (3):433–439

Abdel-Shafy H, Kamel A (2016) Groundwater in Egypt issue: resources, location, amount, contamination, protection, renewal, future overview. Egypt J Chem 59(3):321–362

Abtew W, Dessu SB (2019) The Grand Ethiopian Renaissance Dam on the Blue Nile. Springer International Publishing

Abu-Zeid K (2008) Green water and effective legislation for transboundary water management, pp 89–95

Adamson PT, Rutherfurd ID, Peel MC, Conlan IA (2009) The hydrology of the Mekong River. In: The Mekong, pp 53–76. Academic Press

AlSayyad N (2019) Nile: urban histories on the banks of a river. Edinburgh University Press

Amery HA, Wolf AT (2000) Water, geography, and peace in the middle east. Water in the middle east: a geography of peace, pp 1–18

Berkeley G (1902) The campaign of Adowa and the rise of Menelik. A. Constable

Bewket W (2011) Farmers' knowledge of soil erosion and control measures in the northwestern highlands of Ethiopia. Afr Geograph Rev 30(2):53–70

Biba S (2018) China's 'old and 'new Mekong River politics: the Lancang-Mekong Cooperation from a comparative benefit-sharing perspective. Water Int 43(5):622–641

Blaikie P (1985) The political economy of soil erosion in developing countries. Longman Ltd, New York

Blaikie P, Brookfield H (1987) Land degradation and society. Methuen & Co., Ltd, London

Brown P, El Gohary F, Tawfic MA, Hamdy EI, Abdel-Gawad S (2003) Nile river water quality management study, report No. 67 International Resources Group, United States Agency for International Development/Egypt, and Ministry of Water Resources and Irrigation, Egypt Water Policy Reform Contract No. LAG-I-00-99-00017-00, Task Order 815, p 93
Buckley M (2014) Meltdown in Tibet: China's reckless destruction of ecosystems from the highlands of Tibet to the deltas of Asia. Macmillan
Cascão AE (2008) Ethiopia–challenges to Egyptian hegemony in the Nile Basin. Water Policy 10 (S2):13–28
Cascão AE (2009) Changing power relations in the Nile river basin: unilateralism vs. cooperation? Water Alternatives 2(2):245–268
Cohen JE, Small C (1998) Hypsographic demography: the distribution of human population by altitude. Proc Natl Acad Sci 95(24):14009–14014
Corral Nuñez G, Opazo Salazar D, GebreSamuel G, Tittonell P, Gebretsadik A, Gebremeskel Y, … Van Beek CL (2014) Soil organic matter in Northern Ethiopia, current level and predicted trend: a study case of two villages in Tigray. Soil Use Manag 30(4):487–495
Dinar A, Dinar S, Mckinney DC, Mccaffrey SC (2013) Bridges over water: understanding transboundary water conflict, negotiation and cooperation, vol 11. World Scientific Publishing Company
Dubale P (2001) Soil and water resources and degradation factors affecting productivity in Ethiopian highland agro-ecosystems. Northeast African Studies, 2001, New Series, vol 8, No. 1, Special Issue: Natural resource management, human development, and macroeconomic performance in Ethiopia, pp 27–51
Elhance AP (1999) Hydropolitics in the Third World: Conflict and cooperation in international river basins. US Institute of Peace Press
FAO (1986) Ethiopian highlands reclamation study, Ethiopia, Final Report. FAO, Rome
Ficklin DL, Stewart IT, Maurer EP (2013) Climate change impacts on streamflow and subbasin-scale hydrology in the Upper Colorado River Basin. PLoS ONE 8(8):e71297
Fielding L, Najman Y, Millar I., Butterworth P, Ando S, Padoan M, … Kneller B (2017) A detrital record of the Nile River and its catchment. J Geol Soc 174(2):301–317
Fischer EM, Sedláček J, Hawkins E, Knutti R (2014) Models agree on forced response pattern of precipitation and temperature extremes. Geophys Res Lett 41(23):8554–8562
Gebremedhin B, Swinton SM (2003) Investment in soil conservation in Northern Ethiopia: the role of land tenure security and public programs. Agric Econ 29:69–84
Gunathilake MB, Amaratunga YV, Perera A, Chathuranika IM, Gunathilake AS, Rathnayake U (2020) Evaluation of future climate and potential impact on streamflow in the upper Nan river basin of northern Thailand. Adv Meteorol 2020
Hammond (1985) cited in Fombe LF, Tossa HN (2015) Slope morphology and impacts on agricultural productivity in the Kom highlands of cameroon. Adv Soc Sci Res J 2(9)
Hassan FA (1997) Nile floods and political disorder in early Egypt. In: Dalfes N, Kukla G, Weiss H (eds) Third millennium BC climate change and old world collapse, NATO ASI Series 1, vol 49. Springer, New York, pp 1–24
Hausfather Z (2018) Explainer: what climate models tell us about future rainfall. Carbon Brief
Hoanh CT, Jirayoot K, Lacombe G, Srinetr V (2010) Impacts of climate change and development on Mekong flow regimes. First assessment-2009 (No. H043262). Int Water Manag Instit
Hurni H (1993) Land degradation, famine, and land resource scenarios in Ethiopia. In: Pimentel D (ed) World soil erosion and conservation. Cambridge University Press, Cambridge, pp 27–62
Ibrahim AM (2011) The Nile basin cooperative framework agreement: the beginning of the end of Egyptian hydro-political hegemony. Missouri Environ Law Policy Rev 18:282
Latteman S (2010) Development of an environmental impact assessment and decision support system for seawater desalination plants. CRC press
Li Y, Liu C, Zhang H, Gao X (2011) Evaluation on the human settlements environment suitability in the Three Gorges Reservoir Area of Chongqing based on RS and GIS. J Geog Sci 21(2):346
Ma C, Pan S, Wang G, Liao Y, Xu YP (2016) Changes in precipitation and temperature in Xiangjiang River Basin, China. Theoret Appl Climatol 123(3–4):859–871

Melesse AM, Abtew W, Setegn SG (eds) (2014) Nile River basin: ecohydrological challenges, climate change and hydropolitics. Springer Science & Business Media
Milas SL (2013) Sharing the Nile: Egypt, Ethiopia and the geo-politics of water. Pluto Press
Mitiku H, Herweg K, Stillhardt B (2006) Sustainable land management: a new approach to soil and water conservation in Ethiopia. Mekelle University, Ethiopia; Bern, Switzerland: Center for Development and Environment, University of Bern and Swiss National Center of Competence in Research (NCCR) North-South
Mohamed YA, van den Hurk BJ, Savenije HH, Bastiaanssen WG (2005) Hydroclimatology of the Nile: results from a regional climate model. Hydrol Earth Syst Sci 9:263–278
MRC (2005) Annual flood report 2005. Mekong River Commission. Available at: http://www.mrcmekong.org/flood_report/2005/the_flood2.htm. Accessed on 21 Dec 2020)
MRC (2010) State of the basin report 2010. Mekong river commission, vientiane, lao PDR. http://www.mrcmekong.org/assets/Publications/basin-reports/MRC-SOB-report-2010full-report.pdf. Accessed on 23 Dec 2020
Narasaiah ML (2007) Microcredit and agricultural development. Discovery Publishing House
NBI (2012) State of the Nile Basin 2012, Nile Basin Initiative, Entebbe, Uganda. Produced by Myriad Editions, Brighton, UK, p 255
Negm AM (ed) (2017) The Nile River. Springer International Publishing
Nkemasong N, Neba M (2009) Advanced integrated human geography: concepts and techniques. Agwecams Publishers, Bamenda
Pacini N, Harper DM (2016) Hydrological characteristics and water resources management in the Nile Basin. Ecohydrol Hydrobiol 16(4):242–254
Pankhurst R (1963) The Ethiopian army of former times. Ethiopian Observer 7(2):151
Pratt DJ, Preston L (1998) The economics of mountain resource flows. UNASYLVA-FAO, pp 31–38
Räsänen TA, Someth P, Lauri H, Koponen J, Sarkkula J, Kummu M (2017) Observed river discharge changes due to hydropower operations in the Upper Mekong Basin. J Hydrol 545:28–41
Rosenzweig C, Iglesias A, Yang XB, Epstein PR, Chivian E (2009) Climate change and extreme weather events: implications for food production, plant diseases, and pests. Global Change Human Health 2(2):90–104
Saguye TS (2017) Analysis of farmers perception on the impact of land degradation hazard on agricultural land productivity in Jeldu district in West Shewa Zone, Oromia, Ethiopia. J Agri Extens Rural Devel 9(6):111–123
Salim MG (2012) Selection of groundwater sites in Egypt, using geographic information systems, for desalination by solar energy in order to reduce greenhouse gases. J Adv Res 3(1):11–19
Shahin M (1987) Groundwater resources in Egypt: potentials and limitations. In: Water for the future: hydrology in perspective, p 164. IAHS Publication
Small C, Nicholls R (2003) A global analysis of human settlement in coastal zones. J Coast Res 19 (3):584–599. Retrieved 2 Jan 2021, from http://www.jstor.org/stable/4299200
Smith LC (2020) Rivers of power: how a natural force raised kingdoms, destroyed civilizations, and shapes our world. Penguin, UK
Sonneveld BG, Keyzer MA (2003) Land under pressure: soil conservation concerns and opportunities for Ethiopia. Land Degrad Dev 14(1):5–23
Swain A (2011) Challenges for water sharing in the Nile basin: changing geopolitics and changing climate. Hydrol Sci J 56(4):687–702
Tandon SK, Sinha R (2007) Geology of large river systems. In: Gupta A (ed) Large rivers: geomorphology and management. John Wiley & Sons, London, pp 7–28
Teketay D (2001) Deforestation, wood famine, and environmental degradation in Ethiopia's highland ecosystems: urgent need for action. Northeast African Studies, pp 53–76
Van De N, Douglas I, McMorrow J, Lindley S, Thuy Binh D, Van TT, … Tho N (2008) Erosion and nutrient loss on sloping land under intense cultivation in southern Vietnam. Geographical Res 46(1):4–16

Varis O, Tortajada C, Biswas AK (2008) Management of transboundary rivers and lakes, p 271. Springer, Berlin
Yukimoto S, Koshiro T, Kawai, … (2019) MRI MRI-ESM2.0 model output prepared for CMIP6 CMIP. Earth System Grid Federation. https://doi.org/10.22033/ESGF/CMIP6.621
Zhang L (2000) Social impacts of large dams: the China case. World Commission on Dams

Chapter 8
Benefit-Sharing Framework in the Nile River Basin

Getnet Y. Muluye

Abstract The Nile River is the longest river in the world with eleven riparian countries. The lack of effective institutional mechanism to govern the Nile waters, coupled with increasing competition for scarce water resources, has the potential to generate regional conflicts and political turmoil. The intent of this article is to provide a perspective on cooperative development and equitable benefit-sharing mechanism to govern the Nile waters—as opposed to water sharing or water right—based on the experiences of successful, existing treaties and cooperative initiatives throughout the world. The proposed water governance framework reduces the risk of disputes and conflicts with significant implications in terms of promoting regional integration, cooperation, peace, security and sustainable development. The Nile basin riparian countries would need to change the status-quo of competition and mistrust and focus on nurturing cooperation and trust building to tip the balance from potential conflict to regional integration and sustainable development through basin-wide cooperative programs and initiatives in their pursuit to develop, manage and protect the scarce, shared water resources.

Keywords The Nile river · Benefit-sharing · Cooperative development · Transboundary rivers

Introduction

According to the UN-Water (2008) report, there are approximately 263 transboundary lake and river basins (accounting for nearly 60% of the global freshwater supplies in 145 countries) and 295 international agreements on transboundary waters.

Disclaimer: All the views and opinions presented here are my own and do not represent the views and opinions of any entity whatsoever with which I have been, am now, or will be affiliated.

G. Y. Muluye (✉)
Manitoba Hydro, Winnipeg, Canada

A. M. Melesse et al. (eds.), *Nile and Grand Ethiopian Renaissance Dam*,
Springer Geography, https://doi.org/10.1007/978-3-030-76437-1_8

The Nile River Basin—the longest river in the world with eleven riparian countries—lacks an effective institutional mechanism to govern its waters. The colonial treaties and agreements, such as Egypt and Britain (1929) and Egypt and Sudan (1959), are arguably outdated in the preset context and are generally rejected by the other riparian countries. The 1959 agreement, for example, allocates the entire Nile waters between Egypt and Sudan, with no consideration for upstream countries (Nile Treaty 1959). This agreement contradicts with the principle of "equitable and reasonable use" of shared water resources—the core principle of the 1997 UN-Water Convention (n.d.) and the Nile River Basin Cooperative Framework Agreement (n.d.).

The lack of a strong legal and institutional framework, coupled with rapid demographic growth and urbanization trends, increased socio-economic development and high variability in water resources availability, further aggravate and put extra pressures on already scarce water resources in the region. The increasing competition for scarce water resources in the region could escalate potential for conflicts and political turmoil. Basin countries would need to tip the balance from potential conflict to regional cooperation through cooperative development, management and protection of precious and shared water resources.

Cooperative development and benefit-sharing arrangements could offer a practical solution to govern the finite Nile waters. This mode of transboundary water management reduces the risk of disputes and conflicts in the region and promotes regional integration, transboundary cooperation, peace, security and sustainable development. Incorporating benefit-sharing arrangements in water governance structure could serve as an incentive for basin countries to engage and cooperate. Given that benefit-sharing arrangement considers a multitude of local and national interests, it creates opportunities for a basin-wide integrated water resources management and planning, optimal use of scarce resources and maximization of the overall benefits. Unilateral use and management of water resources would result in sub-optimal benefits and would diminish benefits available to others. The key concept of benefit-sharing arrangement is to focus on maximizing and sharing the benefits instead of focusing on the physical allocation of shared water resources. The UN-Water (2008) and the Nile Basin Initiative (n.d.) are some of the advocates of cooperative development and benefit-sharing arrangements. Increasing number of riparian countries across the world are adapting such water governance model and putting it into practice (Yu 2008; UN-Water 2008).

The creation and implementation of benefit-sharing arrangements for the Nile basin countries will not be easy and straightforward. This requires trust building, political will, commitment, and long-term vision. There is no "one-size-fits-all" approach to solve the Nile basin governance issue. The process might look daunting, cumbersome and time consuming, but its merit is far greater than the alternative.

This chapter presents a high-level cooperative development and equitable benefit-sharing arrangement for the Eastern Nile basin countries (Egypt, Sudan and Ethiopia) considering the basin's geography, hydrology, water consumption, climate variability, physical environment, existing and future projects, socio-economy and geopolitics. Representative examples of successful, existing treaties and cooperative initiatives throughout the world are also given.

Cooperative Development

There are several examples of successful transboundary cooperative development and benefit-sharing experiences, including the Columbia River between Canada and the USA (Stern 2019); the Senegal River between Senegal, Mali and Mauritania (Yu 2008; Mbengue 2014); the Lesotho Highlands Water Project between Lesotho and South Africa (Yu 2008); and the Chukha Hydroelectric Project between Bhutan and India (Dhakal and Jenkins 2013).

The Columbia River Treaty (Stern 2019) is an international benefit-sharing agreement between the governments of Canada and the USA for the cooperative development and operation of the water resources of the Columbia River Basin for the benefit of flood control and hydropower generation. Ratified in 1964, the Treaty is known throughout the world as one of the most successful models of a transboundary water management agreement and is considered as a model of "cooperative development" on an international river system. Under the terms of the Treaty, Canada agreed to provide 15.5 million acre-feet of additional storage on the Canadian side of the basin through the construction of three dams and operate the Treaty dams and reservoirs for optimal power generation and flood control on both sides of the border. In exchange of these benefits, the USA agreed to pay Canada (i) 50% of the estimated value of US flood damages prevented, and (ii) 50% of the estimated power benefits generated in the USA. The Treaty provided enormous economic, social and environmental benefits to the region, including improved conditions for agriculture to vast arid areas in the USA, particularly in eastern Washington, northeastern Oregon and western Idaho.

In the context of African watercourses, the Senegal River is characterized as the most progressive and a pioneering approach to transboundary water cooperation in Africa (Mbengue 2014; Yu 2008). The three countries—Mali, Mauritania and Senegal—through the OMVS (the Senegal River Basin Development Authority)—established effective technical and legal frameworks to jointly develop and manage the river at a basin level to fully realize the available potential, primarily in the areas of hydropower, irrigation and navigation. The costs and benefits resulting from joint developments are distributed across member states based on prescribed arrangements and are subject to periodic adjustments. The OMVS has involved through time to become a model in the governance of transboundary rivers, not only in Africa but also at the global level, in terms of its innovative institutional setups, peaceful regional collaboration and ability to mobilize and attract large amounts of financial resources to fund developmental investments and strengthen its institutions (Mbengue 2014).

India and Bhutan emerge as close friends in an increasingly politically unstable region. One of the major areas of mutually beneficial bilateral cooperation between India and Bhutan is the hydropower sector, which provides a reliable source of affordable and clean electricity to India, generates electricity export revenue for Bhutan and fosters economic integration, peace and security between the two countries (Dhakal and Jenkins 2013). The Government of India has constructed three hydroelectric projects in Bhutan with a total installed capacity of 1416 MW. Recently, the two countries on June 29, 2020, signed a concession agreement for the construction of the 600 MW Kholongchhu (joint venture) hydroelectric project in Bhutan (Ministry of External Affairs 2020, June 29).

There is enormous potential for Egypt, Sudan and Ethiopia to create and share benefits, specifically in the areas of water storage, hydropower and agriculture. Each basin country's strengths and potential for development would need to be considered when exploring and identifying cooperative programs and initiatives to maximize regional benefits on the basis of basin-wide technical studies, stakeholder engagement and relevant interest groups. A newly formed or the current Tripartite Technical Committee could serve to explore cooperative programs and initiatives for these integrated and sustainable development and management of the Eastern Nile Basin—a practical arrangement consistent with the principles of UN-Water (2008) and the Nile Basin Initiative (n.d.) as well as the experiences of successful, cooperative development and management of several transboundary rivers [e.g., the Senegal River (Yu 2008); the Columbia River (Stern 2019)].

Regional Water Storage

Joint regional water storage projects would greatly facilitate regional integration and sustainable development. Improved water storage in the region would increase resilience to climate change, climate variability and droughts and provide better water security in the region, specifically for Sudan and Egypt. For optimal benefits, the regional water storage facilities would need to be operated in a coordinated manner to meet regional demands, such as power generation, agriculture, municipal water use, industrial use, navigation, recreation and ecosystem augmentation. Because of its climate and geographic location, Ethiopia could serve as a regional water storage.

Some of the reasons that favor Ethiopia as a regional water storage include:

- Fewer expected human displacements as storage dams would likely be built in the Blue Nile (Abay) gorge
- Minimal expected damage to monuments, artifacts and archeological remains due to limited development in the Blue Nile (Abay) gorge
- Significantly less evaporation losses due to relatively high altitude and small reservoir surface area in the Blue Nile (Abay) gorge

- Economically attractive due to low construction cost per unit volume of stored water
- Cost-effective and unique opportunity for storage dams to serve as multipurpose reservoirs (e.g., hydropower generation)
- Unique hydrological and geographic advantages in facilitating cooperative operation and management of storage facilities.

Some of the specific benefits of regional water storage include:

- Increased water storage and flow regulation
- Increased resilience to droughts and climate variability
- Increased water security
- Increased hydropower production and agricultural activities
- Improved municipal use, industrial use, navigation, fishing, recreation and ecosystem augmentation
- Decreased risk of flood damage
- Increased economic activity and regional integration.

Hydropower

The Nile River has considerable power generating capacity with the potential to transform the region into a united economic powerhouse. The second in Africa, with the potential to generate 45,000 MW, hydropower investment in Ethiopia would be economically attractive given the enormous untapped potential coupled with the low cost per kilowatt-hour of producing power. To realize this potential, Egypt, in particular, could play several key roles by:

- Providing technical expertise in the development of hydroprojects
- Investing in hydroprojects
- Facilitating external financing
- Facilitating and supporting construction of the necessary infrastructure including international transmission lines to new electricity markets
- Facilitating and serving as an outlet to export electricity to Europe and Asia.

Agriculture

Agriculture in the region has a significant social and economic footprint. The full agricultural potential of the region has not yet been fully developed, and the agricultural system would need to be optimized and modernized. Irrigation potential in the upper Blue Nile (Abay) basin is limited—about 0.8 million hectares, whereas Sudan has vast tracts of arable land for agricultural use. To realize Sudan's irrigation potential:

- Egypt could jointly invest in Sudan to secure its food supply.
- Egypt could provide technical assistance to improve the irrigation system.
- The three countries could create tripartite developmental programs, pool funds and attract investments to ensure food security for themselves and generate extra revenue from exports.

Land and Water Management

The three countries would need to develop and implement basin-wide programs and initiatives in the areas of water conservation, salination, land degradation, drought and desertification.

- The continuing deforestation coupled with overgrazing, poor agricultural practices and unsustainable land and water management in Ethiopia would greatly increase its susceptibility to drought, land degradation and desertification. Inadequate and ineffective land and water conservation strategies and management practices could lead to droughts and desertification in the head waters of the Nile, with significant implications in terms of water security in the region.
- The increasing salination of Egypt's agricultural lands in the Nile Delta continues to require significant amount of water for salt removal. Efficient mechanisms would need to be explored to minimize waste.
- The possibility of lowering the water levels of Lake Nasser from the current high water levels to its original design water level or even lower (without appreciable loss in power production) would need to be considered to reduce surface area of the reservoir and hence evaporation losses. Maintaining Lake Nasser at higher water levels might not be necessary given regulated flow releases from upstream storage dams.
- Considerations should be given to explore alternative water sources than Nile waters, such as groundwater and seawater desalination where practical.
- Considerations should be given to explore opportunities to improve agricultural water use efficiency and water productivity through emerging technologies and system-level approaches.
- Long-term regional developmental projects and policies would need to focus on less-water-intensive farming practices and agro-industries, and high-tech and rapid industrialization (which decreases reliance on agriculture).

Benefit Sharing

Benefit sharing is an incentive and a key instrument for the success of basin-wide cooperative development. There is no "blueprint" framework on transboundary benefit-sharing arrangement. The Global Water Programme at International Union for Conservation of Nature (IUCN 2019) has developed a Six-Step Framework to serve as a roadmap that can be used at different stages of the process:

(i) Stakeholder identification and mapping interests and influence
(ii) Identifying the range of benefits (existing and potential)
(iii) Building benefit-enhancing scenarios
(iv) Quantifying costs and benefits from future scenarios
(v) Negotiating benefits with a win-win approach
(vi) Setting up institutional arrangements and implementation mechanisms.

Different approaches could be applied to share benefits among riparians. A great deal of lessons could be learnt from experiences of existing treaties and agreements, such as the Senegal River Basin (Senegal, Mali and Mauritania); the Lesotho Highlands Water Project on the Orange-Senqu River Basin (Lesotho and South Africa); the Mahakali River Basin (India and Nepal); the Chukha hydropower project on Wangchu River Basin (India–Bhutan); and the Columbia River Basin (Canada and the USA).

Legal Framework

A comprehensive basin-wide legal framework is necessary for stable and reliable cooperation. The legal instrument should be based on the principle of "equitable and reasonable use" and "the obligation not to cause significant harm" consistent with the core principles of the 1997 UN-Water Convention (n.d.) and the Nile River Basin Cooperative Framework Agreement (n.d.).

The term of the agreement should be of a fixed duration that considers planning and operational certainty, as well as adaptable to respond to key components, interests, new information and changing conditions [e.g., the Columbia River Treaty (Stern 2019); the Senegal River Basin Development Authority (Yu 2008)].

Summary

There are several successful cooperative development and benefit-sharing treaties and agreements in different parts of the world with varying degrees of priorities and interests. The Nile basin countries would need to change the status-quo of competition and mistrust and focus on fostering cooperation and trust building to tip the balance from potential conflict to regional cooperation, peace, security and sustainable development through cooperative basin-wide development and management of precious, shared water resources.

Acknowledgements The author acknowledges the organizing committee and the reviewers of this chapter.

References

Dhakal DNS, Jenkins GL (2013) Risk sharing in hydropower development: case study of the Chukha Hydel Project in Bhutan. Water Policy 15(S1):109–125. https://doi.org/10.2166/wp.2013.001

IUCN (2019) Sharing benefits from the river. From theory to practice, p 11. IUCN, Gland, Switzerland

Mbengue MM (2014) A model for African Shared Water Resources: the Senegal River Legal System. Rev Eur Community Int Environ Law, RECIEL 23(1), ISSN 2050–0386, https://doi.org/10.1111/reel.12071

Ministry of External Affairs (2020) Signing of concession agreement for 600 MW Kholongchhu Hydroelectric Project in Bhutan. https://www.mea.gov.in/press-releases.htm?dtl/32794/Signing_of_Concession_Agreement_for_600_MW_Kholongchhu_Hydroelectric_Project_in_Bhutan#:~:text=The%20Concession%20Agreement%20for%20the,of%20India%20and%20Lyonpo%20Dr. Accessed on 23 Jan 2021

Nile Basin Initiative (n.d.) Cooperative framework agreement. https://nilebasin.org/images/docs/CFA%20-%20English%20%20FrenchVersion.pdf

Nile Basin Initiative (n.d.) https://nilebasin.org/

Nile Treaty (1959) Agreement between the Arab Republic of Egypt and the Republic of the Sudan for the full utilization of the Nile waters, p 453. United Nations Treaty Series 64

Stern CV (2019) Columbia river treaty review. Library of Congress. Congressional Research Service, Washington, D.C.

UN-Water Convention (n.d.) Convention on the law of non-navigational uses of international watercourses (1997). https://legal.un.org/ilc/texts/instruments/english/conventions/8_3_1997.pdf

UN-Water (2008) Transboundary waters: sharing benefits, sharing responsibilities

Yu W (2008) Benefit sharing in international rivers: findings from the Senegal River basin, the Columbia river basin, and the Lesotho highlands water project (English). Africa Region water resources unit working paper 1. World Bank Group, Washington, D.C.

Chapter 9
Management Implication of Understanding Flood Variabilities in Transboundary Rivers for Future: A Case of Wabi Shebele River Basin, Ethiopia

Fraol Abebe Wudineh, Semu A. Moges, and Belete Berhanu Kidanewold

Abstract Floods are among the most severe hydrological extremes, in terms of social impact and potential economic damage. In this study, flood variability and associated impacts and managements are investigated in the Wabi Shebele River Basin. In early twenty-first century, floods indicate increasing trend in magnitude and frequency in the entire basin. For longest period, 1981–2010, the annual maximum flood discharge shows upward trends in upper and middle catchments while downward trends are in eastern and lower catchments of Wabi Shebele Basin. Among these, annual maximum discharge shows a significant increasing trend in middle catchments (i.e. Erer at Hamaro and Gololcha at Wabi junction) and significant decreasing trend in Fafen watersheds at Jijiga and Kebridehar gauging stations. Flood variability and socio-economic damages follow similar trend tendency in the basin. Like variability analysis result in early twenty-first century, the number of peoples affected indicates increasing trend in study area. In such case, one must shift from defensive action against hazards to management of the risk considering the evolution and trends of floods. Due to its nature, floods in transboundary river basin have transboundary consequence which indicates need of cooperation in between riparian countries for Integrated Flood Management (IMF). IMF is approach which adopts the best mix of both structural and non-structural strategies by ensuring a participatory approach and adopting integrated hazard management approaches.

F. A. Wudineh (✉) · B. B. Kidanewold
School of Civil and Environmental Engineering, Addis Ababa Institute of Technology, Addis Ababa University, Addis Ababa, Ethiopia

S. A. Moges
Department of Civil and Environmental Engineering, University of Connecticut, 261 Glenbrook Rd, Storrs, CT 06269, USA

A. M. Melesse et al. (eds.), *Nile and Grand Ethiopian Renaissance Dam*, Springer Geography, https://doi.org/10.1007/978-3-030-76437-1_9

Keywords SWAT · Mann–Kendall test · Quantile perturbation method · Flood risk · Endemism · Expectations · Integrated flood management (IMF)

Introduction

Floods are among the most severe hydrological extremes, in terms of social impact and potential economic damage. Although the application of science and medicine has improved humankind's ability to predict, alleviate and survive flood disasters to some extent, there is high probability to increase human exposure and vulnerability to floods in the future since population growth and urbanization, and social, economic and political processes have been increasing over the globe.

Flood variabilities highly affect the poorest region of the world (World Bank 2006) where insufficient investment was made in the hydraulic and institutional infrastructures to provide a reasonable measure of water security. Changes in climate or human interventions in catchments and river systems may change flood hazard and as a consequence, flood risk. Due to this, most of the time floods are evaluated from a hazard perspective, focusing on hydrologic/hydraulic parameters such as discharge, water level or inundation extent neglecting societal processes, which implicitly means they are assumed to be constant or, if random, a stationary process, for example, (Assefa 2018; Bissolli et al. 2011; WORLD BANK 2006). However, some socio-economic processes, like population growth and economic development, may change at a faster pace than long-term physical changes (e.g. the impacts of climate change on discharge), and exposure and vulnerability to floods can be highly dynamic. Therefore, societal processes need to be addressed within a risk-based approach, where next to the hazard, societal exposure and vulnerability play a decisive role in flood management process.

A critical question raising in recent years in extreme hydrology is how space–time variations in flood hazard that may be related to climate variability and change intersect with the changing nature of the flood exposure and vulnerability. To understand climate-flood linkage, some researchers conducted a robust study on the area. Among these, Merz et al. (2014) contrast the traditional narrow framing of floods with the broader perspective which is emerging from an improved understanding of the climate context of flood generation. Accordingly, they identify perspectives in floods as traditional and emerging perspectives. For instance, they explain the traditional approach for linking atmospheric components to flood analysis has been primarily hydro-meteorological.

However, the emerging view of climate-flood linkages is process driven and seeks to understand and analyse flood events in the context of their long-term history of variation in magnitude, frequency and seasonality within the climate framework of the global and regional atmospheric circulation patterns and processes that drive changing combinations of meteorological elements at the catchment scale. According to current emerging perspectives, floods occur within spatial framework of large-scale circulation patterns and global climate mechanisms.

Therefore, long-term climate trend, catchments characteristics like geology, topography, vegetation and humans have to be disentangled to fully understand floods. In case of data-sparse watersheds like Wabi Shebele River Basin, obtaining long period observed extreme data for variability analysis is difficult. Thus, an alternative approach (hydrological modelling) to generate flow from weather and catchment variables is required.

Ethiopia has seven transboundary rivers, of which the Wabi Shebele River is the one frequently affected basin in the country by flood disaster (MoWR 2003; NDRMC 2018; Amer et al. 2013). Due to its nature, the floods in transboundary rivers basins often have transboundary consequences which indicates need of cooperation in between riparian countries. Recognizing this, Ethiopia has started to promote cooperative development and management of its shared rivers with riparian countries, particularly in the Nile Basin Initiative (NBI). Flood risk management should have taken into account changing hazard, exposure and vulnerability, and the combined application of financial, structural and non-structural measures (Merz et al. 2014). The best way to mitigate floods depends on how well changes in flood risk can be predicted at short and long-time scales.

The aim of the chapter is to investigate variabilities in flood discharge and associated impacts and management of flood risk in Wabi Shebele River Basin, Ethiopia. Annual and seasonal variabilities of floods isare detected as first task, and assessment of associated impacts and management of flood risk is performed as second task in the study.

Wabi Shebele River Basin

Catchment Descriptions

The Wabi Shebele Basin is a transboundary basin located at the Horn of Africa, covering parts of Ethiopia and the Republic of Somalia. It originates from Bale highlands ranges of the Galama to Ahmar of Ethiopia, about 4000 m altitude and drains portions of Somalia before draining into the Indian Ocean. More than 70% of the catchment (202,220 km^2) lies in Ethiopia. The Wabi Shebele Basin in this study is used to represent the catchment that lies in Ethiopia within 4° 45’ N–9° 45’ N latitude and 38° 45’ E–45° 45’ E longitude (Fig. 9.1). The climate of the basin is dependent on altitude and strong latitudinal movement of the intertropical convergence zone (ITCZ) (Awass 2009). The highlands are cool and densely populated while the lowlands are arid and sparsely populated with recorded rainfall of 1213 mm and 268 mm, respectively (Awass 2009; MoWR 2003). While having the largest area coverage, the basin’s annual run-off is estimated as 3.4 BCM (Billion Cubic Metres) which is relatively smaller compared to the major river basins that exist in Ethiopia (Awass 2009). The air temperature of the Wabi Shebele Basin varies with altitude (MoWR 2003). The mean monthly temperature of the basin is 19.9 °C.

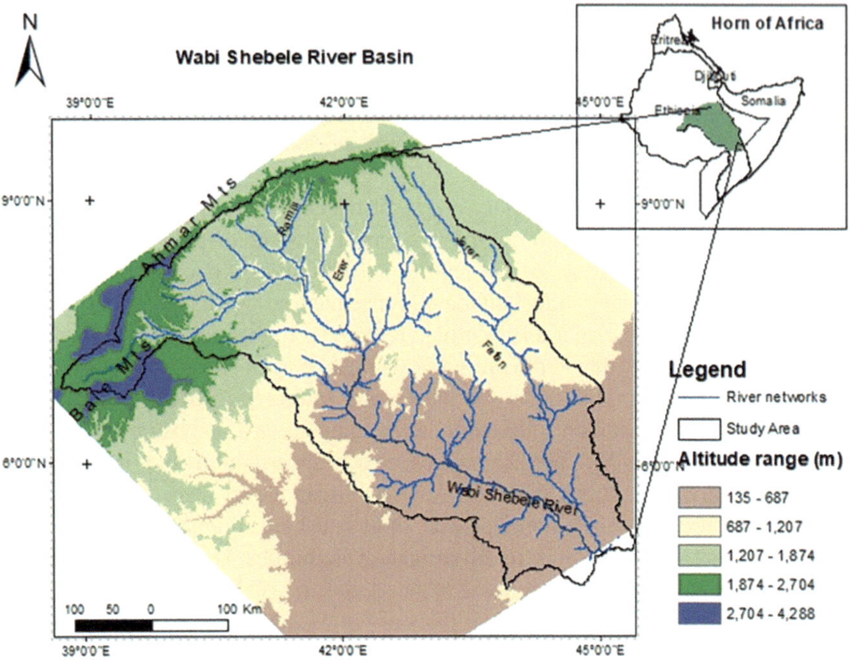

Fig. 9.1 Wabi Shebele River Basin in Eastern Ethiopia

Poorly drained and shallow profile soils are distributed at upstream of the basin, and highly drained soils formed from limestone and gypsum are highly distributed over flat and gently undulating lands of middle and lower parts of the basin (MoWR 2004b). Geologically, the basin falls in three major categories (BCEOM_ORSTOM_EDF 1973; Amer et al. 2013): Precambrian crystalline basement rocks in the northern and valleys of main river and tributary, Late—Paleozoic to Early Tertiary Sedimentary rocks in south-eastern sector of the Ogaden, and sedimentary basin and tertiary to quaternary volcanic rocks in the most north western fringe of the basin. The land use and land cover in the basin are highly dependent on the climatic, topography and edaphic factors (MoWR 2004a). Cultivated land units are dominated in upstream part of the basin, whereas grasses and shrubs are common in the arid and semi-arid areas of the basin, more than 67.8% of the basin land use/cover (MoWR 2004a).

Severity of Floods in Wabi Shebele River Basin

There are two major types of flood events regularly occur in Wabi Shebele River Basin: riverine floods and flash floods. The first type of flood takes place mainly in lower valley of Wabi Shebele Basin when Wabi Shebele River overflows, during torrential rains on upper and middle highland of the basin and localized heavy rainfall in lower Wabi Sheble Basin. This type of flood occurs mainly during first (March–May) and second (October–December) rainy seasons. These floods were reported in April 1995, October 1999, April 2002, May 2003, April 2005, April–December 2006, August 2008, November 2008, March 2010 and April 2016 (MoWR 2003; Tadesse et al. 2016). Floods in Wabi Shebele especially in lower basin floods are favoured by its topography, land cover, run-off from highland, and intensive rainfall condition.

The other type is flash flooding, which occurs from heavy localized rainfall specially in Fafen watershed. In Fafen watersheds, floods are reduced in the channel, and the floods of the tributaries do not directly meet the Fafen River but flow in the alluvial plains in both seasons (March–May and September–November) comparably (MoWR 2003). These floods were reported in October–November 1999, May 2003 and April–June 2005. Flash floods with a rainfall intensity of above 12.5 mm/h can cause high peak discharge, capable of inflicting hazards depending on the type of topography, soil property and effect of rainfall over upstream establishing the occurrence of floods over flood-sensitive areas (Akola et al. 2018).

Data and Methods

Data

We utilize both ground-based station observations and gridded analysis data in this analysis. In hydrologic model, digital elevation model (DEM), physiographic data of paedology, land use and occupation and classes of slopes and meteorological data were used to generate flow. Digital elevation model (DEM) with a spatial resolution of 90 metres is obtained from SRTM GDEM official website.

The soil information was acquired from Food and Agricultural Organization of United Nations (FAO-UNESCO) at a scale of 1:5000000 downloaded from (http://www.fao.org/soils-portal/soil-survey/soil-maps-and-database/FAO-UNESCO-soil-map-of-the-world/en). Land use and land cover data were obtained from the Ministry of Water, Irrigation and Energy.

From the data described above, the hydrologic response units (HRU) were established. After HRU definition, the data from climatic stations located in the study basin were input into the SWAT model. These data are rainfall (mm), maximum and minimum air temperatures (°C), relative humidity (%), solar radiation (KJ/m^2) and wind speed (m/s). These data were obtained from the National Meteorology Service Agency (NMA). The data sets provide daily observations for stations in the basin. Fourteen weather stations were selected with good-quality data, according to a defined criterion and with a minimum of 30 years record length in between 1980 and 2010 (Table 9.1).

Table 9.1 Hydroclimatic data used in SWAT model

Type	Station	Controller	Coordinates*		Altitude (m)	Catchment area (Km^2)
			Lat.	Long.		
Climate	Adaba	NMA	5,43,691	7,73,113	2420	–
	Dodola	NMA	5,19,746	7,72,054	2580	–
	Kofele	NMA	4,79,397	7,81,976	2620	–
	Merero	NMA	5,38,334	8,22,503	2940	–
	Hawassa	NMA	4,43,083	7,81,757	1750	–
	Robe (Arsi)	NMA	5,68,694	8,68,627	2400	–
	Sinana	NMA	6,24,339	7,77,736	2400	–
	Deder	NMA	7,67,132	1,03,2669	2350	–
	Jara	NMA	6,61,100	8,04,664	1960	–
	Haromaya	NMA	8,32,842	10,40,832	2125	–
	Gursum	NMA	8,73,588	10,34,746	1900	–
	Jijiga	NMA	9,15,127	10,33,159	1775	–
	Gode	NMA	1,00,6927	6,54,440	295	–
	Degehabour	NMA	1,00,1635	9,11,616	1070	–
Hydrology	Wabi @ Dodola	MoWIE	7,75,521	5,03,682	–	1040
	Maribo @ Adaba	MoWIE	7,73,692	5,36,818	–	192
	Wabi @ Legahida	MoWIE	8,81,015	7,09,436	–	19793
	Erer Nr. Babile	MoWIE	10,21,721	1,97,816	–	494
	Wabi @ Gode	MoWIE	6,54,138	3,41,331	–	124108
	Fafen @ Jijiga	MoWIE	1,03,4208	2,58,365	–	731

*UTM Zone 37 N

Discharge measured data used for model calibration and validation were accessed from hydrology department of the Ministry of Water and Energy, Ethiopia (MoWRIE), as given in Table 9.1. According to the Ministry of Water and Energy, the measurements of river levels follow the guidelines of the World Metrological Organization (WMO) (MoWR 2003).

Hydrological Model

Soil and Water Assessment Tool (SWAT) is a continuous time, spatially distributed model designed to simulate water, sediment, nutrient and pesticide transport at a catchment scale on a daily time step. In this study, the model was used to generate flows. The model is driven by metrological data like precipitation, temperature, relative humidity, solar radiation and wind speed and physiographic data of paedology, land use and occupation and classes of slopes.

It uses hydrologic response units (HRUs) that consist of specific land use, soil and slope characteristics. The HRUs are used to describe the spatial heterogeneity in terms of land cover, soil type and slope class within a watershed (Neitsch et al. 2005). The model estimates hydrologic parameters such as evapotranspiration, surface run-off and peak rate of run-off, groundwater flow and sediment yield for each HRUs unit. The hydrologic cycle simulated by SWAT is based on the water balance as shown in Eq. (9.1).

$$\mathrm{SW}_t = \mathrm{SW}_o + \sum_{i=1}^{t}\left(R_{\mathrm{surf}} - Q_{\mathrm{surf}} - E_a - W_{\mathrm{seep}} - Q_{\mathrm{qw}}\right) \tag{9.1}$$

where SW_t is the final soil water content (mm water), SW_o is the initial soil water content in day i (mm water), t is the time (days), R_{day} is the amount of precipitation in day i (mm water), Q_{surf} is the amount of surface run-off in day i (mm water), E_a is the amount of evapotranspiration in day i (mm water), W_{seep} is the amount of water entering the vadose zone from the soil profile in day i (mm water), and Q_{gw} is the amount of return flow in day i (mm water).

The water in each HRU in SWAT is stored in four storage volumes: snow, soil profile (0–2 m), shallow aquifer (typically 2–20 m) and deep aquifer. Surface run-off from daily rainfall is estimated using a modified SCS curve number method, which estimates the amount of run-off based on local land use, soil type and antecedent moisture condition. Peak run-off predictions are based on a modification of the rational formula (Chow et al. 1963). The watershed concentration time is estimated using Manning's formula, considering both overland and channel flow. The SCS curve number is described by Eq. 9.2.

$$Q_{\mathrm{Surf}} = \frac{\left(R_{\mathrm{day}} - 0.2S\right)^2}{R_{\mathrm{day}} + 0.8S} \tag{9.2}$$

where Q_{surf} is the accumulated run-off or rainfall excess (mm), R_{day} is the rainfall depth for the day (mm), and S is the retention parameter (mm). The retention parameter is defined by Eq. 9.3.

$$S = 25.4\left(\frac{100}{\mathrm{CN}} - 1\right) \tag{9.3}$$

The SCS curve number is a function of the soil's permeability, land used and antecedent soil water conditions.

For climate, SWAT uses the data from the station nearest to the centroid of each sub-basin. Calculated flow, sediment yield and nutrient loading obtained for each sub-basin are then routed through the river system. The soil percolation component of SWAT uses a water storage capacity technique to predict flow through each soil layer in the root zone. Lateral sub-surface flow in the soil profile is calculated simultaneously with percolation. Groundwater flow contribution to total stream flow is simulated by routing a shallow aquifer storage component to the stream (Moriasi et al. 2007; Schuol and Abbaspour 2007). The model computes evaporation from soils and plants separately. Potential evapotranspiration can be modelled with the Penman–Monteith (Monteith 1965), Priestley–Taylor (Priestley 1972) or Hargreaves methods (Hargreaves and Samani 1985), depending on data availability. In this study, the Penman–Monteith method was used to determine potential evapotranspiration.

Defining Extreme Event

In this paper, six extreme hydrologic indices: annual maximum discharge (AMAX), Peak over threshold (third quartile) frequency (POTF), peak over threshold (third quartile) magnitude, seasonal maximum discharge for winter (SMW), seasonal maximum discharge for spring (SMSp) and seasonal maximum discharge for summer (SMSu) are used to define extreme high discharges.

In extreme value analysis, ensuring independence of samples is a initial task. In this study, the time interval approach is used to ensure the independence of flow discharges. Time intervals between 5 and 14 days between successive peaks, 5 days for catchments <10,000 km^2 and 14 days for catchments $\geq$10,000, are used in this study. This approach is reported as a strong flood-frequency estimations approach (Keast and Ellison 2013; Malamud and Turcotte 2006).

Flood Trend Detection

In this study, two distribution-free (nonparametric, e.g. rank-based) tests: Mann–Kendall trend test to detect trends in flood discharges and quantile perturbation Method (QPM) approach to see temporal variabilities in extreme discharges were used. In practice, there is a continuum between "trend" and "change".

Mann–Kendall (MK) test: The MK test statistic, S is defined as (Kendall 1975; Mann 1945), and the test is conducted by applying Eqs. 9.4–9.9.

$$S = \sum_{i=1}^{n-1} \sum_{j=i+1}^{n} sgn(X_j - X_i) \tag{9.4}$$

where X_j are the sequential data values, n is the length of the data set, and:

$$sgn(\theta) = \begin{cases} 1 & \text{for } \theta > 0 \\ 0 & \text{for } \theta = 0 \\ -1 & \text{for } \theta < 0 \end{cases} \tag{9.5}$$

When $n \geq 8$, the statistic S is approximately normally distributed with mean and variance given by Mann (1945) and Kendall (1975):

$$E[S] = 0 \tag{9.6}$$

$$V(S) = \frac{n(n-1)(2n+5) - \sum_{i=1}^{n} t_1 l(l-1)(2l+5)}{18} \tag{9.7}$$

where t_1 is the number of ties of extent l. The standardized test statistic Zmk is computed by

$$\text{Zmk} = \begin{cases} \frac{S-1}{\sqrt{\text{Var}(S)}} & \text{for } S > 0 \\ 0 & \text{for } S = 0 \\ \frac{S+1}{\sqrt{Var(S)}} & \text{for } S < 0 \end{cases} \tag{9.8}$$

The standardized MK statistic Zmk follows the standard normal distribution with a mean of zero and variance of one under the null hypothesis of no trend. A positive Z-value indicates an upward trend, while a negative one indicates a downward trend. The Kendall rank coefficient is often used as a test statistic to establish whether two variables may be regarded as statistically dependent. Under the null hypothesis of independence of X_i and X_j, the sampling distribution of "tau" has an expected value of zero. The value of "tau" ranges from $-1(100\%$ negative association, or perfect inversion) to $+1(100\%$ positive association or perfect agreement). A value of zero indicates the absence of association. The p-value of the MK statistic S of sample data can be estimated using the normal cumulative distribution function:

$$p = 0.5 - \emptyset(|Z|) \text{ where } \emptyset(|Z|) = \frac{1}{\sqrt{2\pi}} \int_0^{|Z|} e^{\frac{t^2}{2}} \mathrm{d}t \tag{9.9}$$

If the P-value is small enough, the trend is quite unlikely to be caused by random sampling. At the significance level of 0.05, if $p \leq 0.050$, then the existing trend is assessed to be statistically significant.

Quantile Perturbation Method (QPM) is a statistical analysis used to study trends and multi-time period oscillation patterns in hydroclimatic extremes (Ntegeka and Willems 2008; Onyutha 2016; Tabari et al. 2014; Taye and Willems 2012). The method has two concepts: (i) the frequency aspect which focuses on how often an event (quantile) may occur and (ii) the perturbation aspect which determines the changes in the extremes for a particular return period. It is rank-based empirical statistical method which uses ranks of time series to detect frequency and trends of extreme flow series. It uses the given series directly (i.e. without rescaling) to obtain quantile anomalies. The selected sub-period (block period) is a subseries taken from the total time series representing the period of interest. To select appropriate value of block length in between 5- and 15-years intervals, Tabari et al. (2014) recommend applying QPM to extreme time series to different block of years and select the one which shows a high variability at a given time interval. To check the significance of perturbation factor in extreme quantiles, confidence interval is computed using nonparametric bootstrapping method is performed. Perturbation values that fall outside of 95% CI (confidence intervals) are identified as statistically significant perturbation value.

Results and Discussion

Calibration and Validation of the Hydrological Model

In data-sparse watersheds, like Wabi Shebele River Basin, developing a representative hydrological model (e.g. in generating the observed streamflow) is very challenging but it is a prerequisite to accurately assess variabilities in extreme flows. In this study, we used a combination of data sets to calibrate and validate the hydrological model. In addition to the field-based ground stations, some weather data like relative humidity, solar ration and wind speed data (from CFRS data sets) were used to improve the hydrological model accuracy during both calibration and validation.

For model calibration and validation, the observed daily and monthly stream flow data were used from 1988 to 2000 with three years warming period. To evaluate the model performance, three parameters have been used, namely R^2 and NSE and P bias. NSE (Eq. 9.10) is a normalized statistic, ranges from $-\infty$ to 1, used to indicate the relative value of residual variance compared to the variance of the observed data and values close to one show a perfect match of the modelled with the observed data (Nash and Sutcliffe 1970). R^2 (Eq. 9.11) is the square of the correlation coefficient between the observed and modelled data and values close to one show the ability of the model to accurately predict the observed values.

$$\mathrm{NSE} = 1 - \frac{\sum_{i=1}^{N} (X_i - Y_i)^2}{\sum_{i=1}^{N} (X_i - \bar{X})^2} \tag{9.10}$$

$$R^2 = \left[\frac{\sum_{i=1}^{N} (X_i - \bar{X})(Y_i - \bar{Y})}{\sqrt{\sum_{i=1}^{N} (X_i - \bar{X})^2} \sqrt{\sum_{i=1}^{N} (Y_i - \bar{Y})^2}} \right]^2 \tag{9.11}$$

$$P\mathrm{bias} = \frac{\sum_{i=1}^{N} (Y_i - X_i}{\sum_{i=1}^{N} X_i} * 100 \tag{9.12}$$

where x and y are observed and modelled streamflow with their respective means, and N is the number of data pairs. Table 9.2 shows gauging stations, sub-basin area, location, average annual flow and model out statistical evaluation in the calibration and validation runs.

Table 9.2 Evaluation of model performance

Stations	Area (km^2)	Location		Average Annual Flow (Mm^{-3})	Calibration			Validation		
		Lat	Long		R^2	NSE	Pbias (%)	R^2	NSE	Pbias (%)
Wabi at D/ Bridge	1040	7.01	39.02	230.9	0.74	0.74	−3.0	0.74	0.74	−2.1
Maribo	192	7.00	39.20	100.2	0.50	0.62	−18.5	0.35	0.47	−28.7
Robe	169	7.51	39.38	48.5	0.47	0.40	−30.3	0.42	0.39	−20.8
Wabi at L/Hida	19,793	7.58	40.54	1848.5	0.64	0.61	−0.87	0.62	0.65	−0.27
Erer	494	9.14	42.15	87.5	0.43	0.42	−0.19	0.11	−0.4	−53.6
Jijiga	731	9.21	42.48	35.4	0.15	0.49	5.3	0.02	0.16	−59.1
Wabi at Gode	124,108	5.56	43.33	4523.2	0.40	0.20	−29.4	0.16	0.01	−37.6

Trends and Variabilities in Flood Discharge

The MK test applied at each site for the period 1981–2010 showed a non-significant decreasing trend in upper and lower catchments. For the longest period, 55% of the stations indicated weak to strong decreasing trends in annual maximum discharge. However, the watersheds in middle basin indicated a majority of significant increasing trend in annual maximum discharge with $p < 0.05$. Too see multi-temporal changes in annual maximum discharge, Mann–Kendall trend tests were analysed at 5-, 10-, 15-, 25- and 30-year intervals.

From Fig. 9.2, it is evident that blue colours are more frequent than red colours, meaning that positive trends are more frequent than negative, especially for the most recent period since 1996 in all stations. Second, negative trends (red colours) appear more significantly in eastern Fafen catchments and lower stations of Wabi Shebele River before 2000 (as shown on the Fafen @ Jijiga, Fafen @ Kebridehar, Wabi @ Gode and Wabi @ Burkur). In general, most of upper and middle catchments indicate weak to significant increasing trends and decreasing tendency in lower and eastern catchments in annual maximum discharge during the past 30 years in Wabi Sheble River Basin. The years, 1980s and 2000s, are the decades with weak to significant increasing trends when flood discharge occurred, whereas in 1990s, most stations indicate decreasing trends in annual maximum discharge in the study area.

At sample gauging stations, the result of MK trend test for the long-term period (1981–2010) reveals positive trends of peak over threshold (third quartile) frequency (POTF) for 55% of gauging stations (Table 9.3). Among these significant increasing trends in POTF, it is observed over middle catchments (Erer watershed at Hamaro, and Golocha) and the rest 45% of stations indicates negative trends, mainly present in Fafen watershed and Wabi Shebele River at Gode and Burkur.

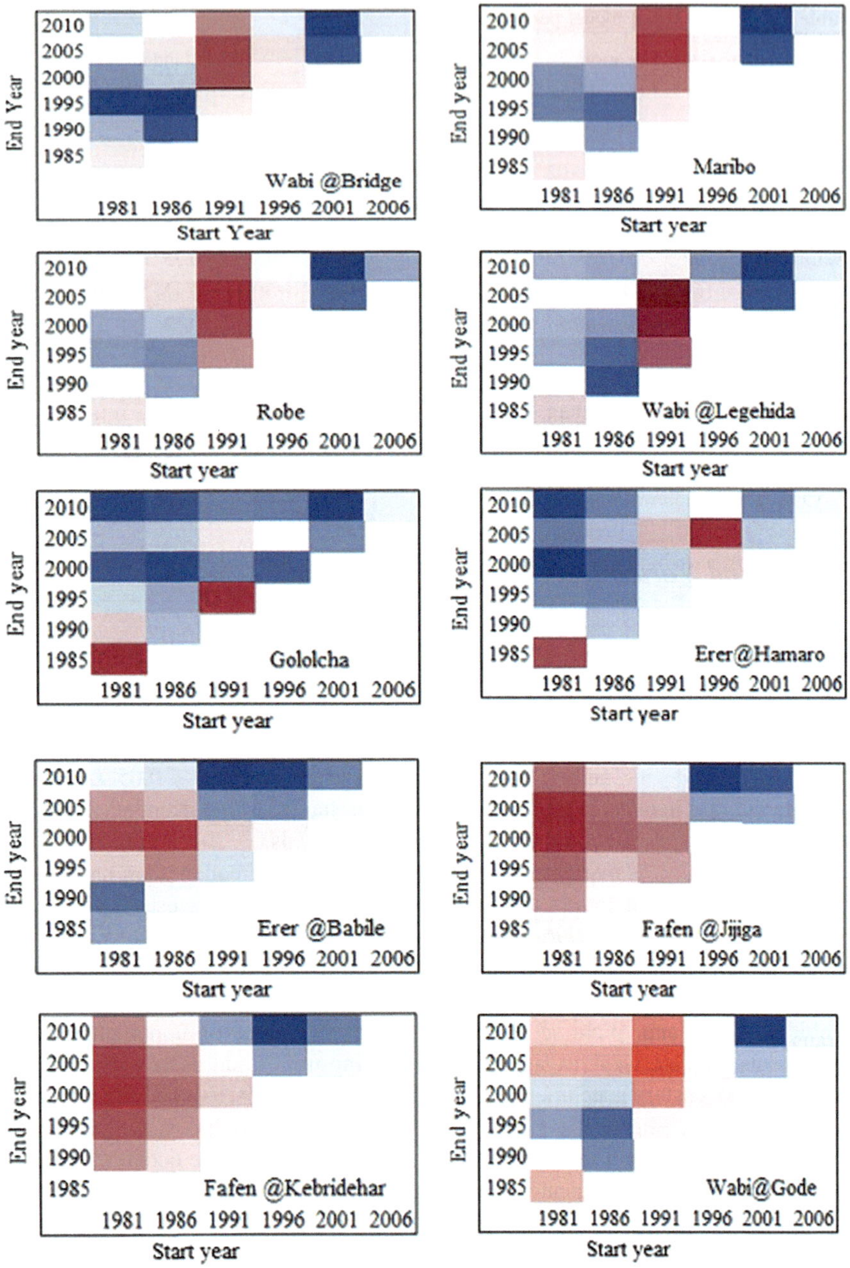

Fig. 9.2 Multi-temporal trend analysis for the annual maximum discharge (AMAX) for river catchments in Wabi Shebele River Basin. Blue and red cells correspond to positive and negative tau values, respectively (the darker the colour the more significant the trend)

Table 9.3 Characteristics of highest positive (+ve) and negative (-ve) anomaly in seasonal extreme discharges

Station/ River name	Highest +ve anomaly (%)	Season of occurrence	Year	Highest -ve anomaly (%)	Season of occurrence	Year
Wabi at Dodola	135	Spring	1989	−29.7	Summer	1983
Maribo	177	Spring	1987	−60.5	Spring	2004
Robe	193.2	Spring	1987	−62.7	Spring	2004
Wabi at Legehida	114.2	Spring	1987	−49.1	Spring	2004
Gololcha	116.5	Spring	1987	−49.2	Spring	2004
Erer at Hamaro	455.5	Winter	2006	−57.4	Spring	2004
Erer at Babile	296.8	Winter	2006	−50.5	Spring	1997
Fafen at Jijiga	177	Spring	1981	−65.7	Winter	1996
Fafen at Kebridehar	357.6	Summer	1984	−65.2	Spring	1996
Wabi at Gode	439.6	Summer	1990	−61.5	Summer	2001
Wabi at Burkur	438.7	Summer	1990	−61.1	Summer	2001

From Fig. 9.3, the QPM analysis using peak over threshold (third quartile) indicates that most of extreme flows varies with a confidence with high oscillation patterns in the entire the basin. Extreme flows vary above reference line (above mean) in upper, middle and lower valley of Wabi Shebele River stations, whereas it varies below reference line in eastern catchments.

The multi-temporal analysis in seasonal flood discharge indicates similar patterns in summer and winter, and differences in spring are observed throughout the basin (Fig. 9.4). In eastern catchments, spring flood discharge indicates weak increasing trend which indicates decreasing trend in summer and winter season for the last 30 years. Darker colours show statistically significant trends, and they are more frequent in summer and less in winter in eastern catchments. For 1990s, upper and lower Wabi Shebele Basin flood discharge indicates significant decreasing trends in all seasons and weak decreasing trend in eastern catchments. In all catchments, flood discharge indicates increasing tendency in 2000s. This pattern is stronger in summer and winter when 72% of stations showed significant increasing trends for the period 2001–2010. The same result was previously observed in the Wabi Shebele River Basin, where a significant increase in spring, summer and winter floods was identified (IWMI 2015). Table 9.2 shows the characteristics of highest anomaly in seasonal extreme discharges with the season and occurrence year.

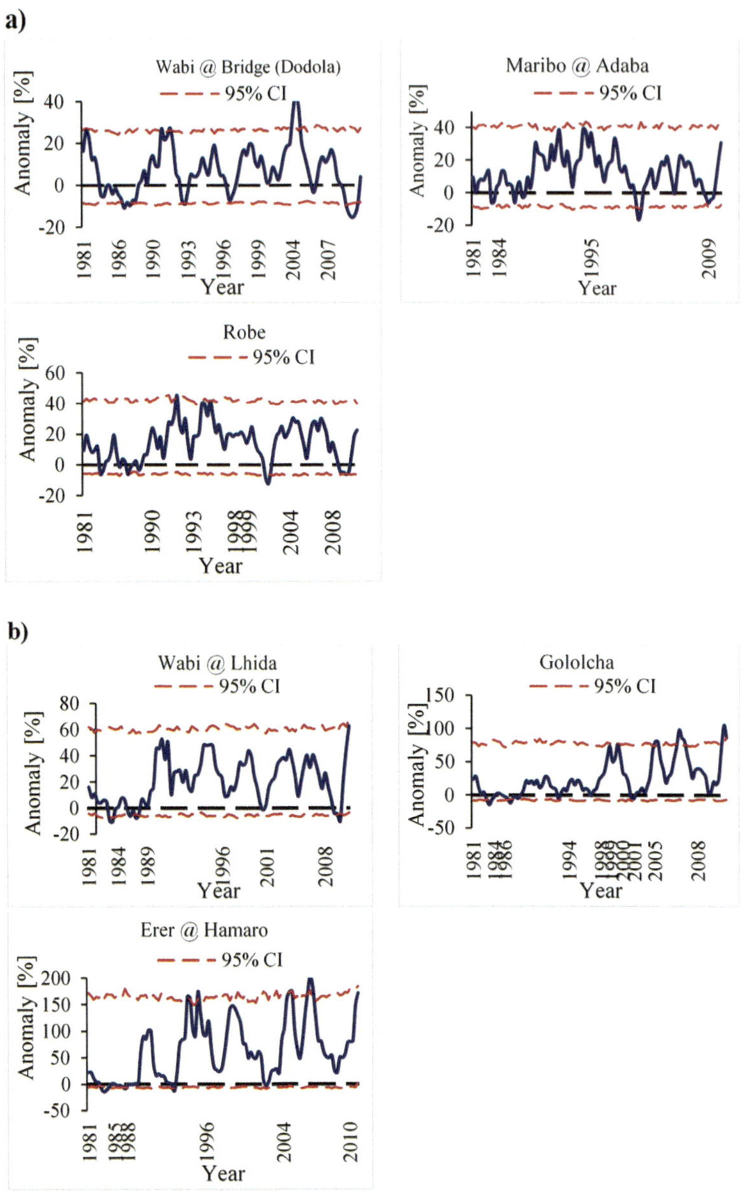

Fig. 9.3 Temporal variability in extreme flood discharge using QPM with 95% CI (confidence interval) in Wabi Shebele River Basin under four categories: **a** upper catchments, **b** middle catchments, **c** eastern catchments and **d** lower catchments

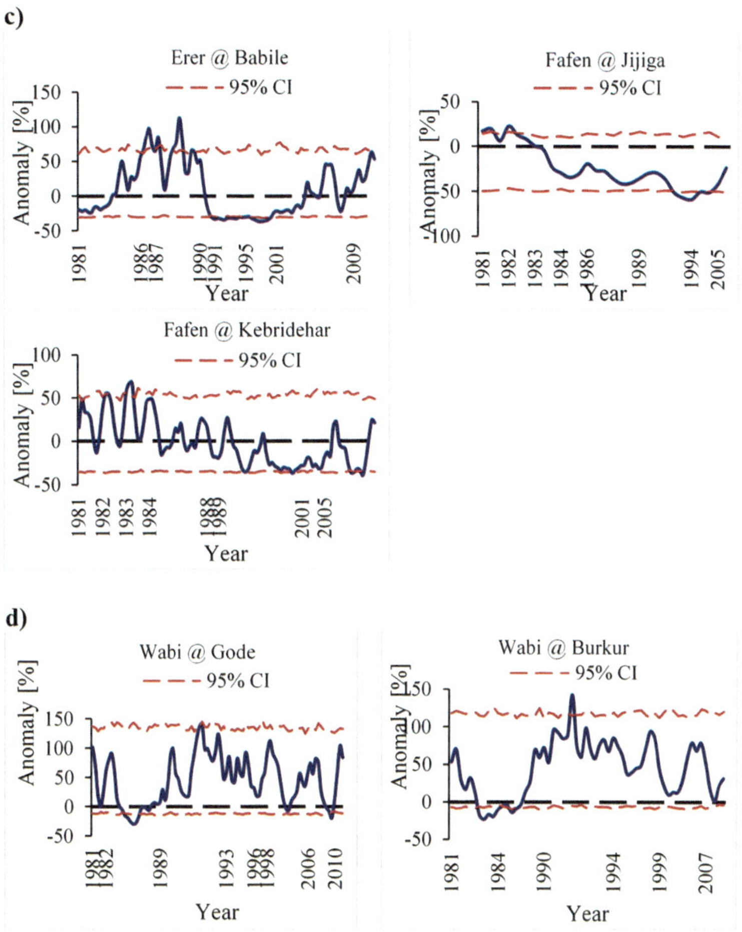

Fig. 9.3 (continued)

Discussion

The presented results revealed that there has been an increasing trend in the flood magnitude and frequency over early twenty-first century over entire the basin. Similar result is reported in literatures that the magnitude and frequency of floods indicate increasing trend in Wabi Shebele River Basin since 2000 (IWMI 2015; MoWR 2003; Tadesse et al. 2016). The positive Kendall's Z-values indicate increasing trend within analysis period (Table 9.4). For the period 1981–2010, the annual maximum flood discharge shows upward trend in the upper and middle catchments, while downward trend is in eastern and lower catchments of Wabi Shebele River Basin. The annual maximum stream flow for middle catchments (i.e.

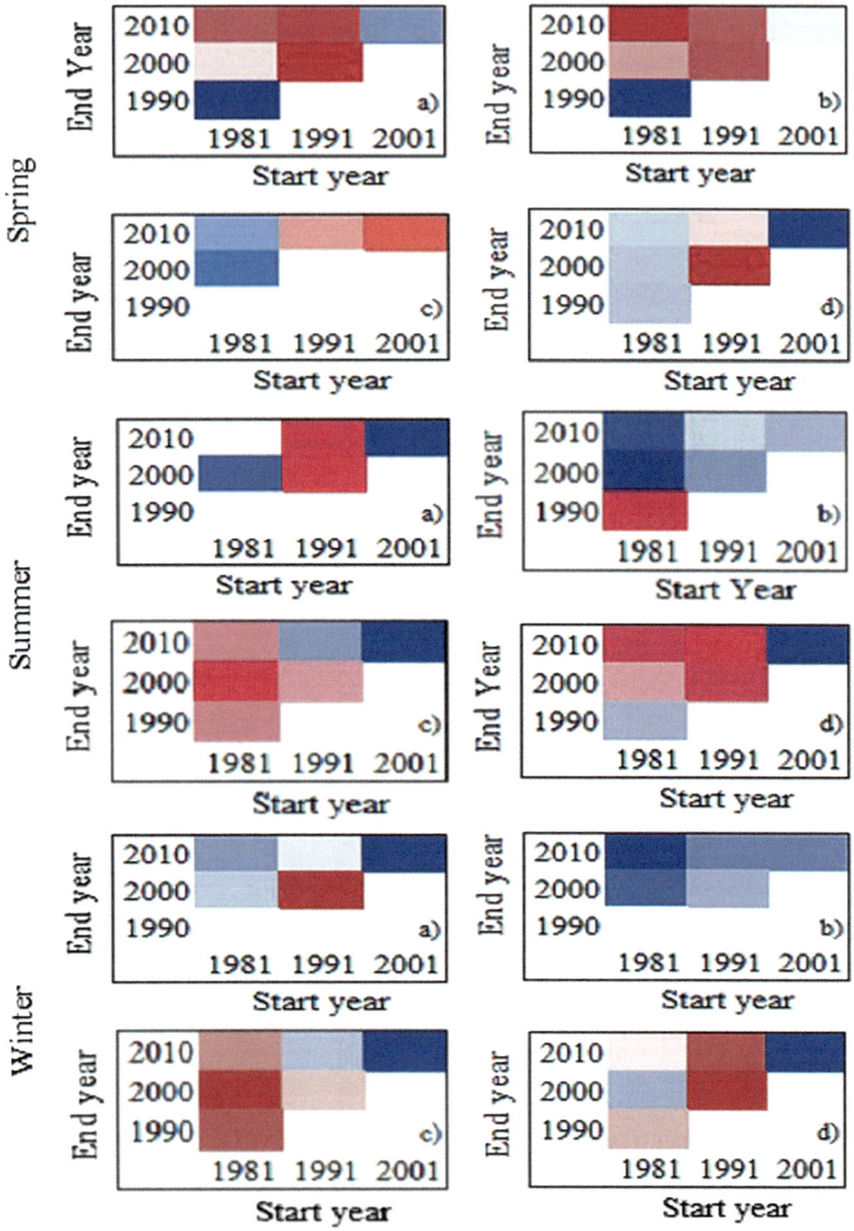

Fig. 9.4 Multi-temporal trend analysis of seasonal maximum discharges for **a** upper catchments, **b** middle catchments, **c** eastern catchments and **d** lower catchments. Legends are the same as shown in Fig. 9.2

Table 9.4 Mann–Kendall trend test summary extreme flood discharges in Wabi Shebele River Basin

Extreme Indices	Mann–Kendall Statistics	Upper catchments			Middle catchments			Eastern catchments			Lower catchments	
		Wabi @ Dodola	Maribo	Robe	Wabi @ Legehida	Erer @ Hamaro	Gololcha @ junction	Erer @ Babile	Fafen @ Jijiga	Fafen @ Kebridehar	Wabi @ Gode	Wabi @ Burkur
AMAX	Test statistics (Z)	0.32	−0.18	−0.07	0.57	2.32	2.00	0.32	−2.64	−2.32	−0.82	0.41
	p-value (two-tailed)	0.75	0.86	0.94	0.57	0.02	0.05	0.75	0.01	0.02	0.41	0.45
	MK Stat (S)	19.0	−11.0	−5.0	33.0	131.0	113.0	19.0	−149.0	−131.0	−47.0	−43.0
	Kendall's tau	0.04	−0.03	0.09	0.08	0.30	0.26	0.04	−0.34	−0.30	−0.11	−0.10
	Alpha	0.05	0.05	0.05	0.05	0.05	0.05	0.05	0.05	0.05	0.05	0.05
	Trend	Not significant	Not significant	Not significant	Not significant	Increasing	Increasing	Not significant	Decreasing	Decreasing	Not significant	Not significant
POTF	Test statistics (Z)	1.471	0.969	1.149	0.997	2.41	2.046	−0.895	−3.207	−2.478	−1.208	−1.206
	p-value (two-tailed)	0.141	0.333	0.251	0.319	0.016	0.041	0.371	0.001	0.013	0.227	0.228
	Alpha	0.05	0.05	0.05	0.05	0.05	0.05	0.05	0.05	0.05	0.05	0.05
	Trend	Not significant	Not significant	Not significant	Not significant	Increasing	Increasing	Not significant	Decreasing	Decreasing	Not significant	Not significant
SMW	Test statistics (Z)	0.77	1.14	1.25	2.36	3.96	2.89	0.57	−2.36	−1.64	−0.04	−0.29
	p-value (two-tailed)	0.44	0.25	0.21	0.02	0.00	0.00	0.57	0.02	0.10	0.97	0.78
	MK Stat (S)	44.0	65.0	71.0	133.0	223.0	163.0	33.0	−133.0	−93.0	−3.0	−17.0
	Kendall's tau	0.10	0.15	0.16	0.31	0.51	0.38	0.08	−0.31	−0.21	−0.01	−0.04
	Alpha	0.05	0.05	0.05	0.05	0.05	0.05	0.05	0.05	0.05	0.05	0.05
	Trend	Not significant	Not significant	Not significant	Increasing	Increasing	Increasing	Not significant	Decreasing	Not significant	Not significant	Not significant

(continued)

Table 9.4 (continued)

Extreme Indices	Mann–Kendall Statistics	Upper catchments			Middle catchments			Eastern catchments			Lower catchments	
		Wabi @ Dodola	Maribo	Robe	Wabi @ Legehida	Erer @ Hamaro	Gololcha @ junction	Erer @ Babile	Fafen @ Jijiga	Fafen @ Kebridehar	Wabi @ Gode	Wabi @ Burkur
SMSp	Test statistics (Z)	0.82	−1.71	−1.82	−1.57	−1.61	−1.57	−2.61	−2.14	−2.53	0.57	0.57
	p-value (two-tailed)	0.41	0.09	0.07	0.12	0.11	0.12	0.01	0.03	0.01	0.57	0.57
	MK Stat (S)	47.0	−97.0	−103.0	−89.0	−91.0	−89.0	−147.0	−121.0	−143.0	33.0	33.0
	Kendall's tau	0.11	−0.22	−0.24	−0.21	−0.21	−0.21	−0.34	−0.28	−0.33	0.08	−0.08
	Alpha	0.05	0.05	0.05	0.05	0.05	0.05	0.05	0.05	0.05	0.05	0.05
	Trend	Not significant	Not significant	Not significant	Not significant	Not significant	Not significant	Decreasing	Decreasing	Decreasing	Not significant	Not significant
SMSu	Test statistics (Z)	0.39	−0.18	0.00	0.54	2.21	1.68	0.00	−2.25	−2.21	−0.82	−0.89
	p-value (two-tailed)	0.70	0.86	1.00	0.59	0.03	0.09	1.00	0.02	0.03	0.41	0.37
	MK Stat (S)	23.0	−11.0	−1.0	31.0	125.0	95.0	−1.0	−127.0	−125.0	−47.0	−51.0
	Kendall's tau	0.05	−0.03	0.00	0.07	0.29	0.22	0.00	−0.30	−0.29	−0.11	−0.12
	Alpha	0.05	0.05	0.05	0.05	0.05	0.05	0.05	0.05	0.05	0.05	0.05
	Trend	Not significant	Not significant	Not significant	Not significant	Increasing	Not significant	Not significant	Decreasing	Decreasing	Not significant	Not significant

Erer at Hamaro and Gololcha at Wabi junction) shows a positive significant trend because the computed *p*-values in both watersheds are lower than the significance level (α 0.05) in the region. However, significant decreasing trend in annual maxima is observed in Fafen watershed at Jijiga and Kebridehar gauging stations. Seasonal trend analysis reveals similar trend and pattern with annual maximum stream flow almost in all stations during past 30 years.

Extreme discharge variability analysis using peak over threshold (3rd quartile) based on QPM showed significant increasing trend in early 1990s and 2000s and decreasing trends in 1980s particularly in upper and middle catchments (Table 9.5). Over eastern catchments, 1980s is the decade in which significant increasing trend is observed and decreasing trend in 1990s and 2000s. The lower Wabi Shebele River stations indicate general decreasing trend in analyses period, 1980–2010. In seasonal extreme analysis, significant anomaly occurrence season varies with catchment (Table 9.3). In upper and middle catchments (Wabi at Dodola, Maribo, Robe, Wabi at Legehida and Gololcha watersheds), spring season is the season in which the highest extreme variabilities are occurred. Similarly, in eastern catchments (Erer watersheds), the highest extreme discharge anomalies occurred in winter season and in lower Wabi Shebele catchments (at Gode and Burkur stations) during summer season. The years: seond half of 1980s, first half of 1990s and second half of 2000s are the years of positive anomalies (significant increasing trend) in flood discharges, whereas the years: first half of 1980s, second half of 1990s and first half of 2000s are the years of significant negative anomalies occurrence in all seasons. It is known that the Wabi Shebele Basin is characterized by two rainfall regimes (NMA 1996): the area characterized by a quasi-double maximum rainfall pattern with a small peak in April and maximum peak in August which covers the west–east highland of the basin (bimodal type I); and the area dominated by double maximum rainfall pattern with peaks during April and October covers the south-eastern low-lying areas of the basin (bimodal type II).

Associated Impacts and Management of Flood

Associated Impacts of Flood Variability

In countries where hydrological variability is high and investments to achieve water security are inadequate, variability is a constant economic risk to small investors (such as farm families) and large ones (such as industries) as well as to the nation. The seasonality of streamflow and cost of extreme weather events have been shown a rapid upward trend in recent decades throughout the world including Ethiopia (Allamano et al. 2009; Bates et al. 2008; Tadesse et al. 2016). As shown in Fig. 9.5, the number of peoples with flood disaster increased in the 2000s in Wabi Shebele Basin, similar to flood discharge trend analysis (Section 'Hydrological Model') which shows increasing trend in early twenty-first century. Similarly, there were

CTable 9.5 Summary of QPM analysis in annual extreme flow

Upper catchments			Middle catchments			Eastern Catchments			Lower catchments		
Sub-basin	Magnitude of highest anomaly (%)	Time	Sub-basin	Magnitude of highest anomaly (%)	Time	Sub-basin	Magnitude of highest anomaly (%)	Time	Sub-basin	Magnitude of highest anomaly (%)	Time
Wabi at Dodola	−10.9	1986	Wabi at Legehida	−6	1986	Erer at Babile	+96.9	1986	Wabi at Gode	−29.4	1983
	+27.3	1991		−7.9	1987		+112.3	1988		−19.4	2009
	+43.9	2004		−10	2009		+65.1	1989	Wabi at Burkur	−23.4	1983
	−15.4	2009	Gololcha	−14.6	1983		−34.5	1992		−14.2	1985
Maribo	−16.5	2000		−11.8	1987		−36.7	1998		+142.3	1991
Robe	−6.03	1982		+81	2005	Fafen at Jijiga	+19.4	1981			
	−6.3	1987		+98.4	2006		+22.5	1982			
	+45.3	1991		+105.4	2010		-58.6	1994			
	−12.3	2001	Erer at Hamaro	-14.2	1983	Fafen at Kebridehar	+68.4	1983			
				−8.5	1987		−37.02	1998			
				+165.3	1994		−39.6	2009			
				+174	1995						
				+203.6	2006						

studies that showed high temporal and spatial variation in hydrologic parameters, precipitation and discharges in Wabi Shebele Basin (MoWR 2003; Seleshi and Zanke 2004). There is a perception in peoples prevailing that flooding in Ethiopia is mainly linked with torrential rainfall (NDRMC 2018; Tadesse et al. 2016) which may not be always true. Short-time rainfall, even a less-than-a-day rainfall event, can also create floods (Bissolli et al. 2011; Liu et al. 2017).

The satellite image (Fig. 6a, b) showed that the extent of flooding in April and May 2005, which was very stressful to the locals and damaged properties (http://earthobservatory.nasa.gov) in south-eastern Ethiopia. As of 5 May 2005, 154 people have been reported dead in the wake of severe flooding along the Wabe Shebele in south-eastern Ethiopia (Tadesse et al. 2016). The United Nations Office for the Coordination of Humanitarian Affairs (UNOCHA) reports that at least 100,000 people have been affected by the flooding (https://www.un.org/press/en/2005/afr1145.doc.html). In addition to flood direct impacts, the swollen river has also extended the reach of crocodiles and water snakes. The UNOCHA reports that 19 of the deaths were caused by crocodiles. The floods have also brought diseases, like malaria and diarrhoea, to the region. On 4 May, skies were clear when the Moderate Resolution Imaging Spectroradiometer (MODIS) flew overhead on NASA's Terra satellite to capture the top image.

According to UNOCHA reports, heavy rains pounded down over the Ahmar Mountains and the desert-dry plain to their south and east on 23 April and 24 April 2005. Rivers flowing out of the mountains overflow far beyond their banks on 27 April 2005. The lower Wabi Shebele Basin is the region very affected by drought and that may have contributed to the flooding, because hard and sunbaked ground cannot easily absorb heavy rain, so the water tends to run-off, filling depressions and riverbeds. After two days of heavy rainfall on 23 April and 24 April 2005, the drought-shrunken Shebele River banks swelled and resulted in large flood events that swept away more than 35 villages in the region. The general environmental and economic impacts of too much rain, as well as of extreme variability in rainfall, are shown in Table 9.6.

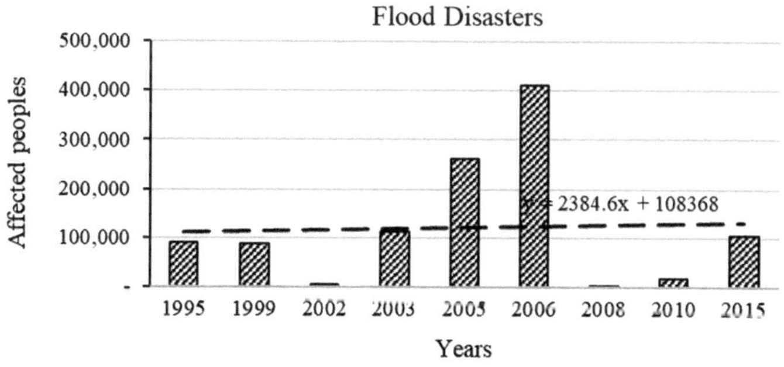

Fig. 9.5 Flood disasters in Wabi Shebele River Basin since 1995 (Assefa 2018; Tadesse et al. 2016)

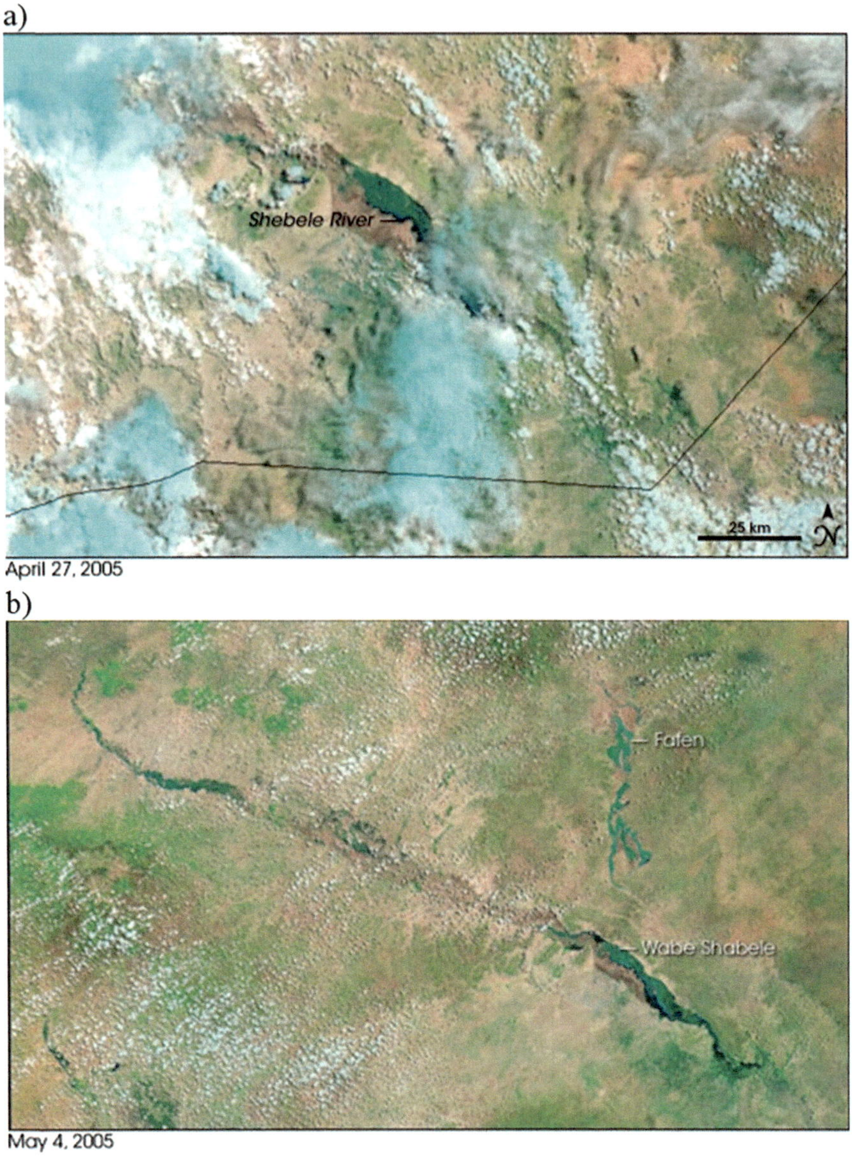

Fig. 9.6 Satellite pictures showing Wabi Shebelle valley floods; **a** 27 April 2005; **b** 4 May 2005. Water is blue and blue-green, vegetation is bright green, bare ground is pinkish tan, and clouds are light blue

These high numbers of economical and human losses indicate the importance of flood risk management practice in the basin. Also, there is now a pressing need for decision-makers to better understand the ongoing changes in hydrologic extremes in order to make preparations for the possibility of changing conditions.

Table 9.6 Historical flood events in Wabi Shebele River Basin, 1980–2019: determined from different sources

Date	Disaster event	Causalities Reported	Source
July 1993	Flooding: 120,000 affected	Heavy rain	http://floodobservatory.colorado.edu/Archives/index.html
1995	Flooding: 89,902 affected, 27 deaths	Heavy rain	http://floodobservatory.colorado.edu/Archives/index.html
October 1997–February 1998	El Niño related flooding	Brief torrential rain	http://www.fao.org/docrep/004/w7832e/w7832e00.HTM
1999	Flooding: 85,789 affected, 34 deaths.	Brief torrential rain	http://floodobservatory.colorado.edu/Archives/index.html
2003	Floods: 119 people died	Heavy rain	ERCS: http://ifrc.org/where/country/check.asp/countryid=65
2005	Flooding: 103,000 affected, 177 deaths	Heavy rain	ERCS: http://ifrc.org/where/country/check.asp/countryid=65
2006	Flooding: 410,132 affected, 132 deaths	Heavy rain	http://floodobservatory.colorado.edu/Archives/index.html
May 2008	Flooding: 11 deaths, 52,000 people abandoned, 164 hectares farmland washed away.	Heavy seasonal rains	http://www.irinnews.org/report/81526/ethiopiathousands-displaced-by-floods-in-Somalia-region
2010	Flooding: 16,000 affected	Heavy rain	http://floodobservatory.colorado.edu/Archives/index.html
2015	Flooding: 105,000 affected	Heavy rain	http://floodobservatory.colorado.edu/Archives/index.html

The recent study of World Bank (World Bank 2006) on environmental analysis for Ethiopia revealed that the country's environmental challenges involve complex cross-sectoral linkages. The two key environmental-development linkages identified by the study relate to the challenges posed by Ethiopia's hydrology are the lack of integrated water resources management and the "land degradation-food insecurity-energy access-livelihood" nexus. The latter includes unsustainable agricultural land management practices and heavy reliance on biomass energy. The consequences of both deforestation and degraded soil structure include less infiltration of rainfall, which diminishes groundwater recharge; more run-off, which contributes to erosion and siltation, and reduced water storage capacity in the soil, which makes crops less able to withstand drought.

Endemism and Expectations of Flood Variability

If the effects of flood extremes are not well recognized, expectations of high variability and endemic floods by itself can affect economic performance and potentially the structure of the economy. The expectation of variability and the unpredictability of rainfall and run-off will make economy actors focus on minimizing their downside risks rather than maximizing their potential gains (World Bank 2006). Because they think as they could lose everything in a single flood and farm families will not invest in land improvements, advanced technologies, or agricultural inputs and thus constraining agricultural output and productivity gains. Lack of such investments can lead to land degradation and desertification, which result in a vicious circle of reducing production and deteriorating assets.

Understanding and mitigating the full impact of hydrological variability on economic performance require a better understanding of the role of expectations and the incentives created by entirely rational risk aversion. In risk assessment, three elements are involved, i.e. vulnerability, exposure and threats. Risk assessment of flood-prone areas in Ethiopia is a challenge task, due to major shortage of adequate and reliable water and soil data. However, some studies (Assefa 2018; NDRMC 2018) done on flood risk assessments in Ethiopia indicate severity of flood risk in the country. According to these studies, the following areas have been recognized as flood-prone areas: parts of Oromia and Afar regions lying along the upper, middle and downstream plains of the Awash River; parts of Somali region along the Wabi Shebele, Genale Dawa Rivers; low-lying areas of Gambella along the Baro and Akobo Rivers; downstream areas along the Omo and Bilate Rivers in southern Ethiopia and the extensive floodplains surrounding Lake Tana and the banks of Gumara, Rib and Megech Rivers. However, National Disasters Risk Management Commission (NDRMC 2018) does not address specifically what should be done on these impacts of floods. A study by Aseffa (2018) proposed structural and non-structural measures to mitigate flood risk in the country. The best-known structural measures are as follows: concrete, earthen or other engineering structures. However, these type measures focus at flood prevention by reducing the amount of discharge running down a river, for example, reservoirs, retention ponds, river training, embankments and flood walls. Flood risk mapping is proposed as a non-structural flood management technique in reducing flood damages in areas frequented by flood.

Flood mitigation actions generally fall into the following categories: preventative measures, property protection measures, natural resource protection activities, emergency services (ES) measures, structural mitigation projects and public education and awareness activities.

Flood Management in Transboundary River Basin

Flood management is a complicated task in river basins controlled by a single authority and becomes more challenging when dealing with transboundary floods. Transboundary floods are floods that originated in one country and then propagate downstream to another country (Bakker 2009). Ethiopia has seven transboundary rivers (Table 9.7), of which the Wabi Shebele River is part Shebele-Juba Basin. Wabi shebele River is one of the frequently affected basins in the country by hydrological extremes (Awass 2009; Amer et al. 2013). Due to the transboundary nature of rivers in the region, the flooding often has transboundary consequences which indicates need of cooperation between riparian countries. To prevent and resolve potential conflicts and avoid severe effects of flooding especially in transboundary waters, countries sharing a water resource need to agree on common rules and procedures of cooperation to jointly manage these water resources (Bakker 2009).

United Nations sustainable flood prevention guides (UN 2000) state that, considering the evolution and trends of floods, one must shift from defensive action against hazards to management of the risk. Flood protection is never absolute; therefore, risk management will be the appropriate method to deal with this challenge. Therefore, a holistic approach based on multilateral cooperation, including interdisciplinary planning for the whole catchment areas including international cooperation in case of transboundary rivers is necessary to take into account the whole river basin.

Table 9.7 Transboundary rivers

River	Total basin area (km^2)	% of basin within Ethiopia	Sharing countries
Nile basin • *Abbay* • *Baro-Akobo* • *Tekeze* • *Mereb*	3,112,369	12	**Burundi, Democratic Republic of Congo, Egypt, Eritrea, Kenya, Rwanda, South Sudan, Sudan, Uganda, Tanzania** *Sudan* *South Sudan* *Sudan* *Eritrea*
Rift Valley Basin • *Gibe-Omo*	637,593	49	**Kenya**
Shebelle-Juba basin • *Wabi Shebele* • *Genale - Dawa*	810,427	46	**Kenya, Somalia** *Somalia* *Kenya, Somalia*

Source http://www.fao.org/nr/water/aquastat/countries_regions/eth/print1.stm

Integrated Flood Risk Management

Integrated flood management (IFM) refers to the integration of land and water management in a river basin using a combination of measures (UN 2009). IFM is holistic approach to address the water cycle as a whole, integrating land and water management. The idea is to adopt the best mix of both structural and non-structural strategies by ensuring a participatory approach and adopting integrated hazard management approaches. IFM requires adopting a river basin approach to planning that involves many disciplines and stakeholders in efforts to reduce flood vulnerability and risk and to preserve ecosystems. Unite Nation Convention on the protection and use of transboundary watercourses and international lakes (UN 2009) proposed four main principles for integrated flood management (IMF):

- *River basin management*: Water management should be based on boundaries of the river basin, not on administrative areas or country borders, thus taking into account a river system as a whole, from source to mouth.
- *Solidarity*: Problems should not be shifted to neighbouring countries or regions. Negative effects between upstream and downstream areas should be prevented, and positive effects should be stimulated.
- *Sustainability*: Integrated water resource management (IWRM) aims at a combination of economic development, ecological protection and improvement of social welfare and justice. In the context of flood management, the principles of sustainable development involve ensuring livelihood and security among different population groups as well as the viability of ecosystems and floodplain functions, including in the long term.
- *Public participation*: Active public involvement in the development and implementation of water management strategies and plans.

Conclusion

In this study, variabilities in flood and its implication in flood risk management are assessed. Nonparametric Mann–Kendal trend test and quantile perturbation (QPM) methods are used to see temporal variabilities in flood discharge in the Wabi Shebelle Basin. The multi-temporal trend analysis performed at 5-, 10-, 15-, 15-, 25- and 30-year intervals starting from 1980 in annual maximum discharge showed increasing trends for most recent periods, since 1996, in all sample stations, while decreasing tendency of flood discharges is observed before 2000s particularly in eastern and lower catchments in the basin. The annual maximum stream flow for middle catchments shows a significant increasing trend and decreasing trends in Fafen watersheds at Jijiga and Kebridehar gauging stations, because the computed p-value in these watersheds is lower than the significance level ($\alpha = 0.05$).

Extreme discharge analysis using QPM shows that there are significant positive anomalies (outside confidence interval) in early 1990s and 2000s and negative anomalies in 1980s particularly in upper and middle catchments. However, in eastern catchments, the 1980s is the decade in which significant positive anomalies were observed and negative anomalies in the 1990s and the 2000s. Significant extreme changes in seasonal flood discharges vary with catchments: in upper and middle catchments during spring season; in eastern catchments during winter season and in lower catchments during summer season.

Wabi Shebele River Basin is one of identified flood-prone areas in Ethiopia. Similar to flood discharge, the impact of floods on human being and resources indicates increasing trend in twenty-first century. In countries where hydrological variability is high and investments to achieve water security are inadequate, variability is a constant economic risk.

Understanding and mitigating the full impact of hydrological variability on economic performance will require a better understanding of the role of expectations and the incentives created by entirely rational risk aversion. Flood risk management in transboundary river basin requires involvement of complex cross-sectoral linkages and more challenging when dealing with transboundary rivers.

Acknowledgements We would like to thank the Ethiopian National Meteorology Service Agency and Ministry of Water, Irrigation, and Electricity for providing the necessary data.

References

Akola J, Binala J, Ochwo J (2018) Guiding developments in flood-prone areas: challenges and opportunities in Dire Dawa city, Ethiopia. OASIS

Allamano P, Claps P, Laio F (2009) An analytical model of the effects of catchment elevation on the flood frequency distribution. Water Resour Res 45:W01402

Amer S, Gachet A, Belcher WR, Bartolino JR, Hopkins CB (2013) Groundwater exploration and assessment in the eastern lowlands and associated highlands of the Ogaden Basin Area, Eastern Ethiopia: Phase 1 Final Technical Report

Assefa TH (2018) Flood risk assessment in Ethiopia. Civil Environ Res 10:35–40

Awass AA (2009) Hydrological drought analysis—occurrence, severity, risks: the case of Wabi Shebele River Basin, Ethiopia 217

Bakker MHN (2009) Transboundary river floods and institutional capacity 1. JAWRA J Am Water Resour Assoc 45:553–566. https://doi.org/10.1111/j.1752-1688.2009.00325.x

Bates BC, Kundzewicz Z, Palutikof J, Shaohong W (2008) World meteorological organisation (WMO), United Nations Environment Program (UNEP), Intergovernmental Panel on Climate Change, Climate change and water [Electronic resource]: IPCC Technical paper VI. IPCC Secretariat, Geneva

BCEOM_ORSTOM_EDF (1973) Wabishebelle survey

Bissolli P, Friedrich K, Rapp J, Ziese M (2011) Flooding in eastern central Europe in May 2010—reasons, evolution and climatological assessment. Weather 66:147–153. https://doi.org/10.1002/wea.759

Chow VT, Maidment DR, Mays LW (1963) Applied hydrology. Technical Correspondence School, N.Z. Dept. of Education, Wellington (N.Z)

Hargreaves GH, Samani ZA (1985) Reference crop evapotranspiration from temperature. Appl Eng Agric 1:96–99
International Water Management Institute IWMI (2015) Drivers of hydrological dynamics in the Bale Eco-Region
Keast D, Ellison J (2013) Magnitude frequency analysis of small floods using the annual and partial series. Water 5:1816–1829
Kendall MG (1975) Rank correlation methads. Charles Griffin, London, UK
Liu Y, Yuan X, Guo L, Huang Y, Zhang X (2017) Driving force analysis of the temporal and spatial distribution of flash floods in Sichuan province. https://doi.org/10.3390/su9091527
Malamud BD, Turcotte DL (2006) The applicability of power-law frequency statistics to floods. J Hydrol Amsterdam 322:168–180
Mann HB (1945) Nonparametric tests against trend. Econometrica
Merz B, Aerts J, Arnbjerg-Nielsen K, Baldi M, Becker A, Bichet A, Blöschl G (2014) Floods and climate: emerging perspectives for flood risk assessment and management. https://doi.org/10.5194/nhessd-2-1559-2014
Ministry of Water Resources MoWR (2003) Wabi Shebele river basin integrated master plan study project, vol VII, Water resources, Part 2 Hydrology, Part. Ethiopia
Ministry of Water Resources MoWR (2004a) Wabi Shebele River Basin integrated development master plan, land use/land cover study, part. Ethiopia
Ministry of Water Resources MoWR (2004b) Federal democratic republic of Ethiopia, ministry of water resources, Wabi Shebele River Basin, Integrated Development Master Plan Study Project, Volume III—Natural Resources, Part 7—Soils (No. II), Addis Ababa, Ethiopia
Monteith JL (1965) Evaporation and environment. Symp Soc Exp Biol 19:205–234
Moriasi DN, Arnold JG, Van Liew MW, Bingner RL, Harmel RD, Veith TL (2007) Model evaluation guidelines for systematic quantification of accuracy in watershed simulations. Trans ASABE 50:885–900
Nash JE, Sutcliffe JV (1970) River flow forecasting through conceptual models, Part I—a discussion of principles. https://doi.org/10.1016/0022-1694(70)90255-6
National Meteorological Service Agency NMA (1996) Climatic and agroclimatic resources of Ethiopia. NMSA Meteorological research report series (Ethiopia)
NDRMC (2018) Federal Democratic Republic of Ethiopia National Disaster Risk Management Commission, Early Warning and Emergency Response Directorate
Neitsch SL, Grassland S, WRL Blackland Research Center (2005) Soil and water assessment tool: theoretical documentation: version 2005. Grassland, Soil and Water Research Laboratory; Blackland Research Center, Temple (Texas)
Ntegeka V, Willems P (2008) Trends and multidecadal oscillations in rainfall extremes, based on a more than 100-year time series of 10 min rainfall intensities at Uccle. WRCR Water Resour Res, Belgium, p 44
Onyutha C (2016) Statistical analyses of potential evapotranspiration changes over the period 1930–2012 in the Nile River riparian countries. Agric For Meteorol 226–227:80–95. https://doi.org/10.1016/j.agrformet.2016.05.015
Priestley CHB (1972) On the assessment of surface heat flux and evaporation using large-scale parameters. Mon Weather Rev 100:81–82
Schuol J, Abbaspour KC (2007) Using monthly weather statistics to generate daily data in a SWAT model application to West Africa. Ecol Modell 201:301–311
Seleshi Y, Zanke U (2004) Recent changes in rainfall and rainy days in Ethiopia. Int J Climatol 24:973–983
Tabari H, AghaKouchak A, Willems P (2014) A perturbation approach for assessing trends in precipitation extremes across Iran. J Hydrol 519:1420–1427. https://doi.org/10.1016/j.jhydrol.2014.09.019
Tadesse T, Haigh T, Wall N, Shiferaw A, Zaitchik B, Beyene S, Berhan G, Petr J (2016) Linking seasonal predictions to decision-making and disaster management in the greater horn of Africa. Bull Am Meteorol Soc 97, ES89–ES92. https://doi.org/10.1175/BAMS-D-15-00269.1

Taye MT, Willems P (2012) Temporal variability of hydroclimatic extremes in the Blue Nile basin: temporal variability of hydroclimatic extremes. Water Resour Res 48. https://doi.org/10.1029/2011WR011466

UN (2000) Meeting of the parties to the convention on the protection and use of transboundary water resources and international lakes

UN (2009) Transboundary flood risk management: experiences from the UNECE Region, New York

World Bank (2006) Ethiopia: managing water resources to maximize sustainable growth. The World Bank Agriculture and Rural Development Department

Chapter 10
Exploring Technologies for Sustainable Transboundary Water Resource Management in the Era of Climate Change: A Case for the Nile River Basin Riparian States

Bertha Othoche

Abstract This book chapter explores technologies for sustainable transboundary water resource sharing and management in the era of extreme weather and climate patterns in and across the Nile River Basin riparian countries. Climate change which has been attributed mainly to anthropogenic forcing has become a major challenge to sustainable water resources managements globally. The adverse impacts of changing weather patterns have resulted in global efforts toward mitigation and adaptation through the United Nations Organization (UNO) satellite organizations. These include the World Meteorological Organization (WMO), the United Nations Environmental Programme (UNEP) and the Inter-Governmental Panel to Climate Change (IPCC), among others. Treaties, Conventions and Conference of Parties (COPs) that address extreme weather and climatic conditions include the United Nations Framework Convention on Climate Change (UNFCCC) and the Kyoto Protocol. The landmark declarations of the COP 2015 Paris Agreement in which nations committed to address anthropogenic causes of climate change as well as adaptation and mitigation funding were a step toward this end. Most models predicting future emission pathways show increasing temperature trends while others predict further variability in rainfall patterns. Such trends will impact on the environment including water resources and especially water catchment areas, watersheds, river drainage systems, wetlands as well as impacting on community livelihoods. The latter will translate into changing production systems, especially in agriculture thereby resulting in large numbers of climate refugees across the globe. The Nile Basin, which covers over 10% of Africa's landmass in 11 countries including Ethiopia, Sudan, South Sudan, Egypt, Rwanda, Tanzania, Uganda, Burundi, the Democratic Republic of Congo (DRC), Eritrea and Kenya, is a development engine across the riparian states. A combined population estimated at 257 million, which is 53% of the Nile Basin countries depend in one way or the

B. Othoche (✉)
Pwani University, Box 195, Kelifi 80108, Kenya
e-mail: b.othoche@pu.ac.ke

A. M. Melesse et al. (eds.), *Nile and Grand Ethiopian Renaissance Dam*,
Springer Geography, https://doi.org/10.1007/978-3-030-76437-1_10

other on the waters of the Nile. The predicted extreme weather and climate scenarios will therefore spell doom in the form of degradation of water towers feeding the Nile, famine, hunger and migration. The Nile Basin riparian countries should therefore jostle for climate change mitigation and adaptation funds and adoption of climate smart technologies and in particular geospatial technologies with special reference to geographic information system (GIS) and satellite technologies to be supported by Information Communication Technologies (ICTs). It is with this background that this book chapter intended to explore appropriate technologies for possible application in management of the Nile Basin waters. The objectives of this book chapter therefore included to (i) establish the current trends and impacts of weather variability and climate change in the Nile Basin, (ii) establish current technologies applied in management of the Nile Basin waters and (iii) explore sustainable potential strategies and technologies to address pertinent issues arising from man–environment interactions. The chapter uses desktop research method to mine adequate data on the current trends and weather patterns as well as current and potential technologies that could offer sustainable solutions to water resources in the Nile Basin. The data is then presented in the form of themes, narratives and graphs. The chapter contributes toward enhancing knowledge on extreme weather patterns, river basin hydrology and appropriate technologies.

Keywords Transboundary resources · Nile river basin · Drainage systems · Riparian states · Climate change · Weather variability · Climate forcings · Adaptation · Mitigation

Introduction

This chapter explores technologies for sustainable transboundary water resource sharing in the era of extreme climates in and across the Nile River Basin riparian countries. Climate change has been attributed to both natural and anthropogenic forcings. Manifestations of weather variability and climate change include changes in temperature conditions, changes in precipitation patterns; increased frequency of occurrence of drought and floods hence, the hotter and longer spells of dry periods will be harmful; higher and increased frequency of rainstorms; sea level rise, as well as frequent floods in areas where the changes will result in an increase in precipitation. On the other hand, impacts of climate change include changes in the hydrological cycle that influences the flow of water in rivers, soil and underground storage reservoirs; higher temperatures resulting in higher rates of evapotranspiration that eventually affect rain fed and irrigation agriculture; higher precipitation resulting in floods that result in environmental degradation, siltation of water resources hence scarcity of water for domestic and irrigation purposes; changes in temperature conditions that result in emergence of pests and diseases with cost implications for their management; as well as higher water temperatures that directly affect aquatic plants and animals which in turn indirectly affect livelihoods.

For the Nile Basin, it is projected that continuous warming of the Mediterranean Sea where the Nile empties its waters will affect the Nile Delta in Egypt thereby reversing the gains and defeating the logic of continuous negotiations on the use of the Nile waters. Various projections on climate change have been made and exist in literature (IPCC 2007).

Global Efforts

The adverse impacts of changing weather patterns have resulted in global efforts toward mitigation and adaptation through the United Nations Organization; World Meteorological Organization (WMO); the United Nations Environmental Programme (UNEP) and the Inter-Governmental Panel to Climate Change (IPCC) are notable world organizations. Treaties, Conventions and Conference of Parties (COPs) also address extreme weather and climatic conditions. Others include the United Nations Framework Convention on Climate Change (UNFCCC) and the Kyoto Protocol. The landmark declarations of the COP 2021 (Paris Agreement) in which nations committed to address anthropogenic causes of climate change, as well as adaptation and mitigation were a step toward this end. Nations also committed to adaptation and mitigation funding as a step toward this end. Regional bodies include the Inter-governmental Authority on Development (IGAD).

Climate Models

Most models predicting future emission pathways show increasing temperature trends. Such trends will impact on the environment including water resources, the hydrological cycle and community livelihoods. This will in turn cause famine and poverty thereby resulting in large numbers of climate refugees. For Africa and specifically the Nile Basin riparian states, all models show that most parts of continent will reach the 1.5 °C limit by 2023. Some parts of the region have already seen a temperature increase of 2.5° (IGAD Climate Prediction and Application Center (ICPAC 2020). For rainfall, the models are predicting an upward trend. This will result in floods that have adverse effects on settlements along the Nile Basin through displacements and destruction of infrastructure and property, wash away crops resulting in great wastage. Floods will also result in the increase and/or emergence of vector and waterborne diseases. On the other hand, livestock sector could also be affected through livestock deaths and increased incidence of animal diseases. Both floods and drought will affect availability and access to clean water for domestic purposes.

The Nile Basin

The drainage basin of the Nile (Fig. 10.1) covers 3,255,000 km^2, about 10% of the area of Africa. The Nile River is the world's longest river (6670 km), and its catchment extends over ten East African countries. This basin covers over 10% of Africa's landmass in 11 countries including Ethiopia, Sudan, South Sudan, Egypt, Rwanda, Tanzania, Uganda, Burundi, the Democratic Republic of Congo (DRC), Eritrea and Kenya. The Nile flow is controlled by a number of manmade structures between Lake Victoria (covering parts of Kenya, Tanzania and Uganda) and the Ethiopian Highlands through Sudan to the High Aswan High Dam in Egypt and several planned facilities in the middle reaches (Ethiopia and Sudan). This basin is a development engine across the Nile riparian states.

A combined population estimated at 257 million, which is 53% of the Nile Basin countries depend in one way or the other on the waters of the Nile. The predicted extreme weather and climate scenarios will therefore spell doom in the form of famine, hunger and migration. Extreme weather patterns manifested in the form of floods would push further the poverty levels and cause degradation of water towers feeding the Nile River. As negotiations on the sharing of the Nile waters continue, the riparian states should include clauses to ensure that this precious resource will be there tomorrow and that the current efforts will not be in vain.

Uses of the Nile Waters

The Nile Basin has several water uses including water supply for agricultural, industrial and domestic use; as well as water for power generation. Agriculture accounts for at least 80% of all water consumption in the Basin. Better and more integrated use of water resources is important including adoption of improved irrigation techniques. The most common method remains flood irrigation, which has been found to be inefficient and wasteful.

Trends and Impacts

Current Trends: Weather Variability and Climate Change in the Nile Basin

The trend of weather patterns in the Nile Basin riparian states varies according to geographic location and variations in topography or relief. Generally, observed trends of weather patterns in the riparian states in the Nile Basin show that the basin receives annual average rainfall of about 650 mm, though regional disparities exist. Rainfall is slightly on the upward trend except for Kenya which shows a downward

Fig. 10.1 Nile River Basin. *Source* https://commons.wikimedia.org/wiki/File:River_Nile_map.svg

trend among the riparian states in the Nile Basin. Probability of exceedance of rainfall is greater in Ethiopia, Sudan and South Sudan and lower in Kenya and Uganda. Rainfall patterns in both the White and Blue Nile Rivers show greater variability with outliers indicating extreme weather events of drought and floods. Water scarcity is common due to temporal and spatial variations in rainfall. Maximum, minimum and mean temperatures are on upward trend in all the selected riparian states in the Nile Basin. There are marked anomalies in rainfall and temperature trends in all the five selected riparian states in the Nile Basin. This is in line with ICPAC (2020) predictions for temperature increase of 2.5° for the region.

Increasing Rainfall and Rising Temperatures in the Nile Basin Riparian States

A number of studies have established the rising temperature conditions. Increasing rainfall has been confirmed (Ogega et al. 2020). The mean annual temperature over the basin varies from a minimum of 10 °C over Ethiopian and equatorial highland areas to about 30 °C in the central Sudan. The mean annual temperatures in the equatorial lakes region have little variations throughout the year and range between 16 and 27 °C depending on locality and altitude. In the Sudan plains, the minimum temperature occurs in January and the maximum in May or June, when it rises to a daily average of 41 °C in Khartoum. From the IPCC Fourth Assessment Report (IPCC 2007) and the country report of Ethiopia (NMA 2007), there is a general consensus on the rise of temperature. In particular, higher water temperatures have been reported in lakes in response to warmer conditions (Bates et al. 2008). The minimum temperatures over Ethiopia show an increase of about 0.37 °C per decade, which indicates the signal of warming over the period of the analysis 1951–2005 (NMA 2007). Within the Nzoia catchment that drains into Lake Victoria, located in Kenya, there is an increasing trend of 0.79 °C per decade in the lowlands and 0.21 °C per decade in the highlands (Githui 2008).

Efforts by Riparian States and Transboundary-Level Adaptation Measures

Adaptation to climate change in transboundary water resources is an urgent requirement (Goulden et al. 2009). The Nile Basin riparian states participate in global discussions on climate change mitigation. They also participate on adaptation discussions; preparation of National Action Plans (NAPA) to prepare for adaptation; preparation of policies on climate change adaptation and mitigation; and on preparation of National Appropriate Action (NAMA). Examples include Ethiopia's Climate Resilient Green Economy initiative to sustain low emissions of greenhouse gases and aim to acquire funding for sustainable and green development through the Clean Development Mechanism. Other Nile riparian states have near similar plans.

Preparing for climate change requires taking action at local, national, basin-wide and global levels. This will involve coordinated inter-state transboundary-level adaptation to include coordinated reservoir operation, strengthening inter-basin agricultural trade, interconnecting power grids, developing joint mechanisms for soliciting climate change adaptation funds, operating joint hydro-meteorological monitoring programmes, as well as conducting joint research.

The Nile Basin as a Fragile Ecosystem and Vulnerable to Climate Change

Two-thirds of the physical environment in the Nile Basin consists of arid and semi-arid lands. It is upon this environment that there is overdependence of the riparian states population on natural resources and primary production systems such as agriculture, fishing and forestry. These primary production systems are sensitive to climate change. Economic diversification into less climate sensitive sectors is also low in most Nile Basin riparian states. The high population growth rate causes pressure on natural resources resulting on their depletion and increases in the number of disasters especially hydro-meteorological disasters such as floods and drought. The lower Nile Basin states also rely heavily on Nile-fed irrigation for agriculture; hence, any change in the flow of the Nile is likely to have adverse effects on these states. There is also high dependency on hydro-power for energy across the basin.

The Future of the Nile Basin Riparian States

Global warming will translate into higher temperatures and very dry and hot years; hence, warming of the region will continue (Coffel et al. 2019). Severe wet and dry years will occur in quick succession. The hot and dry years have become more common over the past four decades in the Upper Nile Basin and this trend is likely to continue. These hot and dry conditions will be similar to those that have resulted in crop failures, food shortages and humanitarian crises in the region over the past decades. By the late twenty-first century, the frequency of these hot and dry years may rise between a factor of 1.5 and 3. In the past, hot and dry years occurred about once every 20 years; but this increase in frequency means that in future, a hot and dry year could occur once every six to 10 years, making them a common experience for people in the region. In addition to becoming more frequent, they will also become more severe. Temperatures during heat waves in the region could rise by between 2 and 6 °C, putting far more stress on people, animals and crops than it occurs today.

Impacts of Weather Variability and Climate Change in the Nile Basin

Impacts of climate change on the hydrology of different landscapes have been documented (Mujumdar et al. 2012; Bates et al 2008; Conway 1996, 2005; Conway and Hulme1993). Increasing rainfall in the Nile Basin has and will result in increasing frequency of floods and displacement of people depending on existing hydrological structures in each member state. Floods damage existing structures and wash away crops and livestock, and this results in famine, hunger and therefore food insecurity. Floods and drought cause an increase in the number of pests and diseases that will eventually affect livelihoods and environmental health in general. Drought has become not only frequent but also prolonged in a greater part of the riparian states and increasing temperatures are a threat to existing livelihoods. Poor water management is a problem in the basin. For example, an average of up to 30% of the region's rainfall is lost before use. Egypt, with rainfall close to zero, except for along the Mediterranean coast and some parts of the Sinai Peninsula, gets almost 90% of its water needs from the river. Among the Nile Basin riparian states, Burundi, Uganda and Tanzania receive relatively high rainfall, along with Democratic Republic of Congo and Rwanda, which also have abundant water resources. Just a 10th of semi-arid Kenya is within the basin, but Nile waters support about 40% of Kenya's population. Ethiopia and Eritrea both have high rainfall, but it is typically seasonal and lasts for only four months of the year. The Nile River flows through six of the world's poorest countries, home to over 300 million people, the majority of whom live in rural areas. Estimates of food insecurity show that approximately 24–25.4 million people will face acute food insecurity requiring urgent action in 2020.

Generally, climate change poses a threat to livelihoods among communities in the Nile Basin riparian states. Climate change has been described as the most pervasive and threatening crises currently facing the global community. It is an emergency that requires adequate and immediate action. This is because a 2 °C increase in temperatures would cause more harm to humanity and natural ecosystems as this would result in more and increased frequency of heat waves, would wipe out the world's coral reefs which harbor valuable marine species and cause damage to livelihoods.

Existing and Potential Technologies in Access and Distribution of Water with Special Reference to Kenya as a Nile Basin Riparian State

Technologies are required to address changing weather patterns in relation to hydrology of the Nile valley. These technologies would be in the form of water-related service provider and end user tools. This is because as climate patterns change, water resources will become scarce; hence, one way of water management would be geared toward conservation of water and water distribution. Large-scale projects would require the use of geospatial technologies, satellite technology and ICT, among others.

The need for appropriate technology in mitigation and adaptation to climate change cannot be understated. The following observation by Dina Zayed confirms the need for adoption of climate smart technology for sustainability in the Nile riparian states:

> Agriculture accounts for at least 80 percent of all water consumption in the basin. Experts call for better and more integrated use of water resources, saying many countries have been slow to adopt improved irrigation techniques. The most common method remains flood irrigation, which has been found to be inefficient and wasteful, (Dina Zayed 2011).

Desktop search method was used to mine data on technologies currently used by the Nile Basin riparian states (Kenya) in relation to management of climate change and weather variability. There are a variety of technologies that are applied in water resource management in Kenya. Examples include technologies for water resources management; technologies for climates mart agriculture; technologies for soil management; and technologies for value chain addition for sustainable livelihoods in the era of changing weather patterns. It means that appropriate technologies will be required in almost all areas or sectors related to water resources in the Nile Basin as most of these areas affect directly or indirectly the waters of the Nile River and vice versa. There will be need to capacity build on scientific methods associated with modeling of climate change, weather variability, adaptation and mitigation. Simple ways or methods of disseminating climate change information and special ways of creating awareness on climate change will be necessary to empower local communities with adequate knowledge on climate change. Technologies on water resources management, environmental conservation, and afforestation/reforestation and pasture management for sustainable water resources will be required. Technologies in construction of small and large water reservoirs to increase water storage capacity will go a long way in ensuring water availability for various uses. Other technologies will be required in the protection of catchment areas and integrated river basin and flood management minimize impact of floods; primary production systems and especially agricultural production; harnessing of renewable energy resources, green technology and diversification of energy sources to minimize reliable on climate sensitive hydro-power; enhancing food storage capacity; trade and value chain addition; soil moisture conservation and irrigation;

mainstreaming climate change adaptation and mitigation in all national development sectors; as well as increasing research into crop varieties, including disease- and drought-resisting varieties, high-yielding varieties, quick-maturing varieties.

It is important to note that technologies required in adaptation cut across various areas and requires multidisciplinary, interdisciplinary and transdisciplinary approaches. These approaches exist in literature (UNEP 2011). Some of these technologies are discussed between the lines. Information communication technologies (ICTs) provide opportunities for information exchange and dissemination of new ideas. Water management in the era of climate variability in the Nile Basin riparian states and prompt weather alerts will be required to ensure that appropriate information is disseminated to local communities and other stakeholders for swift action. On the other hand, geospatial technologies will be significant in climate smart adaptation and mitigation mechanisms and efforts. This will involve and include the application of GIS (GIS and GPS agriculture); application of remote sensing (Satellite imagery); drone and aerial photography/imagery; climate smart agricultural techniques such as precision farming; indigenous knowledge techniques; water resources onsite reuse technologies, green infrastructure and Big Data; farming software and online data that involves merging datasets, among others. Technologies for sustainable water use and management would include a variety of technologies to include sprinkler and drip irrigation; fog harvesting; rainwater harvesting; soil management; slow-forming terraces; conservation tillage; integrated nutrient management. Technologies for sustainable crop management would include crop diversification and new varieties; biotechnology for climate change adaptation of crops; ecological pest management; seed and grain storage; sustainable livestock management; livestock disease management; selective breeding via controlled mating; sustainable farming systems; mixed farming; agro-forestry. Also required are technologies for capacity building and stakeholder organization that would be required for different stakeholders to include community-based agricultural extension agents; farmer field schools; forest user groups; water user associations, among others. Soil management technologies would also be required. Such technologies include sow-forming terraces, conservation tillage and integrated nutrient management.

Technology adoption will require funding. The funding will go into establishing climate smart technologies. The 2015 Paris Agreement agreed on a formula to fund adaptation and mitigation strategies in developing countries. Negotiations and mechanisms should therefore be put in place to enhance existing strategies and to introduce new sustainable technologies. The likely sources of funding for the Nile Basin riparian states for technological upgrade include the United Nations Development Program (UNDP), the Global Environmental Facility (GEF); the AUDA-NEPAD; the African Climate Change Fund through African Development Bank; the Global Green Grants Fund as well as the funding from the National and Regional Governments.

Water Resources Management Technologies Focusing on Supply/End User Interface in Kenya

Water resources management involves planning and management aimed at optimal use of water resources. Due to the changing climatic conditions, there will be a need to adapt and establish alternative management strategies especially in the Nile Basin for sustainable utilization of water resources in the Nile Basin (Fawcett et al. 2012). Water is required by the growing population in the Nile Basin for irrigation, domestic use, for industrial development and for recreation, among others. Globally, it is estimated that 70% of the world's waters is used for irrigation, with 15–35% of irrigation withdrawals being unsustainable. Management of water resources in the Nile Basin riparian states will have to consider issues related to access to water. Different countries in the Nile Basin riparian states have varying levels of access to water based on their rural and urban populations. For example, in Kenya, 58% of Kenyans had access to at least basic drinking water sources in 2015. This means that 42% of the population do not access basic drinking water.

Water resources management in Kenya is guided by Water Act 2002. The constitution of Kenya 2010 also stresses on the need for watershed area conservation. The National Water Master Plan and the Master Plan for Conservation and Sustainable Management of Watershed areas involved participatory planning. It was also realized that the achievement of the Vision 2030 could only be possible through implementation of effective water resources planning in the face of increasing demands and climate change while conserving sustainable watershed.

Kenya has also recognized the role of geospatial technology in the Kenya Vision 2030. All sectors of the economy have embraced geospatial technology including the water sector, agriculture, tourism, among others. The country has adopted Integrated River Basin development programmes to enhance water volume in rivers in the country. Most of Kenya's rivers drain into Lake Victoria which in one way or the other affect the flow of the Nile waters. Kenya has embarked on afforestation programmes, protection of wetlands and swamps as well as integrated soil management strategies. Satellite weather applications and monitoring are ongoing. Examples of management programmes include the transboundary integrated water resources management in the Mara River Basin Management Project jointly management by Kenya and Tanzania. Kenya has also embarked on country-wide catchment-level water resources management and integrated basin planning for sustainable water resource management. These programmes integrate geospatial technologies including aerial surveys, satellite technologies, the use of unmanned aerial vehicles (UAVs) or drones, among others. Geographic information system (GIS) and GPS technologies have been used in water management across the water sector. Geospatial technologies are employed in water resource management in many other parts of the world. For example, these technologies have been used in gravity-based water supply management at the village level. It has been observed that satellite-derived data in conjunction with GIS and ground inventory data are necessary in delineation of gravity-based water supply management in different parts of the world (Surindar 2019). This approach could be adopted by the Nile Basin riparian states to manage their water resources in the era of climate change and weather variability.

Kenya has enhanced the use of ICT in water resources management aimed at reaching the local people in both urban and rural areas. These include Maivoice, the WATEX system; M-Maji; Majidata; Ufahamu; Onkesean; Water flow; Huduma; mWater; Banki Ya Maji, among others (Hilda et al. 2012). The Water Resources Management Board (WASREB) is responsible for enhancing and regulating innovations in technology adoption for sustainable utilization of water resources. These technologies employ the use of ICT, the satellite technology; mobile applications; and also involve the use of open data. These technologies could be replicated in the Nile Basin riparian states to enhance water security, safe water distribution and access in the era of climate change. Most of the ICT platforms are web based or mobile phone based. The basic client/server connections also host ICT applications.

Challenges

Challenges exist in developing, initiating and strategies associated with climate change. In the Nile Basin, challenges for water sharing are related to the changing geo-politics and changing climate (Ashok 2011). Such challenges will definitely influence the adoption of technologies for climate smart technologies; hence, this will depend on geopolitical and other factors. Such factors would include political good will; the inter-state conflicts; difficulty in securing funds; issues related to resistance to change; inadequate research; low technologies. Other factors include water scarcity and shortages, pollution and water protection; water and conflicts at local, national, and regional levels.

Summary/Conclusions And Recommendations

Conclusions

Technologies are a key to future survival of the populations in the Nile Basin riparian States. Every effort should be put in place to ensure most up-to-date technologies are adopted in line with experiences from different states.

Recommendations

Recommendations include the need to establish appropriate policies; support from the international community; a prioritized "no-regret" measure is to expand water storage infrastructure in the Nile region; sourcing for funds.

References

Ashok S (2011) Challenges for water sharing in the Nile Basin: changing geo-politics and changing climate. Hydrol Sci J 56(10)

Bates BC, Kundzewicz ZW, Wu S, Palutikof JP (eds) (2008) Climate change and water. IPCC Secretariat. Technical Paper, Geneva

Conway D (1996) The impacts of climate variability and future climate change in the Nile Basin on water resources in Egypt. Int J Water Resour Dev 12:277

Conway D (2005) From Headwater Tributaries to International River Basin: adaptation to climate variability and change in the Nile River Basin. Global Environment Change 15:99–114

Conway D, Hulme M (1993) Recent fluctuations in precipitation and runoff over the Nile Sub-Basins and their impact on main Nile discharge. Clim Change 25:127–151

Dina Zayed (2011) Reuters, Nile Basin Initiative, World Bank. In: Blair E, Roddy M (eds) United Nations Environmental Program, United Nations Educational, Scientific and Cultural Organization, 2018

Coffel ED, Keith B, Lesk C, Horton RM, Bower E, Lee J, Mankin JS (2019) Future hot and dry years Worsen Nile Basin water scarcity despite projected precipitation increases. https://doi.org/10.1029/2019EF001247

Fawcett et al (2012) Flexible strategies for long term sustainability under water uncertainty. Build Res. 40(5):545–557. https://doi.org/10.1080/09613218.2012.702565. S2CID 110278133

Githui FW (2008) Assessing the impacts of environmental change on the hydrology of the Nzoia catchment, in the Lake Victoria Basin. Doctoral dissertation, Vrije Universiteit Brussel

Goulden M, Conway D, Persechino A (2009) Adaptation to climate change in international river basins in Africa: a review. Hydrol Sci J 54(5):805–828

Hilda et al (2012) Technology in solving society's water problems in Kenya. iHub Research

ICPAC (2020) Climate forecasting. ICPAC

IPCC (2007) Fourth assessment report: climate change 2007. IPCC, A Geneva. (AR4). Core Writing Team

Mujumdar PP et al (2012) Climate change impact on hydrology and water resources. ISH J Hydraul Eng

NMA (2007) Climate change national adaptation program of action (NAPA) of Ethiopia. NMA, Oxfam International, Addis Ababa

Ogega OM, Koske J, Kung'u JB, Scoccimarro E, Endris HS, Mistry MN (2020) Heavy precipitation events over East Africa in a changing climate: results from CORDEX RCMs. Clim Dyn 55:993–1009

Surindar W (2019) Application of geospatial techniques for gravity-based drinking water supply management. Arch Photogrammetry Cartography Remote Sens 31:75–83. ISSN 2083–2214, eISSN 2391–9477. https://doi.org/10.2478/apcrs-2019-0006

UNEP (2011). Technologies for climate change adaptation—the water sector. Energy, climate and sustainable development. UNEP

Chapter 11
Water–Energy–Food (WEF) Nexus Modelling Application to Estimate WEF Investment Portfolio in Ethiopia: A Case Study Applicable to Future Cooperative Investment in the Nile Basin

Megersa Tesfaye, Semu A. Moges, Assefa M. Melesse, and Zeleke Agide

Abstract Resources utilization without due regard to the environment is recognized as one of the reasons for the collapse of past civilization across the planet. The current growth-based economic model of development is believed to be unsustainable and the world needs a new sustainable development path. United Nation (UN) general assembly acknowledges the development path that does not commit to the principle of sustainability is a recipe to future disaster. The Nile basin is one of the regions of the world with water energy and food insecurity is driven by rapid socio-economic growth, investment backlog and climate change uncertainty. Regional WEF nexus understanding and cooperative regional planning and development are required for sustainable development path in the basin. This paper attempts to demonstrate the WEF nexus modelling application to sustainable WEF development and investment. The national target development of achieving lower-middle-income economic level, the national potential resources and scenario analysis method is implemented. The preliminary results indicate that Ethiopia needs at least about 113 billion US$ of total capital investment between 2021 and 2030 to achieve national targets of lower-middle-income consumption in water, energy and food. This case study demonstrates the importance of utilizing nexus models for sustainable development in shared river basins, like the Nile basin, to facilitate cooperation and joint investment.

Keywords WEF · Water · Energy · Food · Intensification · Sustainable development

M. Tesfaye (✉) · Z. Agide
Ethiopian Institute of Water Resources, Addis Ababa University, Addis Ababa, Ethiopia

S. A. Moges
School of Civil and Environmental Engineering, Connecticut University, Storrs, CT, USA

A. M. Melesse
Department of Earth and Environment, Florida International University, Miami, FL, USA

A. M. Melesse et al. (eds.), *Nile and Grand Ethiopian Renaissance Dam*,
Springer Geography, https://doi.org/10.1007/978-3-030-76437-1_11

Introduction

The 2030 Agenda for Sustainable Development adopted by all United Nations Member States in 2015 is a universal call to action to end poverty, protect the planet and ensure shared peace and prosperity (UN 2015). The Sustainable Development Goals (SDGs), each with several quantifiable underlying targets and related data metrics, contains 17goals to be accomplished by 2030. Three of the SGDs specifically recognize the need to deliver access sustainably through dedicated goals on each theme: SDG 6 (clean water and sanitation), SDG 7 (affordable and clean energy and SDG 2 (zero hunger) (UNDP 2015). Yet, a significant percentage of the global population is still a long way from attaining water, energy and food security. Globally, 3.5 billion people lack access to safe drinking water and do not have adequate sanitation, 1.3 billion people lack access to electricity and close to one billion people are undernourished (Neil and Philip 2012). This issue is severe in sub-Saharan Africa.

Ethiopia is one of the most populous countries in sub-Saharan Africa in a similar situation. Despite considerable progress in the past years, these trends are insufficient to meet the national target of water, energy and food security. Still, only 57% of the national population have improved water supply (Parker et al. 2016). In 2012, only 23% of the total population was connected to the national grid (Mondal et al. 2016). Using the threshold of 2550 kilocalorie (kcal) per adult equivalent per day, 40% of the national population were food insecure and undernourished (Mohamed 2017). This reflects in the national per-capita water, electricity and food consumption rates, very low compared to the internationally adopted value for basic human needs. At the same time, national population is likely to increase to 139.6 by 2030 and to 190.9 by 2050 (UN 2017b), while development and consumption patterns are likely to change, leading to even higher water, energy and food demands per-capita. Ethiopia aims to intensify efforts to attain water, energy and food security under this development gap, formulating a 10-year prospective development plan for the year 2019/20 to 2029/30 aligned to the 2030 agenda and SDGs. There is a combined challenge of compensating for the unmeet current demands and meeting additional future demands. In walking towards addressing these challenges, it is critical to understand what resources are required, their adequacy and understanding of the costs connected with meeting these targets. As a pre-condition for assessing the financial mechanisms and sources for achieving the targets, the costs of meeting the targets need to be better understood.

Many methodologies have been used to estimate investment needs on a global, regional and national scales to attain socio-economic goals. Using integrated assessment model, (Guy and Varughese 2016) estimate the cost of meeting the 2030 Sustainable Development Goals targets on drinking water, sanitation and hygiene and (Pachauri et al. 2013) estimate investment cost on pathways to achieve universal household access to modern energy by 2030. Though integrated assessment models are useful for understanding how environmental objectives can be achieved, they are not primarily designed for estimating investment needs and therefore do not produce budgets that can be tied directly to inputs and outputs. Moreover, (FAO et al. 2015) estimate agricultural investment to achieve zero hunger by 2030 using incremental capital output ratio (ICOR) estimates method. Though simple to apply, this method suffers from several conceptual and practical limitations. It simply extrapolates the past into the future, which is a poor guide for the structural change in the economy, especially in developing countries.

Besides, all of these methods in common lack to simulate or account for water, energy and food system interactions both now and in the future. Water, energy and food are interdependence, and the production and consumption chain of these resources are intricately related (Bizikova et al. 2013; Ringler et al. 2013; Rasul 2014) and assessment of these resources should ideally treat them as such. Besides these approaches are also lack a methodological component to calculate the potential resource and service requirements to meet socio-economic goals in a growing economy. An approach that calculates resource requirements and simulates important interactions within the goals and targets of water, energy and food production system to meet services and demands with the constraint imposed by the physical and economic environment is therefore required. Climate, Land, Energy and Water System (CLEWS) framework is a novel prototype to simulate, or account for the CLEW system interactions both now and the future and calculate the resource and service requirements with the lowest net present cost to meet socio-economic goals within a growing economy.

This paper aims to develop WEF nexus model for sustainable development and investment in the water energy and food sector in Ethiopia to fulfil the target goal of achieving lower-middle-income economic country by 2030. The study uses CLEWS (Climate, Land, Energy and Water Strategies) WEF nexus modelling framework to evaluate and understand the linked development pathways. The model is introduced by an international atomic agency for the first time in 2009 (IAEA 2009) and later by a multi-united nations agency application to Mauritius (Howells et al. 2013). CLEWS modelling framework has been applied in several studies at different scales. At the global scale, the model was applied to provide useful insights about the relationship between water, energy, climate, land and material use (Weirich 2013). At a national scale, the framework was applied to provide a robust and integrated framework for assessing nexus interlinkages and future scenario (Welsch et al. 2014). At urban scale, the model was applied successfully to a case study of New York City for the investigation of water and energy system intervention (Rogner 2017).

The paper first presents the study area context, including data type and data sources. Modelling approaches and tools are then discussed including CLEWS framework setup and scenario description. Finally results, discussion and conclusion are provided.

Methods and Models

Description of the Study Area

This study is a national scale study of Ethiopia, which is located in the horn of Africa (eastern part of Africa) (Fig. 11.1). About 75% percent of Ethiopia's landmass is categorized as dryland, experiencing moisture stress during most days of the year and having only 45–120 days of growing season per year (Giorgis 2014). In contrast, most of the water resources are concentrated in the few areas of the country. Ethiopia has 12 major river basins, which comprises four major drainage systems (the Nile Basin, the Rift valley Basin, the Shebelle Juba Basin and the northeast cost Basins). The Nile basin part of the country (32%) generates about 70% of the national surface water resources and the rest of the country generates about 32% of the water resources. As a result, the water energy and food resources of the country is concentrated in the Nile basin and few other small river basins.

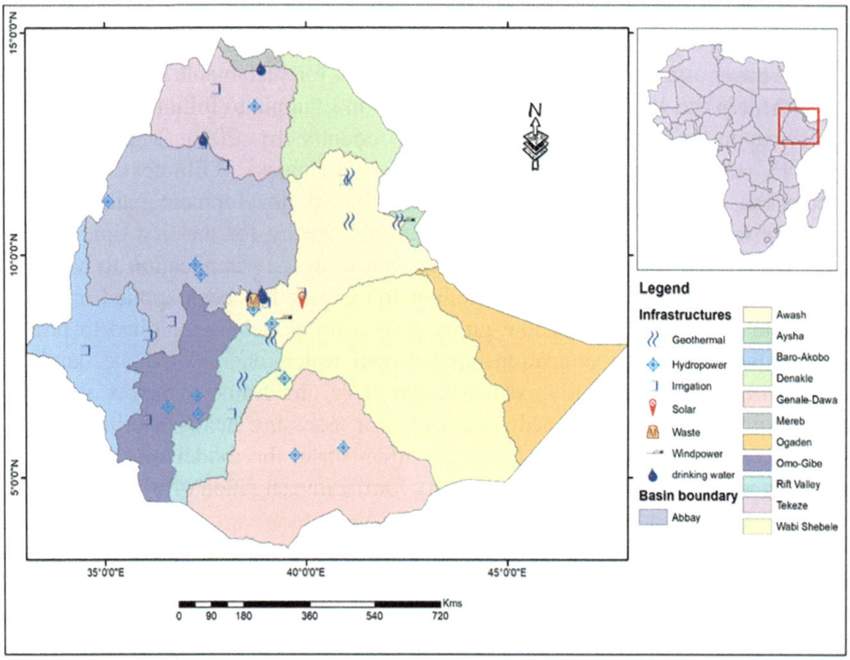

Fig. 11.1 Location map of Ethiopia with major water and renewable energy resource infrastructures

CLEWS Model Structure

CLEWS nexus framework (Fig. 11.2) is a tool for simultaneous consideration of water, energy and food security. It consists of mapping of the relations with the CLEW system, and between it, and other important resources and the economy. It is often promoted as strengthen national capacity for conducting integrated assessment in support of coherent national sustainable development plans. The model structure developed consists of demand projections and a database of water, power supply and land-use technologies that are characterized by economic, technical and environmental parameters, and information regarding the existing capital stock and its remaining life span. Every resource cost and quantities were defined in the model. Furthermore, the model is restricted by so-called constraints used to reflect, among others, operational requirements, government policies, or socio-economic realities. All parameters entered in the modelling framework are time-dependent and can be adjusted over the study horizon to represent a variety of potential futures. Although at its early stage of development, the implementation of the CLEWS nexus system has begun to zoom in exploring various geographical scales:

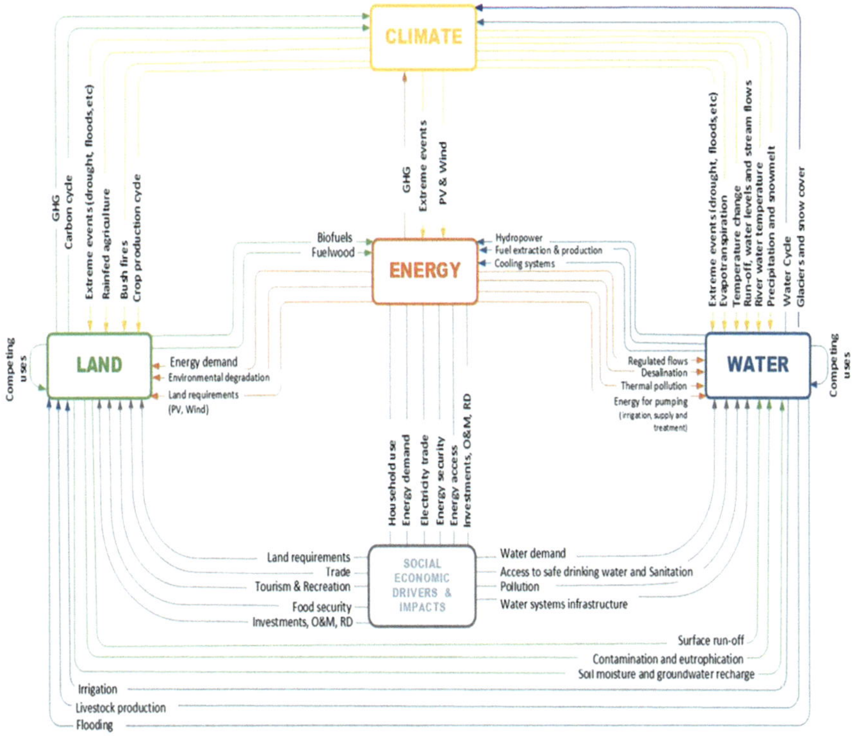

Fig. 11.2 CLEWS framework. *Source* OPTIMUS

from global (Weirich 2013) to regional (Smajgl and Ward 2013) to *national* (Hermann et al. 2012; Sattler et al. 2012; Sridharan et al. 2019).

Open Source energy Modelling System (OSeMOSYS)

For the analysis, a national CLEWS model was developed and created in (OSeMOSYS (Howells et al. 2011). OSeMOSYS is a dynamic, bottom-up, multi-year energy system model applying linear optimization techniques for long-term modelling of energy systems. It determines the optimal investment strategy and product mix of technologies and fuels required to satisfy an exogenously defined energy. It builds on an open-source programming language (GNU Mathprog) and solver (GLPK (GNU programming kit)). The objective in the used version of OSeMOSYS calculates the lowest NPC (net present cost) of an energy system to meet given demands for energy carriers, energy services or their proxies. Technical, economic and environmental implications associated with the identified least-cost system can be easily extracted from the model results. It is structured into blocks of functionalities to calculate what investment to make, when, at what capacity and how to operate them, to meet given final demands and policy together at the lowest cost (Howells et al. 2011). This enables OSeMOSYS to be applied in domains other than energy, such as water systems. Technical constraints, economic restrictions and/or environmental targets may also be imposed to reflect policy considerations.

CLEWS Framework Setup and Application

OSeMOSYS was used to integrate land, energy and water system and their interaction with climate. OSeMOSYS is a flexible, open-source tool for energy modelling, which has been a high degree of flexibility and is capable of accommodating other "sectors", such as water and land and including climate change impacts. To represent national CLEWS framework in OSeMOSYS, existing and future resource flows and activities (technologies) were identified. Resource flow often represents primary resources that contribute to water supply, electricity generation and food production, that is, the maximum resource capacity that could be exploited by each technology. Technology represents elements of a modelling system that converts a resource from one form to another, uses it and supplies it. Resource technology interaction was then identified. This involves the identification of water to energy and food flow, energy to water and food flow, food to energy flow.

The identified resource interactions were quantified using a parameter input to activity ratios and output to activity ratios. For technology (activities) that is to produce a resource flow (water, energy and food), a nonzero output to activity ratio is added. Similarly, for a technology that uses a resource, a nonzero input to activity ratio is entered. A technology (activity)-resource flow relationship is then formed. Further, activities (technologies) were characterized by their techno-economic features and constrained by their physical and financial limits. The level of input to activity ratio and output to activity ratios was determined by technical characteristics of each technology.

The model calculates the resource and service requirements with corresponding annual investment and operation costs to meet socio-economic targets (national targets to achieve middle-income country status). At the same time, the tool should simulate important interactions within the CLEWS system to meet water, energy and food-related service demands, within constraints imposed by the physical and economic environments. Accuracy of exogenous parameters provided to the model, such as technical and economical is of paramount relevance to obtain reliable results. The model considers Ethiopia as a single region economy, in the time horizon of between 2014 and 2050. The model was calibrated based on data range from 1993 to 2030 sourced from FAO, while the future years until 2050, demand has been derived from scenarios data.

Data Type and Sources

This work draws on data from several sources including Ministry of Water, Electricity and Irrigation, National Meteorological Agency, the Central Statistical Agency of Ethiopia, different international databases and published literature.

Water Data

Historical water consumption patterns were used to project water demand under business-as-usual scenario as well as for model calibration and are obtained from (MoFED 2006, 2010,2016). Further, average middle-income-per-capita water consumption rate was used to project national water demand as the country vision is to become middle-income status by 2030 and sourced from (Koontanakulvong 2018). To meet these demands, Ethiopia depends on three water sources: precipitation, surface water and groundwater resources. Maximum national water resources potential was used to constrain the model and defines the real system limitations of water supply infrastructure to provide and supply water over a year to meet an exogenously defined demand and was obtained from (Kidanewold et al. 2014) for surface water resource potential, and for groundwater resources potential, it was sourced from (Moges 2012). Further costs of water supply infrastructures were obtained from Addis Ababa Water and Sewerage Authority (AAWSA) for municipal water supply and irrigation water supply it was obtained from (Inocecio et al. 2007).

Energy Data

National electricity generation system has four tiers, primary resources, power generation technologies, transmission and distribution infrastructures and final demand. Historical electricity consumption rate was obtained from Ethiopian Electricity Utility (EEU), and it was used to project electricity demand for business-as-usual scenario. Middle-income countries electricity consumption rate was obtained from (IEA 2014) and was used to project electricity demand for as Ethiopia targets to achieve middle-income countries status. To meet electricity demand for both business-as-usual scenario and middle-income countries consumption level, the national power generation system draws on renewable energy resources, power generation technologies and transmission/distribution technologies. All technical and economic characteristics for power generation and transmission/distribution technologies were sourced from Brinckerhoff (2014).

Land-Use Data

Total national land stock and land-use share, as well as national food supply data, come from Food and Agricultural Organization (FAO) of the United Nations Statistical Division (Hanna and Max 2013). Agricultural production statistics were sourced from the Central Statistical Agency (CSA) of Ethiopia.

Scenario Descriptions

Business-As-Usual Scenario (BAUScen)

This scenario is considered a starting point for analysing a water, energy and food system. This scenario is based on a set of assumptions based on historic projections of water, energy and food access and consumption rates. The overall assumption in this scenario takes into account the continuous access rate and per-capita consumption of water, energy and food at the same rate for the period of 2014–2050. The growth of the total demand is derived mainly by population and urbanization growth. Table 11.1 presents the evolution of water, energy and food access and consumption rate under business-as-usual scenario.

Table 11.1 Summary of considered indicators for BAUScen

Sectors	Indicator	2014	2030	2040	2050
Water	Water (l/c/day)	35	57	71	85
Energy	Electricity (kWh/c)	69.72	160	268	451
Food	Food (kcal/c)	2137	2253	2328	2406

National Vision 2030 Scenario (NV2030Scen)

Ethiopia aims to spur economic structural transformation and sustain accelerated growth towards the realization of the national vision to become a low-middle-income country by 2030. National strategic plans indicate the country will attain 100% coverage of water supply and electricity by 2030. This scenario stipulates Ethiopia will satisfy its water, energy and food requirements by 2030 and fulfils the 2030 SDG goals in water, energy and food. It is also assumed that the country graduates from a low-income country to lower-middle-income country status by 2030 as planned and continue to grow to a medium middle-income country by 2040 and higher-middle-income country by 2050. Key features of this scenario are the per-capita water, electricity and calorie consumption of the country is assumed to match the current average consumption rates of lower-middle, medium and higher-middle-income country by 2030, 2040 and 2050, respectively. The evolution of water, energy and food demand based on this scenario is represented in Table 11.2.

Agricultural Intensification Scenario (AGIScen)

Ethiopia is not on a track to meet the food security targets based on BAUSen and NV2030Scen. Based on the result obtained from the above two scenarios, the land is the limiting factor and the base of this scenario is, therefore, the availability of resources to satisfy the national vision of achieving middle-income status. This

Table 11.2 Summary of considered indicators for NV2030Scen

Sectors	Indicators	The base year (2014)	Target by 2030	Gap relative to 2030	Target by 2050
Water	Percentage of the population safely managed drinking water service	57.3% (World development Indicator 2012)	100%	42.7%	100%
	Per-capita water consumption (l/c)	35	130.52	173	222.82
Energy	Percentage of population with access to electricity	26.6% (World development Indicator 2012)	100%	73.4%	100%
	Per-capita electricity consumption (kWh/c)	69.72	760	690.28	3469
Food	Prevalence of undernourishment	32% (World development Indicator 2012)	0%	32%	0%
	Per-capita calorie consumption(Kcal/c)	2137	3180	1043	3180

scenario assumes the NV2030Scen of achieving middle-income country continues but considers enhancing agricultural productivity as a means of sustainable food production and protection of the ecosystem. In addition to the NV2030Scen assumptions above, this scenario sets out an ambitious vision of doubling of agricultural productivity (both rainfed and irrigation) through implementing agricultural intensification and technology. The current average agricultural productivity of Ethiopia is one of the lowest in the world. It is assumed that the average crop productivity will be enhanced to double from the current 2.3 tones/ha.

Results

Water Supply

Extending basic water access rate to current unserved (i.e. achieving 100% access rate) and enhancing per-capita water consumption rate to average lower-middle-income countries level will cost 5.87 billion US$ or 0.65 billion US$ annually from 2021 to 2030. This accounts for 5.64% of the current National Gross Domestic Product (GDP). This adds up to almost 12.84 times the current investment levels of 457 million US$ in the water sector. Compared to BAUScen, 4.84 Billion US$ additional capital investment is needed. The total expenditure needed to attain higher-middle-income country consumption rate from 2030 to 2050 is about 25.05 billion US$ which is 26.85 billion US$ greater than BAUScen. Tables 11.3 and 11.4 present the total spending required for achieving the national vision of water security under BAUScen and NV2030Scen. The difference in investment is largely explained by the current lower access and consumption rate of water compared to middle-income country status.

Electricity Supply

Investment required to achieve electricity security is shown in Tables 11.5 and 11.6 for BAUScen and NV2030Scen respectively. Projected national electricity demand

Table 11.3 Cost of water supply under BAUScen (Billions$)

Year	Investment cost	Operation and maintenance cost	Total cost
2020	0.079	0.04	0.12
2021–2030	1.03	0.43	1.46
2031–2040	1.33	0.62	1.95
2041–2050	1.87	0.86	2.73
Total	4.30	1.94	6.25

Table 11.4 Cost of water supply under NV2030Scen (Billions$)

Year	Investment cost	Operation and maintenance	Total cost
2020	0.31	0.14	0.45
2021–2030	5.87	2.69	8.56
2031–2040	11.12	5.11	16.23
2041–2050	13.93	6.41	20.34
Total	31.23	14.35	45.58

Table 11.5 Cost of electricity supply (BAUScen) (Billion US$)

Year	Generation cost	Transmission and distribution cost	Operation and maintenance cost	Total cost
2020	0.36	0.14	0.003	0.50
2021–2030	19.71	7.88	0.19	27.78
2031–2040	22.38	8.95	0.22	31.55
2041–2050	7.01	2.80	0.07	9.88
Total	49.46	19.77	0.48	69.71

calls for installing of 31.22 GW, of 21.22 new electricity generation capacity worth an aggregate investment of 69.70 billion US$, 48.92 US$ for generation capacity expansion and 19.56 billion US$ for transmission and distribution as well as 1.22 US$ for operation and maintenance from 2021 to 2030 to achieve lower-middle-income consumption rate. An investment of 7.74 billion US$ per year will be needed to provide electricity access for all. In general, the total investment cost of NV2030Scen is higher than the BAU by 151% to achieve lower-middle-income consumption target. In 2050 (from 2031 to 2050 inclusive), a total cost of 753.24 billion US$, 566.67 billion US$ investment cost to expand generation capacity to 334.22 GW and 172.78 billion US$ for transmission and distribution to attain upper-middle-income electricity consumption rate. Annually this translates to an average of 37.66 billion US$. Figure 11.3a, b shows total cost of electricity supply for BAUScen and NV2030Scen, respectively.

Table 11.6 Cost of electricity supply under NV2030Scen (Billion US$)

Year	Generation Cost	Transmission & distribution cost	Operation & maintenance cost	Total cost
2020	2.47	0.98	0.06	3.51
2021–2030	48.92	19.56	1.22	69.70
2031–2040	137.71	55.08	3.44	196.23
2041–2050	431.96	117.70	7.35	557.01
Total	621.06	193.32	12.07	826.45

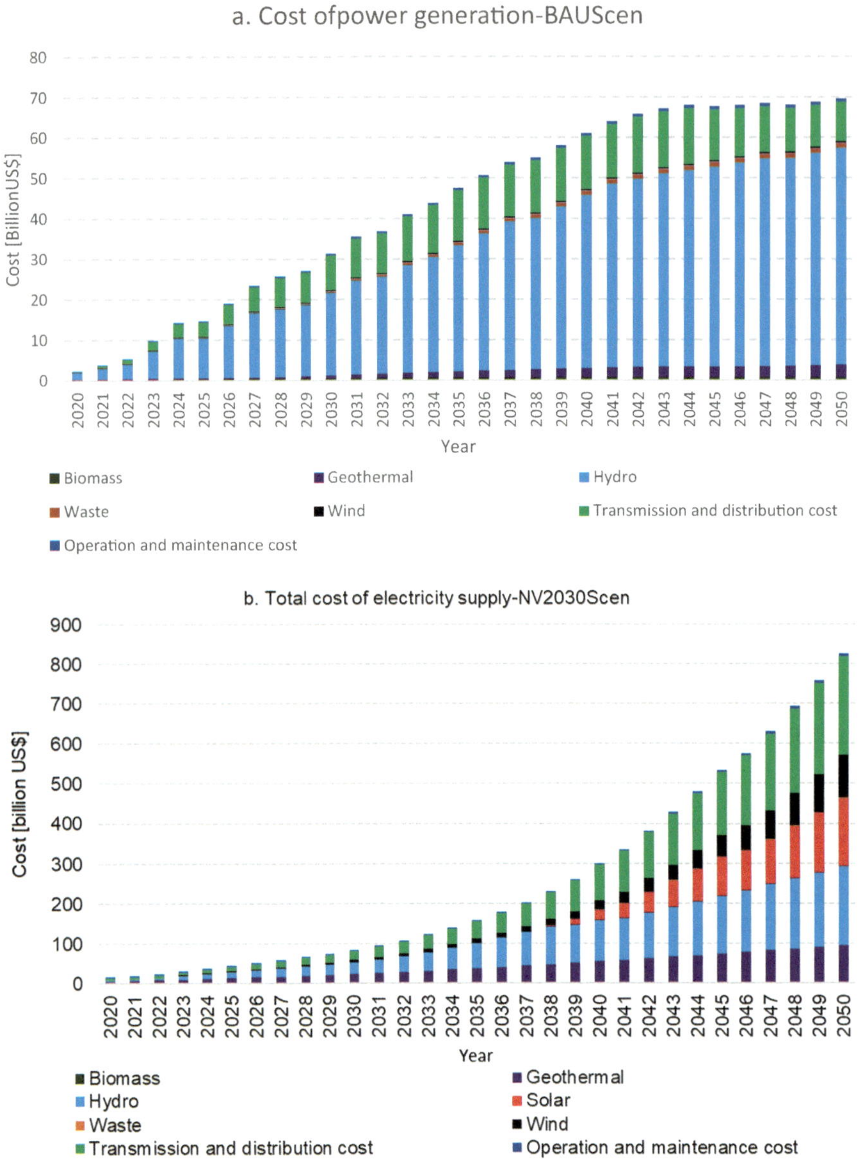

Fig. 11.3 a Total cost of electricity supply for BAUScen. **b** Total cost of electricity supply for NV2030Scen [**b**]

Food Supply and Land Resources

Figure 11.4 presents the land required to support national food demand as defined by scenarios. As indicated in the figure, the current food consumption rate

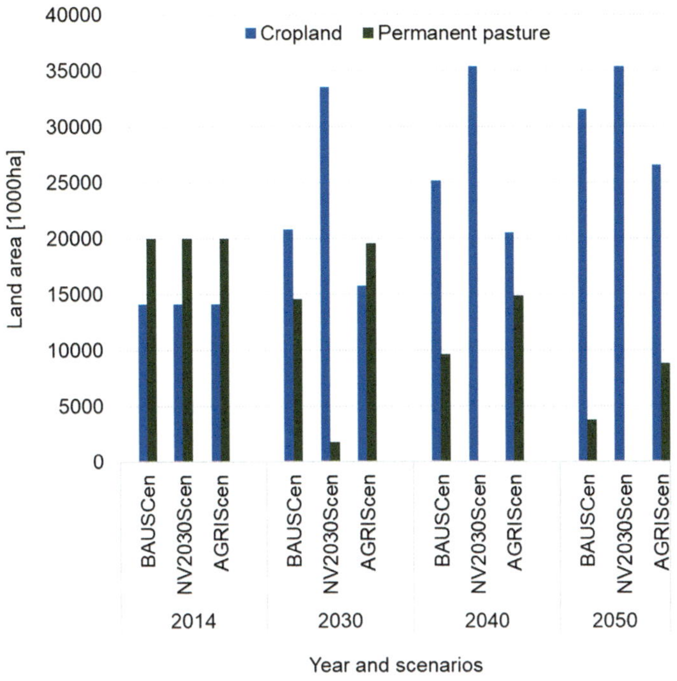

Fig. 11.4 Agricultural Land-use competition under different scenarios

(BAUScen) is meeting at the cost of ecosystem intervention. However, in NV2030Scen, land resources cannot support food demand to meet the national goal of attaining middle-income consumption rate (measured by per-capita calorie consumption).

With doubling crop productivity from the current production rate, cropland significantly reduced to leave space for other land uses (Fig. 11.4). Using crop intensification data from (Abraham et al. 2014), potential investment costs of intensification for AGRIScen show that crop intensification investment cost is projected to be 12.7 US$ billion by 2030. Annually this equals 1.4 US$ billion to increase crop productivity to double by 2030. In 2014, irrigated land accounted for about 280,000 ha, 4.83% of national irrigated area of about 5.8 million ha. Total irrigated area in AGRIScen is projected to increase to 65% of national potential to reach 3.8 million ha 2030. Using cost data on water project from (Innecnio et al. 2007) the average cost to achieve this level of improvement to the irrigation system amounts to about 2.22 US$ billion per year from 2021 to 2030. The total cost of intensification is, therefore, 34.9 US$ billion by 2030, 3.87 US$ billion annually.

Discussion

This study provides indicative figures on the water, energy and land resources and associated costs of attaining the national target of achieving middle-income consumption level until 2030. Developing strategies to accelerate progress in this area requires an understanding of resource requirements and the costs connected with meeting these targets. The study applies the CLEWS nexus framework using OSeMOSYS as a modelling tool. From 2021 to 2030, 113.16 billion US$ needs to be spent to attain WEF security, of which 8.56 billion US$ is for water, 69.70 billion US% for electricity (48.92 billion US$ for production and 19.56 billion US$ for transmission and distribution) and 34.9 billion US$ for agricultural intensification.

The investment required to cover current unmet clean water demand Ethiopia is estimated to invest about 5.87 billion US$ per annum until 2030, where 0.65 billion US$ needs to be invested annually. This accounted for 0.62% of current national GDP which is within the range of capital investment required for sub-Saharan African for water security which ranges from 0.29% to 1.0% of the Gross Regional Product (RGP) (Guy and Varughese 2016). The modelling result is only off to the national estimates for meeting the water and sanitation target from 2021 to 2030 by about 0.05 Billion US$. Annual investment costs estimated by the Ministry of Water, irrigation and Energy (MoWIE) are 0.6 billion US$ on annual basis (MoWIE 2013), 0.05 billion US$ lower than our estimate. This discrepancy is likely attributed by the difference per capital water access considered. We considered the average of the lower-middle-income water consumption per capital, whereas the government used local water consumption rate.

To address 100% access and to attain average lower-middle-income consumption level for electricity, (NV2030Scen), the total investment cost of 48.92 billion US$ is required by 2030. This value is higher than the estimated annual cost of 1.4 billion US$ by national electrification program for first phase implementation (MoWIE 2017). The total investment cost of attaining lower-middle-income country (NV2030Scen) is higher than the current utilization trend (BAUScen) by 151%, indicating huge investment.

Several factors, including those that are limiting food availability—domestic food production due to low productivity among others, influence National Food Security. National Food Security cannot be guaranteed with the current mode of production. Land expansion-based agricultural development may induce conflicts with pastoralists and forest ecosystem. By assuming increasing productivity to double in 2030 through agricultural intensification will guarantee food security as well as reduces the burden on both the pasture and forest resources.

The financial requirement for Ethiopia to meet its cost for food security is around 34.9 billion US$ from 2021 to 2030. The cost includes a financial requirement for agricultural intensification and investment in irrigation expansion. The estimated cost only partially includes investment required for inputs of production to double agricultural productivity for the country to transition to sustainable agricultural practices. Additional investment may be needed for overall transition. Improving agricultural productivity will entail significant research and development costs, which have been included in the cumulative research and development costs. Further, the costs of implementing and scaling up agricultural intensification techniques, urban agriculture, agroforestry and horticulture have also not been included. The continuous shrinking of land for agricultural land due to land demand for industries, infrastructure and cities may further increase the cost of food security.

Ethiopia, therefore, needs to significantly enhance investment and capacity building plan over the coming 10 years to attain sustainable water, energy and food security, and ecological integrity through 2050. Delayed investment in water, energy and food resources will likely aggravate socio-economic challenges and destruction of ecosystem integrity. Expanding water, energy infrastructural development and investing in agricultural inputs to double agricultural productivity guarantees national security and ecosystem integrity. Moreover, building institutional and human resource capacity as well as enabling environment has an impact on achieving the national target. The methodology applied in this study is transferable to Nile basin.

Conclusion

This study calculates the finance required in achieving water, energy and food security for Ethiopia. This study has shown that the current level of financing water and energy resources development, as well as agricultural production financing, cannot cover the capital costs of achieving water, energy and food security targets. Compared to the current investment level, the capital investment required to achieve water, energy and food security target is significantly higher than the current level of financing. Ethiopia, therefore, needs to promote a marshal plan type of investment and capacity building plan over the coming 10 years to develop and attain sustainable water, energy and food security and sustains ecological integrity through 2050. Delayed investment in water, energy and food resources may likely aggravate socio-economic challenges and destruction of ecosystem integrity. Expanding water, energy infrastructural development and investing in agricultural inputs to double agricultural productivity guarantees national security and ecosystem integrity. Moreover, building institutional and human resource capacity as well as enabling environment has an impact on achieving the national target. The methodology applied in this study is transferable to Nile basin.

References

Abraham B, Hailu A, Tareke B, Sue E, Biksham G, Ram B, Khadka, Yang SK (2014) The system of crop intensification : reports from the field on improving agricultural production, food security, and resilience to climate change for multiple crops, 1–12

Bizikova L, Roy D, Swanson D, Venema David H, McCandless M (2013) The water-energy-food security nexus: towards a practical planning and decision-support framework for landscape investment and risk management. Int Inst Sustain Dev. http://www.iisd.org/sites/default/files/pdf/2013/wef_nexus_2013.pdf.

Brinckerhoff (2014). Ethiopian power system expansion master plan study, vol 3. Generation planning report 2014. Ethiopian Electric Power Corporation

FAO, IFAD and WFP (2015) Achieving zero hunger; the critical role of investments in social protection and agriculture. Rome, FAO

Guy H, Varughese M (2016) The costs of meeting the 2030 sustainable development goal targets on drinking water, sanitation and hygiene. World Bank Group. Technical paper

Hannah R, Max R (2013) Land use. Published online at ourworldIndata.org. Retrieved from: http://ourworldindata.org/'land-use' [Online Resource] (2019)

Hermann S, Welsch M, Segerstrom ER, Howells M, Young C, Alfstad M, Rogner HH, Steduto P (2012) Climate, Land, Energy And Water (CLEW) interlinkages in Burkina Faso : an analysis of agricultural intensification and bioenergy production 36:245–262

Howells M, Hermann S, Welsch M, Bazilian M, Segerström R, Alfstad T, Gielen D, Rogner H, Fischer G, Velthuizen van H, Wiberg D, Young C, Roehrl Alexander R, Muller A, Steduto P, Ramma I (2013) Integrated Analysis of climate change, land-use, energy and water strategies. Nat Clim Change 3(7):621–626. https://doi.org/10.1038/nclimate1789

Howells M, Rogner HH, Strachan N, Heaps C, Huntington H, Kypreos SH (2011) OSeMOSYS: the open source energy modeling system an introduction to its ethos structure and development. Energy Policy 39(10):5850–5870. https://doi.org/10.1016/j.enpol.2011.06.033

IEA (2014) Energy statistics, the world bank data. www.iea.org/state/index.asp/

IAEA (International Atomic Energy Agency) (2009) Annex VI: seeking sustainable Climate Land Energy and Water(CLEW) strategies. Nucl Technol Rev

Inocencio A, Kikuchi M, Tonosoki M, Maruyama A, Merrey D, Solly H, de Jong I (2007) Costs and Performance of irrigation projects: a comparison of sub-Saharan Africa and other developing regions. Colombo, Srilanka: Int Water Manage Inst 8199. (IWMI) Research Report 109)

Giorgis K (2014) Agricultural and pastoral technologies and practices for climate change adaptation in dryland areas of Ethiopia. In: Integrated Drylands Management in Ethiopia: proceeding of a high-level policy forum workshop, Samara, Afar National Regional State, and UNDP Country Office, Addis Ababa

Kidanewold B, Seleshi Y, Melesse MA (2014) Surface water and groundwater resources of Ethiopia : potentials and challenges of water resources development chapter 6 surface water and groundwater resources of Ethiopia : potentials and challenges of water resources development, no. February. https://doi.org/10.1007/978-3-319-02720-3

Koontanakulvong S (2018) Water security index concept Thailand's water security situation in the context of world and ASEAN, no. August. https://doi.org/10.13140/RG.2.2.23972.32648

MoWIE (2013) የአሥር ዓመት መሪ የልማት ዕቅድ

MoWEI (2017) Light for all, national electrification program. Implementation of the Road map and Financing Prospectus. Ethiopia: Ministry of Water, Irrigation and Electricity

MoFED (2006) A Plan for Accelerated and Sustained Development to End Poverty (PASDEP). Ministry of Finance and Economic Development (MoFED)

MoFED (2010) Growth and Transformation Plan (GTP) 2010/11 to 2014/15.Ministry of Finance and Economic Development (MoFED)

MoFED (2016) Growth and Transformation (GTP) 2014/15 to 2019/20.Ministry of Finance and Economic Development (MoFED)

Moges S (2012) AgWater solutions project case study agricultural use of ground water in Ethiopia: assessment of potential and analysis of economics, policies, constraints and opportunities

Mohamed AA (2017) Food security situation in Ethiopia : a review study. Int J Healthy Econ Policy 2(3):86–96. https://doi.org/10.11648/j.hep.20170203.11

Mondal H, Bryan E, Ringler C, Rosegrant M (2016) Ethiopian power sector development. Renewable based universal electricity access and export strategies. Renew Sustain Energy Rev 75:11–20. https://doi.org/10.1016/j.rser.2016.10.041 (July 2016)

Neil Coles and Philip Hall (2012) Water, energy and food security: technology challenges of thinking in a nexus perspectives centre of excellence for ecohydrology. The University of Western Australia path, Australia

Pachauri S, Ruijve J, Van B, Nagai Y, Riahi K, Vuuren D, Hammond A, Nakicenovic N (2013) Pathways to achieve universal household access to modern energy by 2030 024015. https://doi.org/10.1088/1748-9326/8/2/024015

Parker H, Mosello B, Calow R, Quattri M, Kebede S, Alamirew T (2016) Water strategies for Ethiopia's new development era. A thirsty future, Aug 2016

Rasul G (2014) Food, water, and energy security in South Asia: a nexus perspective from the Hindu Kush Himalayan region☆. Environ Sci Policy 39:35–48

Ringler C, Bhaduri A, Lawford R (2013) The nexus across water, energy, land and food (WELF): potential for improved resource use efficiency? Curr Opin Environ Sustain 5(6):617–624

Rogner H-H (2017) Connecting the resource Nexus to basic urban service provision—with focus on water-energy interactions in New York City. Sustain Urban Areas. https://doi.org/10.1016/j.scs.2017.02.007

Sattler S, Macknick J, Yates D, Lopez A (2012) Linking electricity and water models to assess electricity choices at water-relevant scales 045804. https://doi.org/10.1088/1748-9326/7/4/045804

Sridharan V, Ramos PE, Zepeda E, Boehlert B, Shivakumar A, Taliots C, Howels M (2019) The impact of climate change on crop production in uganda-an integrated system assessment with water and energy implications. Water 11:1805. https://doi.org/10.3390/w11091805

Smajgl A, Ward J (2013) The water-food-energy Nexus in the Mekong Region. Assessing development strategies considering cross-sectoral and transboundary impacts. https://doi.org/10.1007/978-1-46114-6120-3-2

United Nations Development Program (2015) The 2030 agenda for sustainable development. https://www.undp.org/content/undp/en/home/sustainableDevelopment.goals.Html.2015

UN (2015) Transforming our world: the 2030 agenda for sustainable development. Resolution adopted by the general assembly on 25 September 2015. Available from: http://www.un.org/ga/search/view_doc.asp?symbol=A/RES/70/1&Lang=E

UN (2017b) United Nations, department of economic and social afairs, population division. World population prospects: the 2017 Revision, Online Edition. http://creativecommons.org/licensses/by/3.0/igo/

Weirich M (2013) Global resource modelling of the Climate, Land, Energy and Water (CLEW) nexus using the open-source energy modelling system internship report

Welsch M, Hermann S, Howells M, Rogner HH, Young C, Ramma I, Bazilian M, Fischer G, Alfsad T, Gielen D, Le Blanc D, Rohrl A, Steduto P, Muller A (2014) Adding value with CLEWS—modelling the energy system and its interdependencies for Mauritius. Appl Energy 113:1434–1445. https://doi.org/10.1016/j.apenergy.2013.08.083

World Bank (2012) World development indicators 2012. World development indicators. Washington, DC. https://openknowledge.worldbank.org/handle/10986/6014 License: CC BY 3.0 IGO

Chapter 12
Water Management Priorities for Sustainable Socio-economic Development

Muluneh Imru

Abstract This chapter presents real-time water management system operation practices that can be partially adopted for use in Ethiopia with some adaptations to suit realities in the country. State-of-the-art concepts, methods, and procedures based on current practices in the USA, South Florida Water Management District (SFWMD), are illustrated. The historical evolution of the South Florida Water Management System is described. The current state of water management is explained to provide ideas of what can be used in Ethiopia. Objectives of water management in Ethiopia and USA can be similar. The question is if opportunities exist to utilize information on advances in water management concepts, methods, and processes. It is important to identify practices in the developed world that can be adapted to the needs and realities of developing countries. In most countries, water management evolves through similar stages. The initial stages of socioeconomic development result in undesirable consequences such as deforestation, soil erosion, and ecological degradation. When unintended adverse consequences become apparent, societies wake up to nature's call for help. Historically, South Florida was swampy land. From the late nineteenth to the mid-twentieth centuries, Florida swamps were drained to make land available for agriculture and industrial development. Cities were built and agriculture expanded on drained land. Transportation and other industrial infrastructure substantially replaced natural wetlands. Such endeavors were followed by undesirable environmental outcomes including degradation of natural habitat, floods accompanied by loss of human and animal life, as well as recurrent water shortages. In the late 1940s and 1950s, national efforts were geared toward flood control, water supply, and navigation. In the 1970s and subsequent decades, water management policy centered around water quality and ecology stressing the idea of fishable and swimmable rivers. The current water management practice in the USA focusses on ecosystem restoration while continuing to meet other goals including flood control, water supply, navigation, and recreation. Within the context of water management in Ethiopia, the recommendation is to learn from experiences of nations ahead in the water management evolution and to minimize adverse environmental outcomes. Ethiopia can adopt and

M. Imru (✉)
112 Lancaster Way, Royal Palm Beach, FL 33414-4356, USA

A. M. Melesse et al. (eds.), *Nile and Grand Ethiopian Renaissance Dam*, Springer Geography, https://doi.org/10.1007/978-3-030-76437-1_12

enhance integrated water management with a clear focus on food, ecology, energy, and related development endeavors while promoting mitigation of undesirable environmental consequences.

Keywords Real-time water management · Remote operation · Alternative renewable energy · Water management policy · Flood control · Irrigation · Hydropower · Environmental degradation · Ecosystem restoration

Introduction

This chapter presents water management practices Ethiopia can adopt based on current applications in the USA, particularly in the South Florida Water Management District (SFWMD). SFWMD has flood control, water supply, water quality, and natural systems management responsibilities over 16 counties extending from Orlando to the Florida Keys (Fig. 12.1). For the purpose, SFWMD uses various tools and processes in concert to fulfill the needs of data acquisition, transmission, display, processing, and archival. Extensive infrastructure is in place that enables daily water routing strategies and helps achieve defined goals at high levels of effectiveness. Water control stations, associated weather stations, remote data collection units, and communication loops and lines constitute the infrastructure enabling real-time operation of such a large-scale water management system remotely.

Background

Information Infrastructure

The Supervisory Control and Data Acquisition (SCADA) system links various functions and devices between remote field sites and the Operation Control Center (OCC). Typically, the linkage is between the remote data acquisition device, the telemetry transmission network, and data receiving servers, display terminals, and archival databases (Fig. 12.2). The other function of the SCADA system is transmitting operational commands from OCC to remote sites. Commands include opening and closing gates, starting and stopping pumps, and raising and lowering weirs. Commands initiated in OCC are implemented at remote sites spread out over hundreds of miles (Imru and Damisse 2004). Water management decisions are made remotely or in some cases message is passed to onsite staff operating structures or large pump stations.

Data acquired from remote sites get processed and archived in servers in SFWMD headquarters. The process of data acquisition, verification, and archival triggers automatic discharge computation using a software application called

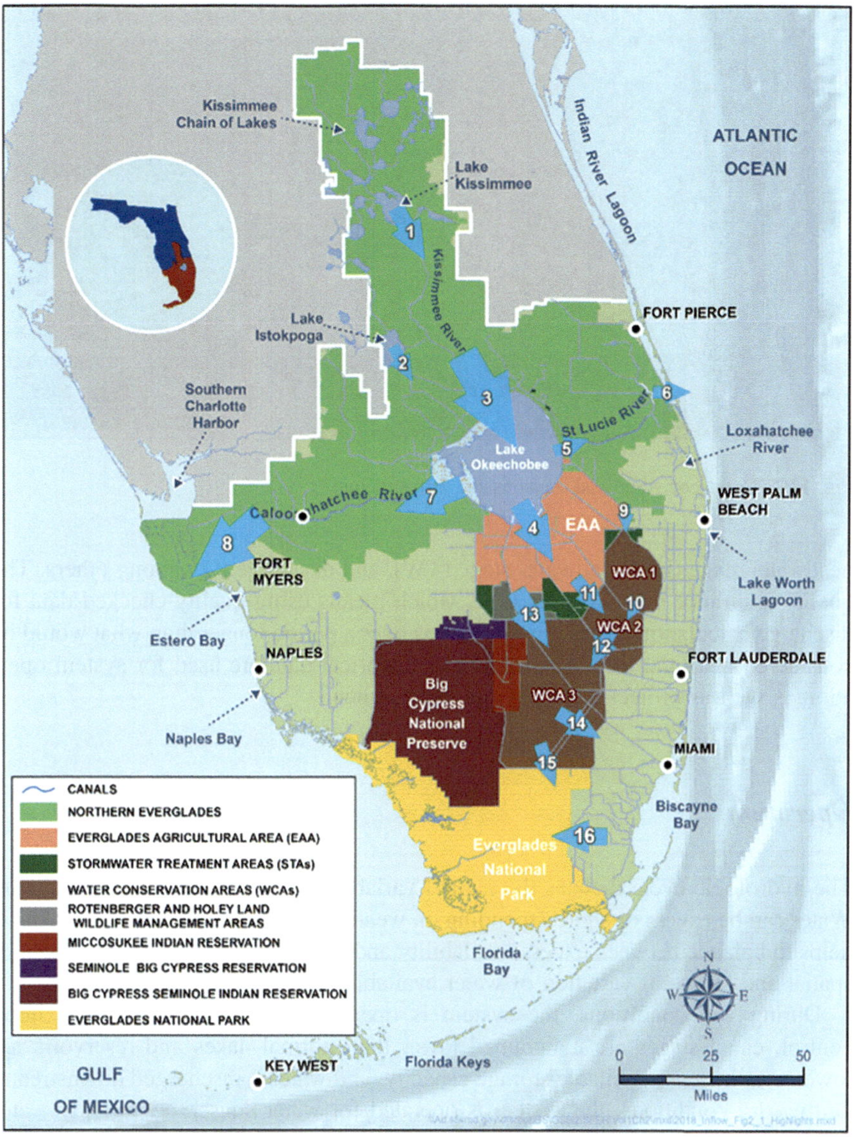

Fig. 12.1 South Florida water management district and major features (South Florida Water Management District 2018)

FLOW that is embedded in the system. FLOW is linked to the corporate hydrometeorological database called DBHYDRO, which is configured to provide static parameters such as structure dimensions and discharge coefficients as needed. Data acquisition, display, and flow calculation can happen in real time. For water management operations, hydrodynamic data available in real time include

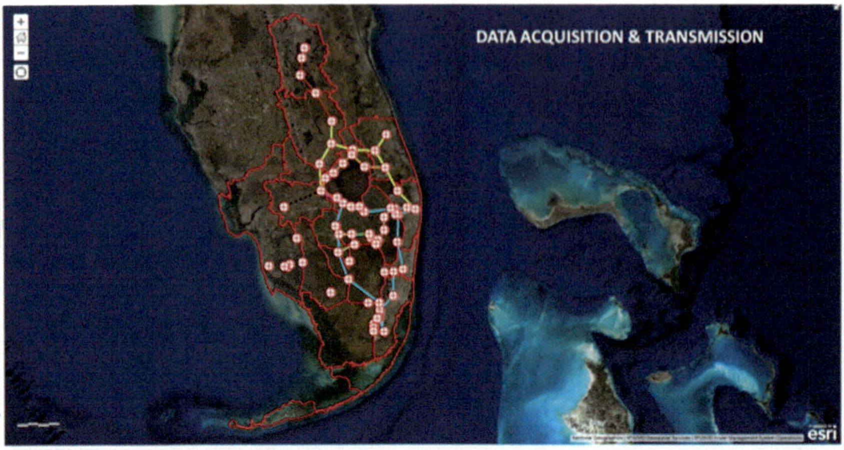

Fig. 12.2 Data acquisition and transmission infrastructure

headwater stage (HW), tailwater stage (TW), and discharge (Q) among others. The quality assurance process takes time, which means using quality-checked data for discharge calculation and archival happens over a period longer than what would be considered near real time. Real-time and historical data are used for system operation as well as project planning and management.

Operation

The hydrologic cycle dictates the natural variability of water flow at any location. Water can be excess or short depending on weather conditions. Human intervention helps to balance between excess availability and shortage of water. Management of spatial and temporal variation of water availability serves various purposes.

During wet conditions, the system is operated for flood control. For flood control, canal stages are maintained lower than normal, lakes and reservoirs are lowered to increase available storage capacity, and water is discharged downstream. During dry conditions, the system is operated for water conservation. Only water needed for downstream use is released while attempts are geared toward conserving fresh water as far upstream as possible. When conditions are dry, stages of lakes, reservoirs, canals, and wetlands are maintained high while trying to provide water supply downstream for agriculture, urban, and ecosystem needs. Efforts are made to keep fresh water inland, maintaining stages higher than the ocean surface water level, always to recharge ground water and protect against saltwater intrusion. Protecting freshwater inland is a critical need. Figure 12.3 shows lakes, shallow reservoirs, and constructed wetlands in the central region of the water management system.

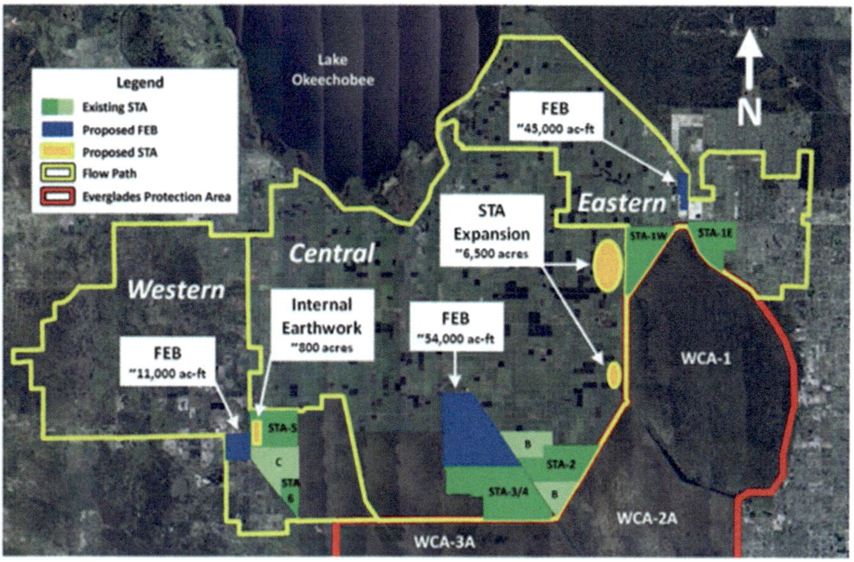

Fig. 12.3 Regional water control infrastructure (central area—partial)

The idea of keeping as much water as far upstream as possible could be considered a particularly good concept for Ethiopia using any means available in the country. Lakes like Tana can be maintained high using a suitable regulation schedule. A regulation schedule could help keep the lake high enough to store extra water for drier times, while simultaneously low enough to keep the beautiful city, Bahr Dar, dry and convenient for its residents and visitors. Water can then be released for groundwater recharge, ecosystem enhancement (in the form of afforestation and soil conservation), agriculture, and urban water use. Results of effective water management include food production in excess of local consumption, well-maintained ecosystem, abundant and healthy marine and other wildlife, adequate urban water supply, and minimized risks of flooding and drought. These attributes account for most of Florida's appeal to residents and visitors. Efficient water management utilizes concepts, processes, and tools effectively in conjunction with environmental data.

In south Florida's case, remote access to instantaneous data has enabled real-time remote operation. This is made possible with the help of the Operation Decision Support System at the OCC where real-time weather conditions and system hydraulic state are displayed on multiple monitors to a 24-h monitoring staff (Fig. 12.4). Decisions are executed mostly via the OCC workstations shown in the figure, while in some cases the decisions are communicated to field staff in remote locations for execution onsite.

Environmental data collection is done at various stations distributed throughout the region. There are large numbers of stations for collecting data on weather, surface and ground water stage, water quality, and ecology. Data from the extensive

Fig. 12.4 South Florida water management district operation decision support system (Sylvester 2015)

network of stations are archived in various specialized databases. Data from weather stations include rainfall, wind speed, temperature, humidity, and solar radiation. Most stage and water quality data collection devices are installed in conjunction with water control sites. Ecological data are generally related to water management areas. Ecological data include vegetation, fish, and wildlife. Water quality is a major aspect of environmental monitoring for the purpose of regulating and managing nutrients and other pollutants. One of the nutrients of concern is phosphorus that adversely affects ecological balance in water bodies and wetlands. To reduce and control nutrients including phosphorus SFWMD uses constructed wetlands called storm water treatment areas, STAs (Fig. 12.3). Information on invasive species is collected and studied, and suitable control measures are implemented.

It is understandable that Ethiopia has hydrology and meteorology data collection networks. It is not obvious how current such information is or if there is a process in place for electronic data collection, transmission, and archival. The country needs an enhanced process for data collection and archival using interconnected, centralized as well as distributed databases. For the purpose, utilizing experiences of developed countries can help. It is likely that there is adequate technology and skill in the country. The question may be how to create a system of coordinated use of the resources for improved water management that can produce desired socio-economic outcomes like in the developed countries. The following sections describe some aspects of data use in water management to meet food production, ecosystem enhancement, and socio-economic development goals.

Hydrometeorological and Topographic Data

Water management heavily relies on hydrometeorology and topographic data. So, data acquisition, transmission, display, processing, and archival are critical functions for water management. Important data include topography, land use, rainfall, hydraulic structure (control structure) dimensions, and water surface elevation (stage) among others. Data on topography, land use, and control structure information are relatively static while water-related data are dynamic. There is variation in time scale of acquisition between various data types. Dynamic data such as water surface elevation require more frequent acquisition and transmission compared to land features or control structure information. Flow can be derived from measured data types and physical parameters of conveyance and control facilities such as canals, weirs, culverts, and pumps.

Remotely located weather stations collect meteorological data including temperature, wind speed, humidity, solar radiation, evaporation, and rainfall. The number of parameters measured varies from one station to another. Some stations have only rain gauges and measure only rainfall. Based on rain gauges, the area covered by SFWMD is divided into rain areas. Areal average rainfall amounts are recorded for each rain area based on readings from several gauges in the area. Radar rainfall estimates are also archived parallel to gauge observations.

Stage records are acquired at representative gauging stations. Water management is characterized by control of stage and discharge. In many cases, stage recorders are associated with locations where stage and discharge are controlled. Such locations have structures including gated culverts, spillways, weirs, and pumps that control stage and discharge (Fig. 12.5).

Data Transmission, Display, and Archival

The Supervisory Control and Data Acquisition (SCADA) system effects acquisition, transmission, and display of data between remote sites and central data servers and workstations. Multiple individual gauges at remote locations are connected to a nearby data polling device called remote terminal unit (RTU). From each RTU, data transmission to central servers occurs via communication loops connecting multiple telemetry towers (Fig. 12.2). The SCADA system also serves to transmit operational changes from the Operations Control Center (OCC) to remote stations where implementation of the commands is initiated and completed within minutes (Fig. 12.4). Data collected from the remote sites is stored in a corporate database, DBHYDRO and made available for public access through the web (Fig. 12.6). Figure 12.7 is an illustration of flow and stage data for a spillway.

Fig. 12.5 Water control structures **a** culvert, **b** spillway, **c** pump station, and **d** weir

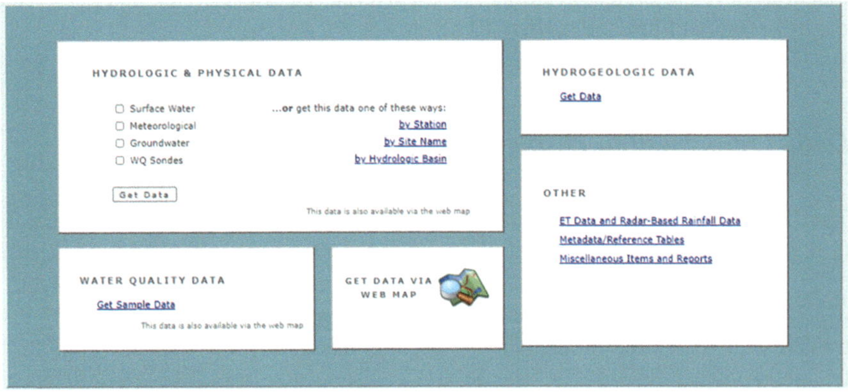

Fig. 12.6 Database display to extract archived data

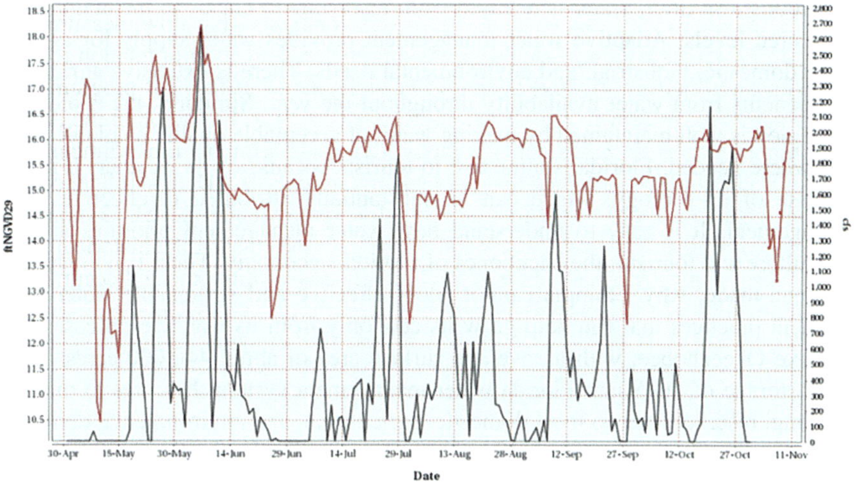

Fig. 12.7 Archive DBHYDRO database daily stage and flow data graphic display for a spillway G302

Water Management Goals

Flood Control

Flood control is a critical water management function that deals with mitigating flood risk. South Florida has a flat topography with coastal areas subject to flooding partly due to high tide. When conditions get wet and flooding is likely due to oncoming storms, operation of water control structures switches to flood control mode. Stages are reduced, in reservoirs to create storage for excess water and in canals to increase conveyance capacity. Reservoirs are operated to store water during wet conditions when there is excess and make controlled releases as needed to meet demand downstream. Preparations for flood control start before a storm arrives in an area. How far ahead the preparation starts depends on the expected intensity and area coverage of the storm. It also depends on level of automation of control structures to be engaged for the operation. If a structure is operable remotely and automated, preparation may be a few hours before the storm, whereas if local manual operation is required preparation can take from one to five days in advance depending on location and accessibility.

Water Supply

In South Florida, to meet water demand, canal stages are kept high enough for users to abstract water efficiently. When water demand is expected to rise, i.e., when conditions turn drier, water is released from reservoirs to canals to maintain stages in desired levels. Adaptive water management provides water supply for agricultural, domestic, industrial, and environmental needs. There is extensive agriculture that benefits from water availability throughout the year. Similarly, the ecology of the region is well maintained by making water of acceptable quality available when and where needed. Florida is attractive to tourists because of its weather. It is also because of its ecology, which can be substantially attributed to effective water management. It is easy to understand how water management priorities support agriculture and tourism, the backbones of Florida's economy. There is no reason, if not man-made, why Ethiopia cannot adopt effective and sustainable water management practices that can help grow its economy from its water resources.

Lake Okeechobee, with an average surface area of about 440,000 acres, is the main storage of the South Florida water management system. It serves to mitigate flooding, reserve water to meet domestic, agricultural, and environmental demands. It is operated based on a regulation schedule, various operational ranges, and guidelines for adaptability (Fig. 12.8).

Water Quality

Clean water requirements, based on human and natural system needs, define water quality standards of discharges going into freshwater bodies as stipulated in the

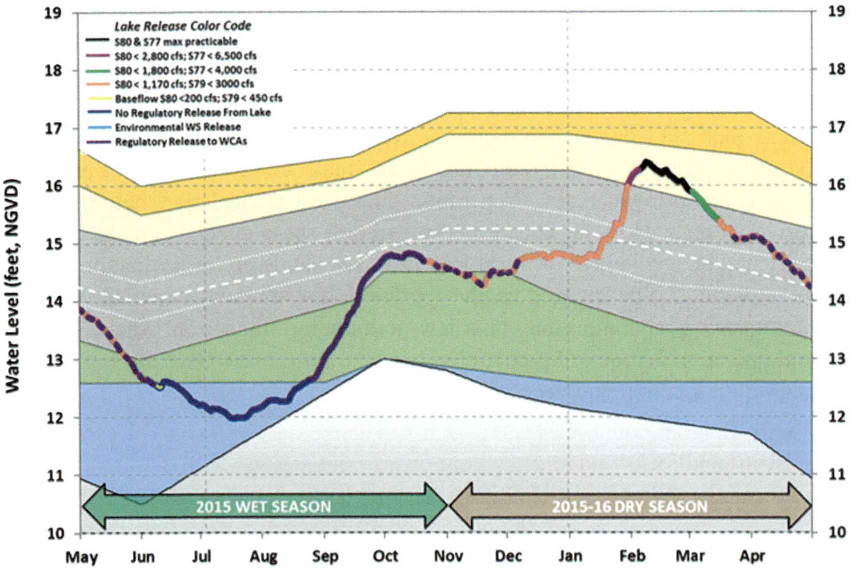

Fig. 12.8 Lake Okeechobee stage hydrograph, operation zones for water management decisions

Clean Water Act (CWA) of 1972. The Act states, "The Clean Water Act (CWA) establishes the basic structure for regulating discharges of pollutants into the waters of the USA and regulating quality standards for surface waters. The basis of the CWA was enacted in 1948 and was called the Federal Water Pollution Control Act, but the Act was significantly reorganized and expanded in 1972." This is something Ethiopia can learn from, adopt, adapt to realities in the country, and implement. A similar act in Ethiopia can help reduce such grave environmental consequences as invasion of Lake Tana by water hyacinth, or unknown contamination of Rift Valley lakes due to industrial discharges from Addis Ababa, Akaki, Debre Zeit, and Nazareth.

Navigation

South Florida's waterways are used mostly for fishing, recreation, and shorter travel for smaller boats from Gulf of Mexico in the west to the Atlantic Ocean in the east. Lake Okeechobee is a major reservoir surrounded by a dyke with gated outlet control structures discharging into major canals leading to the Atlantic Ocean on the east, to the Gulf of Mexico on the west, and to the Everglades on the south. The lake is connected to the ocean by large canals which serve many purposes including navigation. Those large canals are also used for lake stage regulation, urban and rural water supply, recharging aquifers, and protecting against saltwater intrusion among others.

Recreation

Commercial and recreational boating and fishing are widespread in Florida's waters. Boating is regulated as is fishing like other popular activities such as hunting, driving, or aircraft flying. Regulations focus on safety and conservation of human and animal life, as well as protection of the environment. Bird watching is another popular recreational activity many people engage in. The South Florida ecosystem attracts a large variety of avian species including migratory birds. Birds and other wildlife are protected by laws established by acts of congress and bilateral agreements such as the Migratory Birds Treaty Act of 1918 and the Endangered Species Act of 1973.

Enabling Features

Water management objectives are facilitated using important features including reservoirs and lakes for storage, open channels for conveyance, pumps and gravity structures for control of flow and stage.

Lakes and Reservoirs

Storage is a crucial feature of water management. It helps to regulate spatial and temporal variations of water availability. Weather condition dictates rainfall occurrence in space and time. Rainfall in turn influences water availability. In South Florida, like in Ethiopia, the rainy season extends from late May or early June to late September or early October. The remainder of the year is dry. Storage can be facilitated using lakes (natural) or man-made reservoirs. Man-made reservoirs are created by building dams at appropriate locations on water ways. Excess water can be stored in lakes and reservoirs to be released when water availability is lower. An outflow control structure such as a gated culvert or spillway at the outlet of a reservoir can be operated to dictate the amount of water discharged downstream. How much water stays in the reservoir depends on the balance between inflows and outflows. Daily operations determine reservoir stages by varying relative magnitudes of inflow into and outflow from a reservoir. The process of managing stages and flows into and out of the reservoir with time constraint is reservoir operation. The temporal variation of reservoir stage is a regulation schedule. The regulation schedule is used to plan water storage and release depending on weather conditions, flood risks, and water demand following prescribed guidelines. As downstream water demand increases during dry conditions, usually in the dry season, releases are made from the reservoir making more storage available as the rainy season approaches. A typical regulation schedule shows reservoir stage rising through the wet season and falling through the dry season. A reservoir is normally operated to provide maximum available storage space at the end of the dry season and maximum stored volume at the end of the wet season available for gradual release to meet water demand during dry conditions.

Canals and Streams

Conveyance facilities include open canals, natural streams, and closed conduits. Closed conduits can provide for pressurized or free surface flow conditions. Such facilities help to move water from source to points of use.

Stage and Flow Control Means and Methods

Pumps and gravity structures control stages and discharges. Gravity structures include gated culverts, spillways, and weirs. Gate openings determine flow magnitude and stage upstream and downstream of the structure. Flow through control structures can be quantified using standardized equations. Flow through pumps can be quantified by Eq. 12.1 (Imru and Wang 2003, 2005).

$$Q = A\frac{N}{N_o} + BH^C\left(\frac{N_o}{N}\right)^{2C-1} \tag{12.1}$$

where Q is flow rate, N is actual pump speed, N_o is design pump speed, H is head across structure: A, B, C are coefficients and exponent determined through calibration.

Flow over a weir can be determined using the following equation (Imru 2001; Ci and Imru 2006).

$$Q = C_w B_e H^{1.5} \tag{12.2a}$$

where Q is flow rate for a free flow weir, C_w is weir coefficient, B_e is effective weir or spillway length, and H is head across structure. Head across structure is the difference between water level upstream of structure (H_w) and downstream of the structure (T_w). Flow through free flow weirs and spillways can be quantified with Eq. 12.2a and for submerged conditions with Eqs. 12.2b and 12.2c.

For submerged weir, flow can be estimated as follows, using a submergence coefficient (C_s) where H is head water depth above sill crest and h is tail water depth above sill crest, C_w is weir coefficient, and B_e is effective weir/spillway length.

$$Q = C_s C_w B_e H^{1.5} \tag{12.2b}$$

$$C_s = \left(1 - \left(\frac{h}{H}\right)^{1.5}\right)^{0.385} \tag{12.2c}$$

Although culverts do not lend themselves to accurate flow estimation, the following equations have been used to approximate discharge through standard configuration culverts (Eqs. 12.3a–12.3c).

$$Q = C_d A\sqrt{(2gh_i - h_o - h_f)} \tag{12.3a}$$

In Eq. 12.3a, C_d is discharge coefficient, A is flow cross-sectional area in culvert, h_i is head water stage, h_o is tailwater stage, and h_f is friction loss in the culvert barrel.

Using field measurement data h_f can be estimated as follows.

$$h_f = Q\frac{2L}{K_i K_o} \tag{12.3b}$$

where K is conveyance, i and o indicate inlet side and outlet side of barrel, respectively

$$K = \frac{1.486}{n} R^{2/3} A \tag{12.3c}$$

where A is flow cross section, R is hydraulic radius, and n is Manning's roughness coefficient.

Water Management in Ethiopia for Competing Objectives

While there are many problems in Ethiopia, the most depressing attribute which has tarnished the country's image for over four decades is hunger. Ethiopia needs to take food security as its primary objective. Most of the population is engaged in agriculture which relies primarily on rainfall. Sometimes rainfall can be unpredictable in terms of temporal and spatial distribution. That is the reason for farmers not being able to produce enough food at times and millions of people being at risk of hunger every year. There is no alternative to water for agricultural production. Since rainfall is unpredictable, irrigation is critical for sustainable food production and fighting hunger.

While conditions in the country suggest that food production should be the number one priority objective of water management in Ethiopia, attention should also be given to other important purposes including soil conservation, afforestation, sanitation and health, energy production, and related industrial development. Integrated water resources management (IWRM) is possible with food security as the central goal. If a reservoir is built for irrigation as its primary purpose, other purposes including energy production can be achieved using the same reservoir. A multipurpose reservoir in a basin is useful and effective if built as far upstream as possible to achieve the maximum irrigation benefits.

In some cases, energy production and agriculture may be in competition for the same water sources and force making choices. If there is a competition between irrigation and energy production, in Ethiopia's case irrigation should be the overriding priority. In such a situation, other energy alternatives such as wind, solar, and geothermal sources can be considered for energy production. Energy production can adversely affect water availability for irrigation if Ethiopia commits to selling electricity to foreign countries. While integrated water resources management is a widely accepted concept, if water availability dictates making difficult choices, for Ethiopia, the following can be considered a reasonable order of priorities in time.

- Food production (irrigation)
- Afforestation (ecology)
- Soil conservation (ecology)
- Domestic/industrial water supply (rural, urban)
- Hydropower

Hydropower development can be implemented in concert with the other water resources objectives listed above provided power generation facilities are located upstream of the other water consumptive endeavors. This is reasonable considering that hydropower and irrigation can benefit from the same water source if relative locations are carefully selected. An irrigation project needs to be placed downstream of a hydropower facility if in the same water course, because the water that generates energy can be used for irrigation without being reduced by the upstream power plant. Irrigation downstream of a hydropower plant can benefit from flow regulation, i.e., reduction of variation in discharge, which is characteristic of power generation. For integrated water resources management in Ethiopia, priorities can follow temporal and spatial sequencing of land and water projects for beneficial and sustainable socio-economic outcome.

IWRM priorities

Temporal Priorities $\rightarrow$ Food $\rightarrow$ Ecosystem $\rightarrow$ Energy
Spatial Priorities $\rightarrow$ Energy $\rightarrow$ Food $\rightarrow$ Ecosystem (for land with healthy vegetation upstream)
Spatial Priorities $\rightarrow$ Ecosystem $\rightarrow$ Energy $\rightarrow$ Food (for degraded land upstream)

The suggestion here is that food security (irrigation) should be the priority in time. In location (spatial priority), a hydropower generation facility needs to be placed upstream of irrigation and ecosystem projects if in the same watercourse. In some cases, it can be more beneficial or even necessary to have afforestation and erosion control projects upstream of hydropower reservoirs. Erosion control upstream of a reservoir reduces sediment transported into the reservoir improving its service life and maintaining healthy aquatic life.

It is especially important to pay attention to the quality of water flowing into a reservoir or a lake. The invasion of Lake Tana by water hyacinth is an example of severe adverse effects resulting from ignoring what the water flowing into the lake contains. It is apparent that runoff from upstream agricultural land, where a lot of fertilizer was being applied, has been flowing into the lake carrying nutrients with it. This has been going on as long as nutrient loaded water was going into the lake. The nutrients accumulating in the lake promoted the vigorous growth of water hyacinth to the detriment of fish and other native aquatic life.

Invasive plants growing in reservoirs and lakes as a result of nutrient loaded water flowing in are not unique to Lake Tana. It has happened in many parts of the world including in South Florida, USA. In South Florida, scientists have applied mechanical, chemical, and biological means and methods to mitigate the problem. There is literature on the experiences to date, successes, limitations, intended, and unintended consequences of each method employed (Abtew and Dessu 2019).

Lake Tana has another critical water management issue since the building of the Chara-Chara Dam as part of thc Tana-Bcllcs Projcct. Thc Chara-Chara Dam has severely diminished flexibility of managing Lake Tana water levels. Heavy rain in mid-September revealed the problem. At the time of design and construction in the 1980s, the inadequacy of the gates installed at Chara-Chara was suspected and recommendations were made to increase the number of gates. Initially, two outflow gates were planned. During review meetings at the time, the recommendation was to double the outflow capacity by increasing the number of gates from two to four if possible or to three at least. The turnkey contractor for the project resisted and nobody could force that company. During the mid-September 2020 flooding, the residents insisted that the solution was opening the Chara-Chara gates and were making an emotional appeal to visiting government officials at that time. Unfortunately, no one heeded the request of the area residents affected by the flood. A partial relief would have been to increase outflow from Tana via Chara-Chara as well as the hydropower route to Belles. That solution would have possibly reduced the length of time people had to suffer due to their homes and farms being flooded. The long-term solution to avoid similar incidents at the headwater of Tana is to increase the number of outflow gates at Chara-Chara Dam. This solution will also improve the situation of the Tis-Issat Falls. Tis-Issat Falls has been adversely affected by the Tana-Belles Project/Chara-Chara Dam. Its majestic beauty has been destroyed. Concerns were raised at the time of Tana-Belles Project design and construction, which were ignored by the responsible government officials at the time.

Ethiopia's Irrigation and Energy Potential

Irrigation and power potential in Ethiopia, according to a presentation by the Minister of Water, Irrigation and Energy at a workshop on "Prospective Development Plan (2013–2022)", held in the summer of 2020, is shown in Table 12.1. Annual average rainfall was indicated as 840 mm.

In Ethiopia, water plays critical roles in the production of food as well as energy. In most cases, it is possible to utilize water resources for both purposes without adverse effects to either one. However, some water utilization approaches can cause competition between the two objectives. That calls for caution in order to make sure that hydropower generation does not adversely affect irrigation or vice versa.

All medications carry warning labels to make sure users know and understand related side effects. The side effect should not be worse than the ailment the medication is supposed to treat. Similarly, it is important to understand consequences of proposed solutions. GERD is considered a major solution to Ethiopia's energy problem. Caution is warranted considering the possibility that Ethiopia can severely limit its right to use Blue Nile water for food production and ecological objectives. Ethiopia can limit its water use for irrigation if:

Table 12.1 Irrigation and energy potential in Ethiopia

Irrigation	7.5 million ha
Hydropower	45,000 MW
Solar power	5.5 kwh/m^2/day (2 GWH/ha/year)
Geothermal	10,000 MW
Wind (7 m/s, 50 m alt.)	1000 GW+

- any agreement it signs directly or indirectly stipulates that projects upstream of GERD need prior consent of riparian countries (Egypt and Sudan) or if such a precedent is created.
- Over-commits to deliver electricity, generated in the Blue Nile and other transboundary rivers, for export to foreign countries. Overuse of water for domestic power consumption can also strain the ability to develop irrigated agriculture and environmental projects, of course depending on relative locations.

To resolve competition for water between food and energy, it may be a good idea to look at alternatives such as wind, solar, geothermal, and gas for energy production and/or place hydropower facilities upstream of irrigation.

Another important consideration is that Ethiopia should first meet its domestic energy needs before venturing into exporting to foreign countries. It should generate electricity from wind and solar sources, among others, in the order of 40,000 MW or more in combination with the presently operating capacity for the current population, then, if it can produce in excess of that, think of exporting abroad. There is a lot of brainwashing narrative originating from riparian countries that misguides Ethiopia from using its water resources for increased food production to combat hunger. Ethiopia's sustainable economic growth will depend primarily on irrigated agriculture and use of wind, solar, and other renewable energy sources for power generation to support industrial growth. It is time to take the blindfold off and clearly visualize priorities to effectively use water resources for sustainable socio-economic development of the country.

Potential of Great Ethiopian Renaissance Dam (GERD)

GERD is the biggest water resources development project Ethiopia has ever undertaken. It has appealed to the collective psyche and enjoyed a broad support of Ethiopians. Some pros and cons of GERD for Ethiopia and downstream countries are listed below.

Pros for Ethiopia

- Provides a platform for Ethiopians to unite and rally around.
- It has become a symbol of national pride.
- Revealed Ethiopia's ability to use its own resources to undertake such a gigantic project despite strong opposition and lack of support from foreign powers.
- It showed that Ethiopia has no need to yield to unfair demands of foreign countries.
- Brought riparian countries to recognize and negotiate with Ethiopia.
- It created unity between the government and the people as a critical common cause.

Cons for Ethiopia

- Potential conflict between energy production on the one hand and food security and ecological objectives of consumptive use characteristics on the other. Once people in Ethiopia get used to meeting their energy needs from GERD, there is no going back. So, the entire annual flow will be dedicated to energy production leaving no opportunity to use the water source for other poverty alleviation efforts of irrigation in upstream locations.
- Uncertainty of water use agreements with riparian countries and their powerful allies who tend to force terms biased against Ethiopia's water use rights.

Pros for downstream countries

- Additional reservoir upstream at no cost to themselves.
- Regulated flow (discharge variation minimized) at Ethiopia's cost.
- Undesirable sediment load minimized.
- Flood risk reduced.
- All Blue Nile water guaranteed to flow downstream in order to meet Ethiopia's power generation needs.

Cons for downstream countries

- Flow controlled by upstream country, which will be a new uncertain experience.
- Having to recognize and negotiate with Ethiopia unlike in the past.
- Limitation on bilateral 1959 Blue Nile water sharing agreement (Degefu 2003) that had excluded Ethiopia.

A graphic representation of GERD, Fig. 12.9a, (not to scale) below shows its configuration with elevations and corresponding storage volumes. Figure 12.9b shows GERD main dam outlets and operational levels.

The hydropower potential of GERD can be estimated using discharge Q, unit weight of water γ, and static head H_p, with the equation below (12.4). Table 12.2 shows hydropower potential estimates based on Eq. 12.4. Equation for hydropower potential (Roberson et. al. 1997).

$$P = \gamma Q H_p \tag{12.4}$$

Table 12.2 indicates estimates of power generation capacity of GERD for various combinations of water static head in meters (m) and annual volume (in bcm) of Blue Nile at the location of the project. The estimates are calculated for static head ranging from a maximum of 140 m (corresponding to maximum water surface elevation at 640 m) to a minimum static head of 40 m (corresponding to water surface elevation at 540 m). It is not likely to operate at 40 m static head. If it operates at a static head of 140 m, it will be in rare cases. The left side of the table shows installed capacities in megawatts (MW), and on the right side, corresponding annual power generation in gigawatt hours (GWH) are shown. The annual volume used for the estimates ranges

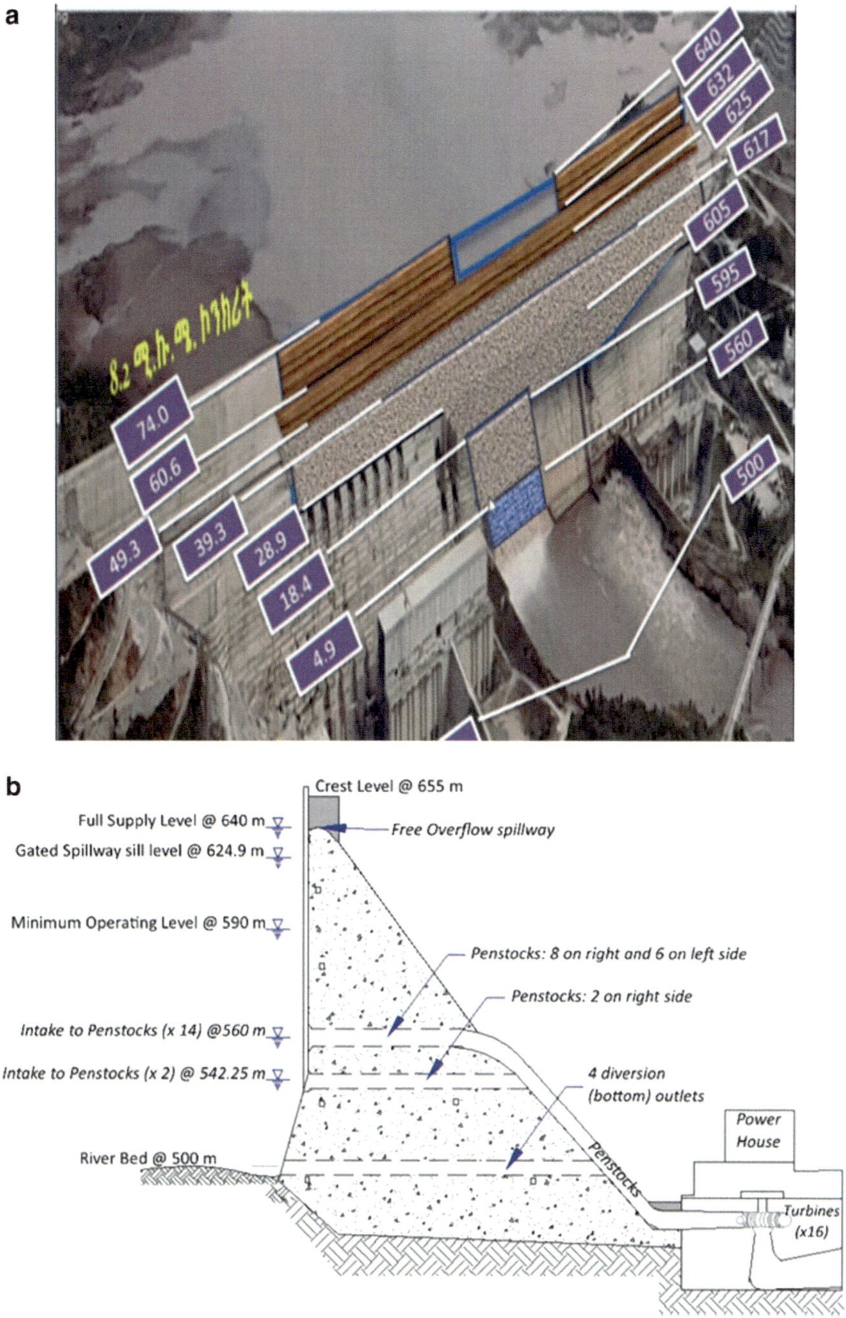

Fig. 12.9 **a** Grand Ethiopian Renaissance Dam volume (billion cubic meter (bcm), left) and corresponding water levels (m a.s.l.) (*Source* Semu Moges). **b** GERD main dam outlets and operational levels (Abtew and Dessu 2019)

Table 12.2 GERD hydropower generating capacity MW (GWH/YR) for Blue Nile annual volumes (bcm)

Static Head (meters)	Blue Nile Annual Volume through GERD (bcm)									
	70	60	50	40	30	70	60	50	40	30
	POWER CAPACITY (MW)					ANNUAL ENERGY (GWH)				
140	3108	2664	2220	1776	1332	27222	23333	19444	15556	11667
130	2886	2473	2061	1649	1237	25278	21667	18056	14444	10833
125	2775	2378	1982	1585	1189	24306	20833	17361	13889	10417
120	2664	2283	1903	1522	1142	23333	20000	16667	13333	10000
110	2442	2093	**1744**	1395	1046	21389	18333	**15278**	12222	9167
100	2220	1903	**1585**	1268	951	19444	16667	**13889**	11111	8333
95	2109	1807	**1506**	1205	904	18472	15833	**13194**	10556	7917
80	1776	1522	1268	1015	761	15556	13333	11111	8889	6667
70	1554	1332	1110	888	666	13611	11667	9722	7778	5833
60	1332	1142	951	761	571	11667	10000	8333	6667	5000
50	1110	951	793	634	476	9722	8333	6944	5556	4167
40	888	761	634	507	381	7778	6667	5556	4444	3333

from 70 bcm (extremely rare) to 30 bcm (possible during droughts). The likely capacity ranges, the annual volume of the Blue Nile at GERD can sustain, are shaded in the table. The shaded values indicate that 50 billion cubic meters of annual volume at a static head of 110 m can sustain about 1700 MW of installed capacity generating about 15,000 GWH/year. Similarly, the same annual volume at 100 m and 95 m static head can sustain 1600 MW and 1500 MW, generating 14000 and 13,000 GWH/year, respectively. The table is populated with installed capacity and annual energy estimates for various combinations of annual volume and static head. Though these are only estimates, they give a good idea of how much energy can be expected if the whole annual volume is dedicated to generating electricity only, with no allowance for water consuming endeavors such as agricultural, environmental, domestic, and industrial water supply. None of the estimates in the table shows that the annual volume of the Blue Nile at GERD can sustain an installed capacity

Table 12.3 Land use by electricity source in Acres/MW produced (*Source* https://wattsupwiththat.com/2017/08/09/the-footprint-of-energy-land-use-of-u-s-electricity-production/)

Electricity source acres per megawatt produced	
Coal	12.21
Natural gas	12.41
Nuclear	12.71
Solar	*43.5*
Wind	70.64
Hydro	315.22

anywhere close to 6000 MW. This calls for viable planning based on realistic expectations paying special attention to land and water needs of various economic development endeavors. Table 12.3 depicts land requirement to generate a megawatt of energy by different power sources.

Integrated water management can serve other goals including flood control, ecosystem enhancement, fishing, recreation, and inland navigation. It can be assumed that Ethiopia's government attempts to plan and achieve such goals. It is not clear, however, if the government is making any plans for flood control. The government's flood control effort, if any, is not visible. There has been a lot of flooding which caused damage to crop and claimed human and animal lives in the past, an example being the 2020 summer Awash River flooding.

Conclusion

Water management planning and implementation determines achievement of national goals including flood control, irrigation, environmental health, water supply, and energy generation.

Ethiopia's image of poverty and hunger needs a lot of effort for enhancement. Government should and can formulate policies that would effectively combat hunger. Policies related to water management, prioritizing irrigated agriculture for increased food production, can significantly reduce hunger and poverty in the country. Ethiopia can learn from relevant experiences of developed nations, such as the USA, to practice sustainable water management practices in its socio-economic development effort.

In cases like GERD where there will soon be competition for water between hydropower and irrigation, energy production should shift focus toward alternative renewable energy based on solar, wind, and geothermal sources. Ethiopia should meet its domestic needs before exporting energy to foreign countries. Ethiopia's desire to export energy, if based on hydropower, will severely diminish its ability to increase food production and to combat hunger.

Flood control needs to be a major goal of water management in Ethiopia. Ethiopia's water resources plan does not appear to give enough attention to flood control as a crucial goal. Ethiopia's economy can be helped to sustain water management priorities and goals by using financial savings that can result from rooting out corruption and illicit financial flows noted in Kukutschka and Martinez (2018).

References

Abtew W, Dessu S (2019) The grand Ethiopian renaissance dam on the Blue Nile. Springer, New York

Ci Y, Imru M (2006) Rating development for culverts under SR9336 in the Everglades National Park, South Florida Water Management District, West Palm Beach, Florida

Degefu G (2003) Historical, legal, and development perspectives. New York

https://wattsupwiththat.com/2017/08/09/the-footprint-of-energy-land-use-of-u-s-electricity-production/

Imru M (2001) Flow monitoring in stormwater treatment area No. 6, Technical publication EMA #399, South Florida Water Management District, West Palm Beach

Imru M, Wang Y (2003) Flow rating analysis procedures for pumps, EMA #413 South Florida water management district. West Palm Beach, Florida

Imru M, Damisse (2004) Real-time flow monitoring in a large-scale water management system, world water & environmental resources congress 2004. Salt Lake City, Utah

Imru M, Wang Y (2005) Quantifying and communicating pumped stormwater flows for real-time flood management. In: International conference on energy, environment and disasters (INCEED 2005), Charlotte, North Carolina

Kukutschka R, Martinez B (2018) Illicit financial flows in Ethiopia, transparency international, u4.no, www.Transparency.org

Migratory Bird Treaty Act of 1918 https://www.fws.gov/laws/lawsdigest/migtrea.html

Roberson JA, Cassidy JJ, Chaudry MH (1997) Hydraulic engineering. Wiley, New York

South Florida Water Management District (2018) The 2018 South Florida environmental report highlights. https://apps.sfwmd.gov/sfwmd/SFER/2018_sfer_final/2018_sfer_highlights.pdf

Sylvester S (2015) Presentation at the UNESCO-IHE modeling lecture series at South Florida water management district. May 26–29, 2015

Water Quality Act of 1972. www.epa.gov/laws-regulations/summary-clean-water-act

Chapter 13
Stage-Based Filling of Grand Ethiopia Renaissance Dam (GERD): Flexible, Adaptive, and Cooperative Approach

Belete Berhanu, Yohannes Gebretsadik, and Seleshi Bekele Awulachew

Abstract The initial filling of the Grand Ethiopian Renaissance Dam (GERD) with its storage capacity of 74 BCM has been the concern of downstream countries since its construction in 2011. The three countries (Egypt, Ethiopia, and Sudan) have been negotiating to establish filling and annual operation rules and guidelines within the umbrella of equitable and reasonable water use without causing significant harm, despite no water-sharing agreement of the riparian countries. They also try to use adaptive, cooperative principles as affirmed by signing the declaration of principles in 2015. Researchers formulated various year-based filling strategies considering different hydrological scenarios—however, the traditional fixed year filling plan, non-flexible by its nature, is unable to tackle the problem. The situation motivated the authors to think and to come up with an innovative concept called stage-based filling. Stage-based filling is flexible, adaptive, and cooperative by its nature. Stages neither refer to the flow conditions nor refer to any water-sharing arrangements. They also do not use a fixed filling year or minimum release and can be accelerated or decelerated based on hydrological variability and cooperation. Based on these concepts, there are five filling stages of GERD; the first stage entirely depends on the construction phase, and the others are more flexible to adapt and cooperate with the changes. Unlike previously proposed filling schedules, GERD's stage-based filling is accepted and adopted by the three countries to solve their concerns. Furthermore, the countries use multiple modeling approaches to evaluate the filling schedule and its impacts according to their perspectives, like using the year of filling, amount of release during filling, energy production, and water level in the reservoir.

B. Berhanu (✉)
Addis Ababa Institute of Technology, Addis Ababa, Ethiopia
e-mail: belete.berhanu@aait.edu.et

Y. Gebretsadik
Nile SECC, Entebbe, Uganda

S. B. Awulachew
Ministry of Water, Irrigation and Energy, MoWIE, Addis Ababa, Ethiopia

A. M. Melesse et al. (eds.), *Nile and Grand Ethiopian Renaissance Dam*,
Springer Geography, https://doi.org/10.1007/978-3-030-76437-1_13

Keywords Stage-based filling · Adaptive GERD filling · Cooperative GERD filling · Flexible GERD filling · GERD · Egypt · Ethiopia · Sudan · Nile river basin · Blue Nile River

Introduction

The "initial reservoir filling" is a deliberate impoundment to meet project purposes and is a continuing process as successively higher pools are attained for flood control projects. It is one of the critical phases of hydroelectric power plant construction. The first filling of the reservoir is always considered the final phase of the construction and the beginning of the discharge structures' operation and the powerhouse (Dalmora and Grube 2019). It is also the first test of the dam and all associated hydro-mechanical equipments to perform the function for which it was designed. The filling rate should be practiced to the extent feasible to allow as much time as needed for a predetermined investigation program, including the observation and analysis of instrumentation data (U.S. Army Corps of Engineers 1979). Moreover, studies indicate that almost two-thirds of all failures and one-half of dam incidents occur during the first filling or the first 5 years of reservoir operation (Foster et al. 1998; FEMA 2015). Therefore, the initial filling of a reservoir is often the primary concern that needs careful planning and management.

Similarly, the first filling of Grand Ethiopian Renaissance Dam (GERD) with its storage capacity of 74 BCM has been a concern of Nile riparian countries, particularly Sudan and Egypt, since the beginning of the dam's construction in 2011. However, the issues of concern were raised quite differently from other historical concerns of high-rise dams, which were more concentrated on the embankment dams' failure (Biswas 2012). The main concern on GERD's first filling is water security (El-Nashar and Elyamany 2018). As of today, there does not exist an all-inclusive, comprehensive water-sharing agreement among the basin countries. Except for the 1959 agreement, which was entierly shared the Nile water for the two downstream countries, Sudan and Egypt. It does not consider the equitable rights of other eight upstream countries, including Ethiopia; the country generates 86% of the Nile of water (Whittington and McClelland 1992; McCartney and Girma 2012).

All the upper riparian countries, particularly Ethiopia, repeatedly have been notified that the 1959 Egypt–Sudan agreement is neither fair nor a base for Nile water issues since its commencement (Kukk and Deese 1996). Tanzania also appeals to it and shows a willingness to discuss and formulate equitable water use rights for all riparian countries (Seaton and Maliti 1976; Mutiti NAB 1976). Moreover, the other riparian countries, Kenya, Rwanda, Burundi, and Uganda endorse Tanzania's stand and discuss among the riparian countries to have an equitable water-sharing agreement (Okoth-Owiro 2004). With all the pressure raised by the upper riparian countries, the Cooperative Framework Agreement (CFA) on the Nile River water had been negotiated by all riparian countries for a decade until Egypt and Sudan dropout at the stage of its signing and signed by other seven countries. Although it did not quantify the "Equitable Water Right," for the riparian countries, the CFA outlines principles, rights, and obligations for cooperative management and development of the Nile Basin water resources and envisages the establishment of a permanent institutional mechanism, the Nile River Basin Commission (NRBC) for its implementation (CFA 2011). Moreover, people quoted the two (GERD's inauguration and CFA) as tools to change the Egyptian hegemony and signal a remarkable rise in Ethiopia's potential to exploit the Nile water (Tawfik and Rawia 2015). Therefore, the first filling of GERD becomes more complicated than other high-rise dams filling. It is not simple technical and construction issues. It has multiple dimensions, including political, legal, and social–economic perspectives.

With all the above concerns, the three countries (Egypt, Ethiopia, and Sudan) signed the declaration of principles (DoP) on GERD in 2015 that promotes equitable and reasonable Nile water uses and cooperation among the parties with all perspectives, including the formulation of the filling and annual operation guideline and rules (DoP 2015). Although the DoP seems to bring a breakthrough for the countries in terms of providing the basis for discussion regarding GERD's filling and operation, it brought its own challenges in interpretations and, consequently, the implementation of the principles to formulate the agreement. Following the signed DoP, the three countries established the Trilateral National Committee (TNC) to implement International Panel of Experts (IPoE) recommended two studies (IPoE 2013). After two years of negotiation for developing the technical proposal, selecting the appropriate consultant, and evaluating the consultant's inception report, the process was stacked again since the three countries cannot establish an agreed baseline for hydrological simulation and environmental impact assessment. It again adds another complication to the technical procedure of the first filling of GERD.

With the direction of the Heads of state meeting on January 27, 2018, the National Independent Scientific Research Group (NISRG) was established by contributing five water resources modelers from each country under the auspicating of the three water affair ministers to come with an innovative solution on the first filling and annual operation of GERD. After two on brief meetings on the idea searing from all the three countries NISRG national members, the Ethiopia NISRG team come up with the idea of stage-based filling in the third meeting with its flexible, cooperative, and adaptive nature, which these chapters focus to share it to the scientific community.

Materials and Methods

Grand Ethiopian Renaissance Dam (GERD)

The GERD is located on the Blue Nile River, which rises in the Ethiopian highlands and joins the White Nile at Khartoum to form the Nile River. Geographically,

GERD is located at 350 12′11.45″ E and 1103′25.79″ N. Its salient features are summarized in Fig. 13.1 and Table 13.1.

GERD is one of the smart reservoirs that use state-of-the-art technologies in its construction. Its safety and quality of construction are approved with different levels of studies. Remarkably, the International Panel of Experts endorsed it as one of the safest dam constructions in this world (IPoE 2013). Its two levels of operation at lower head with two turbines and higher head with other turbines allow generating energy while in the construction phase unlike the global experience

Fig. 13.1 Major physical features of GERD

Table 13.1 Salient features of GERD

Catchment area	172,250 km^2
Mean annual flow 10,000 years return peak flood Probable maximum flood (PMF)	1547 m^3/s 26,860 m^3/s 38,750 m^3/s
Full supply level (FSL) Minimum operating level (MOL) Bottom outlets levels Total storage volume	640 m asl 595 m asl 560 m asl 74.01 Bm^3
Main dam type Maximum height above foundation Crest elevation Crest length	Roller compacted concrete (RCC) dam 145 m 645 m asl 1780 m
Saddle dam type Height Crest length	Rock fill dam with bituminous surface sealing 45 m 4800 m
Total installed capacity	6000 MW
Average annual energy generation	15,692 GWh/yr

different phases for construction and power generation (operation). Its storage capacity also provides additional opportunity to accommodate the maximum historical flow of Blue Nile at GERD construction site, 72 BCM, which can mitigate most of the basin's flood occurrence. Moreover, its contribution to the region's socioeconomic development, such as the provision of sufficient hydroelectric power, changing the recession agriculture in Sudan flood plain to irrigated agriculture, reducing sediment loads on the lower head Sudanese dams, its motivation for significant development in Fishing and recreation, made it a win–win project in the basin.

Data Used

One hundred fifty-five years (1901–2015) flow data of Blue Nile at Sudan border was used as the analysis base. The two hydrological gauge stations, Abaya at Sudan border station from the Ethiopia side and El-Diem gauge station from the Sudanese side, support each other to provide this long-term flow data. The Blue Nile flow is characterized by extreme seasonal and inter-annual variability following the temporal and spatial variability of rainfall in the basin. The annual flow varies from 21.1–71.7 BCM with a long-term mean annual flow of 49.6 BCM at Sudan border station (currently at GERD site). Similarly, the mean monthly flow varies from 142.3 m^3/s in April to 5837.2 m^3/s in August (Fig. 13.2). Typically, four months (July–October) account for 82% of the annual total, and the four low flow months (February–May) contribute only 4% (Awulachew et al. 2008). It contributes about 60% of the annual flow of the Nile measured at Aswan in Egypt.

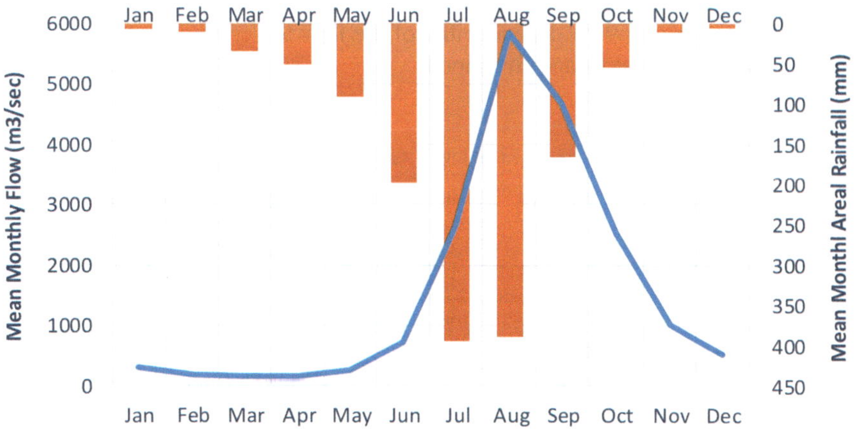

Fig. 13.2 Mean monthly Blue Nile flow and mean monthly areal rainfall at GERD site

Models Used

A multiple modeling approach was applied to evaluate the proposed stage-based filling approach's implication using deterministic and stochastic models. Although there is no agreement among the three countries national experts regarding using one or more selected standard modeling frameworks, each national NISRG team used their choice of models and approach to evaluating the implications of the proposed stage-based filling of GERD.

Results and Discussions

Stages of Filling

The stage-based filling of GERD holds five stages (Table 13.2). The first stage has two phases, and it is highly interrelated with the construction phases. The first phase of the first stage retains 4.9 BCM, and the second phase retains an additional 13.5 BCM of water, and totally at the end of the first stage, 18.5 BCM water is retained behind the reservoir is commonly known as dead storage level. Despite its restriction with the construction process, the first stages also induce a time-based flexibility to accommodate the hydrological changes. The other four stages are more flexible than the first stage in adapting the hydrology change and cooperating with downstream systems. Although the stages were established to address specific target water levels, they are fully flexible to decelerate or accelerate based on the current year's hydrological conditions and water storages in the basin's reservoirs. With multiple of iterative analysis (more than 200 iterations) for different historical and forecasted flow conditions, the retained water after the first stages varies from 0.5 to 24.4 BCM with a 0.57 coefficient of variation. Moreover, despite the needs of the stage-based filling approach, the minimum limit of the potential release is set as 31 BCM, by taking 97% dependability of the flow to describe its sensitivity for downstream environments and livelihoods.

Table 13.2 Guide table for stage-based filling of GERD

Filling stages	GERD target water level (masl)	Retained water (BCM)	Cumulative retained water (BCM)	Ranges of downstream releases based on variable hydrology and stage (BCM)
1	565	4.9	4.9	31–67
	595	13.5	18.4	31–59
2	608	10.5	28.8	31–62
3	617	10.4	39.3	31–62
4	625	10.0	49.3	31–62
5	632	11.3	60.5	31–62

Implications on Filling Years

One of the key features of the stage-based filling, which encapsulates the adaptive and cooperative principles approach, is adaptive to hydrologic conditions, making it better over a fixed year-based strategy. It is essential to show the logical justification and comparative advantage over the "fixed year"-based filling mechanism. The authors have investigated the resilience of the proposed mechanism under different scenarios of future changes both in climate and changes as a result of development. Simulation results for ensembles formulated in combination with historical and forecasted flows, considering climate and development changes, were applied to see the impact on possible ranges of filling years and the probability of occurrences. Accordingly, the mode of iterations gave the filling year other than the first stages varies from 2 to 5 years (Fig. 13.3), which shows the possibility of the entire filling years from 4 to 7 years. The analysis also shows one extreme iteration (condition) that can extend the first filling up to eleven years. The probability analysis also justifies that the filling can be completed within 4–5 years with a 51% chance with a 2.5% chance extending for above 10 years (Fig. 13.4). Generally, the authors believe that the filling years 4–7 are reasonable and consistent with the concept of equitable and reasonable use without causing significant harm. Notably, it compares with High Aswan Dam's filling years, which took six years (1971–1976) (Abtew and Dessu 2019) with its double storage capacity, and the filling year of GERD becomes more rational and equitable.

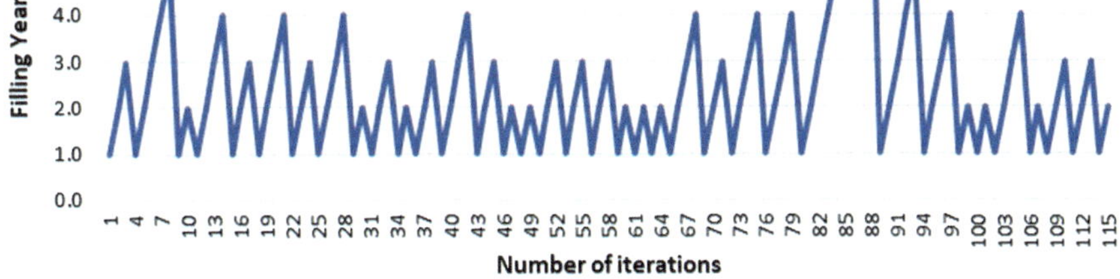

Fig. 13.3 Possible filling years of GERD after the first stage with multiple iteration

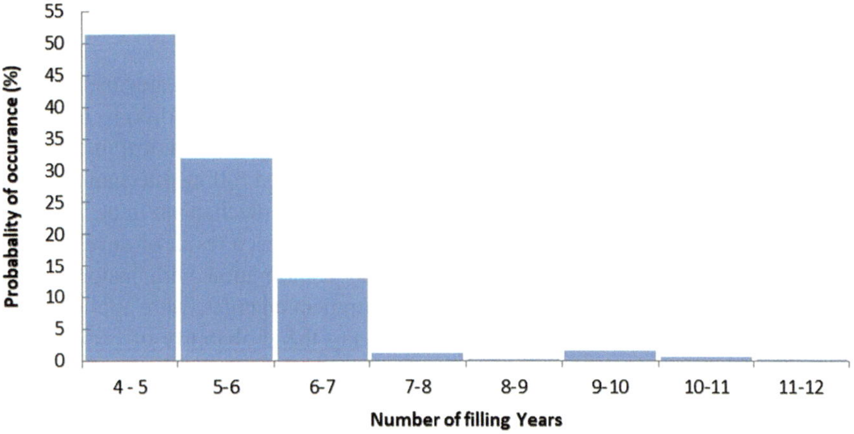

Fig. 13.4 Probability of chance to fill GERD with different filling years

Implication on Energy Production

The hydropower systems' energy production depends on the amount of water released through the turbines and the net head from the reservoir water level. Although filling the dam as quickly as possible allows Ethiopia to achieve a high net head and thus higher energy generation, the adaptive filling approach could, in theory, increase the time of filling (and thus relatively reduced energy generation) if drought conditions occur in the inflow. The tradeoff between the two approaches could be as high as 60%, which Ethiopia can make under the cooperative and adaptive stage-based filling approach (Table 13.3). Moreover, the delay of filling for many years reduces the energy generation, which has financial implications—changing the potential three-year filling period to 4–7 flexible filling years with the acceleration or declaration of stage-based filling lead for the loss of 0.67–2.03 billion USA Dollars, which accounts for up to 30–40% of the investment cost of the project.

Table 13.3 Optimal energy generation in GWH with different reduction factors

	Reduction fraction						
Minimum optimal operating Levels m a.s.l	0%	20%	40%	50%	60%	80%	100%
625	14,462	14,491	14,529	14,546	**14,559**	14,055	12,571

Bold indicates the maximum optimal energy generation possible attend at 60% reduction

Conclusions

This chapter presents the concept of stage-based adaptive and cooperative strategy which can be a useful tool particularly in transboundary setting where comprehensive water-sharing agreement does not exist among the countries such as the Grand Ethiopian Renaissance Dam (GERD) with its storage capacity of 74 BCM.

The initial filling of a large-scale reservoir is often the challenging part of project implementation, and it is even more complicated, as in the case of a transboundary river. Downstream countries can raise a variety of concerns in different dimensions, such as the safety of other downstream water infrastructures and water security as a result of the temporary reduction of flow. In this regard, the root of GERD's first filling concerns is the water use hegemony built by the 1959 Egypt–Sudanese Nile water-sharing agreement without consultation and addressing other upstream riparian countries' rights. Significantly, Ethiopia has objects since 86% of the Nile water contributes from its territory. Without an agreed water-sharing formulation, the three countries (Egypt, Ethiopia, and Sudan) have been negotiating to establish rules and guidelines for initial filling and annual operation within the framework of equitable and reasonable water use without causing significant harm. They also try to use adaptive, cooperative principles, which the traditional fixed year filling approach never able to attain. Therefore, the proposed stage-based filling approach becomes the only alternative to do it.

The authors applied multiple hydrological models to evaluate the framework and its various implications. And justify the rationale of the 4–7 years stage-based filling with acceleration and deceleration ranges of 0.5–24.4 BCM retained water with a multiple iterative deterministic analysis and about 96% probability of occurrences with a stochastic analysis. Similarly, the stage-based filling induces 60% reduction factor and cumulative financial loss of 0.67–2.03 billion USA Dollars from energy generation taken as opportunity costs of its cooperation and flexibility nature. Finally, all parties endorsed the stage-based filling with all its implications as flexible, adaptive, and cooperative.

References

Abtew W, Dessu S (2019) Chapter 7: grand ethiopian renaissance dam reservoir filling. In: The grand Ethiopian renaissance dam on the Blue Nile. Springer, New York. https://doi.org/10.1007/978-3-319-97094-3_7

Awulachew SB, Mccartney M, Steenhuis TS, Ahmed AA (2008) A review of hydrology, sediment, and water resource use in the Blue Nile Basin. International Water Management Institute, Colombo, Sri Lanka

Biswas AK (2012) Impacts of large dams: issues, opportunities, and constraints, chapter 1. In: Tortajada C et al (eds) Impacts of large dams: a global assessment, water resources development, and management. Springer-Verlag Berlin Heidelberg 2012. https://doi.org/10.1007/978-3-642-23571-9_1

CFA (2011) Agreement on the Nile River basin cooperative framework, unpublished online shared document. https://nilebasin.org/documents-publications/30-cooperative-framework-agreement/file. Accessed on 30th Nov 2020

Dalmora CT, Grube R (2019) The first filling of the Laúca hydroelectric power plant reservoir. Int J Hydro 3(3):217–223. https://doi.org/10.15406/ijh.2019.03.00183

DoP (2015) Declaration of principles between The Arab Republic of Egypt, the Federal Democratic Republic of Ethiopia, and the Republic of the Sudan on the Grand Ethiopian Renaissance Dam Project (GERDP). Horn Affairs. 25 March 2015. http://hornaffairs.com/en/2015/03/25/egypt-ethiopia-sudan-agreement-on-declaration-ofprinciples-full-text/

El-Nashar WY, Elyamany AH (2018) Managing risks of the grand Ethiopian renaissance dam on Egypt. Ain Shams Eng J 9:2383–2388. https://doi.org/10.1016/j.asej.2017.06.004

FEMA (2015) Evaluation and monitoring of seepage and internal erosion interagency committee on dam safety (ICODS) FEMA P-1032 /May 2015. https://www.fema.gov/sites/default/files/2020-08/fema_p1032_eval_monitoring_seepage_internal_erosion.pdf. Accessed 13 on Jan 2021

Foster M, Spannagle M, Fell R (1998) Report on the analysis of embankment dam incidents. UNICIV Report No. R-374, School of Civil and Environmental Engineering, The University of New South Wales

International Panel of Experts, IPoE (2013) Final report on Grand Ethiopian Renaissance Dam project (GERDP), 31st May 2015. http://www.internationalrivers.org/files/attachedfiles/international_panel_of_experts_for_ethiopian_renaissance_dam-final_report_1.pdf. Accessed on 1 June 2016

Kukk CL, Deese DA (1996) At the water's edge: regional conflict and cooperation over fresh water UCLA. J Int Law Foreign Affairs 1:21–64

McCartney MP, Girma MM (2012) Evaluating the downstream implications of planned water resource development in the Ethiopian portion of the Blue Nile River. Water Int 37(4):362–379. https://doi.org/10.1080/02508060.2012.706384

Mutiti NAB (1976) State succession to treaties in respect of newly independent African Sates. E.A. L.B, Nairobi/Kampala/Dar es Salaam

Okoth-Owiro A (2004) The Nile treaty: state succession and international treaty commitments: a case study of the Nile water treaties, Konrad Adenauer Stiftung and Law and Policy Research Foundation, Nairobi, Kenya

Seaton, Maliti (1976). Tanzania Treaty Practice

Tawfik A, Rawia M (2015) Revisiting hydro-hegemony from a benefit-sharing perspective: the case of the Grand Ethiopian Renaissance Dam, Discussion Paper, No. 5/2015, ISBN 978–388985–669–2, Deutsches Institut für Entwicklungspolitik (DIE), Bonn

U.S. Army Corps of Engineers (1979) Engineering and design: initial reservoir filling plan. CECW-EH-Y Washington, DC 20314–1000 ETL 1110–2–231, Technical Letter, No. 1110–2–231 30 March 1979

Whittington D, McClelland E (1992) Opportunities for regional and international cooperation in the Nile Basin. Water Int 17(3):144–154. https://doi.org/10.1080/02508069208686134

Chapter 14
The Grand Ethiopian Renaissance Dam: Evaluation of Filling and Operation Plans and Negotiations

Wossenu Abtew

Abstract The history of the Nile water use for agriculture has been associated only with Egypt and Sudan. The upstream countries have mostly managed with rain-fed agriculture with limited consumptive use withdrawal from the Nile River and tributaries. Now, global water stress has increased along with the water demand created by rapid population growth and climate change. According to the World Economic forum, at the 2017 rate, there will be a 40% gap between global water supply and demand by 2030. According to the UN in 2019, 2 billion people live under water stress and 2.2 billion do not have access to safe water. As a result, economically strong countries without freshwater supply have been observed acquiring land and water direct and indirectly for food and fodder production for their growing population. The Nile River basin, especially the Eastern Nile, is currently facing water conflict triggered by Ethiopia's Grand Renaissance Dam, an assertion of upstream water right. The historical uneven social, economic, and military development between Europe and Africa resulted in the colonization and domination of Africa. Through long struggle and sacrifice, colonialism was driven out of Africa. But attempts are being made to keep alive colonial remnant water treaties and agreements by former colonies that benefit from them. Egypt has been aggressively frustrating upstream water projects by blocking funding mechanism and other coercive ways. The Nyerere doctrine on state succession to colonial treaties concluded that the former colonies are not bound by colonial water treaties. By 2011, Ethiopia initiated the self-financed Grand Ethiopian Renaissance Dam (GERD) with storage capacity of 74 billion cubic metre and design power generation of 6000 MW. The dam is located close to Sudan and within three years of completion. Dam filling and operation plan negotiations have sparked conflict mainly between Ethiopia and Egypt. The cause for dragging the negotiations is the implicit position of Egypt that no upstream country has right for a share of the rivers that flow through their territories. Egypt's GERD dam filling and operation proposals have ingenuously embedded minimum water guarantee issues camouflaged

W. Abtew (✉)
Water and Environment Consulting LLC, 6153 Terra Rosa Circle, Boynton Beach, FL 33472, USA
e-mail: wabtew@gmail.com

A. M. Melesse et al. (eds.), *Nile and Grand Ethiopian Renaissance Dam*, Springer Geography, https://doi.org/10.1007/978-3-030-76437-1_14

with droughts and prolonged droughts terms. A series of meetings was conducted between Egypt, Ethiopia, and Sudan to agree upon dam filling and operation plans with mediators and without mediators. The meetings end mostly with Egypt withdrawing when agreement was reported close to finish and the whole process to start all over again with new demands. Sudan is upstream of Egypt and downstream of Ethiopia and initially supporting the project but later was influenced to put challenges to the dam filling. Two of the negotiations on the GERD, the Washington D.C. and the African Union negotiations, are presented with implications of some of the proposals. The Declaration of Principles (DOP) signed on 23 March 2015 after several meetings was the first attempt to move the model of relationship from hostility to cooperation. This chapter analyses global and Nile Basin water stress and put in context the negotiations on GERD. Upstream and downstream transboundary relationships are analysed, and major issues of water share, dam filling, and dam operation are presented. Egypt's extended use of Nile water with trans-basin and trans-continental water transfer to the Toshka Valley and Sinai Peninsula is creating or could create water shortage to the traditional Egyptian farmer, caused by water use policy. Water policy may favour hard cash earning large farms, usually owned by foreign investors, at the expense of the traditional small irrigation farms. Water shortage at the small farms created by Egypt's water policy that favors big farms and transfer of water out of the Nile Basin can be blamed on upstream countries and dams specially on the GERD and Sudan.

Keywords Global water stress · Transboundary rivers · Nile Basin · Blue Nile Basin · Eastern Nile · GERD · GERD negotiations · Egypt · Ethiopia · Sudan · Washington DC · African union · Water transfer · Water share

Introduction

Concern for the prevailing and emerging global freshwater supply shortage to satisfy the needs of the growing global population and the environment is rapidly growing especially in developing countries. In July 2010, the United Nations General Assembly recognized access to water and sanitation as a human right issue. It was recommended that safe, acceptable, and affordable water for personal and domestic use between 50 and 100 l per person per day be available at a cost not more than 3% of household income (United Nations 2010). Unfortunately, rapid population growth and climatic and environmental changes have stressed the limited freshwater supply with regional variation. In light of this, the shared use of transboundary waters presents challenges. The history of the Nile water use for agriculture has been associated only with Egypt and Sudan. The upstream countries have mostly managed with rain-fed agriculture with limited consumptive use withdrawal from the Nile River and tributaries. As population growth pushes demand for water, food and energy, upstream Nile countries have started planning and implementing water projects. The GERD on the Blue, under construction since

2011, is the major upstream project. Egypt and Sudan have raised questions on the dam filling and operation in fear of losing their unfettered use of the Nile water. This chapter presents the history of negotiations, the major proposals, and the intricate link of water sharing and GERD filling and operation plans. The GERD project objective of generating optimal hydroelectric power is dependent on reservoir filling and operation plans.

Global Water Condition

According to the World Economic Forum, at the 2017 rate, there will be a 40% gap between global water supply and demand by 2030 (World Economic Forum 2017). According to the UN in 2019, 2 billion people live under water stress and 2.2 billion do not have access to safe water. Global population growth demands more water for agriculture, industry, domestic use, and the environment. Modelling results show that an additional 1.8 billion people will be under water stress by 2050 (Schlosser et al. 2014). Modelling global annual water demand showed that in 2000 there were 1.7 billion people suffering from moderate to severe water stress. But monthly water demand analysis showed that the number is 40% higher as a result of seasonal distribution of freshwater availability (Wada et al. 2011).

Global water condition has started water conflict in all corners of the globe. From the west, to the middle east, and to the far east, numerous water conflicts occur on daily basis with scales from farmer to farmer, state to state, and country to country. To mention some examples, the Colorado River and the western states of the USA have perpetual water conflict as the river flow is decreasing and water demand is increasing. In the east, Afghanistan's Kamal Khan Dam on the Helmand River upstream of Iran has raised water rights issues with Iran (https://www.lowyinstitute.org/the-interpreter/afghanistan-and-iran-water-treaty-water-dispute). In the west, an example of an acute water conflict is the violent farmers' protest and death in Chihuahua state of Mexico in September 2020. Local farmers took control of the La Boquilla Dam and clashed with Mexican National Guard troops. The La Boquilla Dam is a masonry arch-gravity dam on the Rio Conchos in Chihuahua, Mexico, built in 1910 to provide hydroelectricity, irrigation, and flood control. The clash is caused by the Mexican government action to release water to the USA through the Rio Grande to fulfil its overdue water treaty obligations. By the 1944 water treaty, USA is to send Colorado River water to Mexico and Mexico replenishes the Rio Grande River that irrigates south Texas. Farmers in Chihuahua, a desert state, could not accept the treaty requirements and are complaining that their current president is diverting water to the USA, while their area is in drought. The Los Angeles Times reported that farmer Alejandro Aguilar, 57, said "*we will defend our water until the end. We will not end our fight because this liquid is vital to our culture*" (McDonnell 2020).

The Nile Basin and Water Conflict

The value of water has been increasing steadily as population growth created water demand and increased per capita usage associated with development. A very good summary of global water stress and the transition of water into a commodity is presented by Kacziba and Glied (2020).

> The power factor of water resources hinders the development of effective international cooperation and transforms water into a commodity that can only be controlled by those who have political, military, and economic advantages. The politicisation of water, a global trend which is apparent in all our studies, intensify water stress and increase the possibility of a climate change crisis, which may have larger effects than the current the COVID-19 emergency.

The Nile River basin, especially the Eastern Nile, is currently facing a water conflict triggered by the ongoing construction of Ethiopia's Grand Renaissance Dam (GERD), an assertion of upstream water right. Although several dams are built on the Nile River (Fig. 14.1), but the GERD is being built at a time of highest population and water stress concern in the basin and surrounding regions. The value of water has rose high enough to bring in speculators and out of basin interests. To address rising populations' food and water demand, land and water direct and indirect acquisition has started by out of basin interests (Arafaat and El Nour 2019). Figure 14.2a shows population growth in the three Eastern Nile countries from 1900 on, major infrastructures on the Nile River, and the increase in value of water. The Nile water has also reached commodity status drawing interest from many corners. The intercontinental transfer of the Nile water into Sinai and out of basin transfer to Toshka Valley and Gulf countries interest in large-scale farms to feed their growing population shows the increase in water value. Global land grab in many parts of the world to fulfil faraway growing populations is well documented (Pierce 2012). The Nile water may be auctioned to foreign investors with hard cash. As a result, the traditional Egyptian farmer may lose hydraulic head and available water. The decrease in hydraulic head and water volume may be blamed on upstream countries projects, and the dry small farms could successfully be used for misinformation of Nile water use. In July 2020, Ethiopia completed the first year of GERD filling (4.9 billion cubic metre, bcm). Although it was a very wet year with extreme flooding in the Sudan, Egyptian academic published that the dam filling disrupted drinking water supply systems and hydropower production facilities in Sudan (Helal 2020). The initial filling of the GERD has reduced flood in the Nile River system by 4.9 bcm positively contributing to the 2020 summer Sudanese flood impact reduction.

The rate of population growth in the Nile Basin countries is alarmingly high (Fig. 14.2b). Assuming no human or natural intervention, in the coming 20 years, Nile Basin population will increase by over 275 million. Most of the population depends on land and water resources only. This will result in decline of the Nile River flow increasing water conflict. Although the level of dependence on water varies from country to country, Ethiopia, the source of most of the Nile flow is

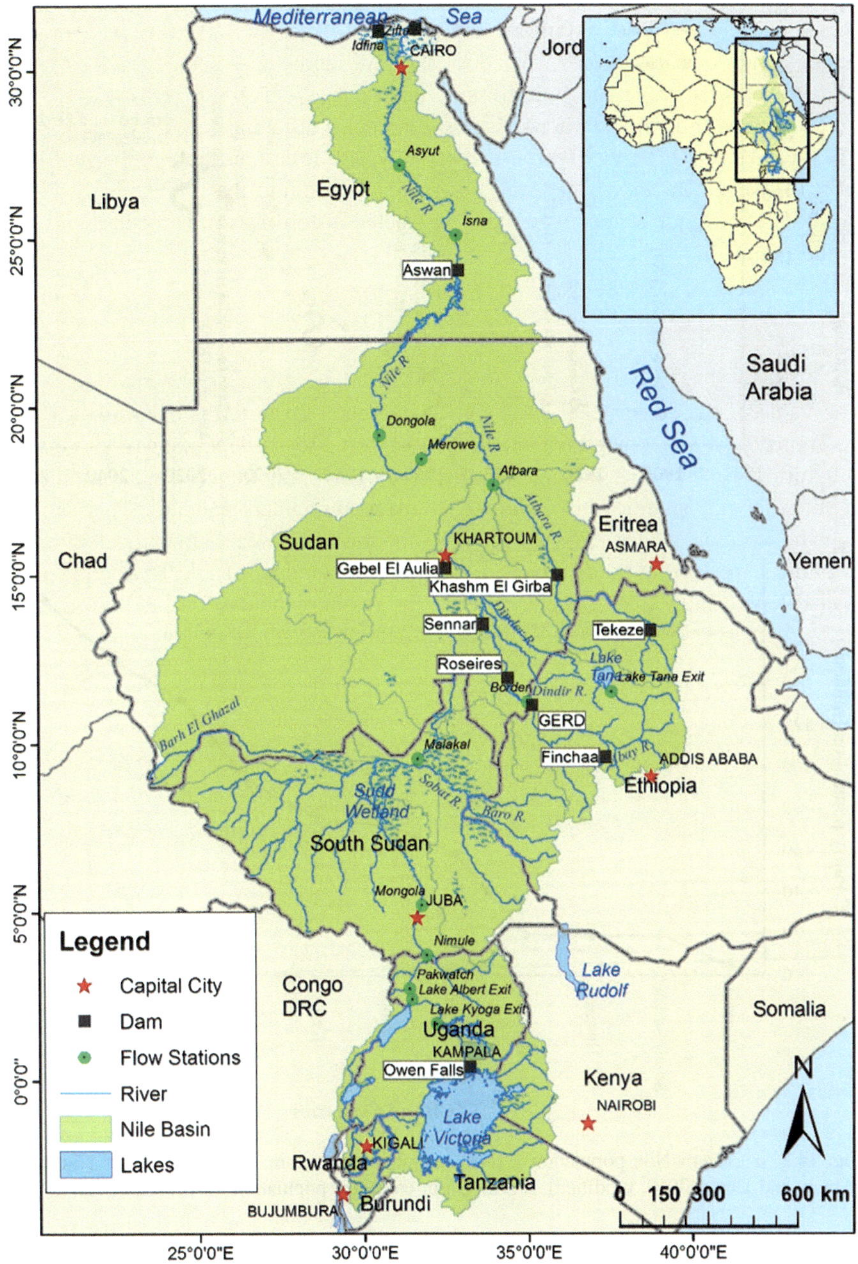

Fig. 14.1 Major dams on the Nile River and the location of the GERD (Abtew and Dessu 2019)

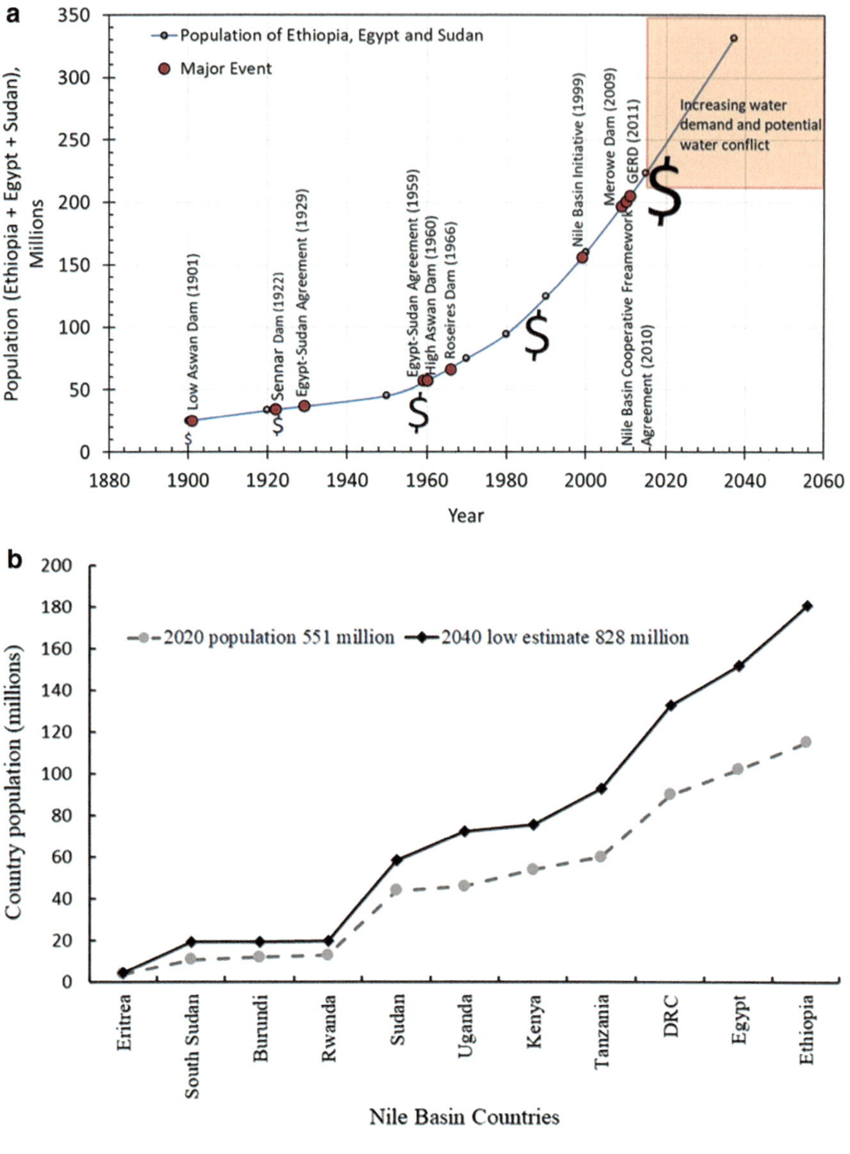

Fig. 14.2 **a** Eastern Nile population growth, major infrastructure, and the value of Nile water (Abtew and Dessu 2019, modified), **b** Nile Basin countries population growth, 2020–2040

totally dependent on land and water resources. Ethiopia's loss of access to the sea resulted in limits to economic growth and access to additional food sources, reduced cost of import and export, forcing it to exclusive dependence on soil and water resources. Ethiopia's contributions to Nile flow will progressively decrease.

Meanwhile, water stress in upstream countries has forced them to assert water rights and start dispersed and point abstraction of Nile water. Institutional developments include the Nile Basin Initiative in 1999, Cooperative Framework Agreement in 2010, Lake Victoria Basin Commission, and other institutions including small farmers water use associations. Structural changes in the basin include rain harvesting, small-scale runoff storage, dams on tributary rivers, and the GERD, a major dam. Although total volume of water may not be high, there are reports showing Blue Nile tributary streams drying during the dry season from gravity and pumping for small farm irrigation. Water use conflicts do arise from upstream and downstream users of stream diversions with new conflict resolution local institutions arising (Deneke 2014). The hydropower dam, The Grand Ethiopian Renaissance Dam, is the biggest of the upstream projects (Abtew and Dessu 2019).

The Grand Ethiopian Renaissance Dam

The GERD is the first major dam on the Blue Nile (Abay) River in Ethiopia (Fig. 14.3). It is the biggest project in the upstream Nile Basin that demonstrates upstream Nile water control and water use rights. It is located about 20 kms from Sudan border. The dam is a roller compacted concrete dam with maximum height of 155 m and width of 1800 m. On the side, a 5-km saddle dam of 50 m height is built to raise a lower land feature to the desired elevation. GERD outlets and operating levels are shown in Fig. 14.4. The dam is currently 75% completed. After nine years of unsuccessful negotiations to agree on dam filling plan, the dam filled in July 2020 to water surface elevation of 560 m a.s.l. with 4.9 bcm storage. By the design of the dam, the height of the main dam at the centre is raised during the dry season (November through May), while dry season flow is passed through the bottom outlets. When the wet season starts in June, flow rates are likely higher than the capacity of the bottom outlets leaving no choice but filling the dam to its current level and start overflowing. Since the inception of the GERD, Egypt has raised concern on dam filling and operation using its current water use status as a reference. In the following section, models of relationships between upstream and downstream countries in a transboundary basin are presented followed by the chronology of major negotiations and implications of respective proposals.

Downstream and Upstream Countries Relationship in Sharing Transboundary Waters

International Watercourses (General Assembly of the United Nations on 21 May 1997; entered into force on 17 August 2014) are the reference norm mostly used on transboundary river water sharing issues. As eloquently put in Article 5,

Fig. 14.3 Grand Ethiopian Renaissance Dam (Awulachew 2017)

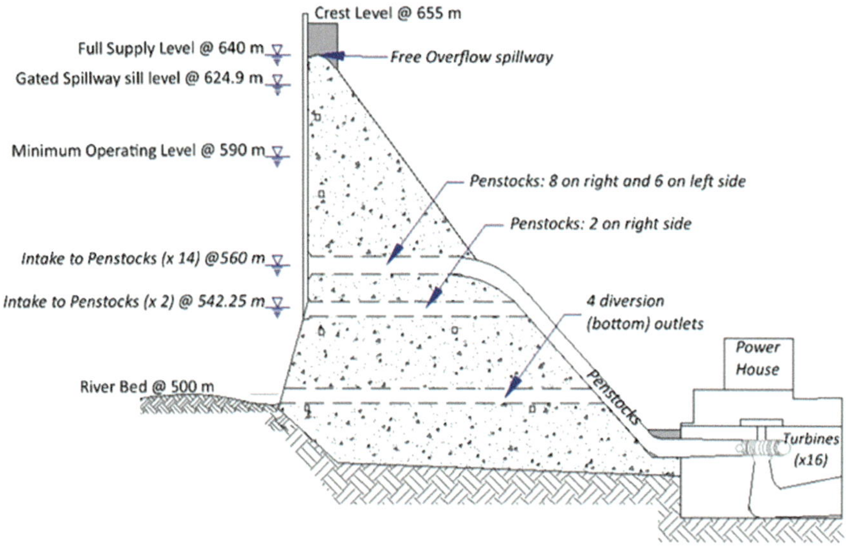

Fig. 14.4 GERD main dam outlets and operational levels (Abtew and Dessu 2019)

> Article 5, 1 "*Watercourse States shall in their respective territories utilize an international watercourse in an equitable and reasonable manner. In particular, an international watercourse shall be used and developed by watercourse States with a view to attaining optimal and sustainable utilization thereof and benefits therefrom, taking into account the interests of the watercourse States concerned, consistent with adequate protection of the watercourse*".

The premise of this noble concept is that "*Affirming the importance of international cooperation and good neighborliness in this field*" as stated in the introduction of the 1997 UN Convention. But, from greed and threat of current and future freshwater scarcity, upstream and downstream countries in some basins decide not to cooperate. In the Nile River Basin, there are 11 countries which can be classified as shown in Table 14.1.

In transboundary basins, there are two models of relationship between upstream and downstream countries, hostility or cooperation (Fig. 14.5). If upstream and downstream countries do not wish to cooperate but deal with hostility, there is the potential for water conflict. Hostile relationship includes undermining upstream security and peace to discourage economic and military growth. Blocking of international funding for water projects has been commonly practiced in the Nile Basin. Hostility also includes refusal to share transboundary water. Since the 1800s, Egypt had unsuccessful military expeditions to Ethiopia, threats, and destabilizations efforts have been recorded (Chesire 2010). This model of relationship currently is displayed by the endless negotiations on GERD filling and operation plans without any resolution. Egypt is not willing to any equitable share of Nile water and put great effort to sustain the status quo by internationalizing the water conflict and inviting out of basin entities to create undue pressure on Ethiopia. In a hostility model, there is no trust, and no agreement can be reached even when one side cooperates. The second model of relationship is cooperation. Examples of successful cooperation on a transboundary water and benefit sharing include Brazil and

Table 14.1 Nile Basin countries and, relative location along the Nile, and percentage of area in basin

Country	Relative location in basin	Percentage of country area in Nile Basin (FAO 2000)
Burundi	Upstream	46
Democratic Republic of Congo	Upstream	1
Egypt	Downstream	33
Eritrea	Upstream	21
Ethiopia	Upstream	32
Kenya	Upstream	9
Rwanda	Upstream	83
South Sudan	Upstream/Downstream	78
Sudan	Upstream/Downstream	
Tanzania	Upstream	13
Uganda	Upstream/Downstream	98

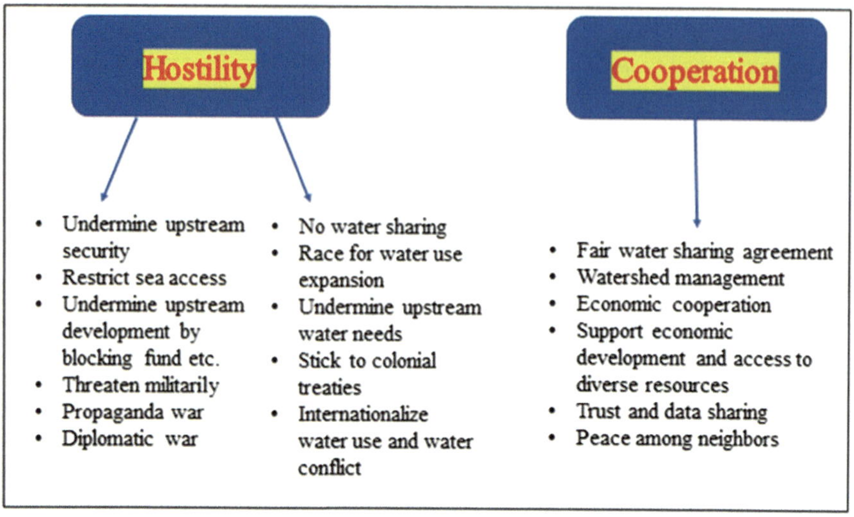

Fig. 14.5 Models of downstream and upstream countries relationship in the Nile Basin

Paraguay on the Parana River, USA and Canada on the Columbia River, and Guinea, Mali, Mauritania, and Senegal on the Senegal River.

The relationship on the GERD filling and operation plan combines both hostility and cooperation models. Ethiopia's dam site selection close to the border of Sudan and far from large track of irrigable lands should give confidence that the dam is a hydropower dam. Ethiopia also provided all necessary design reports of GERD to Egypt and Sudan. It agreed for the establishment of international panel of experts (IPoE), which presented their findings and recommendations which Ethiopia has been implementing. It agreed for the establishment of tripartite technical committee as recommended by IPOE to address downstream concerns. From the series of negotiations, one can conclude that Ethiopia follows the cooperation model and has bent back to accommodate downstream concerns. Egypt on the other hand follows the hostile model as evaluated from historical records from 1800s to current including military invasions in the late 1800s and instigating and supporting insurrections in Ethiopia (Ram 2009; Chesire 2010). The policy of interference in Ethiopia's internal affairs and destabilization, blocking international funds for water projects, military threats, propaganda war, and diplomatic pressure on Ethiopia is caused by Nile water (Hassen and Al Rasheedy 2007; Hassen 2017). Egypt's hostility model is based on objective of not sharing Nile water with upstream countries except Sudan. Egypt's failure to join the Cooperative Framework Agreement of 2010 is clear indication for not choosing the cooperation road. Sudan is both upstream and downstream country in the Nile Basin (Table 14.1). Its stand has been changing from time to time, with first supporting the GERD project and later siding with Egypt apparently from pressure from multiple corners. Sudan is the most to benefit from the GERD from flood control, regulated irrigation water supply, and sediment reduction that extends its dams life. As global water stress grows, the biggest Nile water conflict will be between Egypt and Sudan as all Nile water passes through Sudan to reach Egypt and the two countries are bordering. Egypt will pressure Sudan to reduce its water use once the GERD negotiations get some resolution or no resolution. Through long struggle and sacrifice, colonialism was driven out of Africa. But attempts are being made to keep alive colonial remnant water treaties and agreements by former colonies that benefit from them. Egypt has been aggressively frustrating upstream water projects by blocking funding mechanism and other coercive ways. The basis of Egypt's Nile water domination is based on colonial treaties and agreements which should not have relevance at this age. Upstream countries have stated that they are not bound by colonial power treaties. The Nyerere doctrine on state succession to colonial treaties concluded that the former colonies are not bound by colonial water treaties (Okidi 1982).

Negotiations Between Ethiopia, Egypt, and Sudan on GERD Filling and Operation

Since the inception of the GERD in 2011, negotiations between Ethiopia, Egypt, and Sudan have been undergoing on GERD design, filling, and operation. Ethiopia agreed on the formation of the international panel of experts (IPoE) to review the dam design with intention of building confidence between the three countries. The IPoE, after reviewing design documents and visiting the dam site, produced a report on 30 May 2013 with certain recommendations (IPoE 2013). Two experts from each country and four international experts formed the IPoE (Table 14.2).

Series of meetings were held between Ethiopia and Egypt and Sudan from 2013 to 2015 mostly on the implementation of the IPoE recommendations and

Table 14.2 International panel of experts

Origin	Name	Expert
Egypt	Dr. Sherif Mohamady Elsayed	
Egypt	Dr. Khaled Ahmed	
Ethiopia	Eng. Gedion Asfaw	
Ethiopia	Dr. Yilma Seleshi	
Sudan	Dr. Ahmed Eltayeb Ahmed	
Sudan	Eng. Deyab Hussein Deyab	
International	Dr. Bernard Yon	Environmental expert
International	Mr. John D. M. Roe	Socio-economic expert
International	Mr. Egon Failer	Dam engineering expert
International	Mr. Thinus Basson	Water resources and hydrological modeling expert

culminated on signing of the Declaration of Principles (DOP) on 23 March 2015. The DOP is the first attempt to move the model of relationship from hostility to cooperation. The ten Articles in the DOP are indicators:

1. Principles of cooperation
2. Principle of development, regional integration, and sustainability
3. Principle not to cause significant harm
4. Principle of equitable and reasonable utilization
5. Principle to cooperate in the first filling and operation of the dam
6. Principle of confidence building
7. Principle of exchange of information and data
8. Principle of dam safety
9. Principle of sovereignty and territorial integrity
10. Principle of peaceful settlement of dispute.

Since the signing of the DOP, Egypt has continuously shown lack of will to cooperate by continuously changing positions, withdrawing from negotiations, rejoining and attempting to put Ethiopia in a trap to sign obligations that will maintain its current water hegemony. On 15 May 2018, the three countries agreed to form an independent scientific group of 15 to evaluate GERD impact on Nile flow and various scenarios of GERD filling. Continuing meetings of the three countries were held from capital to capital reaching a critical level when GERD negotiations moved to Washington D.C. in November 2019, to continue negotiations under the sponsorship of the USA Treasury Department and the World Bank. The goal was to sign agreement after four monthly meetings with implicit promise of Ethiopia getting fund. The negotiations should have been based on Ethiopia's dam filling and operation plan presentation and downstream countries presenting their concerns. Although the official agenda was GERD filling and operation, but in reality, there were two agendas. The first unstated agenda being securing Egypt's water share, guarante minimum flow, under the disguise of dam filling and operation with assumed drought conditions. The second being a stated agenda of GERD filling and operation plans. A high-speed and high-pressure negotiations were conducted in four meetings between November 2019 and February 2020. It is clear that the draft presented for discussion in Washington D.C. was produced by Egypt after intensive hydrologic and dam operation modeling as demonstrated by details of the plan and also later submitted by Egypt to UNSC. By the end of February 2019, the fourth meeting, Ethiopia realized that only Egypt's interests were being promoted with the mediators taking Egypt's side and returned back to Ethiopia excusing itself from the negotiations for consultation reasons. Egypt by itself signed the draft while Sudan refrained. Egypt's proposal at the D.C. meeting is discussed in a later section. On February 28, 2020, the US Treasury Department released a statement on the D.C. negotiations failure and stated that Ethiopia should not start filling the GERD before the three countries come into agreement. The first-year filling of 4.9 bcm resulted later in July 2020 before agreement is reached dictated by the design and construction phase of the dam. In March 2020, COVID-19

(Coronavirus) pandemic became rampant globally, and the USA and every country were consumed by the alarming spread and casualty from the virus making everything else secondary. Ethiopia's temporary relief from US pressure may be at least partly attributed to the plague.

In June 2020, Egypt took the GERD conflict to the Arab League where the League showed support for Egypt and Sudan on Nile water issues with Resolution 8524 of 24 June 2020. The Arab League stressed that Egypt and Sudan's water security is part of Arab national security. In the same month, Egypt took the GERD issue to the United Nations Security Council. All three countries wrote letters to the Security Council stating their positions. Ethiopia's position was that this matter does not belong in the Security Council. Egypt's letter reinstated the D.C. proposal that Ethiopia rejected and made the matter an international security concern. On 29 June 2020, the Security Council heard from the three countries and others on the matters of the Nile and GERD with no resolution, and the matter was pushed to the African Union (AU). The AU has already expressed its interest on mediating the issue in Africa under its auspices. Outcome of AU led negotiations is presented in a later section.

Washington D.C. Negotiation and Egypt's Proposal

The D.C. plan with Annexes A, B, C, D (Permanent Mission of the Arab Republic of Egypt to the United Nations 2020) makes GERD a water supply dam to fulfil the interests of Egypt with a maximum water level of 625 m a.s.l than the design 640 m a.s.l. The D.C. proposal mainly is centred on minimum quota (guaranteed flow) of the Blue Nile flow out of the GERD during filling and normal operation with drought terms not commonly used—drought, prolonged drought, and prolonged dry years. The main feature of the proposal is minimum GERD guaranteed outflow with release compensation scheme from the reservoir when inflows are below 37, 39, and 40 bcm (Table 14.3). Under the so-called drought conditions, the reservoir must release an annual minimum of prescribed flow dependent upon the inflow amount and the reservoir stage during filling. During filling and normal operation, for inflow between 20 bcm and 37 bcm, additional water must be released from storage (Annex A, Fig. 14.6). As shown in Fig. 14.6, when inflow to the GERD is 20 bcm at water level of 618 m a.s.l., 29.3 bcm must be released with 9.3 bcm added from GERD storage. Additional release from the reservoir pushes water level down to below 608 m a.s.l when evaporation and seepage losses are accounted. Even more is demanded when the 4-year average GERD inflow is below 40 bcm during "prolonged dry years"; 50% of GERD storage above 603 m a.s.l. is ought to be released for the next four years regardless of flow conditions during those four years. In the dam operation phase, annual minimum outflow from GERD of 37 bcm during drought, minimum of 4-year average of 39 bcm during prolonged drought, and 5-year average of 40 bcm during prolonged period of dry years are obligations. A spreadsheet (Annex A) that determines the amount of annual release and release from storage, based on reservoir water level and annual inflow into the

Table 14.3 Egypt's minimum flow guarantee during GERD filling and operation in the D.C. plan

Drought	Filling		Long-term operation	
	Inflow (Q_i)	Minimum release (Q_r)	Inflow (Qi)	Minimum release (Q_r)
Drought (annual)	$Q_i < 37$ bcm	$Q_r = Q_i$ + From Reservoir (Annex A)	$Q_i < 37$ bcm	$Q_r = Q_i$ + From Reservoir (Annex A)
Prolonged drought (4-yr average)	Q_i (4-yr Avg.) < 37 bcm	$Q_r = Q_i$ + 62.5% of storage above 603 m a.s.l., the following 4 years*	Q_i (4-yr Avg.) < 39 bcm	$Q_r = Q_i$ + 100% of storage above 603 m a.s.l., the following 4 years*
Prolonged period of dry years	Q_i (4-yr Avg.) < 40 bcm	$Q_r = Q_i$ + 50% of storage above 603 m a.s.l., the following 4 years	Q_i (5-yr Avg.) < 40 bcm	$Q_r = Q_i$ + 100% of storage above 603 m a.s.l., the following 5 years**

Q_i = inflow into reservoir, and Q_r is reservoir release in a year (July 1 to June 30)
*minimum annual release in a year from reservoir is 1/8 of total
**minimum annual release in a year from reservoir is 1/10 of total
Releases from reservoir are obligated even if year is wet
Releases from drought, prolonged drought, and prolonged dry periods are additive
Annex A (Fig. 14.6)

GERD, was provided for the filling period, Annex A (Fig. 14.6). Annex C (Figs. 14.7, 14.8) is prescribed releases during filling for non-drought flows based on GERD water level and annual inflows. For GERD inflows of 37–70 bcm, release requirements during filling are depicted in Fig. 14.7, and retention in reservoir without considering evaporation and seepage losses is shown in Fig. 14.8. The maximum amount of water that the GERD can retain for filling is 17.9 bcm at the maximum inflow of 70 bcm and the low stage of 595 m a.s.l (Fig. 14.8). Following year, if flows fall below 37 bcm for a year, reservoir levels will be forced to go down due to releases, and an estimated 2 bcm loss due to evaporation and seepage is shown later. As shown in Figs. 14.7 and 14.8, when the most extreme high inflow of 70 bcm occurs at 623 m a.s.l. water level, the prescribed release is 67.5 bcm. With 2 bcm losses due to evaporation and seepage, the net retention is half bcm. If the reservoir is at 625 m a.s.l, with extreme high inflow of 70 bcm, no water is retained. This practically limits the highest water level far below the design level of 640 m a.s.l.

The hydrologic year is defined as July 1 to June 30 where annual inflows into the GERD are determined. The proposal has a revision time of 10 years, and the term "flow" is defined as total volume of water entering the GERD. If a drought condition is followed by another drought condition, the releases from storage are additive. The drought classifications are based on inflows to the GERD. Normally, droughts are classified by rainfall in the basin. When drought is classified on inflow to the GERD, Ethiopia's water share and upstream withdrawals are considered as drought with penalty to drain from the reservoir. The effect of this plan in limiting the GERD from fulfilling its objective of optimal hydropower generation is discussed in a later section.

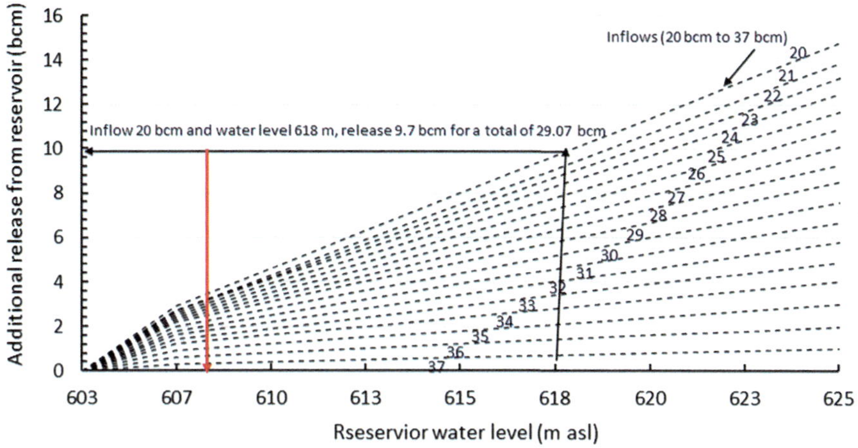

Fig. 14.6 Additional release from reservoir for inflows into the GERD from 20 to 37 bcm at different water levels (Data source: Annex A)

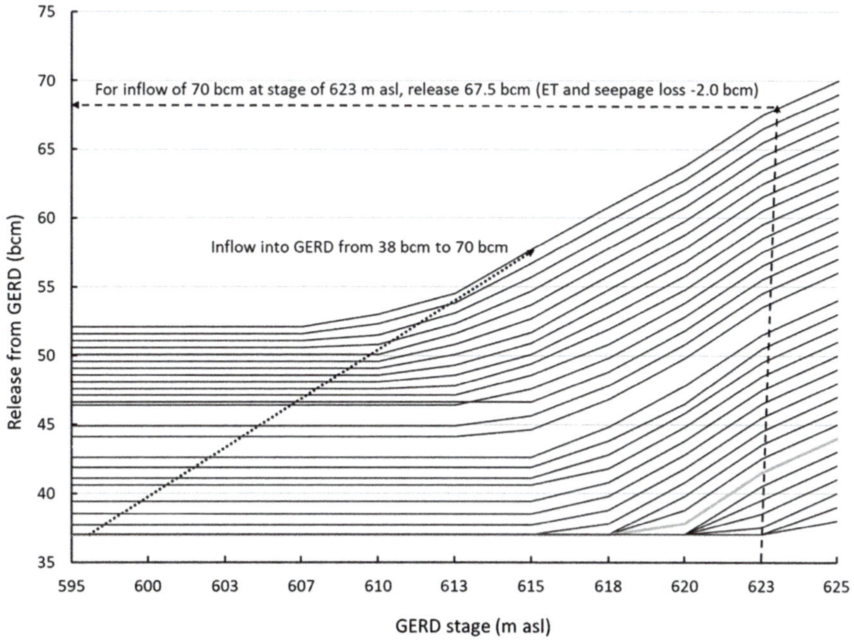

Fig. 14.7 GERD inflows and prescribed releases during filling at given water levels (Annex C)

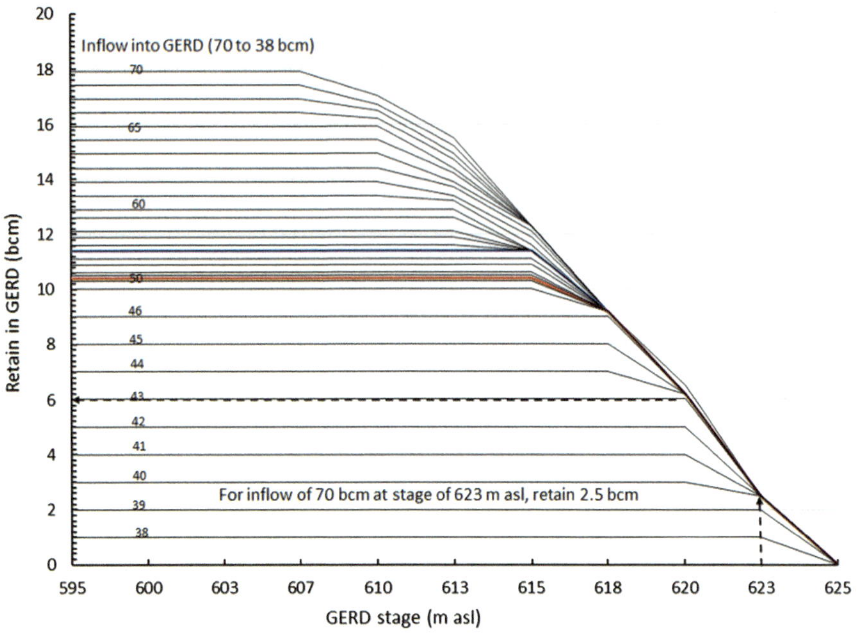

Fig. 14.8 Annual retentions in GERD for inflow of 38–70 bcm (Annex C)

How Does Drought Came into GERD Filling and Operation Plan

From, the three countries, Ethiopia is the one with most impacted from drought, famine, mass death, and it is under perpetual food shortage due to its lack of water storage for irrigation. Table 14.4 shows number of deaths from drought by country from 1900 to recent time. Egypt is not shown in the figure, but Ethiopia is mentioned two times, 1983 and 1973, with 100 s of thousands of human losses. Egypt has large storage in the High Aswan dam (132 bcm) to overcome flow variations including drought. In the D.C. plan, no mention of consideration for drought impact in Ethiopia was made. Drought was used for minimum guaranteed flow determination for Egypt without mentioning its storage capacity of the Aswan Dam.

The history of drought and famine in Ethiopia is internationally known. Hunger defined the country's image for many years. Thousands of people died of starvation during severe droughts in the 1970s and 1980s. Ethiopia has a chronic food shortage requiring the presence of the World Food Organization in the country. Recurrent droughts have been common. Food security should be a paramount national goal for Ethiopia. Seasonal rainfall has sustained agriculture for centuries. With climate change and population growth, rain-fed agriculture only cannot meet the growing food demand. That is why irrigation has to be a national priority. Seventy percent of Ethiopia's surface water is in the Nile Basin. Equitable use of this water for irrigation, energy production, and other services is a national security issue.

Table 14.4 Historical human losses from droughts since 1900 (World Economic Forum 2019)

Country	Death from Famine	Year
China	3,000,000	1928
Bangladesh	1,900,000	1943
India	1,500,000	1942
India	1,500,000	1965
Soviet Union	1,250,000	1900
China	1,200,000	1921
Ethiopia	300,000	1920
Sudan	150,000	1983
Ethiopia	100,000	1973

During El Nino climate pattern, drought occurs in the Blue Nile Basin (Abtew et al. 2009). Recently, the 2015/2016 El Nino year resulted in drought in Ethiopia with 2016 reported as the worst drought in 50 years with millions needing food assistance (Speckhard 2016). The 2002–2003 drought in Ethiopia affected 13.2 million people (de Waal et al. 2006).

The Washington D.C. Plan Impact on Ethiopia

The Washington D.C. plan can be summarized as an ingenious way of inserting Egypt's minimum water guarantee in GERD filling and operation plans. It has put undue burden on Ethiopia to pay water from the reservoir for inflow variation that is mostly influenced by the climate (Fig. 14.6). The scheme extends the filling period of the dam and makes it a water supply dam than a hydropower dam that generates optimal and dependable power for marketing. At the same time, it combined upstream abstractions of Ethiopia's consumptive use, current and future, into drought because the drought classification is on inflow to GERD not rainfall in the basin. It implicitly reduced the dam water level from 640 to 625 m a.s.l.

The Washington D.C. Plan GERD Filling (Annex A, B, C)

The filling of the GERD is determined by Annex A, B, C. Annex B limits the filling plan to two years with the remaining filling years regulated by Annex A and Annex C (Figs. 14.6, 14.7, Table 14.5). The maximum reservoir stage is limited to 625 m a.s.l (Figs. 14.6, 14.7, 14.8). Figure 14.7 shows Annex C annual releases during filling for non-drought years based on inflow to GERD and GERD water level. Figure 14.8 shows Annex C annual GERD retentions during filling for non-drought years based on inflow to GERD and GERD water level. Maximum inflow of 70 bcm has zero retention at stage of 625 m a.s.l showing that Annex C keeps the reservoir far lower than the design full-service level of 640 m a.s.l

Table 14.5 GERD stage I filling (Annex B)

Stage I filling (to 595 m a.s.l level of GERD	Incremental retention
Hydrological year 1	4.9 bcm
Hydrological year 2	13.5 bcm (18.5 bcm total)
Definition of drought	31 bcm
Release rule	Lower 31 bcm or flow
Postponement of stage I	If flow less than 31 bcm, stage I will be postponed

effectively limiting its ability to generate the design power. For inflow of 70 bcm at 623 m a.s.l., the retained amount is 2.5 bcm, while seepage and evaporation losses are about 2 bcm.

Table 14.6 shows GERD filling simulation with the D.C. plan. Depending on the beginning of the simulation year and the period of flow data, the number of years to fill the dam changes. Simulation starting from 1979 took 18 years to fill the dam to 626 m a.s.l. As shown in bold in Table 14.6, the criteria to release additional water from the reservoir were met in 1987, inflow was less than 37 bcm, and water level was greater than 603 m a.s.l. The minimum guaranteed flows of 37, 39, and 40 bcm under the various assumed drought conditions are more probable increasing the chance of draining the reservoir (Fig. 14.9). Under other sets of inflow data, it may take longer years to reach to 625 m a.s.l.

The filling simulation for the D.C. plan can be compared to other proposals. Figure 14.10 depicts a previously published suggested filling plan with retaining 20% of any incoming flow in the year. It takes up to eight years to fill to the full-service level of 640 m a.s.l. In this plan, both drought and wet period impacts are equally shared by upstream and downstream countries. Ethiopia's proposal of filling up to 625 m a.s.l. (49.3 bcm storage) in five years, shown in red, is plotted over the simulation results from 20 percent retaining. This was a very reasonable filling plan that should have been accepted and implemented.

The D.C. plan also affects the reservoir during normal operation after initial filling is completed. For illustration purposes, flows of the Blue Nile River at El Diem from 1979 to 1987 were used to simulate normal operation with the D.C. plan (Table 14.7). As shown in Table 14.7, from 1979 to 1981, reservoir was assumed to operate under the D.C. plan, under normal operation (Annex D, Fig. 14.11) and drought plan for annual inflows below 37 bcm (shown in italics), 4-year average below 39 bcm, and 5-year average below 40 bcm (shown in bold). An initial stage is assumed at 625 m a.s.l. In 1982, since inflow was below 37 bcm, incoming inflow plus additional release from reservoir was made (Fig. 14.6 (Annex A)). In 1983, the five-year average was below 40 bcm, and for the next five years, the prolonged period of dry years formula was applied. The storage volume above 603 m a.s.l. was distributed to the following 5 years as additional flow. Even if the following five years were wet years, the penalty of draining the reservoir continues.

Table 14.6 GERD filling simulation with D.C. plan

Year	Inflow	4-yr inflow avg.	5-yr inflow avg.	Rain-ET	Release	Release from reservoir	Retained	Cumulative storage	Stage	Area	Release decision
	BCM	BCM	BCM	BCM	BCM	BCM	BCM	BCM	m a.s.l.	km^2	
1979	38.13			0.19	33.23	0.00	4.90	4.71	571	260	Annex B
1980	42.50			0.47	31.00	0.00	11.50	16.21	597	650	Annex B
1981	42.77			0.38	40.76	0.00	2.00	18.21	590	522	Annex B
1982	34.33	39.43		0.51	34.23	0.00	0.10	18.31	600	700	Annex B
1983	39.53	39.78	39.45	0.57	37.00	0.00	2.53	20.84	604	781	Annex C
1984	29.73	36.59	37.77	0.60	28.60	0.00	1.13	21.98	606	823	Annex C
1985	45.11	37.18	38.29	0.75	37.00	0.00	8.11	30.08	615	1030	Annex C
1986	34.58	37.24	36.66	0.75	35.40	0.00	0.00	29.27	615	1030	Annex C
1987	33.41	**35.71**	36.47	0.70	34.80	0.00	0.00	27.87	612	954	Annex A
1988	64.48	44.39	41.46	0.95	49.70	**1.45**	14.78	42.65	625	1304	Annex A
1989	32.83	41.32	42.08	0.87	36.20	**1.45**	0.00	37.83	621	1192	Annex A
1990	37.99	42.18	40.66	0.79	37.00	**1.45**	0.99	32.33	617	1082	Annex A
1991	45.38	45.17	42.82	0.79	38.80	**1.45**	6.58	32.41	617	1082	Annex C
1992	44.21	40.10	44.98	0.89	37.80	0.00	6.41	32.31	622	1220	Annex C
1993	56.10	45.92	43.30	0.87	53.00	0.00	3.10	30.36	621	1190	Annex C
1994	52.51	49.55	47.24	0.95	46.00	0.00	6.51	36.87	625	1304	Annex C
1995	37.13	47.49	47.06	0.97	37.00	0.00	0.13	37.00	626	1333	Annex C
1996	55.06	50.45	49.20	0.97	56.00	0.00	0.06	37.06	626	1333	Annex C

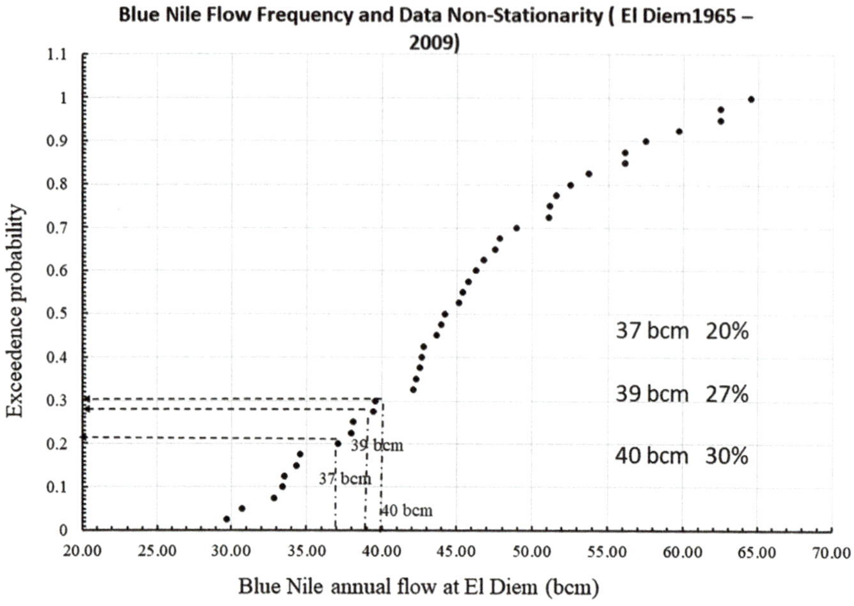

Fig. 14.9 Exceedance probability of Blue Nile flows at El Diem (1965–2009)

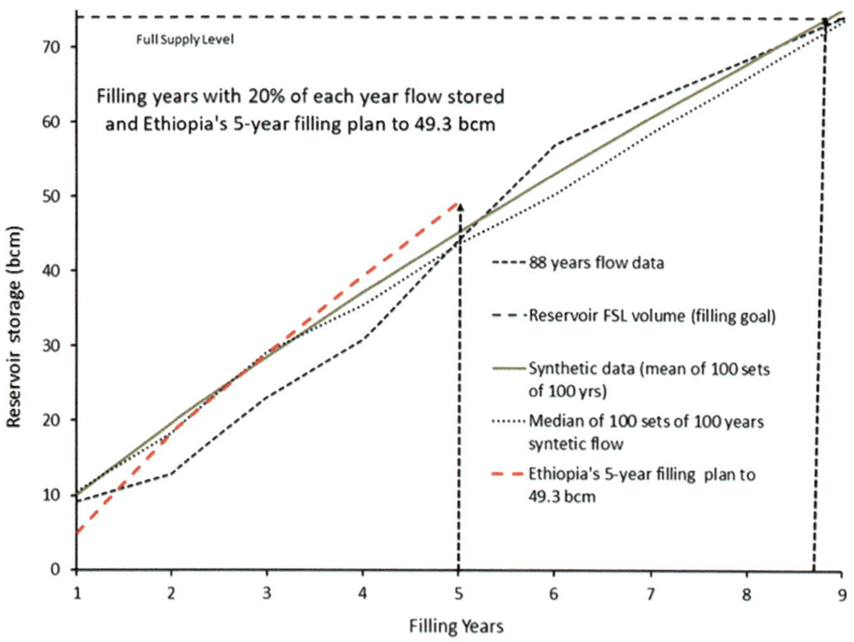

Fig. 14.10 GERD filling simulation with synthetic and observed data; Ethiopia's GERD filling plan overlaid in red

Table 14.7 GERD normal operation with the D.C. plan

Year	Inflow	Release	4-yr avg. inflow	5-yr avg. inflow	Rain-ET	From reservoir	Cum. Storge	Stage	Assumed operation start at 625 m a.s.l.
	BCM	BCM	BCM	BCM	BCM	BCM	BCM	m a.s.l.	Annual operational release (BCM)
1979	38.13	37.00			0.95		42.20	625	Annex D
1980	42.50	42.50			0.95		41.25	625	Annex D
1981	42.77	42.77			0.93		40.32	624	Annex D
1982	*34.33*	37.00	39.43		0.87	2.67	34.10	619	Annex D
1983	39.53	37.70	39.78	**39.45**	0.81		35.13	619	Annex D
1984	*29.73*	32.61	**36.59**	**37.77**	0.75	2.88	28.62	613	5-yr drought (<40 bcm)
1985	45.11	47.99	37.18	**38.29**	0.70	2.88	22.16	606	5-yr drought (<40 bcm)
1986	*34.58*	37.46	37.24	**36.66**	0.65	2.88	15.75	597	5-yr drought (<40 bcm)
1987	*33.41*	36.29	*35.71*	**36.47**	0.64	2.88	9.35	585	5-yr drought (<40 bcm)

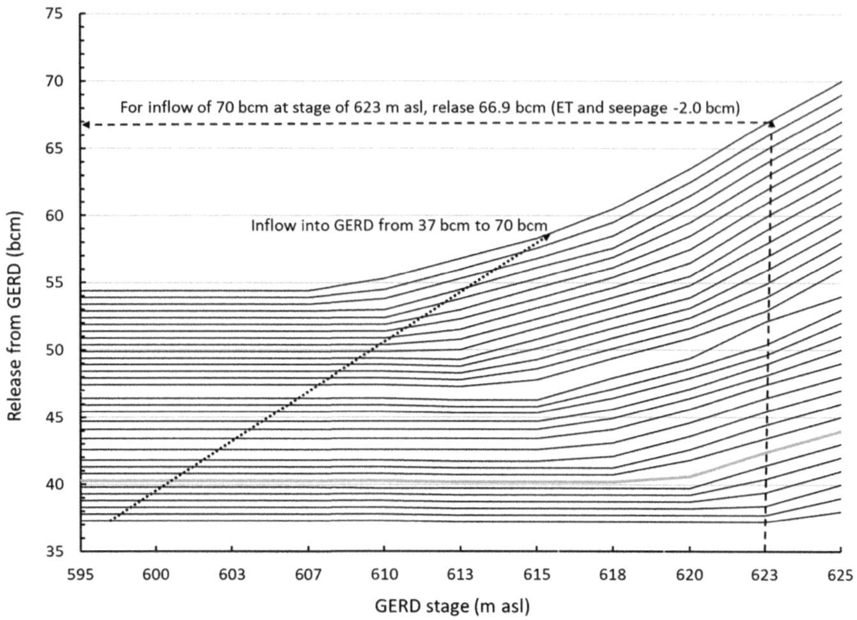

Fig. 14.11 GERD inflows and prescribed releases during normal operation at given water levels (Annex D for flows from 37 to 70 bcm)

The reservoir operation is controlled by the downstream release requirement. For the case period, reservoir consistently declined from 625 to 585 m a.s.l. Since the five-year average is below 40 bcm from 1983 to 1987, the prolonged period of dry years will dictate release for following another five years through 1992. One can imagine when the reservoir will ever fill back to 640 m a.s.l. and if such operation ever allows to generate dependable and marketable power. The normal operation plan (Annex D) combined with the drought mitigation plan is more likely to keep GERD water level at 625 m a.s.l. or lower. An example in Table 14.8 shows that at the wettest inflow of 70 bcm, the reservoir loses storage of about 2 bcm from evaporation and seepage with no retention allowed. This was the time to refill the reservoir when the basin is the wettest. Figure 14.11 shows during normal operation during extreme high flow of 70 bcm at 623 m a.s.l. water level, the prescribed release is 66.9 bcm. The net retention is 1.1 bcm with 2 bcm evaporation and seepage losses. If the reservoir does not significantly increase water level and storage during extreme high inflow periods, it will not function as designed.

Downstream Control Impact on Hydropower Reservoir Operations

The impact of the D.C. plan on Ethiopia is eloquently summarized by Addisu Lashitew (Lashitew 2020) "This proposal is likely to cause many years of delay in

Table 14.8 GERD storage loss during maximum inflow of 70 bcm (Annex D)

Stage m a.s.l.	Inflow (bcm)	Release (bcm)	Area (km^2)	Evap. (bcm)	Seepage (bcm)	Change in storage (bcm)
625	70	70	1304	0.95	1.01	−1.97

the filling period of the dam's reservoir, which has a capacity of 74 bcm, reducing its potential to generate electricity at full capacity. Moreover, this restriction could prevent Ethiopia from starting other projects along the Nile. If a drought or a new upstream project reduces the water flowing into the dam, the minimum flow would have to be maintained by drawing from the GERD's reservoir."

Downstream control impact on hydropower dams has been reported. Deviations from the pre-dam flow regime correspond to high return from power generation indicating that full control of reservoir operation is critical (Kim 2013). Depending on the inflow pattern and reservoir size, hydropower dams can be operated to meet temporally varying power demand. Figure 14.12 depicts types of downstream demand on hydropower reservoir. Simulation results from GERD and High Aswan Dam (HAD) operations during drought periods of 1972 through 1987 under condition of GERD releasing from storage illustrate downstream control impact. The simulation covered period of 1972–1987 as drought operation and 1987–2002 as drought recovery operation. During the thirty years period, GERD mostly stayed below 40 bcm storage and reaching to 20 bcm low storage level in 1983 and taking close to 20 years to reach full-service level (Fig. 7 in Wheeler et al. 2020).

The D.C. plan is based on meeting downstream water needs and not on optimal power generation by GERD. Therefore, if operated under such plan, it may not generate dependable and marketable power and meeting peak power demand. In summary, the D.C. plan fulfilled Egypt's objectives of maintaining its minimum guaranteed flow at the expense of the GERD hydropower production and Ethiopia's Nile water share for irrigation. The plan also would have implicitly restricted other dam developments on the Blue Nile River.

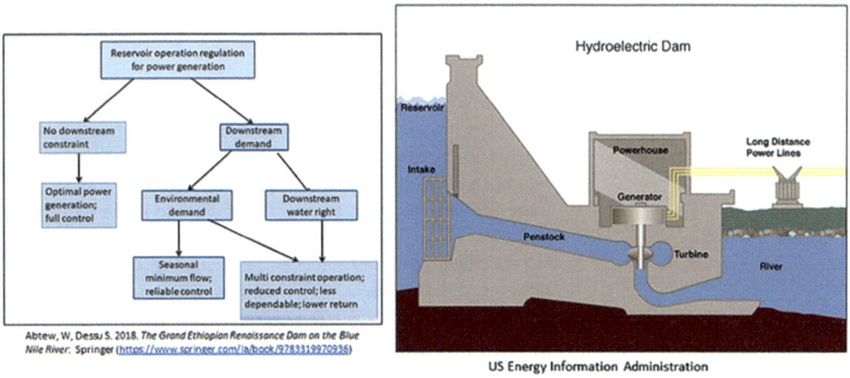

Fig. 14.12 Downstream demand impact on hydropower reservoir operation

African Union Led Negotiations

In June 2020, GERD negotiations restarted under the auspices of the African Union and observers from the European Union and USA. Meetings were conducted with the normal pattern of Egypt pulling back, Sudan asking break and reconvening. The reason is that minimum water guarantee for Egypt is plugged into GERD filling and operation plan negotiation and has made the problem unsolvable without infringing on Ethiopia's equitable share of consumptive use water. Egypt's status quo water share and Ethiopia's future consumptive water share cannot be achieved at the same time. Figure 14.13 depicts author's understanding of Egypt's and Ethiopia's positions in the negotiations.

After series of meetings, the three countries delivered reports on their positions on the GERD negotiation to the Africa Union (AU) Bureau of the Assembly of Heads of State and Government under the chairmanship of Matamela Cyril Ramaphosa, President of the Republic of South Africa. The Bureau had its first extraordinary meeting on 26 June 2020. At its second extraordinary meeting on 21 July 2020, the Bureau released communique after receiving AU experts report on the trilateral GERD negotiations. The Bureau's communique of its agreement on the matter is quoted as follows. "*The meeting of the Bureau of the Assembly of AU Heads of State and Government held extensive discussion on the matter of the first filling and annual operation of the GERD and future development of projects on the Blue Nile River upstream of the GERD. Consequently, the Meeting of the Bureau of the Assembly of AU Heads of State and Government agreed on the process of finalizing on the text of finalizing Agreement on the Filling and Operation of the*

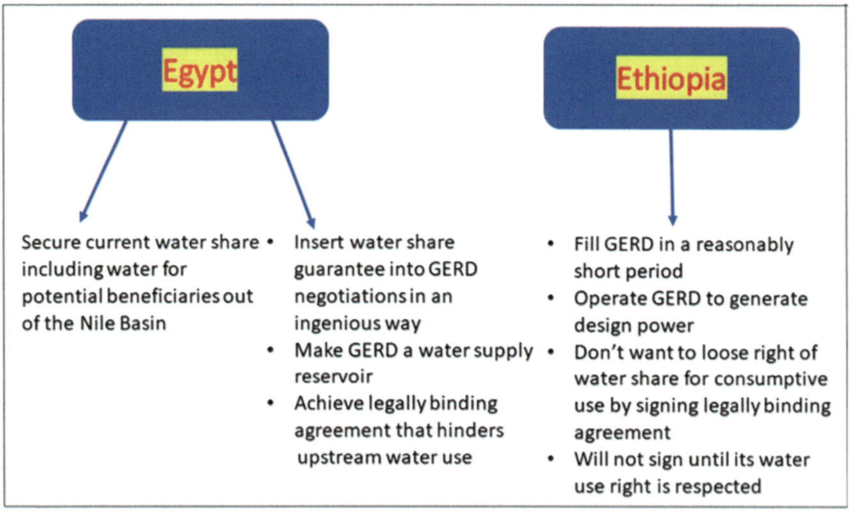

Fig. 14.13 Understanding of Egypt's and Ethiopia's positions in the GERD negotiation

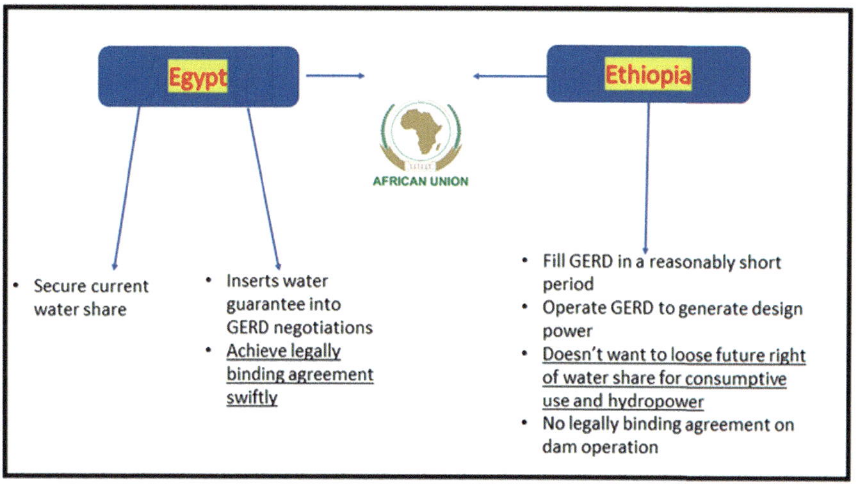

Fig. 14.14 Summary of AU communique on GERD negotiations

GERD, which should include a Comprehensive Agreement on future development on the Blue Nile River" (African Union 2020).

The AU agreed on two issues that satisfy the positions of Egypt and Ethiopia each (Fig. 14.14). Egypt's insistence to get a legally binding agreement on GERD filling and operation was included, while Ethiopia's future water consumptive use right was also covered. Egypt did not secure its minimum water guarantee, and Ethiopia did not avoid legally binding agreement.

The issue of treaty or legally binding agreement on reservoir operation plan is meaningless. Reservoir operation plan development is time taking process and dependent on ever changing climatic factors, power demand, and other factors. If relationship is based on hostility, the plan should minimize the interference of hostile parties and be based on optimal benefit of the reservoir owner. But should preserve international norms of not severely impacting downstream entities unless their dam operation and water use policy severely impacts the upstream reservoir owner. If the relationship is cooperation, coordination can be implemented on optimal operation of dams in series on a river system.

Dam operation plan has to be flexible enough to meet its power demands, maintain dam safety, and deal with climatic extremes such as the 2020 floods in China, Ethiopia, and Sudan. The operation of dams includes short-term weather prediction and long-term climate predictions and incorporates forecasts in dam operation plans. One of the climate predictions relevant for the region is El Nino Southern Oscillation (ENSO). During La Nina years, the Blue Nile Basin gets wet, and drought is likely during El Nino years. Storage or release decisions can be made with a lead time of at least few months (Abtew and Melesse 2014; Abtew and Trimble 2010, Abtew et al. 2009). Sometimes, the predicted climate conditions may not materialize, and water management decisions may turn out to be incorrect. Also, dam safety needs to be paramount in the operation plan with full flexibility in decision making. As a hydropower dam, GERD has to be operated in a way to generate optimal power for meeting needs. Flexibility is required to meet pick power demand. Reservoir or dam operation plans need to leave room for deviations. Deviations are needed to operate outside the operation plan when weather and other factors force. In July 2020, Ethiopia completed the first year of GERD filling (4.9 bcm), few weeks later, extreme flooding started in the Sudan with lives lost and thousands of houses flooded (https://earthobservatory.nasa.gov/images/147288/record-flooding-in-sudan). All these make legally binding agreement on dam operation not a sincere demand but driven by the goal of plugging minimum water guarantee in the dam operation plan.

The negotiations stopped after AU communique to resume for another series of meetings between the three countries with observers for USA, European Union, and AU experts, under the AU auspices. The second tripartite meeting was conducted on 28 and 29 August 2020, and no new outcome was reported by September 2020. On 23 October 2020, the US president, Donald Trump, while on telephone talk with Israel and Sudan leaders, talked about bombing the GERD by Egypt as the GERD blocks flows of the Nile and creates undue burden on Egypt and paused the question why it was not bombed at early stage of dam construction. Many corners were alarmed and expressed objections. The filling of the dam should contribute to deterrence to attack as the risk of downstream flooding is high. The AU sponsored meetings continued in early November 2020 under this background. Earlier, Ethiopia has restricted flights on the dam site region.

Summary

The Nile River Basin, especially the Eastern Nile, is currently facing water conflict triggered by Ethiopia's Grand Renaissance Dam (GERD), an assertion of upstream water right. Dam filling and operation plan negotiations have sparked conflict between Ethiopia and Egypt. The cause for dragging the negotiations is the implicit position of Egypt to insert minimum water guarantee in the dam filling and operation plans to maintain its current hydro hegemony. A series of meetings was conducted between Egypt, Ethiopia, and Sudan with and without mediators. The meetings end mostly with Egypt withdrawing when agreement was reported close to finish and the whole process to start all over again with new demands. Sudan is upstream of Egypt and downstream of Ethiopia and initially supporting the project but later was influenced to put challenges to the dam filling. Two of the negotiations on the GERD, the Washington D.C. and the African Union negotiations, are the latest. The Washington D.C. negotiation failed to address the interests of all sides clearly favouring Egypt. The D.C. plan would have put GERD to be a water supply dam under Egypt's management plan. The filling plan would have extended the period of filling and keep water levels low. The operation plan also has no consideration to the hydropower generation objective of the dam with uncontrolled fluctuation of water levels and drawdown. The AU led negotiations are in progress. As long us Egypt continuously expands its water use by transferring Nile water out of its basin and out of the continent, it does not seem it will respect water rights of upstream countries that will reduce the flow. The rejection of the Cooperative Framework Agreement is another indication. Any effort by upstream countries to use Nile water will be frustrated with propaganda and diplomatic campaign, interference, aid and loan blockage schemes, and corruption. The GERD negotiations cannot reach agreement under Egypt's unilateral interest of maintaining the statues quo with upstream Africans' water share being dedicated to global water markets. It is a matter of time before Egypt restricts Sudan's Nile water use. The best way out of this quagmire is water-sharing agreement between all basin countries based on an international water conventions norms of transboundary water sharing.

References

Abtew W, Dessu S (2019) The grand ethiopian renaissance dam on the Blue Nile. Springer, New York

Abtew W, Melesse AM (2014) Chapter 33: climate teleconnections and water management. In: Melesse A, Abtew W, Setegn S (eds) Nile River Basin: ecohydrological challenges, climate change and hydropolitics. Springer, New York

Abtew W, Trimble P (2010) El Niño southern oscillation link to South Florida hydrology and water management applications. Water Resour Manage 24(5):4255–4271. https://doi.org/10.1007/s11269-010-9656-2

Abtew W, Melesse AM, Dessalegne T (2009) El Niño southern oscillation link to the Blue Nile River Basin hydrology. Hydrol Process 23:3653–3660. https://doi.org/10.1002/hyp.7367

African Union (2020) Communique of the 2nd extraordinary African Union (AU) Bureau of the Assembly of Heads of State and Government video-teleconference meeting on the Grand Ethiopian Renaissance Dam (GERD) held on 21 July 2020

Arafaat N, El Nour S (2019) How Egypt's water feed the gulf. ResearchGate

Awulachew SB (2017) Key water issues in African and Ethiopian perspectives amid water related sustainable development challenges. Water Catalyst Award Lecture. Hamilton, ON, Canada, 21st July 2017

Chesire DK (2010) Control over the Nile: implications across nations/MS Thesis. Naval Postgraduate School, Monterey, California, USA

De Waal A, Taffesse AS, Carruth L (2006) Child survival during the 2002–2003 drought in Ethiopia. Global Public Health 1(2):125–132

Deneke TT (2014) Chapter 24 process of institutional change and factors influencing collective action in local water resources governance in the Blue Nile Basin of Ethiopia. In: Melesse AM, Abtew W, Moges S (eds) Nile River Basin. Springer, New York

FAO (2000) Water and agriculture in the Nile Basin. AGL/MISC/29/2000, United Nations

Hassen HA (2017) Egypt and controlling the Nile: from mythologies to real politics. (http://dehai.org/dehai/assets/dehai/2017/204549-Egypt%20and%20controlling%20the%20Nile-From%20mythologies%20to%20real%20politics.pdf)

Hassen HA, AL Rasheedy A (2007) The Nile River and Egyptian foreign policy interests. African Sociol Rev 11(1):25–37

Helal MS (2020) Ethiopia's power play on the Nile has left the region in a deadlock. Foreign Policy (foreignpolicy.com), 28 Sept 2020

International Panel of Experts (IPoE) (2013) Grand Ethiopian Renaissance Dam Project (GERDP) Final Report. Adds Ababa, Ethiopia, 31 May 2013
Kacziba O, Glied V (2020) Water Stress emerging challenges of global water security. University of Pecs, Hungary
Kim JD (2013) How deregulated markets influence hydro revenue and downstream flow. Hydro Rev. www.hydroworld.com
Lashitew A (2020) Why Ethiopia, Egypt, and Sudan should ditch a rushed, Washington-brokered Nile Treaty. Brookings Institute (https://www.brookings.edu/blog/africa-in-focus/2020/02/18/why-ethiopia-egypt-and-sudan-should-ditch-a-rushed-washington-brokered-nile-treaty/)
McDonnell P (2020) Mexican water wars in dispute about water-sharing with US. Los Angeles Times. Los Angeles Times, 11 Sept 2020
Okidi CO (1982) Review of treaties on consumptive utilization of Lake Victoria and Nile drainage system. Natural Resour J 22(1):161–199 (https://digitalrepository.unm.edu/cgi/viewcontent.cgi?article=2166&context=nrj)
Permanent Mission of the Arab Republic of Egypt to the United Nations Letter to the United Nations Security Council on GERD. (https://www.securitycouncilreport.org/atf/cf/%7B65BFCF9B-6D27-4E9C-8CD3-CF6E4FF96FF9%7D/S_2020_566%20Egypt%20letter%20of%2019%20June.pdf) Accessed 31 Dec 2020
Pierce F (2012) The land grabbers: the new fight over who owns the earth. Beacon Press, Boston, MA Investment
Ram KV (2009) Anglo-Ethiopian relations 1869–1906 a study of British policy in Ethiopia. Concept Publishing Company, New Delhi, India
Schlosser CA, Strzepek K, Gao X, Fant C, Blanc E, Paltsev S, Jacoby H, Reilly J, Gueneau A (2014) The future of global water stress: an integrated assessment. AGU Publications, Research Article. https://doi.org/10.1002/2014EF000238
Speckhard D (2016) How Ethiopia can overcome the worst drought in 50 years. Time
United Nations (2010) Human right to water and sanitation. Resolution 64:192
Wada Y, van Beck LPH, Viviroli D, Durr HH, Weingartner R, Bierkens MFP (2011) Modelling global water stress at the monthly time-scale. Water Resour Res 47:W07518. https://doi.org/10.1029/2010wr009792
Wheeler KG, Jeuland M, Hall JW, Zagona E, Whittington D (2020) Understanding and managing new risk on the Nile with the Grand Ethiopian Renaissance Dam. Nature Commun. https://doi.org/10.1038/s.41467-020-19089-x
World Economic Form (2019) Five droughts that changed the world. https://www.weforum.org/agenda/2019/05/5-droughts-that-changed-human-history/
World Economic Forum (2017) The global water initiative. December 2017, weforum.org

Chapter 15
Scientific Misconduct and Partisan Research on the Stability of the Grand Ethiopian Renaissance Dam: A Critical Review of a Contribution to *Environmental Remote Sensing in Egypt* (Springer, 2020)

Wuletawu Abera, Nigussie Haregeweyn, Yihun Dile, Ayele A. Fenta, Mulatu L. Berihun, Biadgilgn Demissie, Chalachew A. Mulatu, Tewodros A. Nigussie, Paolo Billi, Hailemariam Meaza, Kifle Woldearegay, Assefa M. Melesse, Semu A. Moges, and Lulseged Tamene

Abstract We present factual errors, methodological flaws, wrong assumptions, inadequate data use, misleading conclusions and scientific misconducts committed by a book chapter by Dandrawy and Omran (2020) on the GERD, titled "*Integrated Watershed Management of Grand Ethiopian Renaissance Dam* via *Watershed Modeling System and Remote Sensing*" in environmental remote sensing in Egypt and published at Springer (https://www.springer.com/gp/book/9783030395926). We used direct and indirect fact-checking methods that include consulting of the

---

W. Abera (✉) · L. Tamene
International Center for Tropical Agriculture (CIAT), Addis Ababa, Ethiopia
e-mail: Wuletawu.Abera@cgiar.org

L. Tamene
e-mail: LT.Desta@cgiar.org

N. Haregeweyn · P. Billi
International Platform for Dryland Research and Education, Tottori University, Tottori, Japan
e-mail: bli@unife.it

Y. Dile
College of Agriculture and Life Sciences, Texas a&M University, College Station, TX, USA
e-mail: yihundile@tamu.edu

A. A. Fenta
Arid Land Research Center, Tottori University, 1390 Hamasaka, Tottori 680-0001, Japan

M. L. Berihun
The United Graduate School of Agricultural Sciences, Tottori University, 1390 Hamasaka, Tottori 680-8553, Japan

A. M. Melesse et al. (eds.), *Nile and Grand Ethiopian Renaissance Dam*,
Springer Geography, https://doi.org/10.1007/978-3-030-76437-1_15

design report of the GERD, literature review and re-calculating some parameters todemonstrate the major flaws of the chapter. The main points of concern are (1) absence of hydrological and hydraulic model calibration to assess model input and parameter uncertainty; (2) generation of intensity–duration–frequency curve of the Upper Blue Nile basin based on a single station data, while there are more than 40 meteorological stations within and around the basin; (3) ill-defined and exaggerated topographic parameters such as flow length, slope and time to peak to overestimate flood flow; (4) failure to include routing component in the modeling while estimating flood despite large size of the basin with over 1000 km river hydraulic length; (5) exaggerating the elevation–areacapacity curve of the GERD and significantly overestimating GERD's reservoir (lake) area; (6) unwarranted and misleading conclusions on the structural integrity of the GERD without supporting dam failure analysis and proper hydrologic and flooding assessment. Moreover, the book Chapter has several accounts of scientific misconducts of plagiarism, falsification and fabrications that should not have passed any standard peer-reviewed processes for a highly reputable publisher with the stature of Springer. If such gross failures of scientific veracity and analytical weaknesses continue unaddressed, there is a clear danger that the public credibility of scientists and the forum in which they publish their scientific findings can be compromised. In addition, such publications, which lack scientific foundation, could undermine both scientific integrity and regional peace and security. The latter is predicated upon the realization that the discourse on GERD is a highly sensitive matter in north-east Africa. We thus believe that this critique paper can serve as a basis both for defending scientific integrity in other similar cases in the future and for providing pointers that would put science at the service of society.

Keywords Scientific misconduct · GERD · Nile basin · Ethiopia · Sudan · Egypt

B. Demissie · H. Meaza
Department of Geography and Environmental Studies, Mekelle University, Mekelle, Ethiopia

C. A. Mulatu
Faculty of Civil and Water Resources Engineering, Bahir Dar University, Bahir Dar, Ethiopia

T. A. Nigussie
Institute of Technology, Hawassa University, Hawassa, Ethiopia

K. Woldearegay
Department of Earth Science, Mekelle University, Mekelle, Ethiopia
e-mail: kiflewold@yahoo.com

A. M. Melesse
Department of Earth and Environment, Florida International University, AHC-5-390, Miami, FL, USA
e-mail: melessea@fiu.edu

S. A. Moges
School of Civil and Environmental Engineering, University of Connecticut, Storrs, CT, USA

Introduction

Scientific publications enable the advancement of science through sharing ideas, supplying data, methods, tools and results and provide a foundation on which new scientific discoveries and inventions are built (National Research Council 2003). Scientific publications generally build on one another thus authors have paramount responsibilities for the body of knowledge they generate. In addition, researchers are expected to uphold high scientific standards and be aware of the reputational risks that would follow when deviating from the cardinal principles of honesty, impartiality and the quest for truth when doing scientific research (World Economic Forum 2018). Generally, the scientific community invests in a range of self-correction mechanisms that would enable it produce publications that reflect not only the current state of scientific advancement but also expand the frontiers of scientific knowledge.

While scientific publications have tremendously propelled further scientific discoveries and made the world a better place to live in, it is important to note that a small minority of researchers display "scientific misconducts" due to different motives and reasons (Khadilkar 2018). Scientific misconducts take different forms, and the main ones include mistakes arising from capacity limitations to analyze, synthesize and/or report findings; use of erroneous data sets; plagiarism; fabrication and falsification fueled by tendency to be guided by motives other than scientific excellence (Goodstein et al. 2002; Fleet et al. 2006). Scientific misconducts can have varying degrees of adverse impacts depending on, among other factors, the sensitivity of the subject of the research, its timing and the way the results are presented and communicated.

The Grand Ethiopian Renaissance Dam (GERD) has attracted a significant attention from both the scientific and political communities due to its strong socio-economic and geopolitical importance. There is great interest related to the project whereby a simple Google search using keywords "GERD" and "Ethiopia" for example results in over 3 million hits (on June 24, 2020) in a space of under one minute. There are also several scientific publications coming up, most of which have attempted to provide valuable information for the sustainable management and utilization of the dam both for upstream and downstream users (e.g., Kahsay et al. 2015; Yihdego et al. 2016; Wheeler et al. 2016; Tawfik and Dombrowsky 2017). However, there are several publications on the GERD that are replete with scientific misconducts. These publications are labeled as scientific works while overtly and blatantly disregard basic norms of scientific discovery and research decency. Such publications must be scrutinized using the best available scientific methodology and the oversights, misinterpretations and misrepresentation of facts laid to bare. Otherwise, left unchecked and unchallenged, publications characterized by scientific misconducts not only can belittle scientific credibility but also the reported findings can be used by unscrupulous state and non-state actors to sow the seed of confusion, discord and enmity among communities and countries.

A book chapter—hereafter referred to as the report—by Dandrawy and Omran (2020) published at Springer In: "*Environmental Remote Sensing in Egypt*", titled, "Watershed Modelling System and Remote Sensing on Grand Ethiopian Renaissance Dam (GERD)", illustrates a typical case of gross scientific misconduct. Detailed analysis of the report suggests that the authors are politically motivated and have conflict of interest rather than a genuine effort in communicating scientific facts. Using suspect data, disregarding basic ethics of research and employing fabrication and falsification, the report arrived at the following two main unwarranted conclusions: (1) it is more probable that the GERD will collapse, and (2) this would risk flooding the capital city of Sudan and the White Nile River, covering a total area of 667,228 km^2. Given the fact that these conclusions signal a catastrophic flood disaster for the region and the significant stakes involved in the current state of regional hydro politics, a group of interdisciplinary researchers covering the fields of geology, climatology, geomorphology, hydrology, dam engineering and land use planning and modeling took an initiative to investigate the scientific validity of the research methodology and the robustness of its findings and conclusions. Also, recognizing that *Springer* is a widely acclaimed publication house among the global scientific community, we are concerned that conclusions of the above kind that appear in reputable publications can easily be taken as facts among upcoming researchers, thereby misleading future body of research work.

In the interest of space and time, the review has chosen to focus on very fundamental issues that the scientific community should know as to how scientific credibility is violated and scientific affiliations misuse to arrive at misguided conclusions. This is also aimed to alert the *Springer* about the infringement of scientific ethics to make bold statements and assertions without any supporting and scientifically valid evidence.

Grand Ethiopian Renaissance Dam

The Grand Ethiopian Renaissance Dam (GERD) is a hydroelectric power plant under construction on the Upper Blue Nile (UBN) River (Abay) in Ethiopia, located about 20 km from Ethio-Sudan boarder. The UBN river originates from Lake Tana, the largest lake in Ethiopia and the third largest in the Nile Basin, contributing about 85%, to the Nile system (Conway 2000). The UBN basin has an area of about 176,000 km^2, covering some 17% of the total area of Ethiopia. The topography of the UBN basin ranges from 500 masl in the lowlands at the Sudan border to 4160 masl in the upper parts of the basin. Due to this topographic variation, the climate and rainfall regime of the basin varies significantly.

The GERD is the first major roller compacted concrete (RCC) dam being constructed on the UBN river of Ethiopia. GERD consists of two dams, the RCC dam with 1.8 km crest length and 145 m high gravity dam and the rockfill saddle dam with crest length of 5 km and 50 m height. Three spillways are provided—gated and ungated spillways at the RCC dam and ungated emergency spillway at the right side of the saddle dam. When completed, the dam will be the largest hydroelectric power plant in Africa and the seventh largest in the world (Gebreluel 2014). GERD will be primarily used for hydroelectric power generation and will account 40% of Ethiopia's currently installed generation capacity (Liersch et al. 2017). Detailed discussions on the hydrological system and geopolitical arena around GERD are published in Abtew and Dessu (2019).

Ethiopia is keen on utilizing the hydroelectric power to be generated from the GERD to power homes and its planned agro-industrial as well as other modern establishments. In addition, Ethiopia envisages exporting electricity to downstream countries and others at a reasonable price and earns much-needed hard currency. Despite these, the construction of the GERD has created tensions among some of the riparian countries, especially more so between Egypt and Ethiopia. For instance, Egypt alleges that the dam will impinge on its historic water use rights over River Nile, while Ethiopia maintains that this project is not only beneficial to herself but also is of significant value to downstream countries. Ethiopia, Sudan and Egypt are under discussion on the filling and operation of the dam, with limited breakthrough in some key items. Thus, the GERD, rather than being an example of inter-state collaboration, has become a source of dispute, a hydropolitics flash point. Consequently, any scientific research or journalistic reportage on the hydrology and hydropolitics of the GERD are scrutinized for supportive evidence by the parties involved. Given these facts, when researching into water management issues in connection with the GERD, the research community has an utmost responsibility of presenting scientific evidence in a robust and rigorous manner.

Methodology

In this review, we focused on key flaws of the report by Dandrawy and Omran (2020)—a study document that does not have any scientific credibility and should never have passed a rigorous review process—so that the broader scientific community can pass a fair judgement on the report's worth as a knowledge product. Several direct and indirect fact-checking methods were employed, including review of the design report of GERD, literature review and re-calculating some parameters. For some of the statements that do not have any evidence to corroborate, we have quoted the report and highlighted why those claims cannot be acceptable. For hydrological and hydraulic models, we have reviewed the work from the perspective of model input data, the availability/types of calibration and validation and the validity and consistency of model results and conclusions. When it is necessary, we tried to analyze data and compared with the results in the report to highlight gaps and errors. In addition, we have indicated the problems of the results and conclusions of the report by comparing with other publications and findings.

Major Thematic Areas of the Critique

We have divided the paper into major thematic areas of critique involving model calibration, estimation of design rainfall for extreme events, estimation of peak flow, analysis of dam storage capacity and downstream flood modeling.

On Hydrological Model Calibration

The report used the WMS version 7.1 hydrologic modeling with HEC-1 to estimate flow at the GERD. The report used the curve number (CN) method, which is determined based on soil group and land cover type, to estimate surface runoff. Although the authors presented land use (Fig. 17.7) and soil hydrologic maps (Fig. 17.8) for the UBN basin, they did not use these data to determine the CN number and properly estimate the surface runoff. Moreover, the authors did not provide references about the sources of the soil and land use data. For a basin with a size of about 176,000 km^2, the report just used a single CN value of 63 (page 551). Such wrong use of CN method yields unreasonable estimates of surface runoff, which is a critical component of the streamflow. Moreover, as the CN method is sensitive, calibration and validation should have been done using observed data. In case of limited access to observed streamflow data, the report could have used globally available observed streamflow data such as from the Global Runoff Data Centre (GRDC 2020). Discharge data at El Diem, which is near to the GERD, could also have been used to calibrate and validate the model. Moreover, there are many advances in calibrating ungauged basins to reduce predictive uncertainty of flow estimation that were developed by the research initiative Predictions in Ungauged Basins (PUB) (2003–2013) (Sivapalan et al. 2003). Such initiatives by the International Association of Hydrological Sciences (IAHS) provide insights and soft approaches to calibrate and validate hydrological models in ungauged basins (van Emmerik et al. 2015). In summary, it is incorrect to use a single CN value to estimate surface runoff for large catchments like the GERD. In addition, the report did not make an effort to use appropriate data to run the HEC-1 hydrological model, which requires detailed data on soil, land use, topography, river cross section and roughness. The report also did not use any calibration method to fine-tune model parameters and/or evaluate model outputs against observed data.

On Estimation of Design Rainfall for Extreme Event

The report characterized the UBN basin's rainfall from three stations of Bahar Dar, Combolcha and Gondar mentioning that there is data scarcity, and later on in the report, a single weather station, Debremarkos, has been used to develop intensity–

duration–frequency (IDF) curves. As such, the report made major flaws in its rainfall data generation, which is a core input to the hydrological model. First, the report used rainfall data from only three stations for hydrologic simulation with the hydrological model. These stations do not represent the diverse climate of the UBN basin covering about $\sim 176{,}000\ \mathrm{km}^2$ with extreme spatial and temporal rainfall variability (Abtew et al. 2009; Taye and Willems 2012, 2013; Melesse et al. 2011; Mengistu et al. 2014; Abera et al. 2016; Fenta et al. 2017; Berihun et al. 2019). Second, the report indicated data limitation for their decision to use just three weather stations despite the fact that there are more than 40 meteorological stations within, and some very close to, the UBN basin (http://www.ethiomet.gov.et/). Third, the report used an IDF of a single station for the whole area as large as UBN, which is certainly not representative. Despite strong suggestion by studies such as Tefera et al. (2006) to estimate IDF values within a radius of about 25 km of the respective stations, the report used data from a site, which is 300 km far from GERD. Sivapalan and Blöschl (1998) also found that when IDF was estimated for a basin, the spatial mean was much smaller than the point IDF rainfall. This means when basin size gets larger like the UBN basin, the deviation of a single station IDF from the basin IDF gets larger. Fourth, the IDF curve concept is not applied for large basins as UBN. Its application is limited to design of small structures such as storm drainage system in highway and urban areas, irrigation field drainage, with smaller drainage areas. Fifth, while the IDF for the historical rainfall were generated using highly questionable assumptions, the report assumed that the potential future daily rainfall in the basin will be 100 mm. The authors could have used future climate data based on General Circulation Models (GCMs) to get scientifically justified future rainfall in the basin.

Overall, the report did not use reliable rainfall data or generated IDF curves following robust methodological approaches that produce representative rainfall for the different sub-basins in the UBN basin. The various assumptions and unreliable data therefore led to misleading results on different components of the hydrology such as peak flow and soil moisture estimations. It is therefore concluded that because of the lack of calibration and validation as well as due to inappropriate use of data, the reported results pertaining to runoff estimation are not valid.

For a robust analysis, the report should have used actual observed rainfall data from the Ethiopian National Meteorological Agency and apply spatial interpolation methods to get a spatially representative rainfall for the entire basin. If, for some reason, this option was unavailable to the authors, the report could have used well-studied global data sources (Dile and Srinivasan 2014; Abera et al. 2016; Worqlul et al. 2017; Fenta et al. 2018). These types of data provide gridded daily rainfall that better represents the spatio-temporal rainfall distribution in the basin with reasonable bias. Another option could have been the use of empirically derived areal reduction factors (ARFs) to drive basin IDF from point IDF (e.g., Svensson, and Jones 2010; Srikanthan 1995). Generally, the basin IDF is smaller than the point IDF, and the reduction is much sharper for larger basin and shorter duration events (Sivapalan and Blöschl 1998). The ARFs could in fact be generated by coupling high spatio-temporal remote sensing products such as Global Precipitation

Mission (GPM) and the data available from the few stations that exist in the basin. Advanced stochastic regional modeling could also have been used to produce space–time rainfall fields that can generate different design rainfall information for ungauged areas (Rodríguez et al. 2013; Martins et al. 2017).

On the Use of Unit Hydrograph for Runoff Estimation and Routing

Besides the use of wrong IDF generation, inadequate model setup and absence of calibration and validation of surface runoff estimation, the report did not use appropriate topographic parameters to estimate the remaining runoff and associated processes properly. The HEC-1 is based on digital representation of the basin topography and morphometry such as watershed and sub-basin areas, sub-basin flow length, sub-basin slope and time to peak to estimate channel processes (HEC 1981). However, the report did not give necessary attention to (a) generate such physical parameters, (b) describe the assumptions and conditions used to generate the underlying topographic inputs and (c) the way in which they fine-tuned the model. In addition, the report used a single time-lag value of 20 min to force higher peak flow.

The report used a unit hydrograph (UH) approach to estimate discharge and time to peak. Although the representation of topographic information and parameters to estimate peak flow is a well-established practice in hydrology (Rigon et al. 2016), the report failed to provide information on how the UH and associated morphological parameters were developed in their analysis which are sensitive to threshold area. The idea of UH is that the holding time of non-interacting and identical rainfall injected to the basin at the outlet is equal to probability density function (pdf) of the holding time (Gupta et al. 1980). This channel and hillslope representation are key for determining the flow paths and runoff flow time. The relative flow path length in hillslope and river channel determines the flow propagation processes and peak flow level (Rinaldo and Rodriguez-Iturbe 1996).

To show how flood hydrograph is sensitive to a particular topographic parameter (i.e., threshold area to extract river networks), we have generated a topographic-based UH for the Jemma sub-basin. Jemma sub-basin is selected because studies estimating IDF for Addis Ababa are available (De Paola et al. 2014), and Jemma is one of the nearest basins to Addis Ababa. While maintaining all other variables constant, an increase in a threshold area from a single pixel cell to 3000-pixel cells increases the time to peak from 21 to 41 min (Fig. 15.1). Likewise, the peak can vary from 5500 to 8300 m^3/s. This suggests that a change in one topographic parameter can change the peak flood by 3000 m^3/s and time to peak by 100% to the outlet of Jemma sub-basin. This demonstrates the incorrectly used a single time-lag value of 20 min in order to "force" quick peak flow. Such approaches and cherry picking model parameters and data where the authors were

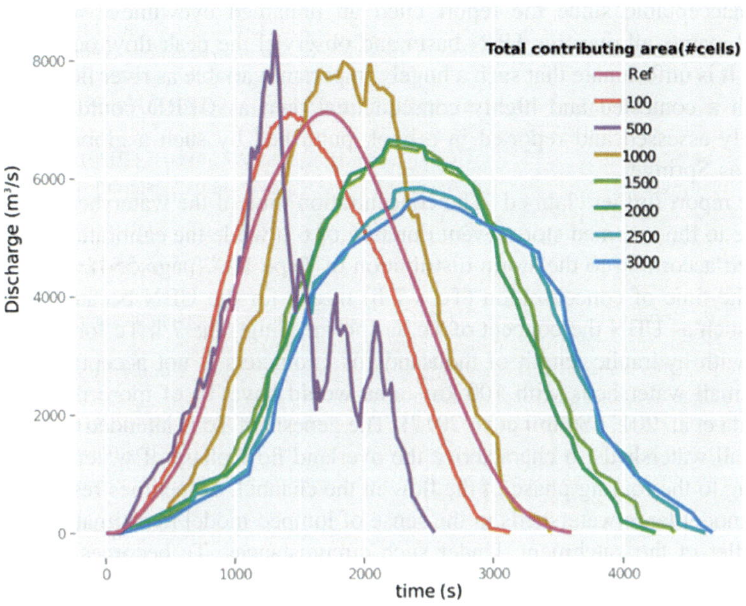

Fig. 15.1 Synthetic unit hydrograph (UH) based on different threshold areas to create stream networks for the Jemma sub-basin of the Upper Blue Nile Basin, Ethiopia

using topographic information that systematically overestimated the peak flow caused faster peak flow to the outlet. A robust approach could have been one that shows the sensitivity of the system to topographic information and that captures the upper and lower boundary of the estimates. Moreover, the authors should have fine-tuned the topographic assumptions and inputs to replicate observed historical records.

On Hydrological Routing and Peak Flow Estimation

Under a subsection *"Hydrologic Model Evaluation", Dandrawy and Omran's report stated that "Depending on the annual hydrograph form at the outlet of the reservoir ..., the flow begins 3 h lag after the storm and then reaches its peak after 7 h. These two findings fully match the data gathered with individuals in the village during site visits (flood incident May 2014). The precipitation began around 6:00 a. m., according to eyewitnesses, while the stream came at 10:00 a.m. and continued for 6–7 h at intensive flow."* (page 554, emphasis added). This is another typical result that critically exposes the methodological flaw of the report. First, there is no way that flow will begin within the same time lag after a storm over the whole of UBN with an area of over 176,000 km^2. The assertion becomes even more strange and unacceptable since the report cited an unnamed eyewitness who watched rainfall events all over the UBN basin and observed the peak flow occurring after 6–7 h. It is unfortunate that such a hugely important variable as river flow dynamics in such a contested and highly consequential dam as GERD could be so amateurishly assessed and reported in a book published by such a global publishing house as Springer.

The report further claimed "The concentration time of the watershed Tc = 7.0 h is close to the assumed storm event duration of 6 h while the calibration of Paris is required according to the storm distribution of Type 1A" (page 554). It is not sure what the time of concentration (Tc = 7 h) means for the UBN because for larger basin such as UBN the concept of Tc has no meaning. The 7 h Tc for such a huge basin with hydraulic length of thousands of kilometers is not acceptable at all as even small watersheds with 100 km^2 area would have Tc of more than 10 h (de Almeida et al. 2017; Salimi et al. 2017). The genesis of Tc is intended to be applied for small watersheds to characterize the overland flow phase of watersheds before entering to the routing phase of the flow in the channel. Sometimes researchers use Tc to model large watersheds in the sense of lumped model to estimate the flow at the outlet of the catchment. Under such circumstances, Tc becomes one of conceptual model parameters to be calibrated. The lack of model calibration in this work invalidates the use of Tc for large watershed, like UBN. Such erroneous outcomes may result from two major methodological drawbacks. First, streamflow routing was not considered indicating that all the runoff productions from the 11 sub-basins of the UBN basin meet at the outlet at the same time. This assumption ignores the time lag that will occur for the flood generated from the most upstream sub-basin (e.g., Lake Tana sub-basin) and the downstream sub-basins such as Dabus or Beles and presumes flood from all over the basin will take the same time to reach the GERD (see Fig. 17.11 of the report). When all sub-basins are assumed to receive the same amount of rainfall at the same time and time lag between them to arrive at the GERD is ignored, it is obvious that the peak flow will be misleadingly overestimated—as stated in the report. For a small Gilgel Abay catchment (1,609 km^2) in the Lake Tana sub-basin, Tassew et al. (2019) reported flood lag time of 227 min using a calibrated HEC-HMS model, whereas the report under review (i.e., by Dandrawy and Omran) estimated 33.9 min for the whole Lake Tana sub-basin at its Abay-Beshilo confluence with a catchment area of 24,652 km^2. Second, the exclusion of the routing process of the HEC-1 model affected the model result since it did not account for channel storage effects. For example, the report did not account for the flood attenuation effect due to storage outflow characteristics of Lake Tana, the largest Lake in the UBN basin. If flood peaks of basins, which have significant storage, are simulated with HEC-1 model without accounting routing, the flood peak will significantly increase (Sui 2005). All in all, the report's authors could have made a simple test using river flow and rainfall data to check, at least, the order of magnitude of their results. In addition, there is an implied suggestion in Dandrawy and Omran's report to calibrate the model Paris, without any indication it was actually applied.

In addition to unreliability and inaccurate model assumptions, the results in the report did not add up. Here, we have used the authors' own assumptions and data to evaluate their results using an inverse approach. These assumptions are (1) a future rainfall of 100 mm, (2) runoff coefficient of 23% (with 77% loss) (page 555), and (3) a catchment area of 176,000 km^2 for the UBN basin (Conway 2000). These data will generate 4,048,000,000 m^3 of water (i.e., 0.1 m rainfall over 176,000, 000,000 m^2 catchment area with 0.23 runoff coefficient). However, the authors' estimated flood value for a future rainfall of 100 mm was 8,492,585,048 m^3, which is twice as large as the results generated using our indirect validation. Further, inconsistencies are also prevalent. For example, for the 50- and 100- year return period, they reported on page 571 that "The peak floods are 157 and 223.6 m^3/s for both return periods with a runoff coefficient of 23% and 31.5%", whereas on page 560 (Table 17.3), they reported peak floods of 11,670 and 44,330 m^3/s, respectively, contradicting flow rates. Such a result shows that the authors did not put effort to check the conflicting results in their report.

On Elevation–Area–Storage Volume Curve

As demonstrated in this critique paper, model inputs (used in the report under review) such as the IDF, CN and other topographic parameters were not accurate, and thereby, generated model outputs using these data (which include peak flood, lag time, and time of concentration) are wrong as shown with cross validation efforts. Likewise, the report brought out wrong results and inferences regarding GERD's storage volume, which was a result of inaccurate model outputs. The report asserted that "the Renaissance Dam's GIS-simulated reservoir at 606 masl (i.e., 100 m maximum water depth) showed that the lake would cover roughly 745 km^2 and that the quantity of water stored would reach [74] billion cubic meters" (pages 561, [74] in parenthesis added from abstract of the report). On the other hand, in their Table 17.7 (page 563), the authors showed that at 606 masl (i.e., 100 m water depth) the reservoir capacity is about 17.5 billion cubic meters. This reveals a difference of about 56 billion cubic meters of storage capacity between what is reported in the abstract and results of analysis. This is another evidence of critical failure of the report. In addition, it is important to note that the report erroneously reported 100 m as the maximum water depth of the GERD reservoir without any reference. Nonetheless, at its maximum capacity of 74 billion cubic meter, the lake area size is not 745 km^2. Based on the reservoir elevation–area–capacity curve that we developed using SRTM DEM (Fig. 15.2), the reservoir volume would reach 74 billion cubic meters at about 645 masl, not at 606 masl; and the corresponding area at 645 masl is about 1800 km^2, not 745 km^2. The maximum depth of water in the reservoir is close to 145 m as the riverbed elevation at the dam site is 500 masl (Abtew and Dessu 2019).

The authors reported incorrect and unrealistic GERD's reservoir (lake) average width as 10,748 km, with a maximum width of about 37,941 km (page 564). This

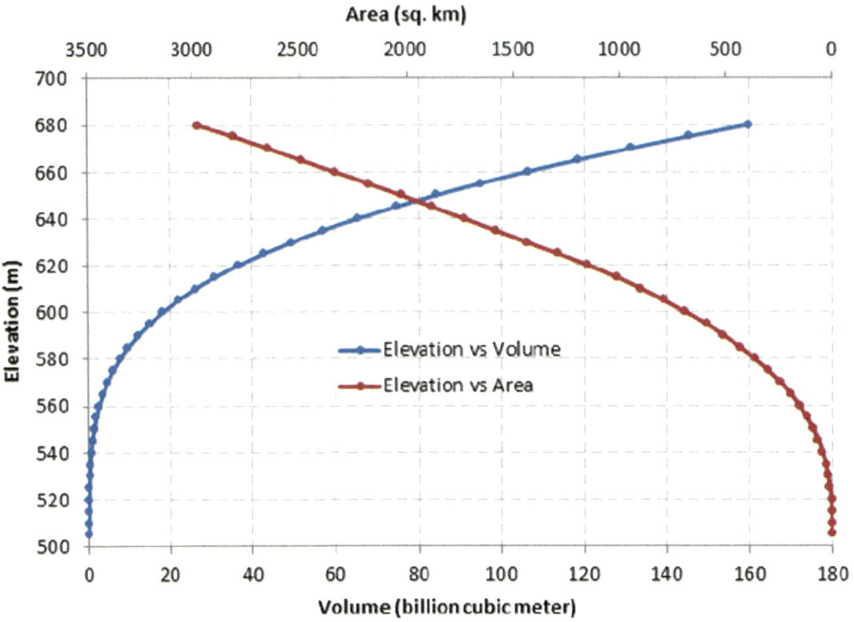

Fig. 15.2 Elevation–area–capacity curve of GERD reservoir produced from SRTM DEM to compare with the values presented in the report

is 7 times the width of Ethiopia. Ethiopia (north to south) is about 1577 km, and west to east is about 1639 km. In addition, Africa extends about 8000 km north–south and 7400 km east–west.

On GERD's Safety Based on Geological, Geotechnical and Seismological Perspectives

The report stated that "*Safety surveys on dams have shown that the Ethiopian GERD Dam's safety factor is only 1.5 degrees from 9 degrees. It is more probable that the GERD Dam will collapse. Experts said the dam was created to collapse. The safety of the dam is very low. Any earthquake will damage this dam and it is a seismic zone adjacent to the African groove*" (page 572). When making such strong statements, the report did not mention any dam safety analysis, evidence or reference.

By virtue of their construction nature, concrete dams resist different forces with the worst loads combinations and thus do not fail due to earthquakes because they are built considering combinations of all possible risk scenarios (Anderson et al. 1998). In addition, no large-size concrete dams constructed in recent years had failed because of several factors such as advancement in dam technology, better investigation of foundation features and quality control on the construction process and materials (Zhang et al. 2016). This indicates that the failure probability of concrete dams such as the GERD, which is being constructed using state-of-the-art technology and where the site was well investigated, is negligible. In addition to this, extensive ground improvement methods such as consolidation and contact grouting were carried out to ensure the highest safety of the dam (Pietrangeli and Rossini 2017).

With regard to the geology of the area, the report is filled with inaccurate information. For instance, on page 536, the report stated that the UBN basin is dominated by basement rocks, which are predominantly acidic to basic rocks, which include quartzite, granites, granodiorite genisses, diorite, metasediments and metavolcanics. However, none of these rocks is basic. The authors of the report also claim that "*Rainwater speeds up in regions with a steep slope and the likelihood of rock erosion and landslides rises (Elmahdy et al.* 2016*; Mohamed* 2016*; page 536)*". This is another misleading and unfounded statement because landslides are not affected by the velocity of overland flow; they can be caused and triggered by other factors such as long, uninterrupted rains for days, unfavorable structural settings, deforestation, river bank erosion, soil characteristics and hillslope gradient. (Popescu 2002).

The report characterizes the rocks of the basin based on large basement information from all over Ethiopia and the Horn of Africa and concludes that the lithology of the GERD site is erodible and undesirable (page 538–539). Such broader assessment not only conceals local variability in the petrographic composition, structure and mechanical resistance but also lacks site-specific information to support the claim of the report. Such claims can only be made after investigating data on cleavage and fractures obtained from preliminary inspection drillings that can enable to ascertain the stability, permeability and mechanical resistance of the dam site. Moreover, according to the finer scale geological map of Merla et al. (1973), the dam site is underlain by Precambrian granites and granodiorites, which also matches with the map Kazmin (1972) produced for the area.

The report claimed that statistical assessment was conducted using rose diagrams and reported that the hilly regions, including the GERD location, are structurally regulated by trend fault zones NNE–SSW, NNW–SSE and ENE–WSW (page 539). However, no ground truthing has been made to confirm if those observations are in fact fractures and to assess their characteristics in terms of design parameters of the dam. The report claimed that "Building such a dam on extensive junctions of geological fractures can trigger tremendous water leakage through the conglomerates and boost the time it takes to fill the reservoir" (page 539). Despite the fact that a dam of considerable economic and international significance would not be built on conglomerates, we used the time function of Google Earth to ascertain that the dam base is made of hard granite rock and also confirmed in Pietrangeli and Rossini (2017).

The report stated that earthquakes will damage the GERD since it is in a seismic zone area adjacent to the African groove. This statement is contrary to the earthquake's records (Fig. 17.22, page 568) and seismic distribution map (Fig. 17.23,

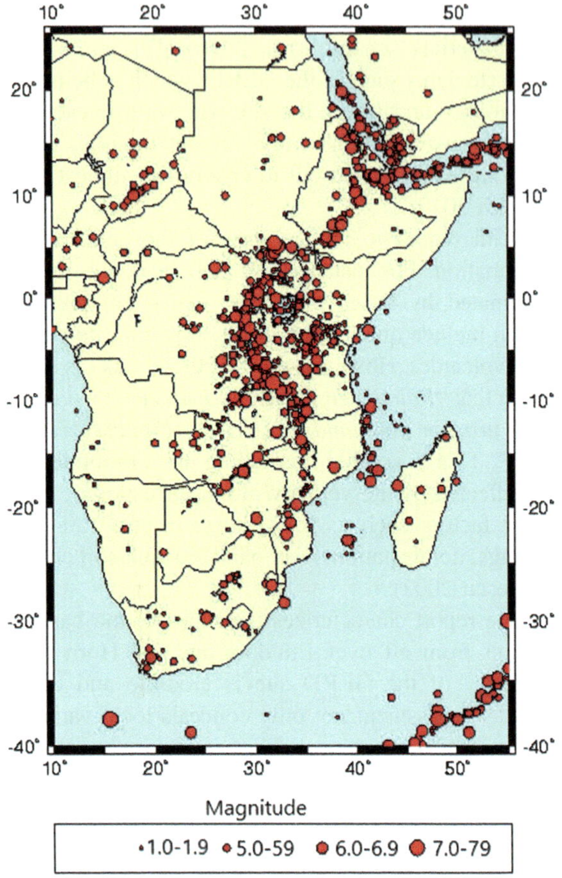

Fig. 15.3 Seismicity of Eastern and South Africa adapted from Turyomurugyendon (1996)

page 569) presented in the same report. As shown on the seismicity map (Fig. 15.3), the GERD is located more than 500 km away from the nearest earthquake site ever recorded. The authors of the report also claimed that three dams failed in Ethiopia (page 569), without making reference to any source. There is no rockfill, or concrete dam that failed in Ethiopia.

In Fig. 15.3, the purple star at the northwestern Ethiopia indicates the dam site, while the reddish dots show previous earthquake incidence. As it is clearly seen from the map, the GERD dam site has not been prone to any earthquake incidence. Across its various sections, the report stated "GERD dam may fail and it may cause flooding to the capital of Sudan" (pages 534, 570–572). However, all these statements are not either scientifically supported or complemented with any dam failure analysis and proper hydrologic and flooding assessment. The report concluded that GERD will fail without dam failure analysis that requires geometric features (location and geometry of failure section such as width, depth, side slope), amount of outflow discharge and time of failure formation (i.e., either gradual or instantaneous dam break) (Veale and Davison 2013; Zhang et al. 2016). In fact, most of the dam breaches in the world are recorded on embankment dams, and breaches on concrete dams are rare. In addition, the report reached to the conclusion of the dam failure without referring to any of the detailed geological, geotechnical and seismological data used for the design of the dam.

Given the justifications provided, it is scientifically unacceptable to report that "GERD is at danger of collapse". The GERD was designed and is under construction with the highest technical standards. As reported by Pietrangeli and Rossini (2017), "seepage through the dam body and foundation after impounding, has been studied and for all the analyzed sections and for all the investigated scenarios, the uplift pressure at roller compacted concrete (RCC) dam/rock contact remains below the correspondent reference case recommended by United States Army Corps of Engineers (USACE) and adopted in the stability analyses of the dam". Pietrangeli and Rossini (2017) also evaluated the seismic behavior by means of linear time history procedure and proved the stability of the dam and its foundation. It should be noted that linear time history analysis involves the direct integration of the equations of motion and is, therefore, the most powerful method available in the literature for evaluating the response of structures to earthquakes (Alembagheri 2016).

On Downstream Flooding

The report stated that one of the main objectives of the study was to map and simulate potential flooded area in the Sudan and Egypt in case of GERD failure (page 535). Likewise, in the conclusion and recommendation sections (page 571), the report claimed that the HEC-RAS program was applied to delineate the flooding zone. Nevertheless, it failed to describe the method employed to delineate the flooding zone and never attempted to conduct any flooding simulation analysis. Rather the report used results from Mohamed and Elmahdy (2017) who reported that the maximum extent of the inundation area below 480 masl in Sudan was 667,228 km^2.

Although available evidence does not point to any possibilities of GERD failure, we conducted dam failure flooding scenario to test the validity of the flooding results given in the report. We conducted the flooding simulation integrating a 30 m resolution Shuttle Radar Topographic Mission (SRTM) DEM with MicroDEM for a reference elevation of below 450 m. We used MicroDEM and 450 m reference elevation to make our results comparable with that generated in the report. Our analysis showed that the flooding area might be about 700,000 km^2 for a reference elevation below 450 m (Fig. 15.4). The report also mentioned that it presented digital floodplain maps that show the volume and depth of flooding in major rivers (page 540). However, although the flooding simulation result revealed that the volume of water inundating Sudan is about 53 trillion m^3, with an average flooding depth of about 75 m, the authors of the report failed to present or omitted the

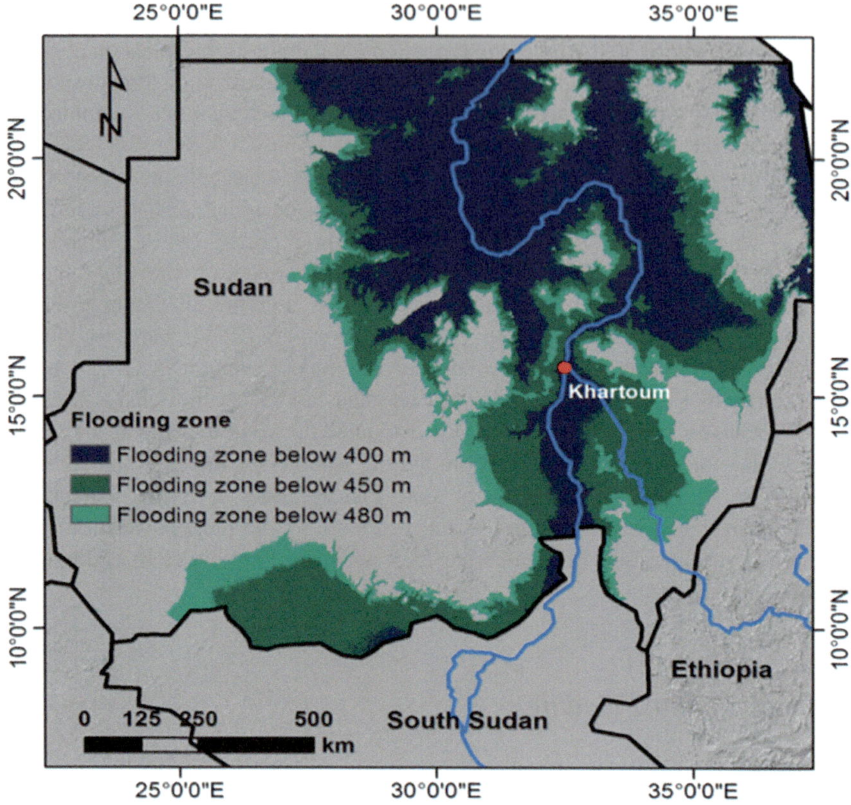

Fig. 15.4 Flooding simulation results at different reference elevations based on a Shuttle Radar Topographic Mission (SRTM) Digital Elevation Model (DEM) and MicroDEM flood basin model

magnitude of flooding in terms of volume and depth. This volume of flooding estimated below 450 m elevation (in the Sudan) is about 700 times higher than GERD's maximum capacity of 74 billion m^3. This is a critical error with misleading results that has the potential to create unwarranted fear on downstream Sudan. The Dandrawy and Omran report has no positive contribution to the ongoing efforts of Ethiopia, Sudan and Egypt on coming to agreement on the filling and operation of the GERD.

Moreover, if one assumes 667,228 km^2 area in the Sudan will be flooded as Mohamed and Elmahdy (2017) reported and quoted in the report, the average depth of flooding water due to unlikely GERD's failure will be about 11 cm. This is a very small depth of flooding that may have little effect in the Sudan. It is unfortunate that the report did not indicate the depth of flooding and its potential effect to support their argument of flooding risk. Moreover, although the report claimed in the introduction that they aim to map potential flooding areas in the Sudan and Egypt (page 535), the report never attempted to estimate any flooding risk in Egypt.

This shows that the main motive of the report was to sensationalize the issue of GERD using emotive words and unfounded results, which, we believe, are aimed at deceiving and destabilizing the Sudanese public with fabricated stories.

Although we applied the MicroDEM model to estimate flooding areas in the unlikely event of GERD failure, the MicroDEM model has its limitation when applied for mega scale dams such as GERD (Guth 2009). For example, the flooding simulation in the MicroDEM assumes that for a given elevation point and reservoir level, all areas below that elevation level will be flooded regardless of existence of natural barriers, or other factors that determine flooding such as river cross section, depth, surface roughness and slope (Guth 2009). Among others, the flooding simulation approach of MicroDEM with regards to GERD is not appropriate because the model is developed to suit flooding simulation due to sea level rise, and its application to simulate flooding in rivers is hardly found in the scientific literature. A Web of Science Database search by fifth of May 2020 with keywords MicroDEM and flooding provided no records of published literature. In addition, the model considers only elevation information and does not account for flow routing and other important river characteristics such as river cross section, depth, surface roughness, slope to simulate flooding in rivers (Guth 2009). Because it is designed for coastline flooding risk analysis, MicroDEM also ignores natural barriers that lie between some lowlands in Sudan and GERD that deter/block flooding (Guth 2009). As such, the chapter erroneously delineates all areas below a given reference elevation as being flooded as is shown by the flooding map in the northeastern part of the Sudan.

Lack of Basic Research Integrity

The report does not even meet some of the basic standards of scientific writing. There are significant plagiarisms in most sections. For example, substantial content was directly taken from Abera et al. (2017). The authors did not paraphrase and/or cite where the contents were taken from. Major claims such as "GERD will probably collapse" are everywhere such as in the introduction and methodology section, before data analysis and result presentations but with no substantial evidence. With regards to some references, the report made substantial misquoting and wrong inferences of referred studies. In some instances, the authors deliberately mention that they have used a given parameter but actually use different value. For example, the flooding simulation by Mohamed and Elmahdy (2017) as reported in Dandrawy and Omran (2020) was done for a reference elevation of below 450 masl, while the latter claimed that it was done for elevation level of less than 480 masl. Nearly, in all sections of the report, there is no justification for selecting a given method/approach whatsoever. In addition, references/sources are not given as to where some very basic data such as soil, land use and geology have come from. This is practically not allowed as data sources should be properly acknowledged or if the report has produced those data, adequate references should have been given about the methods employed.

Mismatch Between Stated Objectives and Content of the Report

The title starts with "*Integrated watershed management of the Grand Ethiopian Renaissance Dam ...*" but there is nothing about watershed management in the content. The report claimed one of its objectives to be "*use the HEC-1 model to create a hydrological modeling rain-flow form; ... and estimate the Upper Blue Nile (UBN) basin's water budget ...; and then to forecast the basin's hydrological reaction to climate change and land-use situations*" (page 535). However, the report neither presented any component of the water budget of the UBN basin except the surface runoff, nor evaluated the hydrological system change due to climate change. The report also claimed that their work "*... is a methodological research, in that it delineates various methodologies to overcome the data scarcity*" (page 535). Nevertheless, it failed to show any approaches used to resolve the claimed data scarcity challenges. To the contrary, the report did not make any attempt to use existing local or global datasets as indicated in this critique chapter. Moreover, the report did not provide methods on how results were derived almost across all of its sections.

Concluding Remarks

Our critical evaluation showed that the book chapter by Dandrawy and Omran (2020) has not followed robust methodological approaches and has not used sound data and assumptions in its analysis. Using direct and inverse methodological approaches as well as existing evidence, we showed that most of the results are wrong and unreasonable. We showed our concern that those unfounded results were used to infer misleading conclusions. In summary, the major drawbacks of the report are (1) use of inappropriate rainfall data for hydrological model setup to simulate peak flood and volume without any model calibration and validation; (2) blatant conclusions about the dam failure due to earthquake without properly evaluating the dam design and safety protocols; (3) use of the MicroDEM model out of context for flooding simulation and wrongly claiming that a potential failure of GERD may flood a large swath of Sudan at unreal average water depth; 4) reaching unverified and erroneous conclusions on downstream impacts of GERD with potential of creating conflict between riparian countries.

It is therefore regrettable to see that a report that transgresses basic scientific standards of avoiding plagiarism, falsification and fabrication was published by a reputable publisher like Springer House. We recommend Springer to revoke the book chapter before this work causes further damage to the thinking processes of young and upcoming researchers and possibly before the work is used to inform discourse of a strategic nature. We also urge the scientific community to keep its guard on the use of corrupt data, model and technical analysis on the Nile Basin, GERD and other tributaries to advance water share interests.

References

Abera W, Brocca L, Rigon R (2016) Comparative evaluation of different satellite rainfall estimation products and bias correction in the Upper Blue Nile (UBN) basin. Atmos Res 178:471–483

Abera W, Formetta G, Brocca L, Rigon R (2017) Modeling the water budget of the Upper Blue Nile basin using the JGrass-NewAge model system and satellite data. Hydrol Earth Syst Sci 21 (6):3145–3165

Abtew W, Dessu SB (2019) The Grand Ethiopian Renaissance Dam on the Blue Nile. Springer, New York

Abtew W, Melesse AM, Dessalegne T (2009) Spatial inter and intra-annual variability of the Upper Blue. Hydrol Process 3082:3075–3082

Alembagheri M (2016) Earthquake damage estimation of concrete gravity dams using linear analysis and empirical failure criteria. Soil Dyn Earthq En 90:327–339

Anderson C, Mohorovic C, Mogck L, Cohen B, Scott G (1998) Concrete dams case histories of failures and nonfailures with back calculations. DSO-05

Berihun ML, Tsunekawa A, Haregeweyn N, Meshesha DT, Adgo E, Tsubo M, Masunaga T, Fenta AA, Sultan D, Yibeltal M, Ebabu K (2019) Hydrological responses to land use/land cover change and climate variability in contrasting agro-ecological environments of the Upper Blue Nile basin, Ethiopia. Sci Total Environ 689:347–365

Conway D (2000) The climate and hydrology of the Upper Blue Nile River. Geogr J 166(1):49–62

Dandrawy ME, Omran ESE (2020) Integrated Watershed Management of Grand Ethiopian Renaissance Dam via Watershed Modeling System and Remote Sensing. In: Elbeih SF, Negm AM, Kostianoy A (eds) Environmental remote sensing in Egypt. Springer, Cham, pp 533–574. https://doi.org/10.1007/978-3-030-39593-3_17

de Almeida IK, Almeida AK, Gabas SG, Sobrinho TA (2017) Performance of methods for estimating the time of concentration in a watershed of a tropical region. Hydrol Sci J 62 (14):2406–2414. https://doi.org/10.1080/02626667.2017.1384549

De Paola F, Giugni M, Topa ME, Bucchignani E (2014) Intensity-Duration-Frequency (IDF) rainfall curves, for data series and climate projection in African cities. SpringerPlus 3 (1):133

Dile YT, Srinivasan R (2014) Evaluation of CFSR climate data for hydrologic prediction in data-scarce watersheds: an application in the Blue Nile River Basin. J Am Water Resour Assoc 50:1226–1241. https://doi.org/10.1111/jawr.12182

Elmahdy S, Marghany M, Mohamed M (2016) Application of a weighted spatial probability model in GIS to analyse landslides in Penang Island, Malaysia. Geomatics Nat Hazards Risk 7:345–359

Fenta AA, Yasuda H, Shimizu K, Haregeweyn N, Kawai T, Sultan D, Ebabu K, Belay AS (2017) Spatial distribution and temporal trends of rainfall and erosivity in the Eastern Africa region. Hydrol Process 31(25):4555–4567

Fenta AA, Yasuda H, Shimizu K, Ibaraki Y, Haregeweyn N, Kawai T, Belay AS, Sultan D, Ebabu K (2018) Evaluation of satellite rainfall estimates over the Lake Tana basin at the source region of the Blue Nile River. Atmos Res 212:43–53

Gebreluel G (2014) Ethiopia's Grand Renaissance Dam: ending Africa's oldest geopolitical rivalry? Wash Q 37(2):25–37

Goodstein D (2002) Scientific misconduct. Academe 88(1):28–31. http://dbbs.wustl.edu/curstudents/Documents/Misconduct_Goodstein_2002.pdf

GRDC (2020) Global Runoff Data Centre (WWW Document). https://www.bafg.de/GRDC/EN/Home/homepage_node.html

Fleet CM, Rosser MFN, Zufall RA et al (2006) Hiring criteria in biology departments of academic institutions. Bioscience 56:430–436

Gupta VK, Waymire E, Wang CT (1980) A representation of an instantaneous unit hydrograph from geomorphology. Water Resour Res 16(5):855–862

Guth PL (2009) Geomorphometry in MICRODEM. In: Hengl T, Reuter HI (eds) Geomorphometry: concepts, software, applications. Developments in Soil Science Series, Elsevier, pp 351–366

Hydrologic Engineering Center (US) & Water Resources Support Center (US) (1981) HEC-1 flood hydrograph package: User's manual. US Army Corps of Engineers, Water Resources Support Center, Hydrologic Engineering Center

Kahsay TN, Kuik O, Brouwer R, van der Zaag P (2015) Estimation of the transboundary economic impacts of the Grand Ethiopia Renaissance Dam: a computable general equilibrium analysis. Water Resour Econ 10:14–30

Kazmin V (1972) Geological map of Ethiopia, 1: 2,000,000 Scale. Ethiopian Institute of Geological Survey, Addis Ababa

Khadilkar SS (2018) Scientific misconduct: a global concern. J Obstet Gynecol India 68(5):331–335. https://doi.org/10.1007/s13224-018-1175-8

Liersch S, Koch H, Hattermann FF (2017) Management scenarios of the Grand Ethiopian Renaissance Dam and their impacts under recent and future climates. Water 9(10):728

Martins D, Kruk NS, Magni NLG, Queiroz PIB (2017) Comparison of two methods for the determination of the intense rainfall equation for Caraguatatuba (SP). DAE J 207(5):34–49

Melesse AM, Abtew W, Setegn SG, Dessalegne T (2011) Hydrological variability and climate of the Upper Blue Nile River basin. In: Nile River Basin. Springer, Dordrecht, pp 3–37

Mengistu D, Bewket W, Lal R (2014) Recent spatiotemporal temperature and rainfall variability and trends over the Upper Blue Nile River Basin, Ethiopia. Int J Climatol 34(7):2278–2292

Merla G, Abbate E, Canuti P, Sagri M, Tacconi P (1973) Geological map of Ethiopia and Somalia, scale 1:2.000.000. Consiglio Nazionale delle Ricerche, Italy

Mohamed M (2016) Mapping of tecto-lineaments and investigate their association with earthquakes in Egypt: a hybrid approach using remote sensing data. Geomatics Nat Hazards Risk 7:600–619

Mohamed M, Elmahdy S (2017) Remote sensing of the Grand Ethiopian Renaissance Dam: a hazard and environmental impacts assessment. Geomatics Nat Hazards Risk 8(2):1225–1240

National Research Council (2003) Sharing publication-related data and materials: responsibilities of authorship in the life sciences. Washington, DC: The National Academies Press. https://doi.org/10.17226/10613

Pietrangeli G, Bezzi A, Rossini C, Masciotta A (2017) Design of Grand Ethiopian Renaissance RCC main dam. https://www.researchgate.net/publication/319317235_Design_of_Grand_Ethiopian_Renaissance_RCC_main_dam

Popescu ME (2002) Landslide causal factors and landslide remediatial options. In: 3rd international conference on landslides, slope stability and safety of infra-structures. CI-Premier PTE Ltd, Singapore, pp 61–81

Rigon R, Bancheri M, Formetta G, de Lavenne A (2016) The geomorphological unit hydrograph from a historical-critical perspective. Earth Surf Proc Land 41(1):27–37

Rinaldo A, Rodriguez-Iturbe IGNACIO (1996) Geomorphological theory of the hydrological response. Hydrol Process 10(6):803–829

Rodríguez R, Navarro X, Casas CC, Ribalaygua J, Russo B, Pougetd L, Rodriguez AR (2013) Influence of climate change on IDF curves for the metropolitan area of Barcelona (Spain). Int J Climatol:440
Salimi ET, Nohegar A, Malekian A, Hoseini M, Holisaz A (2017) Estimating time of concentration in large watersheds. Paddy Water Environ 15:123–132. https://doi.org/10.1007/s10333-016-0534-2
Sivapalan M, Blöschl G (1998) Transformation of point rainfall to areal rainfall: intensity-duration-frequency curves. J Hydrol 204(1–4):150–167
Sivapalan M, Takeuchi K, Franks SW, Gupta VK, Karambiri H, Lakshmi V et al (2003) IAHS decade on predictions in ungauged basins (PUB), 2003–2012: shaping an exciting future for the hydrological sciences. Hydrol Sci J 48(6):857–880
Srikanthan R (1995) A review of the methods for estimating areal reduction factors for design rainfalls. CRC for Catchment Hydrology, Department of Civil Engineering, Monash University
Svensson C, Jones DA (2010) Review of methods for deriving areal reduction factors. J Flood Risk Manage 3(3):232–245
Sui J (2005) Estimation of design flood hydrograph for an ungauged watershed. Water Resour Manage 19(6):813–830
Tassew BG, Belete MA, Miegel K (2019) Application of HEC-HMS model for flow simulation in the Lake Tana basin: the case of Gilgel Abay catchment, upper Blue Nile basin, Ethiopia. Hydrology 6(1):21
Tawfik R, Dombrowsky I (2017) GERD and hydropolitics in the Eastern Nile: from water-sharing to benefit-sharing? In: The Grand Ethiopian Renaissance Dam and the Nile Basin. Routledge, pp 113–137
Taye MT, Willems P (2012) Temporal variability of hydroclimatic extremes in the Blue Nile basin. Water Resour Res 48(3)
Taye MT, Willems P (2013) Identifying sources of temporal variability in hydrological extremes of the upper Blue Nile basin. J Hydrol 499:61–70
Tefera C, Yitaye M, Seleshi Y (2006) Rainfall intensity-duration-frequency relationship for Northern Ethiopia. Zede J 23:29–38
Turyomurugyendo G (1996) Some aspects of seismic hazard in the east and south African region. Unpublished. M. Sc. Thesis, Institute of Solid Earth Physics,University of Bergen, Bergen, Norway, p 80
Van Emmerik T, Mulder G, Eilander D, Piet M, Savenije H (2015) Predicting the ungauged basin: model validation and realism assessment. Front Earth Sci 3:62
Veale B, Davison I (2013) Estimation of gravity dam breach geometry. In: Paper presented at the Multiple use of dams and reservoirs: needs, benefits and risks. Proceedings of the NZSOLD/ANCOLD Conference, Rotorua, New Zealand
Wheeler KG, Basheer M, Mekonnen ZT, Eltoum SO, Mersha A, Abdo GM et al (2016) Cooperative filling approaches for the grand Ethiopian renaissance dam. Water Int 41(4):611–634
World Economic Forum (2018) Code of Ethics. World Economic Forum, Geneva
Worqlul AW, Yen H, Collick AS, Tilahun SA, Langan S, Steenhuis TS (2017) Evaluation of CFSR, TMPA 3B42 and ground-based rainfall data as input for hydrological models, in data-scarce regions: the upper Blue Nile Basin, Ethiopia. CATENA 152:242–251. https://doi.org/10.1016/j.catena.2017.01.019
Yihdego Z, Rieu-Clarke A, Cascão AE (2016) How has the Grand Ethiopian Renaissance Dam changed the legal, political, economic and scientific dynamics in the Nile Basin?
Zhang L, Peng M, Chang D, Xu Y (2016) Dam failure mechanisms and risk assessment. Wiley. https://doi.org/10.1002/9781118558522

Chapter 16
River Flow Monitoring and Data Quality for Equitable Nile Water Sharing

Wossenu Abtew, Semu A. Moges, Menberu Meles, and Muluneh Imru

Abstract Stage and discharge monitoring, data analysis, and interpretation are essential for flood control, hydropower operation, navigation, water allocation, and ecological management. In the context of transboundary rivers, hydrometric measurements are crucial to establish and maintain legal regimes of water allocation and operation of hydraulic infrastructures in the basin. In the case of the Nile, where there is no legal regime or prior comprehensive agreement, there is serious need for accurate data for negotiation and establishment of such legal regimes. Quality of stream flow data is dependent on who does the stream gauging, method of stream gauging, data acquisition, data transfer, and data storage. Long-term historical stream flow data is needed to understand basin hydrology, stream flow trend, and changes. Blue Nile stream flow data obtained from different sources are examined in the context of ongoing negotiation of the filling and operation of Grand Ethiopian Renaissance Dam (GERD). The current negotiation is anchored on the data obtained from El Diem Sudan that gives a long-term mean annual flow of 49 billion cubic meters (BCM). Analysis of a set of Blue Nile flow data from different sources and temporal scales indicates significant variation in the data sets. Variability of flow from different records shades doubt on the reliability of using a single series of historical data for long-term negotiation. This paper considers different issues related to flow that affects the GERD negotiation and highlights the importance of accurate streamflow data, instrumentation, and the need for generating new data sets

W. Abtew (✉)
Water and Environment Consulting LLC, Boynton Beach, FL, USA
e-mail: wabtew@gmail.com

S. A. Moges
University of Connecticut, Storrs, CT, USA
e-mail: semu.moges.2000@gmail.com

M. Meles
USDA, Washington, DC, USA
e-mail: menberu.bitew@usda.gov

M. Imru
South Florida Water Management District, West Palm Beach, FL, USA
e-mail: mimru@msn.com

A. M. Melesse et al. (eds.), *Nile and Grand Ethiopian Renaissance Dam*,
Springer Geography, https://doi.org/10.1007/978-3-030-76437-1_16

for long-term operation and agreement. The paper stresses the importance of data quality, the type and location of hydrometric measuring stations and the impact of associated data error and measurement inaccuracies in the context of allocation and management of transboundary waters in general terms. Finally, the chapter advises the importance of developing a protocol for upstream and downstream data sharing from all monitoring systems in the basin in a timely manner. An example of limitation of model generated streamflow data is illustrated.

Keywords Flow monitoring · Blue Nile River · Grand Ethiopian Renaissance Dam · Data quality · Water sharing · Nile basin

Introduction

In any river basin, the importance of streamflow and water-level data is critical for flood control, hydropower operation, navigation, water allocation, and ecological management. In the case of transboundary rivers, the importance of acceptable quality streamflow data is high. The accuracy of the streamflow measurements in the Blue Nile can be affected by the sizeable contributing area that receives localized highly variable tropical rainfall systems over a dominantly steep slope that facilitates runoff, causing abrupt changes in the amount of flow in a short period of time. Quality of streamflow data depends on monitoring entity, method, data acquisition, and archival. Long-term historical streamflow data is needed to understand basin hydrology and transboundary water sharing negotiations.

The United States Geological Survey (USGS) is one of the prominent institutions that measure streamflow. And the process is called stream gauging in the US. Stream gauging by the USGS started in the late 1800s. The first stream gauging in the USA by the USGS was on the transboundary river, Rio Grande, flowing from the USA to Mexico. The purpose was to evaluate if there was enough water for irrigation development and western expansion in the state of New Mexico (USA). Currently, the USGS operates over 8500 stream gauging stations with continuous records (USGS.gov). USGS has made streamflow measurements at about 37,000 sites at different times. The USGS has developed the standard for stream gauging which has improved the monitoring technology and data processing over the years leading to improved accuracy and reliability of streamflow measurements under a wide range of hydrologic conditions. Stream flow data in the USA is used for a variety of purposes including flood protection, water allocation, design, operation of dams and water control structures which demanded a more accurate and timely data sets. The USGS collects streamflow data and makes available to users, without charge through publicly accessible databases. Such information is used for any water sharing issues at any level. The nationally consistent data is impartially used to settle water sharing issues for interstate and international water compacts and agreements (USGS 2018).

Streamflow measurement generally consists four steps: (1) measuring stream stage or height above an established level at the lowest point in the riverbed called zero level or from an elevation reference datum, (2) determining cross-sectional area from depth and width, (3) measuring mean velocity across the stream cross section at the location, and (4) stage–discharge relation development. The basic procedure to select a cross section for stream gauging is to pick a river section where flow is uniform, not turbulent, and no aggrading and degrading of banks, or no pooling is observed. The accuracy of flow data depends largely on the performance of velocity meters, depth sounding equipment and stage (water level) gauge setup and reading. Corbett (1943) in developing stream gauging manual, predicted that the need for more accurate streamflow data will influence the improvement of equipment. Since then, progress has been made in measurement equipment, and procedures and improvements in flow data quality have been observed.

Similar to other hydrologic systems, errors in the Blue Nile flows could be related to poor stage–discharge relation (Buekham and Dawdy 1970). Other factors that contribute errors to the streamflow observation in the Blue Nile could be: (1) incorrect recording of the stage and time obtained from manual recordings, and data transfer, (2) frequency of recording which is mainly once a day observation for a highly dynamic stage, and (3) the proximity of the recording location from towns/cities. The Blue Nile river is located in a highly incised gorge away from towns and cities. Strict adherence to monitor plan such as following the planned frequency of observation may not be easy due to staff and monitoring program capacity limitations. In the Blue Nile, it is also important to recognize the instability of the riverbanks and the dynamics of hydraulic resistance to flowing water in the alluvial channel. Therefore, future plans to improve the accuracy of the Nile river water measurement may require staff training on observation accuracies, data handling, transfer and archive, frequency of observation, and close follow up of the changes in the hydraulics of the river cross section.

Methods and Technologies in Streamflow Measurements

In both manual and automated streamflow measurement methods, the area and mean velocity of the cross section are required. In the following sections, stream cross-sectional area, average velocity determination and stage monitoring are presented. It is important to mention that remote sensing application to remote stream flow measurement has future in stream gauging although not discussed in this chapter (Samboko et al. 2020).

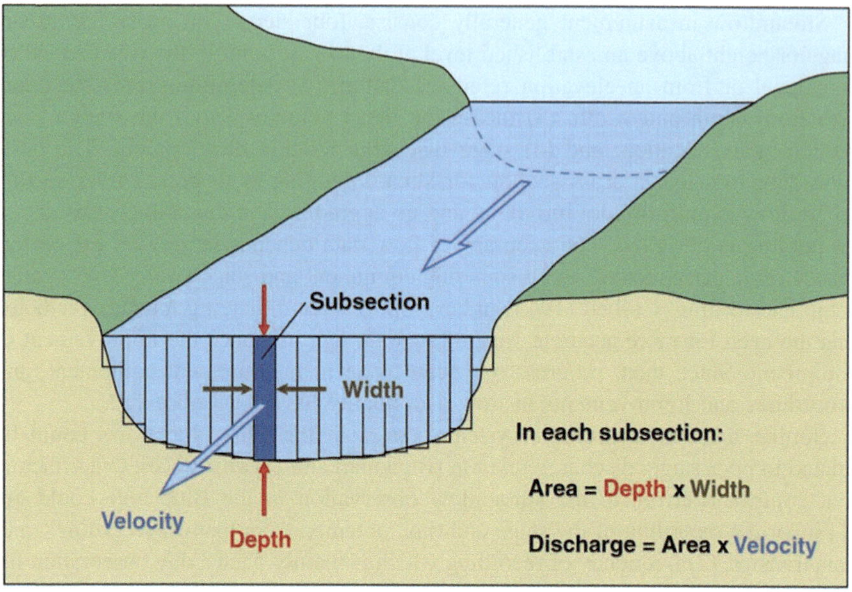

Fig. 16.1 Canal cross section with subsections and predetermined velocity measurement points (USGS)

Stream Cross-Sectional Area Measurement

Area is determined by sectioning a given cross section of river, measuring depth at each subsection, and multiplying the depth by subsection width. The sum of all the subsections area gives cross-sectional area (Fig. 16.1). The simplest method of cross-sectional width is extension of marked tape across the cross section anchored at both ends. Depth can be read with a marked rod at the center of each subsection by wading when stream is shallow or otherwise by boat. One can rarely find the most suitable and accessible cross section for stream gauging. But consideration should be given to look for a cross section with desirable features as uniform depth change, uniform velocity distribution, stable streambed material, minimal turbulence, and the absence of horizontal and vertical angles (Crowell and Mtundu 2000). Advanced methods include the use of digital echo depth finder on a boat and GPS-based RTK technology for cross-sectioning of rivers (Xiao et al. 2009).

Velocity Measurement Methods

Streamflow or the amount of water that flows in a river channel is measured in various ways. Most methods involve average water velocity measurement across a known or measured cross-sectional area (Fig. 16.1; Eq. 16.1). There are several ways of velocity measurement. A simple way, without stream gauging equipment, is to time a floating orange between two cross sections and calculate average velocity from multiple runs. This simple method does not account for horizontal and vertical variations of velocity in the cross section. It can be used to approximate water velocity on the surface. Using this simple method, velocity can be estimated by dividing the distance between the two cross sections by the time it took for the orange to travel. For better accuracy of velocity measurement devices commonly used include Vertical Axis Current Meter, Price AA Current Meter, Electronic Current Meter such as Marsh-McBirney Model 201D, Acoustic Velocity Meter (EG&G Smart Acoustic Current Meter, Acoustic Doppler Current Profiler (ADCP), Ultrasonic Velocity Meter (UVM)). Devices as the Acoustic Doppler Current Profiler (ADCP) do both velocity and cross-sectional area measurement at the same time and output flowrate as product of the two. The equipment is pulled across a stream along a cable or by boat. Multiple runs are averaged to minimize measurement errors.

Acoustic Doppler Current Profiler (ADCP)

An acoustic Doppler current profiler (ADCP) is a hydroacoustic current meter similar to a sonar, used to measure water current velocities over a depth range using the Doppler effect of sound waves scattered back from particles within the water column. The working frequencies of ADCPs range from 38 kHz to several Megahertz. In rivers, the ADCP is used to measure the total water transport. The method requires a vessel with an ADCP mounted over the side to cross from one bank to another while measuring continuously. A cable across the stream anchored on both sides of the bank can be used to pull the device across the stream with a tagline. Measurements can also be made across a bridge by walking along with the ADCP lowered down to the stream water surface with a cable. Discharge is calculated as the dot product between the vector track and the current velocity. Detail data is transferred to a nearby laptop using Bluetooth technology. Generally, four passes are conducted, and results are checked if coefficient of variation is within 5% or additional measurements are performed.

In a project, a StreamPro ADCP was used with Bluetooth USB adapter connection to a laptop. StreamPro ADCP collects complete sets of streamflow measurements in streams or canals in a matter of minutes. Data was collected from a bridge (crossing the canal) or a boat or using a tagline across a canal depending on the flow rate (Fig. 16.2). Data was conveniently acquired using a laptop nearby the bank, equipped with a highly intuitive user interface, WinRiver II Teldyne RD. With this equipment, minimum cell size is 2 cm with up to 30 cells with an upgraded extended profiling range of up to 6 m (Teledynmarine.com, Accessed January 3, 2020).

Systematic errors from ADCP are from methods used for estimating missing parts of velocity profile, improper beam profiler geometry, mispositioning of receiver tracking filters, profiler transmit-filter skew, and others. The systematic errors can be evaluated using other flow meters concurrently (Kinsman et al. 1994).

Fig. 16.2 StreamPro ADCP measuring flow across a canal (SFEC 2020)

Uncertainty in flow measurements with moving boat was studied in Canada with four site measurements and 10 data sets. It was concluded that flow measurement uncertainty sources are unmeasured flow along the cross section in relation to total flow, flow non-uniformity, and operator decisions about instrument programming and choice of measurement cross section (Mueller 2017). A limitation of the ADCP is assumption of flow homogeneity required for an accurate three-dimensional velocity solution. Study showed that flow pattern around the ADCP is influenced by the mere presence of the device inducing negative bias (Mueller 2015). Comparison of flow measurements with Price AA and ADCP methods during the historical flood of the Mississippi River showed that the difference is within expected accuracy of measurements when performed concurrently (O'Brien et al. 2012).

Ultrasonic Method

Velocity is measured using ultrasonic velocity meter (UVM). Using ultrasonic transducers, the flow meter can measure velocity index along the path of an emitted beam of ultrasound, by averaging the difference in measured transit time between the pulses of ultrasound propagating into and against the direction of the flow or by measuring the frequency shift from the Doppler effect. The USGS uses UVMs to measure flow across channels. A study in California reported channel flow velocity index measurement by two transducers placed on opposite channel sides at the same depth. The UVM generates continuous velocity index. In this case, the USGS used acoustic Doppler discharge measurement system (ADDMS) to measure channel cross-sectional mean velocity and develop relationship with UVM index velocity (Oltmann 1993). From the calibrated mean cross-sectional velocity and cross-sectional area, flow rate is computed by Eq. 16.1.

Concerning sources of uncertainty in culvert flow measurement with UVM, Gonzalez Castro and Chen (2005) reported that the quality and reliability of index-velocity-based discharge records depends on uncertainties on rating curve, UVM, and stage measurement.

Stage Monitoring

Staff Gauge for Stage (Water Level) Observation

The most common and simplest method of stage or water-level observation is the staff gauge (Fig. 16.3). A marked rod with elevation readings from a known reference is used to read water levels instreams and ponds. This method requires

Fig. 16.3 Staff gauge for reading stage or water level (Photo by Wossenu Abtew)

manual site visit at the desired frequency and recording observation into a data sheet.

Automated Stage Monitoring Methods

Stage measuring devices range from graphic recorders to digital pressure transducers (Fig. 16.4). With stage recorders, generally a float is lowered to the water level through a stilling well (perforated pipe) and water-level fluctuations are recorded in several ways. Recorders vary by technology as punched-tape recorder, graphic recorder, and digital output that can be transmitted instantaneously. Recent technology includes radar technology with observation out of the stream.

Radar Stage Method

The radar stage sensor is a water-level sensor developed by Forest Technology Systems (FTS). According to its Web site, FTS is a leading manufacturer of remote environmental monitoring solutions including systems, instrumentation and communications technology for the hydrology, fire weather and meteorology industries. The method uses radar to measure the distance between the sensor and the water surface. A radar sensor transmits energy in the direction of the water surface and uses the energy reflected back to the sensor to determine the distance between the

Fig. 16.4 Manual and automatic stage recorders (Photo by Wossenu Abtew)

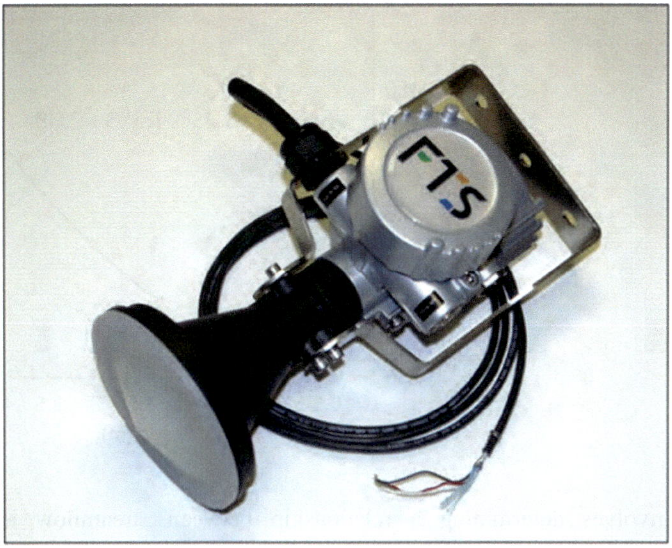

Fig. 16.5 Radar Stage Sensor By Forest Technology Systems, (photographed by Joanne C. Jones, USGS)

radar and the water (Fig. 16.5). By surveying the elevation of the sensor, water surface elevation is derived from the radar measured distance between sensor and water level. Field test measurements of water level with radar and comparison to Sutron Accubar Constant Flow Bubble Gauge showed that the method is suitable for USGS hydrologic data collection (Kunkle 2018).

Streamflow Measurement

Stage–Discharge Method of Flow Determination

The stage–discharge method is commonly used for stream flow monitoring. Once stream cross-sectional area and mean velocity are determined, streamflow or discharge is computed using Eq. 16.1.

$$Q = V_{mc}A_c \tag{16.1}$$

where Q is flow rate in $m^3\ s^{-1}$, V_{mc} is mean cross-sectional velocity ($m\ s^{-1}$), and A_c is stream cross-sectional area in m^2.

Because it is cumbersome to continuously measure streamflow velocity in a river under different stage (water level) and flow rate, the stage–discharge method is used to derive flow rates. This is the most widely used method to gauge streamflow. The

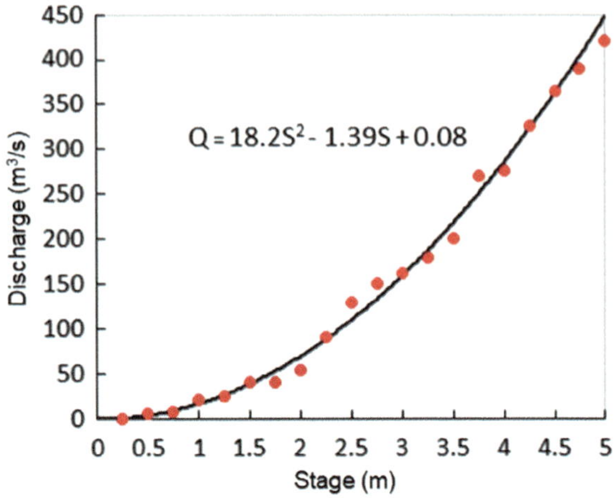

Fig. 16.6 Stage–discharge relationship

concept involves determining a relationship between streamflow and stage. Flow rate can be derived just from stage observation and using the predetermined flow and stage relationship called rating curve development. To develop a rating curve, cross-sectional area and average velocity are concurrently measured at different stages in the stream. Cross-sectional area changes with stage as the depth of water correspondingly changes. Since flow rate is the product of mean velocity and cross-sectional area, multiple velocity and area and stage measurements are conducted under temporal changes of water level. By a process called rating curve development, the relationship of flow rate and stage is determined (Fig. 16.6). A graphic chart or lookup table is developed to readout flow rate just from river stage observation. Stage is manually read from a staff gauge and/or continuously recorded with a water-level recorder (Fig. 16.4).

USGS study shows that the cross section should be divided into 25 to 30 subsections and velocity be measured at 0.2 and 0.8 of depth of stream and averaged to represent the section to address the nonlinear velocity profile through the depth of flow.

Radar Flow Measurement

The flow measurement principle involves a signal with certain frequency transmitted by the radar sensor on to the surface of the water. By detecting reflected signals, velocity is computed using the Doppler principle (https://www.azom.com/article.aspx?ArticleID=15080). Wave formation on the water surface is required for this sensor to measure velocity. Another model is used to translate surface velocity into average cross-sectional velocity. Flow is calculated as in Eq. 16.1.

Fig. 16.7 Nile Basin with major dams and tributaries (Abtew and Dessu 2019)

Wave formation on the water surface is the precondition for the radar technique. The sensor measures the waves' movement and therefore the surface velocity of the water. A single velocity is selectively measured on the water surface. With the help of NIVUS' hydraulic COSP model, it is possible to calculate the average velocity from the selected single velocity. An extra level sensor allows the determination of the wetted area (A). Flow (Q) is calculated as a product of wetted area (A) and the average velocity (V), Eq. 16.1.

Measurement principle is based on radar sensors installed outside of or above the measurement medium. A signal with a certain frequency is transmitted out by the radar sensor. This signal is reflected when it impinges on the water surface. Once the signal is reflected from the water surface, a frequency shift is created. The radar sensor detects the reflected signal, which will be assessed through the Doppler principle.

Acoustic Doppler Current Profiler (ADCP)

An acoustic Doppler current profiler (ADCP) is a hydroacoustic current meter similar to a sonar, used to measure water current velocities over a depth range using the Doppler effect of sound waves scattered back from particles within the water column. The working frequencies of ADCPs range from 38 kHz to several Megahertz. In rivers, the ADCP is used to measure the total water transport. The method requires a vessel with an ADCP mounted over the side to cross from one bank to another while measuring continuously. A cable across the stream anchored on both sides of the bank can be used to pull the device across the stream with a tagline. Measurements can also be made across a bridge by walking along with the ADCP lowered down to the stream water surface with a cable. Discharge is calculated as the dot product between the vector track and the current velocity. Detail data is transferred to a nearby laptop using Bluetooth technology. Generally, four passes are conducted, and results are checked if coefficient of variation is within 5% or additional measurements are performed.

Streamflow rate is measured by ADCP method, by pulling the ADCP equipment across the stream/canal from bank to bank (Fig. 16.2). Measurements can also be made across a bridge by walking along with the ADCP lowered down to the stream water surface with a cable. Boats are also used to pull along the ADCP equipment in deep canals or streams. Discharge is calculated as the dot product between the vector track and the current velocity. Detail data is transferred to a nearby laptop using Bluetooth technology. Table 16.1a and b shows results of stream gauging with and ADCP across a 60 m canal. Four passes were run from the right and left banks, back and forth. The output is averaged and a coefficient of variation of 5% or less within the four measurements is of acceptable quality. In Table 16.1a and b, the standard deviation of the four passes divided by the average gave 3.83% in this event, indicating further measurement was not needed.

Table 16.1 Summary of output from and ADCP four passes across a 60 m canal

(a)

Transect	Start bank	Start time	Total Q (m^3/s)	Delta Q (%)	Top Q (m^3/s)	Meas. Q (m^3/s)	Bottom Q (m^3/s)
b4000	Right	14:58:53	*16.647*	−0.09	0.936	13.534	1.892
b4001	Left	15:10:35	*16.74*	0.47	0.927	13.465	1.978
b4002	Right	15:19:22	*17.408*	4.48	0.957	14.155	2.023
b4003	Left	15:57:15	*15.852*	−4.86	0.897	12.758	1.88
Average			*16.662*	0	0.929	13.478	1.943
Std Dev.			*0.637*	3.83	0.025	0.572	0.069
Std./|Avg. (%)			*3.83*	0	2.68	4.24	3.54

(b)

Transect	Start bank	Start time	Left Q (m^3/s)	Left dist (m)	Right Q (m^3/s)	Right dist (m)	Width (m)	Total area (m^2)	Q/Area (m/s)
b4000	Right	14:58:53	0.122	4.88	0.163	5.49	61.05	224.6	0.074
b4001	Left	15:10:35	0.135	4.88	0.235	5.49	61.04	227	0.074
b4002	Right	15:19:22	0.081	4.88	0.193	5.49	59.68	219	0.079
b4003	Left	15:57:15	0.165	4.88	0.152	5.49	60.53	222.6	0.071
Average			0.126	4.88	0.186	5.49	60.58	223.3	0.074
Std Dev.			0.035	0	0.037	0	0.65	3.37	0.003
Std./| Avg. (%)			27.71	0	19.99	0	1.07	1.51	4.45

Flow Data Availability and Quality

Historical flow data has limitations to be used as baseline data for water agreements due to accuracy issues with earlier period streamflow measurement methods and frequency of discrete observations. The Mississippi River has flow data from 1886 to current with the stage–discharge method. After a thorough review of historical flow data, a study concluded that the pre-1930s flow data does not have sufficient accuracy to be compared with modern flow values (Chester et al. 2013). Factors that affect flow data accuracy are velocity measurement and cross-sectional area determination methods, observation quality, and rating curve fitness to the measured data. Frequency of stage reading and streamflow data derivation frequency as daily, weekly, or other time step creates differences in data quality. Errors associated with data recording, transfer, and data archival, and monitoring system integrity determine data quality. Another factor is data collection period where the mean flow may not be stationary.

In a study of uncertainty in nutrient load, Harmel et al. (2006) compiled different sources of data and calculated 6–19% cumulative probable uncertainty in streamflow data. A study of errors in individual streamflow measurement states that

Table 16.2 **a** Accuracy ratings for single flow measurements the midsection method. **b** Best possible rating of midsection flow measurement

(a)

Rating	Deviation from "True" flow (+ or −) (%)
Excellent	2
Good	5
Fair	8
Poor	>8

(b)

Best possible rating of measurement	Max. % statistical uncertainty (+ or −)
Good fair	1.5
Good	3.5
Fair	7
Poor	>7

sources of errors include cross section (channel width and depth) and mean velocity (instrument error, pulsation error, vertical velocity distribution, oblique flow, stream turbulence). A quasi-quantitative method labels errors for single flow measurements from excellent to poor, Table 16.2a (Sauer and Meyer 1992). When discharge from a rating curve is within a certain percent of an independent "true" measurement, its quality is classified as shown in Table 16.2b (Painter and Loving 2015). This is also USGS flow data quality code (https://help.waterdata.usgs.gov/codes-and-parameters/discharge-measurement-quality-code).

Challenges of Streamflow Measurement in the Nile Basin

The Nile Basin (Fig. 16.7) is one basin where there is no basin-wide water allocation and basin management agreement between riparian states. Historical Nile flow into Egypt varies from source to source and most important is the period of averaging. A maximum of Nile flow of 135 BCM and a minimum of 55 BCM was reported for the period 1870–2006 in the analysis of the impact of Grand Ethiopian Renaissance Dam filling on Lake Nasser's active storage (Ramadan et al. 2015). Annual average flow of 78 BCM with a maximum of 111 BCM and a minimum of 42 BCM, at Dongola, on the Nile, upstream of the Aswan, was reported for the period 1933–2012 (Ahmed et al. 2019). Sadek (2006) reported an average flow into Aswan of 84 BCM. The assumption of stationarity, that mean flow will stay the same through time, may not hold correct. Colorado River has decreased 19% compared to its twentieth century average with models predicting that by 2100, the river flow could fall as much as 55% (Heggie 2020). Transboundary water allocation/sharing based on volumetric flow with reference to historical mean flow poses uncertainty and conflict. Therefore, percentage of prevailing flow-based water allocation\water sharing is a *necessity for* the Nile Basin. Percentage sharing should be used for Nile Basin water share agreements.

Hydrometeorology Monitoring Network in the Blue Nile Basin

The main agency for hydrometeorology monitoring in Ethiopia is the National Meteorological Agency, Ethiopia (NMA-Ethiopia). In the Blue Nile Basin, there are stream gauging stations and meteorology stations both on the main Blue Nile River and tributaries. There are various estimates of the Blue Nile flow at the border between Ethiopia and Sudan. There are various estimates or measurements of Blue Nile River flow in the literature as high as 54.4 bcm (Berhanu 2014) and lower (Table 16.3a). Also, annual mean flow at Dongola varies by the period of record (Table 16.3b).

A study of hydrometeorology monitoring network evaluation and design of upgrade for flood forecasting in the Eastern Nile Basin was performed by Riverside Technologies Inc and submitted to Eastern Nile Technical Regional Office (ENTRO) in 2010. Figure 16.8 shows Blue Nile Basin existing stream gauging and meteorology monitoring networks. Figure 16.9 shows both existing and proposed networks (Riverside Technologies Inc 2010).

Riparian states of the Nile Basin should agree on the UN-supported principle of equitable and reasonable use of shared water resources. For this purpose, riparian states need to agree on flow measurement locations. For water going into and out of the reservoir at GERD, stream flow measurements should be made at selected multiple upstream locations and at least one location downstream. The upstream gauging locations shall be at the entry points of tributaries with meaningful flow along the reservoir.

Table 16.3 **a** Blue Nile flows at El Diem/Border. **b** Nile annual flows at Dongola (Ahmed et al. 2019)

(a)	
Period	Annual mean flow (bcm)
1965–1987	43.22
1990–1996	47.05
2002–2009	50.85
1999–2003	51.92
(b)	
Period	Annual mean flow (bcm)
1933–1972	85.4
1973–2012	70.01

Fig. 16.8 Blue Nile Basin and Surrounding existing hydrometeorology monitoring network (green hydrometric and orange meteorological station), ENTRO, Riverside Technologies Inc 2010

Modeling and Remote Sensing Applications in Blue Nile Basin Flow Estimation

Modeling and remote sensing data are applied with uncertainty in outputs but successfully used for early prediction of famines in East Africa and other regions. To fulfill its global water security interest, the USA has formed an interagency water working group for science and applications, to use models for the analysis of the GERD. According to the March 2019 modeling report on the GERD, understanding the state of global water and implications for US national security requires strategic coordination of the best available science and technical capabilities across the US Government (Globalwaters.org USAID) . The report states that GERD is an example of a water policy issue of interest to the US Government that could benefit

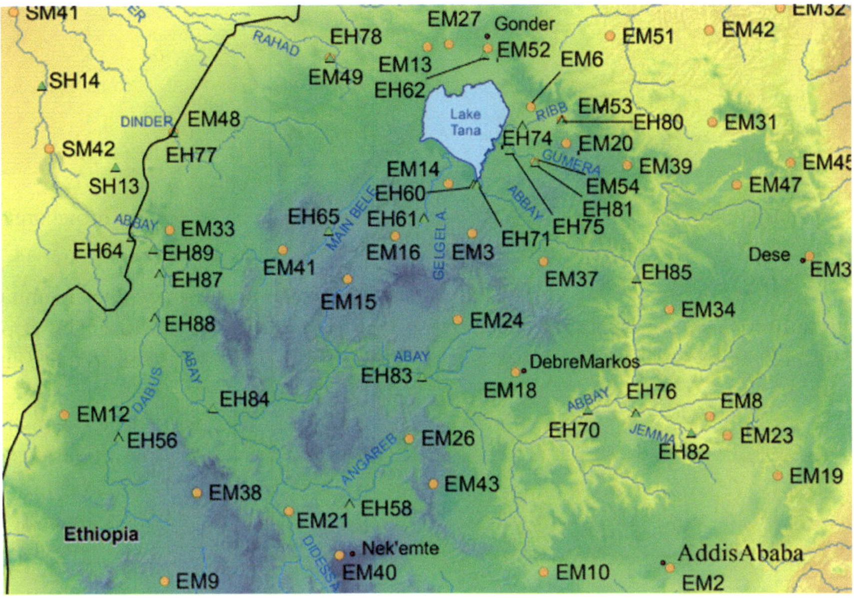

Fig. 16.9 Blue Nile Basin and surrounding hydrometeorology existing and proposed monitoring network (green hydrometric and orange meteorological station), ENTRO, Riverside Technologies Inc 2010

from an integrated modeling approach and inform international negotiations. The US effort started by forming Interagency Water Working Group for Science and Applications Team (ISAT) with resources from NASA, U.S. ACOE and USAF Weather, in 2017.

Figure 16.10 depicts results of six years of model simulated and observed Blue Nile River flow at El Diem by ISAT. Three hydrologic, land and environment, river routing models were applied. HYMAP (Hydrologic Modeling Analysis Platform) is a routing model from NASA. RAPID (FLDAS) is part of the Famine Early Warning System Network (FEWS NET) from NASA and it is a land surface hydrologic model where FLDAS stands for Land Data Assimilation System (McNally et al. 2017). JULES (Joint UK Land Environment Simulator) was originally developed to provide surface boundary conditions for climate models but increasingly used for hydrological simulation and water cycle changes (Zulkafli et al. 2013). The magnitude of differences in modeling and observation data shows appreciation for data quality and flow monitoring. Figure 16.11 is Blue Nile discharge observed (reported) data from 1965–2009. A stark difference from what is in Fig. 16.10 (black line) and in Fig. 16.11 proves the importance of data quality and standardization. In Fig. 16.11, six years of Blue Nile flow data are extracted from a reported flow from 1965 to 2009.

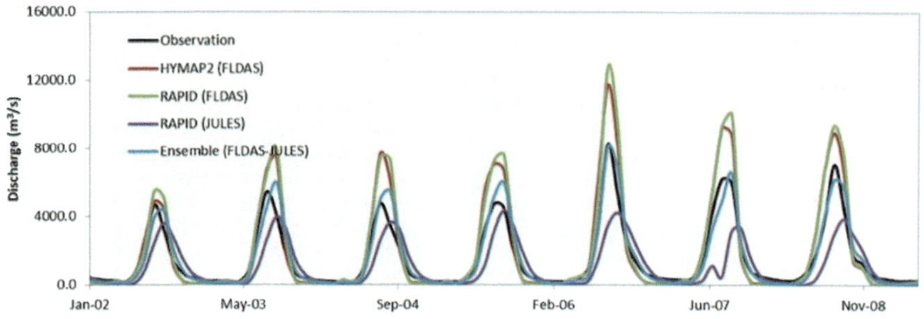

Fig. 16.10 Comparison of six years of observed and model generated Blue Nile River flow at El Diem (https://www.globalwaters.org/GWS-Stories/providing-scientifically-robust-tools-global-water-security)

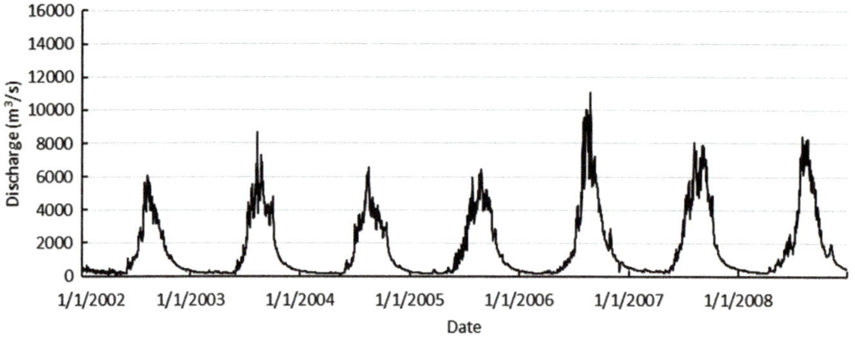

Fig. 16.11 Blue Nile River flow at El Diem from a 1965–2009 observed data source

Conclusion and Recommendation

The importance of baseline streamflow data for basin water agreements is a critical element. The question of how much are the flows in the Nile River and the Blue Nile River needs to be addressed before long-term operation and water allocation agreements reached. Historical flow data may not be of satisfactory quality and any agreement based on such data may not match with current flows. Also, mean flows could change with time and the selection of period of data record to be used for negotiations matters. If agreements are needed soon, percentage of prevailing flow-based approach can be implemented for long-term operation or water allocation/water sharing activities in the basin. The importance of streamflow monitoring, site of monitoring, monitoring entity, method and instrument of monitoring, and length of monitoring for determining current amount of river flow needs critical consideration. Data quality evaluation is also critical with support from water budget analysis and hydrologic modeling.

References

Abtew W, Dessu S (2019) The grand Ethiopian Renaissance dam on the Blue Nile River. Springer. https://www.springer.com/la/book/9783319970936

Ahmed Y, Al-Faraj F, Sholtz M (2019) Assessment of upstream human intervention coupled with climate change impact for a transboundary river flow regime: Nile River basin. Water Resour Manage. https://doi.org/10.1007/s11269-019-02256-1

Berhanu B (2014) Chapter 6 Surface water and groundwater resources of Ethiopia: potential and challenges of water resource development. In: Melessse, Abtew, Setegn (eds) Nile River Basin ecohydrological challenges, climate change and hydropolitics. Springer, New York

Buekham DE, Dawdy DE (1970) Error analysis of streamflow data for an alluvial stream, Gila River Phreatophyte Project, Geological Survey Professional Paper 655-C, USGS

Chester CW, Holmes RR, Biedenham DS (2013) Mississippi River streamflow measurement techniques at St. Louis, Missouri. https://doi.org/10.1061/(ASCE)HY.1943-7900.0000752

Corbett DM (1943) Stream gaging procedure a manual describing methods and practices of the Geological Survey. Water Supply Paper 888, USGS

Crowell ML, Mtundu ND (2000) Guidelines for quality control and quality assurance of hydrologic and meteorologic data. Vol. 2: Data management. South Florida Water Management District, West Palm Beach FL

Globalwater.org. USAID. 2019. Providing scientifically robust tools for global water security. https://www.globalwaters.org/GWS-Stories/providing-scientifically-robust-tools-global-water-security

Gonzalez J, Chen Z (2005) Uncertainty of index-velocity measurement culverts. ASCE/EWRI World Water & Environmental Congress. May 14–19, 2005, Anchorage, Alaska, USA

Harmel RD, Cooper RJ, Slade RM, Haney RL, Arnold JG (2006) Cumulative uncertainty in measured streamflow and water quality data for small watersheds. Trans ASABE 49(3):689–701. https://doi.org/10.13031/2013.20488

Heggie J (2020) Can the Colorado River keep on running? https://www.nationalgeographic.com/science/2020/03/partner-content-colorado-river-preserving-stressed-water-resources/

Kinsman G, Kite J, Mtundu D (1994) Guidelines for the collection of hydrologic and meteorologic data. Volume 1: Field applications. South Florida Water Management District

Kunkle GA (2018) Evaluation of the radar stage sensor manufactured by forest technology systems—results of laboratory and field testing. Open-File Report 2017–1085 U.S. Department of the Interior U.S. Geological Survey

MaCnally A, Arsenault K, Kumar S, Shukla S, Peterson P, Wang S, Funk C, Peters-Lidard CD, Verdin JP (2017) A land assimilation system for sub-Saharan Africa food and water security application. Sci Data 4:170012. https://doi.org/10.1038/sdata.2017.12

Mueller DS (2015) Velocity bias induced by flow patterns around ADCPs and associated deployment platforms. https://doi.org/10.1109/CWTM.2015.7098103

Mueller DS (2017) Quantifying acoustic doppler current profiler discharge uncertainty: a Monte Carlo based tool for moving-boat measurements. Water Surv Can. https://doi.org/10.1061/(ASCE)HY.1943-7900.0001249

O'Brien P, Mueller D, Pratt T (2012) Comparison of acoustic doppler current profiler and Price AA mechanical current meter measurements made during the 2011 Mississippi River Flood. https://doi.org/10.1061/9780784412312.127

Oltmann RN (1993) Hydrodynamics of California's Sacramento-San Joaquin Delta, and the collection of flow data for tracking of pesticides. In: Morganwalp DW, Aronson DA (eds) Proceedings of the technical meeting, vol 2, Colorado Springs, Colorado, September 20–24, pp 695–698

Painter CC, Loving BL (2015) Geological survey quality-assurance plan for surface water activities in Kansas, 2015. Open-File Report 2015–1074, USGS

Ramadan EM, Negm AM, El-Sammny M, El-Sayed AH (2015) Quantifying the impacts of impounding Grand Ethiopian Renaissance Dam on Nasser Lake active storage. Eighteenth International Water Technology Conference. Sharm ElSheikh, 12–14 March 2015

Riverside Technology Inc. (2010) Inception report design of an upgraded data acquisition, communication and flood forecasting systems in the EN countries. Submitted to ENTRO

Sadek N (2006) River Nile flood forecasting and its effect on national projects implementation. Tenth International Water Technology Conference, IWTC102006, Alexandria, Egypt

Samboko HT, Abas I, Luxemburg WMJ, Savenije HHG, Makurira H, Banda K (2020). Evaluation and improvement of remote sensing-based methods for river flow management. Phys Chem Earth 117. https://doi.org/10.1016/j.pce.2020.102839

Sauer VB, Meyer RW (1992) Determination of error in individual discharge measurements. USGS, Open-File Report 92–144

SFEC (2020) Task 3 report: stream gauging and water sediment sampling at four (4) transects and analytical lab results for TSS and TP concentrations (Low Flow Event 12/23/19). Submitted to South Florida Water Management District, West Palm Beach, FL

USGS 2018 Water Resources. USGS Streamgauging Network. USGS.gov

Xiao B, Wan F, Wu CQ, Zhang K (2009) River cross-section surveying using RTK technology: the Yangtze River project case study. Coordinates https://mycoordinates.org/river-cross-section-surveying-using-rtk-technology/all/1/

Zulkafli Z, Buytaert W, Onof C, Lavado W, Guyot JL (2013) A critical assessment of JULES land surface model hydrology for humid tropical environments. Hydrol Earth Syst Sci 17:1113–1132. https://doi.org/10.5194/hess-17-1113-2013

Chapter 17
Soil and Water Conservation Technology and Sediment Retention Assessment

Sileshie Mesfin Leyew, Tesfahun G/Michael, Arega Mullu, and Assefa M. Melesse

Abstract Soil erosion is a common menace to Ethiopian highlands. As a result, mitigation measures were practiced for decades without evaluation of the efforts made on the highlands. Therefore, this field survey was conducted in 34 community watersheds in Amhara region, Ethiopia, to determine the performance of conservation practices with respect to soil erosion and sediment retention. The study used the methodology of biophysical field surveys, and hence, different sites were selected based on agro-ecology, topography, and land use. Different types of structures were identified; vertical intervals, horizontal intervals, dimension of embankments and collection ditches' height, width, and depth were evaluated against the standards provided by ministry of agriculture. Google earth images were downloaded, and structures were digitized for verification using field observations. Stream power index was used to delineate gullies and soil loss with two scenarios, with conservation practices (existing scenario) and without the practices (base case scenario) were estimated, and sediment retained was evaluated. Based on the results, the land affected by gullies was estimated 2% from the total land area. The gullies treated by check dams and plantation of gully sides were estimated about 54.1%. The coverage of the practices in the study watersheds accounts 60.38% and 38.02% by physical and biological measures, respectively. About 42% of the bunds were stabilized with trees and grass, and 16% of the area of community watersheds were delineated for enclosure. The gradients of bunds in humid areas where there is high rainfall, only 27.62% fits the criteria of the standard, but the remaining 72.38% were below the standard. But in most sub-humid areas where the rainfall is lower

S. M. Leyew (✉) · T. G/Michael
Amhara Design and Supervision Works Enterprise, P.o.Box 1921, Bahir Dar, Ethiopia

A. Mullu
Hydraulic and Water Resources Engineering, Debre Markos University, Debre Markos, Ethiopia

A. M. Melesse
Department of Earth and Environment, Institute of Environment, Florida International University, Miami, USA
e-mail: melessea@fiu.edu

A. M. Melesse et al. (eds.), *Nile and Grand Ethiopian Renaissance Dam*, Springer Geography, https://doi.org/10.1007/978-3-030-76437-1_17

than moist humid areas, 91.25% of structures were constructed based on the recommendation. Average annual soil losses of watersheds before conservation were estimated as 77 t/ha/year and reduced to 34 t/ha/year after conservation was practiced. Thus, the amount of soil retained as a result of the practices was estimated as 43 t/ha/year, and hence 56% (0.37 cm) of soil could be engaged on land resource managed areas. From the analysis, it can be concluded that significant soil was retained as a result of implementation of practices.

Keywords Gullies · Soil erosion · Sediment loss · Soil and water conservation · Amhara region · Sediment retention

Introduction

The livelihood of Ethiopians is supported by the agriculture sector, which is widely considered a sector for improving food security and poverty. However, this sector is highly affected by soil erosion, decrease of soil fertility, and land degradation (Grepperud 1996; Pender and Gebremedhin 2006). Furthermore, climate change impact is shown to affect the agricultural sector in Ethiopia. This has led Ethiopia to be highly vulnerable to food insecurity and hence depend on international aid. In Ethiopia, soil degradation can be understood as a direct result of historical agrarian practices in its uplands (Bishaw 2001). In the uplands, the expansions of deforestation, over cultivation, and overgrazing lead to enhanced soil erosion (Hurni and Pimentel 1993).

Watersheds affected by high degradation tend to hasten overland flow, increases sediment detachment and transportation, decreasing soil moisture and base flow. Several studies used terrestrial cover representing tools and methods to comprehend land use differences, record of natural resources and forest as well as recognize the changes in the hydrologic performance of watersheds (Getachew and Melesse 2012; Mango et al. 2011a, b; Wondie et al. 2011, 2012; Melesse and Jordan 2002, 2003; Melesse et al. 2007; Yesuf et al. 2013).

Numerous studies (Aga et al. 2018, 2019; Defersha and Melesse 2012a, b; Defersha et al. 2010, 2012; Maalim and Melesse 2013; Maalim et al. 2013; Setegn et al. 2010; Melesse et al. 2011; Msagahaa et al. 2014; Wang et al. 2008; Mekonnen and Melesse 2011; Setegn et al. 2009; Yesuf et al. 2015) were conducted to comprehend soil erosion and sediment conveyance. Multi-approach studies indicate that watershed procedures and landscape structures regulate the rate of sediment detachment and conveyance.

Similar to the rest of Ethiopian regions, major economic activities of the Amhara region are largely confined to cropping and livestock farming whose misuses are strongly connected to the degradation of land resources. Various soil and water conservation measures have been conducted over the last 40–50 years. Physical soil and water conservation measures such as soil bund, stone bund, hillside terrace, cut of drain, check dams, micro basin construction, and biophysical conservation measures such as river bank plantation, gully side plantation, hedge row cropping, area closure, and tree planting on degraded lands are the main soil, and water conservation measures that were implemented for the last decades under Amhara Bureau of Agriculture (BoA).

In Amhara region, soil and water conservation actions are implemented widely, especially in those 4–5 years in the first million development goals (MDG) period, but their impacts were not appraised. The quantity and quality of the structures were not documented. So that, the social, physical, and biological state of the soil and water conservation structures appraisal were done through selected 34 community watersheds chosen based on the agro-climate zonings (ACZ). Soils surrounded by the construction, gully treated by the check dam, physical and biological soil and water conservation structures were evaluated throughout the Amhara region for selected community watersheds of treated and untreated units with the main objective of assessing the performance and effects of soil conservation practices in respective to soil erosion and sediment retention.

Materials and Methods

Description of the Study Area

The Amhara National Regional State (Fig. 17.1) inhabits a territory prolonged within a geographical coordinate between 9° 29'–14° 0' North latitude and 36° 20'–40° 20' East longitude. It covers an area of 170,152 km^2. It is enclosed by Sudan and Benshangul Gumuze in the west, Tigray in the North, Afar in the East, and Oromia region in the South. Thirty-four small watersheds were selected based on the agro-ecology throughout the Amhara region, and evaluations were conducted on the effect of natural resource conservation on biophysical and socio-economic changes.

Data Set

Base Map Preparation

Base maps using Google earth images were prepared showing treated and untreated watersheds. Observation points were selected based on the prepared base map considering conserved and degraded areas of the watersheds under different slope class.

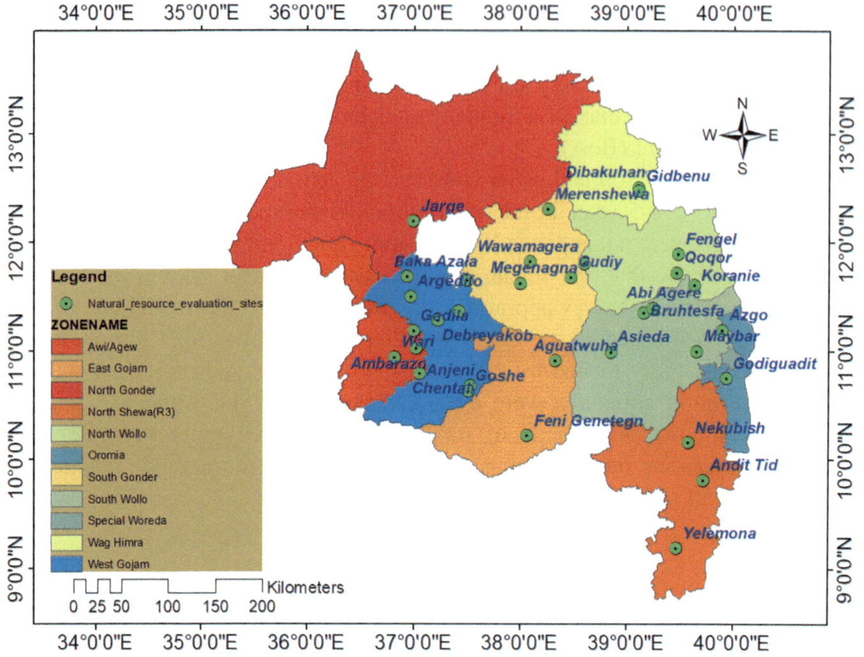

Fig. 17.1 Figure location map of the study area

Preparation of checklists and questionnaires was designed for the assessment of both secondary and primary data on biophysical and socio-economic issues at different levels that enable for gathering preference of farmers and best practices of the watersheds.

Primary Data Assortment

This study was conducted mainly through field observation with the aid of GPS and spatial tools to investigate biophysical data of the watershed. The primary data were also collected from farmers, development agents by using questionnaires and discussions with the concerned experts.

Observation through transect walk and soil and water conservation structures measurements were taken. Field measurements such as vertical interval and horizontal interval between structures, height, depth, and width of structures of embankment and ditch dimensions were taken from field. Key informant interviews were also taken from key persons of each watersheds. Coordinate points were collected to take representative places where soil and water conservation structures were found.

Data Analysis

Assessment of Areas Affected by Gullies

Gullies are formed by concentrated flow of runoff. Area coverage of land affected by gully was assessed with the use of raster calculator as

$$\text{SPI} = \text{LN}(([\text{FlowAccum\_Raster}] + 0.001) * ((\text{Slope\_Raster}]/100) + 0.001) \tag{17.1}$$

Stream power index (SPI) refers to the flow accumulation raster which is the output from flow accumulation analysis and slope raster from slope analysis.

Soil and Water Conservation of Watersheds

Using image downloaded from Google earth, digitization of constructed structures were done by grouping types of structures such as bunds, bunds with stabilized tree types, hill side terraces, trenches, micro-basins, etc. The coverage of structures was digitized in polygon. Attributes were coded for each construction based on the data collected from field.

Evaluation of Dimension of Structures

Soil and water management structures have their own standards to correctly control erosion. If we decrease the spacing between bunds, it may occupy cultivable land, and in the contrary, if we construct wide spacing, erosion may aggravate and the structure may collapse so that appropriate spacing between structures is necessary. Vertical intervals, horizontal intervals, dimensions of different embankments, ridges, and gradients of structures were evaluated with standards. The actual dimensions were collected during the field trip period, and standards of the dimensions were calculated using the formula from guidebooks or from standards obtained in literatures such as participatory watershed management knowledge compiled by the ministry of asgriculture (MoA). Different structures may be appropriate in different agro-ecological zones, so that this variation of appropriateness was also evaluated.

Vertical interval between bunds (VI) was estimated as

$$\text{VI} = \left(b + \frac{S}{a}\right)0.3 \tag{17.2}$$

where S = Land slope in (%); a and b are constants, and $a = 3$ and $b = 2$ for medium and substantial rainfall zones, and $a = 2$ and $b = 2$ for short rainfall zones.

Horizontal spacing in between bunds (HI) is given by

$$\mathrm{HI} = \left(\frac{\mathrm{VI}}{S}\right) \times 100 \quad (17.3)$$

The spacing of check dams is estimated as

$$S = \left(\frac{1.2 \times h}{G}\right) \quad (17.4)$$

where S = the spacing in m, h = the effective height of the check dam (spillway height in m), and G = the gully gradient.

The vertical interval, height, and collection ditch's width and depth were measured at selected points in each structure within the bunds, check dam, micro-basin, and terrace or trench length. The mean height, mean length, and mean vertical interval are computed for each structure and compared with the recommended (standard) value using basic statistics, z-test (when the mean of the population is known and the sample number is greater than 30), and t-test (when the population mean is unknown and sample number is less than 30) to evaluate the quality of the structure.

$$Z = \left(\frac{X - \overline{x}}{\frac{S}{Sqrtn}}\right) \quad (17.5)$$

$$t = \left(\frac{X - \overline{x}}{\frac{S}{Sqrtn}}\right) \quad (17.6)$$

where X is the recommended standard value, $\overline{x}$ is the mean of the data set, S is the standard deviance of the data set, and n is the total number of each data set.

Sediment Analysis

The data collected through different methods were organized and analyzed into different forms in line with the nature of issues supported by theoretical and empirical evidence from literatures. In analyzing the data, both qualitative and quantitative methods were applied.

Soil Erosion Estimation Method

The area was assessed using soil erosion hazard as an indicator for quantitative land degradation. To assess soil erosion threat for the project area, Revised Universal Soil Loss Equation (RUSLE) approach was used.

The land degradation maps were developed on ArcGIS environment by using RUSLE params (erosivity of rainfall; erodibility of soil; slope distance and gradient; land cover; and land management practices) as an input to assess average annual soil loss rate of the area. The mathematical equation can be represented on physical-based models in the ArcGIS environment. Each variable was overlaid to make the overall spatial analysis. Mathematical equation of Revised Universal Soil Loss Equation to guess soil loss before soil and water conservation is

$$\mathrm{A = R.K.LS.C.P} \tag{17.7}$$

Soil loss after SWC was implemented and estimated using

$$\mathrm{A' = R'.K'.L'S'.C'.P'} \tag{17.8}$$

where A = Annual soil loss in tons/ha/yr; R = Rainfall erosivity; K = Soil erodibility; LS = Topographic factor (slope length and gradient factor), C = Soil cover factor, and P = Land management factor.

Soil Loss Before and After Construction of SWC

Soil loss before and after soil and water management practices were implemented is assumed to be calculated by changing slope distance factor, land protection factor, and land management factor.

Rainfall Erosivity factor (*R*) There are different ways of examining the R-factor for different areas. The factor of the development corridor was calculated on the bases of mean annual rainfall data of each station rendering to the calculation given by Hurni (1985a, b), for Ethiopian circumstances based on the available mean annual rainfall (*P*). It is given by a regression equation as

$$R = -8.12 + 0.562 * P \tag{17.9}$$

where R = Rainfall (Erosivity) factor unit less, and P = Mean annual rainfall in mm.

To compute erosivity rate of the basin, the mean value at each meteorological stations was taken. To change this point rainfall into areal rainfall, the Theisen polygon method was used. Areal rainfall of Theisen polygons was area weighted; the values were mapped in the ArcGIS environment.

The formula used to calculate point rainfall into area based is given by the equation:

$$P_{\text{ave}} = \sum_{i=1}^{i=n} \left(\frac{A_i P_i}{A_n} \right) \quad (17.10)$$

where P_{avg} = Average precipitation of the rainfall, A_i = area of Thiessen polygon i, A_n = Sum of Area of "n" numbers, and P_i = yearly mean Precipitation of each meteorological station.

Soil Erodibility (K): Erodibility of soils was estimated from the generated soils map of the project area. To estimate the erodibility value, major soil types based on FAO classification were used.

Slope length and gradient factor (LS): The technique used for computing LS requires length of overland flow for a watershed which is assumed to be the surface flow length till the runoff gets channel such as rills and drainage ways. Stream density was calculated with drainage length divided by area of the watershed, and the length of overland flow is the reverse of twice drainage density.

The slope distance and slope steepness can be used in a single index, which expresses the ratio of soil loss as well-defined by

$$\text{LS} = \left(\frac{X}{22.13} \right)^{0.5} * (0.344 + 0.0798 * s) \quad (17.11)$$

where X is the length of overland flow (before) and length of spacing between bunds (after) SWC and S is slope in percent.

Land cover factor (C): The land cover feature was planned using the land cover map as an input. Each cover value of the project area (Table 17.1) could be synchronized with the adopted C-value in Ethiopian condition (Hurni 1985a, b). The land cover/use map was developed from the Google earth imaginary by using ArcGIS.

Land management practice factor (P): The land management factor (P) in RUSLE is defined as the ratio of soil loss with a specific management practice to the

Table 17.1 C-factor values

Land use	C-factor
Forestland	0.01
Eucalyptus plantation	0.02
Grassland	0.05
Shrub land	0.05
Cropland	0.15
Waste land	0.5
Built up area	0.09
Water body	0

Source (Hurni 1985a, b)

corresponding soil loss with straight row upslope and down slope tillage. Management factor was taken as unity assuming there was no management intervention in the base case scenario.

Soil Loss After SWC

Slope length and gradient factor (LS): It is well known that the slope length is becoming short with the practice of management measures. Length of overland flow is divided in to short slope length because of bunds and terraces are constructed. Therefore, the slope length factors were obtained from spacing between two constructed practices using the above LS formula.

Land Cover factor: Land cover is expected to be changed in each watershed after soil and water conservation strategies were implemented. Land uses were prepared for both before and after soil and water conservation structures implementation, so that the soil loss cover factor could be changed in estimating the soil loss for both scenarios.

Land Management practices factor: It is clearly known that soil loss is minimized after proper soil and water conservation structures were implemented, so that the management factors were determined based on management type for the given land use for Ethiopian condition. It is assumed that the *p*-values (Table 17.2) were given to management factor for those areas treated by bunds (Hurni 1985a, b).

Estimation of Sediment Deposition

The change in between the two scenarios (without SWC and with SWC) was assumed to be sediment deposited between bunds in ton/ha/yr, and this soil contains parallel soil depth increment. The soil depth increment is modeled as in

$$D = \left(\frac{m}{a} * d\right) \qquad (17.12)$$

Table 17.2 Management factors for some land use types

Land use	*p*-value
Water body	0
Cultivated land	0.9
Settlement	0.63
Eucalyptus plantation	0.53
Grazing	0.63
Bare land	0.73

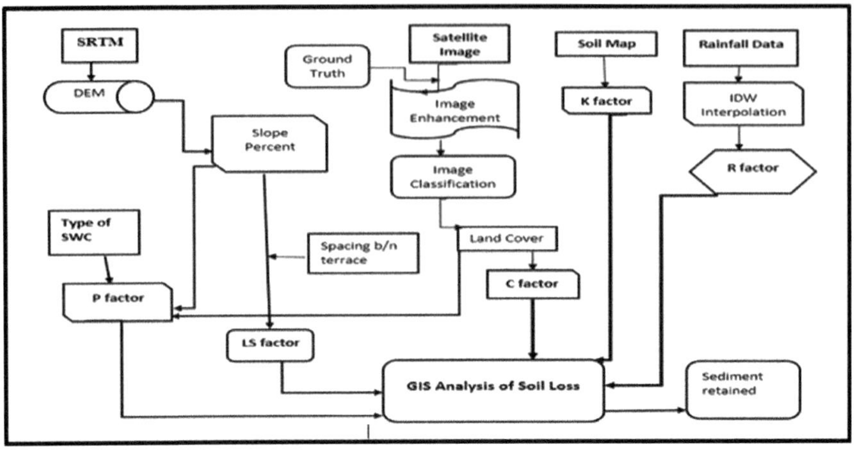

Fig. 17.2 Flowchart of soil loss analysis and sediment retention

where D = soil depth increment, m = mass of soil retained by SWC, a = unit area of grid cell, and d = bulk density of the retained soil taken from laboratory analysis for samples taken from representative samples for physical and chemical analysis of deposited soil. Figure 17.2 shows the methodology flowchart followed in this study.

Results and Discussion

Soil Erosion

Splash, Sheet and Rill Erosion

Splash, sheet, and rill erosions were seen in all watersheds. Earth smoothing and crusting, which is the effect of splash, are the common phenomena of farmlands. Soil deposit at the high side of obstructions and most lately constructed bunds and hill side terraces formed benching. These indicate the presence of the movement of soil as a sheet flow and indication of soil loss as sheet erosion. Rill erosion as a result of wide spacing of bunds and weakly constructed bunds was observed in all watersheds (Fig. 17.3).

Gully and their treatment coverage

From analysis of stream power index (SPI), a land affected by gully erosion was assessed, and from Google earth image and field observations, the gully treated by check dams and plantations was determined. The land occupied by gullies account 2% of the total land area of community watersheds. This implies that this much

Fig. 17.3 Rill erosion formed in poorly constructed soil bund in Dibakuhan watershed (**a**). Earth crusting, smoothing of cultivated land and rills formed in Andit Tid watershed (**b**)

amount of land is affected by gully destruction in the region. From observation and Google earth image, 54.1% of the gullies are treated by check dams and plantation.

Area Coverage of SWC Practices in Community Watersheds

Traditional as well as introduced soil and water management structures are observed in the study watersheds such as contour plowing, traditional ditches, soil bunds, stone bunds, Fanya juu, stone-faced soil bunds, hillside terrace, bench terrace, trenches, micro-basin, eyebrow basin, and percolation pit. Biological measures such as grass strip, area closure, and hedgerow planting were also used. *Trelucern, Susbania susban, Lucinia, Sisal (Erate), Kitikita, and Kundo Berberie (Schieness molle)* were plants, and Vetiver grass, Sendedo, Densho, key Sar, Serdo, Lime, and Guasa are grasses grown in the bunds of the watersheds.

The coverage of soil and water conservation works in the watersheds were 60.38% of the total area covered by physical measures and 38.02% by biological interventions. This implies that 39.62% parts of the region's community watersheds are not covered by physical and 61.98% by biological conservation measures.

Evaluation of the Technical Standards of SWC Structures

Different types of dimensions and spacing of soil and water conservation structures were evaluated with their standards. Those that are the same types of structures data collected from field are grouped together and evaluated with statistical params.

Table 17.3 Comparison of existing with standard values set for Soil bund

Variable	Existing Av	Standard value	*n*	*p*-value	Decision
Height of bund (m)	0.61	0.6	31	0.4588	Ho = accepted
Width (m)	0.71	0.5	10	0.0005	Ho = rejected
Depth (m)	0.29	0.5	10	0.0000	Ho = rejected
Vertical spacing (m)	1.36	1.58	28	0.4071	Ho = accepted

Soil Bund

Soil bund is constructed in slopes ranged from 3 to 33% with 6 to 32-m horizontal distance and 0.7 to 3.45-m vertical interval. As indicated in Table 17.3, the height and the vertical spacing of the soil bund is found to be supporting the hypotheses that there is no a significant difference between the existing and standard one. This implies that the construction of soil bund in watersheds in the region is carried out as per the recommended standards. On the other hand, embankment depth and width of soil bunds are found to be highly significantly different from recommended standard. This implies that the width and the depth of soil bunds are incorrectly constructed in the watersheds throughout the region in the probability of ($p = 0.0005$) and ($p = 0.000$), respectively, at 5% level of significance.

Stone Bund

The mean height of the stone bund is 0.87 m (Table 17.4) with minimum of 0.33 m and maximum of 2.5 m which is observed in old stones that the height is added by frequent maintenance through time. About 25.6% of the bunds have greater heights from the recommended values. The mean value of the stone bund vertical interval is 2 m, but the standard recommended value is 2.13 m.

Statistical analysis also shows that there is no significant difference between height and vertical interval of stone bunds with their recommended standards in 0.05 significance level: the *p*-values are 0.06 and 0.26, respectively, which is greater than 0.05. This implies that height and vertical intervals of the stone bunds were constructed according to the standard in the region.

Table 17.4 Comparisons of existing with the standard values of stone bund

Variable	Existing Av	Standard value	*n*	*p*-value	Decision
Height (m)	0.87	0.76	38	0.06	Ho = accepted
VI spacing (m)	2	2.13	34	0.29	Ho = accepted

Table 17.5 Comparison of existing with standard values of stone-faced soil bund

Variable	Existing Av	Standard value	*n*	*p*-value	Decision
Height (m)	0.629	0.7	39	0.13422	Ho = accepted
Width (m)	0.76	0.5	10	0.0009	Ho = rejected
Depth (m)	0.35	0.5	10	0.000072	Ho = rejected
VI spacing (m)	2.33	2.4	39	0.40389	Ho = accepted

Stone-Faced Soil Bund

The minimum standard of the height of the stone-faced soil bunds is 0.7, and it can extend up to 1 m according to the topography and availability of quality of stones (Table 17.5). The mean height of the stone-faced soil bund is 0.63 m with the minimum of 0.2 m and maximum of 1 m. However, 77% of the height of stone-faced soil bund is less than the recommended value. Height and vertical interval are not significantly different with the standard at 0.05 significance level, and the *p*-values are 0.13 and 0.40, respectively, but width and depth of the ditches are different from standards.

The mean spacing of stone-faced soil bund is 2.33 m, but the mean of the standard value is 2.4 that means the mean spacing is less wide but not so much different from the recommended mean spacing. About 65% of the bund spacing is less than the recommended value. In lower slopes or gentler slopes, the spacing is less wide compared to the standard, but in higher slopes, it is wider than the recommended value of that slope.

The width is wider than the recommended, and the depth is less deep than the standard. The mean values of depth and width are 0.35 and 0.76 m, but the recommended values are 0.5 m for both structures. The depth may be filled with sediment. The height and spacing are correctly constructed in the region, but the width and depth of ditches are not.

Fanya Juu

Fanya juu is applied in cultivated lands with slopes 2–15% which is constructed in recommended slope with 9–25 m horizontal distance and 0.28- 2.67 m vertical interval. Heights and vertical intervals were not significantly different with the recommended standard at 0.05 level of significance since the *p*-value is 0.207 and 0.17, respectively, which is greater than 0.05. However, the mean height of these bunds is greater than the standard. The mean value of heights of fanya juu is 0.667 m, but the recommended standard is 0.6 m. About 64% of the measured height is less than the recommended and 36% is above 0.6 m. The maximum height was 1.96 m which is very old fanya juu in Anjeni watershed, and the minimum height is 0.2 m.

Table 17.6 Comparison of existing with the standard values of Fanya Juu

Variable	Existing Av	Standard value	*n*	*p*-value	Decision
Height (m)	0.66	0.6	25	0.20311	Ho = accepted
Vertical interval spacing (m)	1.65	1.479	24	0.17131	Ho = accepted

Even if there is no difference with *z*-test, the mean spacing between measured fanya juu is 1.65 m, but the recommended value is 1.479 m (Table 17.6). This indicates that the spacing of fanya juu exists in field is higher than the standard. About 63% of the spacing is wider than the standard. In sloppy areas, the spacing is wider than in gentler slopes compared with the recommended value of that slope.

Hill Side Terrace

Hill side terrace was constructed in hill lands and degraded areas to rehabilitate the abandoned lands. It was stabilized with some trees and protected by area closure. Revegetation of natural trees was allowed in hill lands, and plantation was applied in degraded areas together with hill side terrace.

The height and vertical intervals are significantly different from the standard, and the *p*-values (0.042) and (0.007), respectively, are less than the significance level 0.05 (Table 17.7). From the analysis, we can conclude that height and vertical intervals of hill side terrace were not constructed according to the recommendation in the region.

Trench

The value of the constructed trenches was weighed up in terms of minimum.

Trenches were constructed at an interval of 3–5 m depending upon on the suitability of the land. The values of the existing average trench length, width, and depth were not significantly different from recommended standard. This showed that the average height, width, and depth of the already existing trenches in the catchment were up to the MoARD recommended standard (Table 17.8) and can be conclude that they are constructed as per the recommendation in the region.

Table 17.7 Comparison of existing with standards values of hill side terrace

Variable	Existing Av	Standard value	*n*	*p*-value	Decision
Height	0.72	0.6	15	0.04194	Ho = rejected
VI spacing	2.23	3.24	16	0.00779	Ho = rejected

Table 17.8 Comparison of existing with standard values of trench

Variable	Existing Av	Standard value	*n*	*p*-value	Decision
Length (m)	3.03	3- 5	9	0.4519	Ho = accepted
Width (m)	0.52	0.5	9	0.3642	Ho = accepted
Depth (m)	0.43	0.5	9	0.0546	Ho = accepted

Table 17.9 Comparison of existing with standard values set for bench terrace

Variable	Existing Av	Standard value	*N*	*p*-value	Decision
Vertical interval (m)	1.56	2.65	7	0.000125	Ho = rejected
Width (m)	4.5	5.13	7	0.2133	Ho = accepted

Bench Terrace

The spacing between existing bench terraces ranged from 0.85–2.17 m with an average of 1.56 m on 14–50% slope lands. About 86% did not meet the minimum technical standard of vertical interval (Table 17.9). It was significantly different; the null hypothesis is rejected since the *p*-value is less than 0.05 significance level. The analysis shows that it is narrower than the recommended standard.

From the analysis of vertical interval of the bench terraces, we can conclude that bench terrace can be done wider than the existing bench terrace in the region, then it could be possible to get wide farm area possibly to maximize cropping area of high slopes. In Gidbenu watershed, bench terraces were constructed to use the land for youths in order to use for income generation. Hence, there is a possibility to use for the purpose like cropping, afforestation, and apiculture for job creation of youths in the rest watersheds and in the region as a whole.

Check Dam

Check dam are constructed to reduce velocity of runoff and prevent the deepening and widening of gullies. Single brush wood, brush wood together with gabion, gabion check dam, and stone rip rap check dams and in some places like Dibakuhan watershed rock fill dams are also constructed in the study areas. Among collected sample data, 62.5% of the check dams are stone riprap, 18.25% are brushwood, and 18% are Gabion check dams.

Maximum height of check dams is 1.9 m, and the minimum height is 0.3 m. The mean height (1.03 m) is in between the recommended value, 1 and 1.5 m. But 18.75% are constructed above the maximum standard, and 37.5% were constructed below the recommended standard.

Average spacing of check dams is 10.3 m which is greater and wider than the recommended average 9.4 m. The spacing between two check dams was evaluated with *z*-test, and the result showed it is significantly wide with 0.05 significance level. About 37.5% are narrower than the mean and 62.5% are wider. The maximum length between the check dams is 30 m and also 75% of the check dams have no foundation and 81.75% has no drop structure. Besides the drop structure is less than the recommendation. The drop structure must be 1.5 times the height of check.

The width of check dam should be between 1.5 and 3.5 m but the average width is 1.3 m. The value is less than the recommendation which is shown as 87% the check dams' width deviate from the minimum value of the recommendation.

Gully Rehabilitation Problems

The following difficulties can be taken as the main reasons for the catastrophe of most of the gully rehabilitation schemes in Amhara region as summarized from study watersheds.

- Poor attention for upper catchment treatment.
- Poor fitting of check dams which is related to absence of keying the check dam to the floor and side walls of the gully.
- Absence of apron.
- Absence of spillway.
- Deprived maintenance.
- Inappropriate spacing of check dams.
- Structures are occasionally made too high and the water which pools causes unsteadiness of the soil and piping underneath or around the structure.
- Deprived integration among physical and biological measures.

According to the guideline, recommended bund gradient ranged from 0 (level) for the purpose of moisture retention and 0.5–1% for the draining of surplus runoff for the purpose of flow in permissive velocity based on agro-ecology (Hurni 1985a, b). In moist agro-ecological areas, it has to be constructed up to 1% gradient, because the rainfall is high, excess runoff water has to drain. But in this area, only 27.62% fits the criteria of the standard, while the 72.38% violates the recommendation. But in sub-humid areas, it has to be level in most places because the rainfall is lower than the moist areas the analysis show 91.25% was level which fulfills the standard. Figure 17.4 shows gullies formed due to improper land managment.

Biological SWC Measures

Biological measures including protection of existing vegetation or planting of trees and grasses on degraded and stabilization of bunds in cultivated areas were practiced in the study community watersheds (Figs. 17.5 and 17.6). Plantations of up slopes, bund stabilization, gully rehabilitation, and regeneration through plantation of trees and grasses are some activities which were done artificially as biological

Fig. 17.4 Gully formed by improper land management and concentration of runoff in Deresinilih watershed (**a**) and gully formed by improper waterway construction in Asieda watershed (**b**). In these areas, high cattle population and overgrazing constitute a major factor for gully formation

Fig. 17.5 The result of Overgrazing in Baka Azala watershed. Road construction done in steep lands, without adequate delivery for drainage systems, is a main cause of gully erosion

soil and water conservation. Protection of natural vegetation through area closure is practices applied in the study watersheds which are important to regenerate the natural vegetation. Among 34 studied community watersheds, 19 watersheds do have areas delineated for enclosure which have 2657 ha, and it accounts 16.74% from the total area of community watersheds. About 42% of the bunds constructed

Fig. 17.6 Gully formed by road culvert; it is treated by check dam in Abiager watershed

in community watersheds were stabilized with different types of trees and grasses selected according to the agro-ecology (Fig. 17.6). It is is shown that in some parts of the study areas footpath was a contributing factor for gully formation (Fig. 17.7).

Fodder trees, shrubs, and grass used for the stabilization of bunds are *Trelucern, Susbania susban*, Pigeon Pea, and grasses like Vetiver, Phalaris, Elephant grass, etc. The farmers used these plants as a fodder, the seed for sell collected from ripen trees, and fuel wood and cutting grasses for fodder and to cover the house roof in addition to the purpose they give for conservation of the soil. In addition to grasss and vegetation that are used for gully stablization, soil and stone bunds are also used as a measure for soil erosion reduction (Figs. 17.8, 17.9 and 17.10).

Soil Loss and Sediment Retention: RUSLE model was used to estimate the soil loss of the area in two scenarios; before and after soil and water conservation practices with multiplying erosivity, erodibility, slope steepness, slope length, cover factor, and management factors existed in each parcel of land in the study watersheds. Soil losses before SWC assumes no interventions on lands were applied such as up and down ploughing, with no intervention in degraded communal lands and the length of over land flow unchanged. But when SWC practices were applied, the coverage and management of communal lands were changed, and length of overland flow was decreased in cultivated lands and in some communal lands as a result of bunds and hillside terraces, respectively. The slope length and steepness factor, crop cover management factor, and land management factor on

Fig. 17.7 Gully formed by footpath and cattle trafficking line in Gadila watershed

Fig. 17.8 Soil bund in Andit Tid watershed

Fig. 17.9 Stone bund in **(a)** Aguat wuha, and **(b)** Bruhtesfa watersheds

Fig. 17.10 Stone-faced soil bund in Andit Tid watershed

natural resource managed areas were changed, but rainfall and soil erodibility factor remained unchanged.

Average annual soil loss of watersheds before soil and water conservation was estimated as 77 t/ha/yr, and it became 34 t/ha/yr after conservation was practiced. Thus, the amount of soil retained as a result of physical and biological soil and water practices was estimated as 43 t/ha/yr. On average, there was 0.66 cm depth of soil lost in a hectare each year before SWC structures implementation throughout the study watersheds but it decreased to 0.29 cm as a result of management practices. Hence, there could be 0.37 cm per hectare per year depth of soil could be remained as a result of soil and water conservation structures implementation. This indicates that 56% of soil could remain in places on land resource managed areas (Fig. 17.11).

Engdayehu et al. (2016) estimated in single watershed in Debre Mawi learning watershed that there was 55% decrease of soil loss after soil and water conservation intervention reducing from 39 to 17.36 t/ha/year. The difference in soil loss decline was most probably due to SWC measures since it has an important effect in trapping soil and decreasing sediment transport. In general, soil loss was found as above the tolerable soil loss value (0.2–12 t/ha/yr) estimated for Ethiopian condition (Hurni 1985a, b) as cited in (Engdayehu et al. 2016). Figures 17.12, 17.13, 17.14, 17.15, and 17.16 show various types of intervention measures used to reverse the soil erosion process and gully formation.

The analysis shows that soil loss is dependent on the area covered with soil and water conservation measures. It was estimated that 18.9 ton/ha/yr from five among studied watersheds which were on average covered with 93% of the total area with soil and water conservation. It was also seen from the analysis that from five selected study watersheds with 29.08% area treated by soil and water conservation, the soil loss is estimated as 46.86 ton/ha/yr. From these five watersheds, the average

Fig. 17.11 Stabilized fanya juu in Argedifo community watershed

Fig. 17.12 Newly constructed hill side terrace in Nekubish watershed

sediment retention was estimated as 73%. Compared to the average watersheds sediment retention of 56%, it is much greater; therefore, we can conclude that if the whole watersheds were covered with SWC, it can retain considerable amount of sediment, and soil loss can decrease substantially.

Conclusion and Recommendation

Conclusion

In this study, watersheds experience different types of soil erosion. Rills and gullies are the visible form of soil erosion which were commonly observed in study watersheds. It was also common to see stream channel erosion and sediments and boulders movement and color of the runoff water which is muddy. This study showed that there is a decrease in soil loss as a result of soil and water conservation implementation, even though the result was far above the soil formation rate in the Amhara region.

Fig. 17.13 Trench in Deresinilih watershed

Sediment retention in watersheds increases as the coverage of watersheds in soil and water conservation increases. Soil loss of watersheds with and without soil and water conservation was estimated, and the result was 77 ton/ha/yr in base case scenario assuming there was no SWC and 34 ton/ha/yr after implementation of SWC where the watersheds were covered with 60% of physical SWC from 34 watersheds. But when the watersheds area coverage increases to 93% as seen in five selected watersheds, the soil loss decreases to 18 ton/ha/yr, and the soil loss increases to 46 ton/ha/yr as the SWC area coverage decrease to 29% based on the average value of five watersheds.

Physical and biological soil and water conservation in study watersheds were identified, delineation of the coverage of implemented SWCs in the map was done, and the standards of the structures were technically evaluated by direct survey compared with the standard. Contribution of soil and water conservation for erosion control was assessed by RUSLE model.

From the survey of watersheds, it was possible to conclude that physical and biological soil and water conservation were selected based on the agro-ecology and land use requirement. Trees used for hedge row plantation are practiced according to agro-ecology, in moist areas suitable trees such as Trelu-ccrne were observed and adapted and grown well, likewise Susbania susban were grown in humid areas and performed well.

Fig. 17.14 Bench terrace in Godiguadit watershed (**a** and **b**), Side view of Bench terrace (**c**), Front view of Bench terraces Gidbenu watershed (**d**)

Structures have their own standard, and some SWC structures meet the requirements, but some others did not. The height of the SWC structures after compaction was as per the standard, but the width and depth of ditches were not as recommended for Ethiopian condition. Some structures like soil bund and stone-faced soil bund, the width were wider and the depth were less deep. In most types of structures, the mean vertical intervals were as per the standard, but for low sloppy areas, it was narrow, and for high slopes, it was wider than the recommended. The gradient of bunds violates the standard in some places, especially in wet and humid agro-ecology of watersheds. In gully rehabilitation, some conditions which should aggravate the problem such as constructing without foundation, side key, and apron and spill way were observed. The spacing between two check dams also were wide in most gullies treated by this structure.

Biological measures were implemented according to the agro-ecology, but there was no or few protections from cattle in some watersheds, so that those trees that are planted as stabilization of bunds were not grown well if not, they are stunt in their growth. Some managements such as area closure, control grazing, and gully rehabilitation were seen as a good start in rehabilitation of the land, and they become as a source of income by cutting grasses and selling tree seeds as well. In some watersheds, hilly areas were used as a job creation for youths by constructing

Fig. 17.15 **a** Area closure and a man used cut and carry system in Abiager watershed, **b** Hedge row plantation in Aseida watershed, **c** plantation of Eucalyptus trees in Maybar watershed, and **d** protected forest in Godguadit watershed

Fig. 17.16 Grass strip in Deresnilih watershed

bench terracing for maximizing crop land, apiculture, and tree planting. These have again indirect positive impact on regeneration of the land unless it could be used unwisely.

When both physical and biological soil and water conservation structures are done rendering to the standard and suitable agro-ecology, the target, minimizing erosion problem can be achieved, but to achieve the intended objective and for sustainability, technical standards such as vertical intervals, dimension of bunds and ditches, check dam dimensions, and gradients of SWC measures need great attention and close follow up.

Recommendation

Based on the results of the study and field observations of community watersheds, the following recommendations are forwarded.

I. Detailed surveys to map area coverage of soil and water conservation implemented in the region is required.
II. Farmers should be consulted and should be participated in the selection of SWC measures before application on their farm and maintenance should be taken on their own because it develops the ownership and responsibility.
III. Indigenous plants should be practiced as bund stabilization, and reforestation was seen in Gidbenu watershed and planting Kitikita in soil bund structure and regenerated better than exotic plants.
IV. It is better to treat Vertisols farmlands, which tare characterized by their properties of swelling and shrinkage according to the moisture availability, that have gentle slopes with only grass strip and other biological measures. Bunds were cut and damaged since farmers used to construct traditional farm ditches by crossing the constructed SWC structure.
V. If fanyajuu terrace is to be done in gentle slopes, it should be treated for water logging in the upper part, and the mound should be stone faced and slope should be 1%.
VI. Inter-terrace management practices like ploughing in graded contours, strip cropping, cover cropping, mulching or residue management, and planting in rows along graded contours can be applied in addition to physical practices.
VII. If bench terrace is also to be done in loose soils, the embankment should be stone faced, and the width is also to be according to the soil depth since it may be unproductivity if sub soils exposed entirely.
VIII. It is better to avoid using small and round stones when stone bund and stone-faced soil bunds are constructed since they can easily be collapsed. And bunds should not be long enough and must be staggered besides cattle tracks are very important to protect the collapse of bunds.
IX. Some of the water way selected on natural depression do not much as the depth of bund ditches are dipper than the water way depth and farm

boundary were not done along the slope, and this needs special attention because concentration of runoff can damage the land.

X. Accuracy of standards should be given great attention when surveying is held such as layout of gradients, embankments, and ditches dimensions like width, depth, and foundation and side keys on the period of implementation of SWC structures, so that they are constructed as per the recommendation.

XI. Documentation of the data of constructed structures in the community watersheds must be underway in each district since it is very important to evaluate the performance and coverage of SWC.

References

Aga AO, AM Melesse, Chane B (2019) Estimating the sediment flux and budget for a data limited Rift Valley Lake in Ethiopia. Hydrology 6(1). https://doi.org/10.3390/hydrology6010001

Aga AO, Chane B, Melesse AM (2018) Soil erosion modelling and risk assessment in data scarce Rift Valley Lake regions, Ethiopia. Water 10:1684. https://doi.org/10.3390/w10111684

Betrie GD et al (2011) Sediment management modelling in the Blue Nile Basin using SWAT model. Hydrol Earth Syst Sci 15(3):807–818

Bishaw B (2001) Deforestation and land degredation in the Ethiopian highlands: a strategy for physical recovery. Northeast Afr Stud 8(1):7–25

Defersha MB, Melesse AM (2012a) Effect of rainfall intensity, slope and antecedent moisture content on sediment concentration and sediment enrichment ratio. CATENA 90:47–52

Defersha MB, Melesse AM (2012b). Field-scale investigation of the effect of land use on sediment yield and surface runoff using runoff plot data and models in the Mara River basin, Kenya. CATENA 89:54–64

Defersha MB, Quraishi S, Melesse AM (2010) Interrill erosion, runoff and sediment size distribution as affected by slope steepness and antecedent moisture content. Hydrol Earth Syst Sci Discuss 7(6447–6489):2010

Defersha MB, Quraishi S, Melesse AM (2011) The effect of slope steepness and antecedent moisture content on interrill erosion, runoff and sediment size distribution in the highlands of Ethiopia. Hydrol Earth Syst Sci 15:2367–2375. https://doi.org/10.5194/hess-15-2367-2011

Defersha MB, Melesse AM, McClain M (2012) Watershed scale application of WEPP and EROSION 3D models for assessment of potential sediment source areas and runoff flux in the Mara River Basin, Kenya. CATENA 95:63–72

Desta L (2000) Land degradation and strategies for sustainable development in the Ethiopian highlands: Amhara Region, ILRI (aka ILCA and ILRAD)

Engdayehu G et al (2016) Evaluation of technical standards of physical soil and water conservation practices and their role in soil loss reduction: the case of Debre Mewi watershed, North-west Ethiopia. Landscape dynamics, soils and hydrological processes in varied climates. Springer, pp 789–818

Getachew HE, Melesse AM (2012) Impact of land use/land cover change on the hydrology of Angereb Watershed, Ethiopia. Int J Water Sci. https://doi.org/10.5772/56266,Vol.1,4:1-7

Grepperud S (1996) Population pressure and land degradation: the case of Ethiopia. J Environ Econ Manag 30(1):18–33

Helldén U (1987) An assessment of woody biomass, community forests, land use and soil erosion in Ethiopia. A feasibility study on the use of remote sensing and GIS [geographical information system]-analysis for planning purposes in developing countries. Lund University Press

Herweg K, Ludi E (1999) The performance of selected soil and water conservation measures—case studies from Ethiopia and Eritrea. CATENA 36(1):99–114

Hurni H (1985a) Erosion-productivity-conservation systems in Ethiopia

Hurni H (1985b) Soil conservation manual for ethiopia: a field guide for conservation implementation

Hurni H, Pimentel D (1993) Land degradation, famine, and land resource scenarios in Ethiopia. World soil erosion and conservation, pp 27–61

Keeley J, Scoones I (2000) Knowledge, power and politics: the environmental policy-making process in Ethiopia. J Mod Afr Stud 38(01):89–120

Ludi E (2009) Climate change, water and food security, Overseas Development Institute

Masresha B (2014) Evaluation of soil and water conservation measures in Dejiel watersheds, Choke Mountains, East Gojjam Zone of Amhara Region, Ethiopia, AAU

Maalim FK, Melesse AM (2013) Modeling the impacts of subsurface drainage systems on Runoff and Sediment Yield in the Le Sueur Watershed, Minnesota. Hydrol Sci J 58(3):1–17

Maalim FK, Melesse AM, Belmont P, Gran K (2013) Modeling the impact of land use changes on runoff and sediment yield in the Le Sueur Watershed Minnesota Using GeoWEPP. CATENA 107:35–45

Mango L, Melesse AM, McClain ME, Gann D, Setegn SG (2011a) Land use and climate change impacts on the hydrology of the upper Mara River Basin, Kenya: results of a modeling study to support better resource management. Special Issue: Climate, weather and hydrology of East African Highlands. Hydrol Earth Syst Sci 15:2245–2258. https://doi.org/10.5194/hess-15-2245-2011

Mango L, Melesse AM, McClain ME, Gann D, Setegn SG (2011b) Hydro-meteorology and water budget of Mara River basin, Kenya: a land use change scenarios analysis, In: Melesse A (ed) Nile River Basin: hydrology, climate and water use (Chapter 2). Springer Science Publisher, pp 39–68. https://doi.org/10.1007/978-94-007-0689-7_2

Mekonnen M, Melesse A (2011) Soil erosion mapping and hotspot area identification using GIS and remote sensing in northwest Ethiopian highlands, near Lake Tana, In: Melesse A (ed) Nile River Basin: hydrology, climate and water use (Chapter 10). Springer Science Publisher, pp 207–224. https://doi.org/10.1007/978-94-007-0689-7_10

Melesse AM, Jordan JD (2002) A comparison of fuzzy vs. augmented-ISODATA classification algorithm for cloud and cloud-shadow discrimination in landsat imagery. Photogram Eng Remote Sens 68(9):905–911

Melesse AM, Jordan JD (2003) Spatially distributed watershed mapping and modeling: land cover and microclimate mapping using landsat imagery, Part 1. J Spatial Hydrol 3(2)

Melesse A, Weng Q, Thenkabail P, Senay G (2007) Remote sensing sensors and applications in environmental resources mapping and modeling. Spec Issue Remote Sens Nat Res Environ Sens 7:3209–3241

Melesse AM, Ahmad S, McClain M, Wang X, Lim H (2011) Sediment load prediction in large Rivers: ANN approach. Agric Water Manag 98:855–866

Yesuf HM, Alamirew A, Assen M, Melesse A (2013) Spatiotemporal mapping of land cover in Lake Hardibo Drainage Basin, Northeast Ethiopia: 1957–2007. Water conservation: practices. Nova Publishers, Challenges and Future Implications, pp 147–164

Yesuf HM, Alamirew T, Assen M, Melesse AM (2015) Modeling of sediment yield in Maybar gauged watershed using SWAT, northeast Ethiopia. CATENA 127:191–205

Msagahaa J, Ndomba Melesse AM (2014) Modeling sediment dynamics: effect of land use, topography and land management, In: Melesse AM, Abtew W, Setegn S (eds) Nile River Basin: ecohydrological challenges, climate change and hydropolitics, pp 165–192

Nyssen J et al (2010) Impact of soil and water conservation measures on catchment hydrological response—a case in north Ethiopia. Hydrol Process 24(13):1880–1895

Pender J, Gebremedhin B (2006) Land management, crop production, and household income in the highlands of Tigray, Northern Ethiopia: An econometric analysis. Strategies for sustainable land management in the East African highlands, pp 107–139

Renard KG et al (1997) Predicting soil erosion by water: a guide to conservation planning with the Revised Universal Soil Loss Equation (RUSLE), US Government Printing Office Washington, DC

Setegn SG, Srinivasan R, Dargahi B, Melesse AM (2009) Spatial delineation of soil erosion prone areas: application of SWAT and MCE approaches in the Lake Tana Basin. Ethiop Hydrol Process Spec Issue Nile Hydrol 23(26):3738–3750

Setegn SG, Bijan Dargahi B, Srinivasan R, Melesse AM (2010) Modelling of sediment yield from Anjeni Gauged watershed. Ethiop Using SWAT JAWRA 46(3):514–526

Wang X, Garza J, Whitney M, Melesse AM, Yang W (2008) Prediction of sediment source areas within watersheds as affected by soil data resolution, In: Findley PN (ed) Environmental modelling: new research, Chap 7. Nova Science Publishers, Inc., Hauppauge, NY 11788, pp 151–185. ISBN: 978–1–60692–034–3

Wischmeier W, Mannering J (1969) Relation of soil properties to its erodibility. Soil Sci Soc Am J 33(1):131–137

Wondie M, Schneider W, Melesse AM, Teketay D (2011) Spatial and temporal land cover changes in the Simen Mountains national park, a world heritage site in Northwestern Ethiopia. Remote Sens 3:752–766. https://doi.org/10.3390/rs3040752

Wondie M, Teketay D, Melesse AM, Schneider W (2012) Relationship between topographic variables and land cover in the Simen Mountains National Park, a World Heritage Site in northern Ethiopia. Int J Remote Sens Appl 2(2):36–43

Zeleke G, Hurni H (2001) Implications of land use and land cover dynamics for mountain resource degradation in the Northwestern Ethiopian highlands. Mt Res Dev 21(2):184–191

Zhang G et al (2005) Potential effects of climate change on rainfall erosivity in the Yellow River basin of China. Trans ASAE 48(2):511–517

Chapter 18
Conceptual and Practical Approaches to Integrated Watershed Management and Agroforestry to Address Food Security and Environmental Degradation in Lake Tana and the Blue Nile River Basin, Ethiopia

Badege Bishaw

Abstract Forests and trees provide food, fuelwood, and construction materials and many environmental benefits, such as erosion control, clean water and air, biodiversity conservation, and carbon sequestration. The forest covers in Ethiopia and that of the Blue Nile River Basin in particular have been deteriorating at a progressive rate. The situation is even worse in the Angereb watershed. The major causes of deforestation and environmental degradation in Ethiopia are increasing population, increasing demand for farmlands, and increasing demand for fuelwood for cooking and construction materials. The effects of poor farming practices and lack of conservation are the main causes for the siltation of the Angereb dam and the pollution of drinking water from the dam. To address these environmental and livelihood problems, concepts of sustainability and ecosystem-based approaches were applied. We involved different stakeholders, such as the Gondar City Administration, Department of Forestry and Agriculture, Department of Water Affairs, Farmers in the Angereb watershed, University of Gondar, and the Corvallis-Gondar Sister Cities Association in the project. Through this participatory approach, the stakeholders identified the natural resource problems in the Angereb watershed, proposed integrated watershed management involving soil and water conservation, agroforestry, tree planting, and reduced siltation of the dam. Through this project, we improved the tree nursery at Weleka and raised multipurpose indigenous and exotic tree species for planting. Farmers were involved in tree planting and agroforestry practices, such as establishing riparian buffers along the streams, alley cropping for soil and water conservation by planting trees on terraces and agroforests/home gardens to address food security. Through this project, 2.5 million tree seedlings were planted on 560 hectares in the Angereb watershed in ten years. Since 2010, 19 water sources were constructed, including drilled wells,

B. Bishaw (✉)
Oregon State University, Corvallis, OR, USA
e-mail: badege.bishaw@oregonstate.edu

A. M. Melesse et al. (eds.), *Nile and Grand Ethiopian Renaissance Dam*,
Springer Geography, https://doi.org/10.1007/978-3-030-76437-1_18

hand-dug wells, and developed springs to provide clean drinking water for 1700 households in Gondar city and the surrounding villages. We recommend scaling up the ecosystem-based approaches, using lessons learned on agroforestry, tree planting, and soil and water conservation in the Angereb watershed to the larger Blue Nile River Basin. This will address environmental degradation, soil erosion, and food security, which will have great impacts on survival of Lake Tana and extend the longevity of the Grand Ethiopian Renaissance Dam (GERD) by reducing siltation.

Keywords Sustainability · Ecosystem-based · Agroforestry · Participatory approach · Integrated watershed · Angereb dam · Wells and springs · Drinking water · Blue Nile River Basin · Lake Tana · Siltation · Grand Ethiopian Renaissance Dam

Introduction

Forests and trees provide food, fuelwood, and construction materials and many environmental benefits, such as erosion control, clean water and air, biodiversity conservation, and carbon sequestration. Deforestation, soil erosion, and land degradation are serious problems in Ethiopia (Bishaw 2001). The forest covers in Ethiopia and that of the Blue Nile River Basin in particular have been deteriorating at a progressive rate. The major causes of deforestation and environmental degradation in Ethiopia are increasing population, increasing demand for farmlands, and increasing demand for fuelwood to cook food and construction materials (Bishaw 2001; Haileslassie et al. 2008a, b; Gashaw et al. 2014, Kidane and Alemu 2015). The effects of poor farming practices and lack of conservation are responsible for the siltation of many rivers and dams in Ethiopia. The deforestation and soil erosion situations in the Angereb Watershed are worse and affect pollution of drinking water from the dam (Amare 2005; CSCA-Gondar 2020). The social and environmental problems in this watershed are complex and need long-term investment to improve sanitation, education, and livelihoods (Guillozet 2010). To address these environmental and livelihood problems, integrated watershed management and agroforestry were applied as strategy to overcome these problems. This chapter presents the challenges and opportunities faced by the Corvallis-Gondar Sister Cities Association on tree planting and agroforestry practices to address food security, soil erosion, and environmental degradation. It also explores the concepts of ecosystem-based approaches and integrated watershed management to achieve sustainability and proposes scaling up best practices for sustainable land use in Lake Tana and the greater Blue Nile Basin.

Biophysical and Sociocultural Attributes of the Angereb Watershed

The Angereb watershed lies between UTM coordinate of N 1,394,096, N 1,407,336, E 328,073, and E 337,991 in North Gondar Zone, in the Amhara Regional State, Ethiopia (Fig. 18.1). Elevation ranges from 2100 to 2870 m above sea level and mean rainfall 1200 mm peaking in July and August (Guillozet 2010). The watershed covers approximately 7600 hectares and is part of the Blue Nile drainage (Fig. 18.1). The Blue Nile Basin (Abay in Ethiopia) covers wide range of landscapes and climatic zones in Ethiopia and Sudan. In response to those diverse landscapes and climatic zones, different agricultural production systems and socio-economic dynamics have evolved in the basin that responds to changing livelihoods opportunities (Haileslassie et al. 2008a, b, Kidane and Alemu 2015).

The watershed is the primary source of drinking water for the residents of Gondar. The Angereb River feeds the Angereb reservoir, and it is an earthen dam completed in 1986 to create the reservoir (Fig. 18.2). The reservoir was designed with a storage capacity of 5 million cubic meters (Amare 2005; Guillozet 2010; CSCA-Gondar 2020). The reservoir was expected to supply the majority of the city's drinking water until 2021; however, sediment from the surrounding hillsides

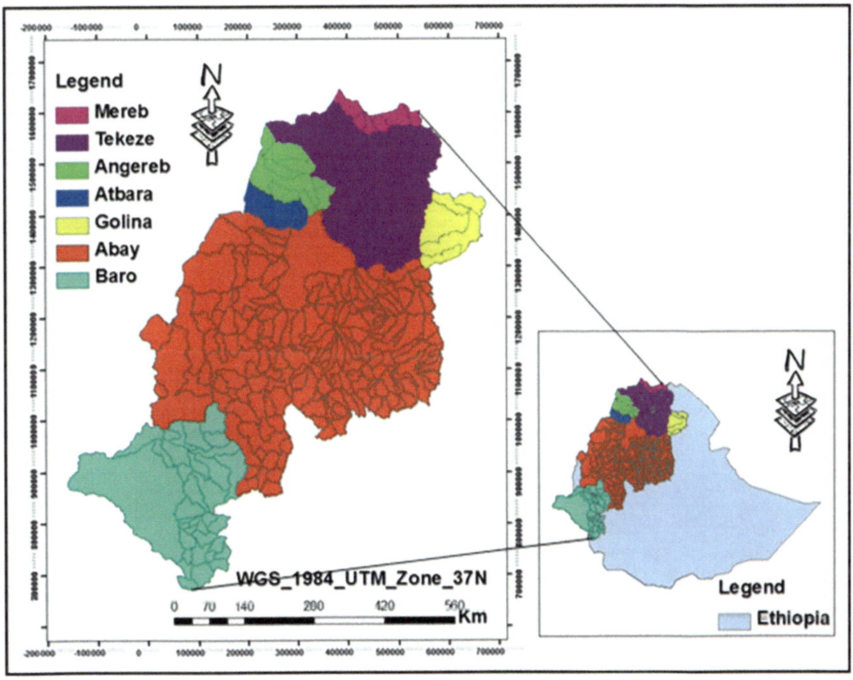

Fig. 18.1 Map of the upper Blue Nile Basin, Ethiopia (Kidane and Alemu 2015)

and the Angereb River began accumulating soon after the completion of the dam. It is well documented that sedimentation problems will significantly reduce the life of the reservoir (Amare 2005; Guillozet 2010; CSCA-Gondar 2020).

Mixed farming practices of crops and livestock production using animal traction and wooden plows over centuries have created highly eroded soils leaving rocks scattered on fields and a reservoir half-filled with sediment (Hailesellasie et al. 2008a, b; Amare 2005; Guillozet 2010). Guillozet (2010) reported that the population of the Angereb watershed was 29,148 (5279 households) of which 57% urban and 43% percent rural. The average family size is about 5.52 persons. Population density within the watershed is about 382 per km^2. Approximately 13,685 persons or 2407 households engaged in farming, covering 70% of the watershed, with average farm size 2.2 ha. Main crop production includes wheat (38%), barley (28%), teff (21%), and horse beans (13%). Livestock is principal source of income and capital with estimates of 7000 cattle, 6300 sheep, 6500 donkeys, 3500 chicken, and 1100 goats. The hillside surrounding the reservoir has been stripped of trees primarily for fuelwood having the area adjacent to the reservoir exposed to soil erosion and entering the water as sediment.

The Corvallis Sister Cities Association-Gondar (CSCA-Gondar) was established in 2005, and the primary goal of this partnership are (1) To promote awareness and understanding between the people of Corvallis, Oregon, and the people of Gondar,

Fig. 18.2 Angereb reservoir (CSCA-Gondar 2020)

Ethiopia, and (2) To undertake activities that are beneficial to the people of both Corvallis and Gondar (CSCA-Gondar 2020). To achieve these goals, the CSCA-Gondar established three working groups: (1) water and watershed, (2) education and (3) health. This chapter will focus on the contributions of the water and watershed-working group.

In 2007, CGCA-Gondar hydrologist and forester made preliminary visit to the Angereb watershed. The hydrologist noted that the water plant intake had been raised approximately 4.5 m (14.8 ft) since the start of the operation of the reservoir and was raised an additional 3.0 m (9.8 ft) in 2014. The intake should be at a depth of 5.0 m (16.4 ft) below surface water. The intake is now only 0.5 m (1.6 ft) below surface water (CSCA-Gondar 2020).

CGSA-Gondar forestry expert visited the watershed with counterparts in Gondar to explore the current land use and farming practices in the watershed. There are mixed cereal-livestock farming practices with eroded farms, big gullies, and striped hillsides denuded of trees. Consultation with counterparts and stakeholders discussed options on restoration of the Angereb watershed to improve the environmental degradation and increase agricultural productivity through an integrated watershed management and agroforestry. They also discussed to improve the Woleka nursery and explored potential seed source for the tree planting and agroforestry practices.

Restoring the Angereb Reservoir and Watershed

The goal of the Corvallis Sister Cities Association-Gondar integrated watershed management project is to improve existing environmental degradation, such as soil erosion and improve farming practices to produce food, fuelwood, and income, while integrating practices and interventions on ecosystem services and develop human capacity to deal with complex systems (Negasa 2020). The objectives of the project were (1) Promote integrated watershed management, soil conservation, and longevity of the Angereb reservoir; (2) Collaborate with farmers, government representatives, and NGOs to develop sustainable practices and enterprises; (3) Promote tree planting through agroforestry to improve environmental function; and (4) Increase the awareness and participation of people who live in the watershed.

Methodology

The methodology applied for this project has two parts: (1) Implementing integrated watershed management and agroforestry practices in the Angereb watershed. (2) Explaining how the concepts of ecosystem-based approaches and sustainability applied in the Angereb watershed. We used participatory approaches and field visits to identify the natural resources problems that affected the livelihoods in the Angereb watershed. We proposed and implemented integrated watershed management and agroforestry as solutions to address these problems.

Participatory Approaches

We involved different stakeholders, such as the Gondar City Administration, Department of Forestry and Agriculture, Department of Water Affairs, Farmers in the Angereb watershed, University of Gondar, and the Corvallis-Gondar Sister Cities Association in the project (Fig. 18.3). Field visits were carried out by experts from the Department of Agriculture and Forestry, Department Water Affairs, Gondar City Administration, and Corvallis Water and Watershed work group to observe the watershed functions and current land use and farming practices in the watershed (Fig. 18.4). We conducted meetings with the urban residents and rural farming communities to identify and prioritize the problems in the Angereb watershed. Through the participatory approaches (Fig. 18.3), the stakeholders identified the following natural resource problems in the Angereb watershed: (1) deforestation, (2) soil erosion, (3) loss of soil fertility, (4) severe water shortage (urban and rural), (5) chronic health problem, and (6) high levels of poverty. These natural resources constraints and lack of coordinated planning were listed by people of the Amhara region as critical to sustainable livelihoods and ecosystem functions. These are wicked problems and have cascading effects on the farming and land use practices in the Angereb watershed. Figure 18.5 reflects the environmental degradation associated with food insecurity and poverty of the region.

Application of Ecosystem Approaches and Sustainability

The ecosystem approach is a strategy for the integrated management of land, water, and living resources that promotes conservation and sustainable use. It is based on the application of scientific methodologies focused on levels of biological organization. The ecosystem approach encompasses the essential processes, functions, and interactions among organisms and their environment. It recognizes that humans, with their cultural diversity, are an integral component of ecosystems. Additional information on ecosystem-based approaches can be found at: https://www.cbd.int/ecosystem/.

The United Nations General Assembly first introduced the concepts of sustainability in early 1980s. The assembly was concerned about world population growth, deforestation, and environmental change. In 1987, the commission published "The Brundtland Report" or known as "Our common Future." In this report, they defined sustainable development: "development that meets the needs of the present without compromising the ability of the future generations to meet their own needs" (Brundtland 1987; Kuhlman and Farrington 2010). The concept of sustainability is complex and important to find the balance between the ecological

Fig. 18.3 Participatory approaches to identify natural resource problems

and human systems. The complexity lies in determining the nature of present and future needs, and who or what has those needs. Keeping with the traditional three-pronged approach to sustainability, "needs" can be generally broken up into those pertaining to the economy, to the environment, and to society. If sustainable development is to provide for the needs of the present as well as the future, economic priorities must shift from production to sufficiency; environmental concerns must focus on resilience; and society (farmers) must benefit equitably from resources.

In order to practice sustainability, one has to address the three pillars of sustainability that include environmental: addresses productivity and ecosystem services, society includes social, culture, ethical and political aspects, and economic market and non-market goods. The three pillars are interconnected and interdependent benefits. Giddings (2002) discuss about weak and strong sustainability and defines strong sustainability, as "a healthy environment is the foundation for heathy society and healthy economy."

To apply the concepts of ecosystem approaches and natural resource sustainability, we applied the following practices: integrated watershed management, (management of land, water and living resources), integrated socio-agro-ecological approach (long-term approach) (Fig. 18.6), integrate farming practices, resource conservation, reservoir maintenance (sustain long-term function), and adaptive management (learning-by-doing). Project implementation was carried out through

Fig. 18.4 Field visit to the Angereb watershed to understand land use and farming practices

agroforestry and tree planting (nursery improvement, tree planting, and building terraces), community participation (participatory approaches, training, workshop, and extension), and strong integration (watershed restoration, water supply, and education).

Result and Discussion

The stakeholders implemented integrated watershed management involving soil and water conservation, agroforestry, and tree planting to reduce siltation of the reservoir and alleviate food security and poverty. Through this project, the tree nursery at Weleka was improved and raised multipurpose indigenous and exotic tree species for planting. The following indigenous species raised at the nursery for planting: Weira (*Olea europaea*), Gesho (*Rhamnus prinoides*), Kentetifa (*Entada abyssinica*), Wanza (*Cordia Africana*), Misana (*Croton macrostachyus*), and Birbira (*Milletia ferruginea*). The following exotic species raised at the nursery for planting: Nim (*Azandirachta indica*), Key bhairzaf (*Eucalyptus camaldulensis*), Yeferenj tid (*Cupressus lustanica*), Nech bahirzaf (*Eucalyptus globulus*), Arzelibanos (*Casuarina equestifolia*), Yeferenj Grar (*Acacia decurrens*), Spatodea (*Spathodea nilotica*), Kundo berbere (*Schinus molle*), Gravilia (*Gravilia robusta*),

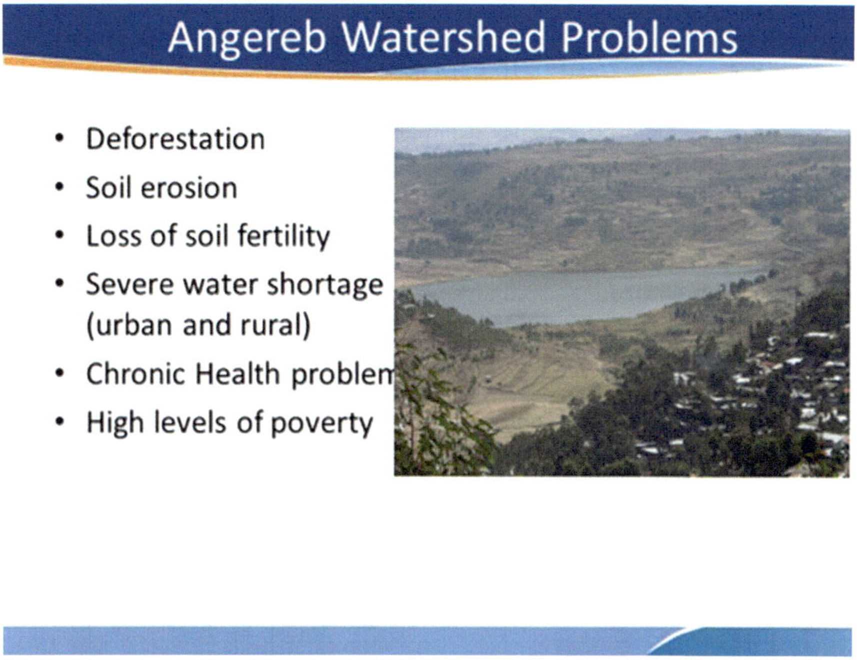

Fig. 18.5 Environmental degradation, food insecurity, and poverty

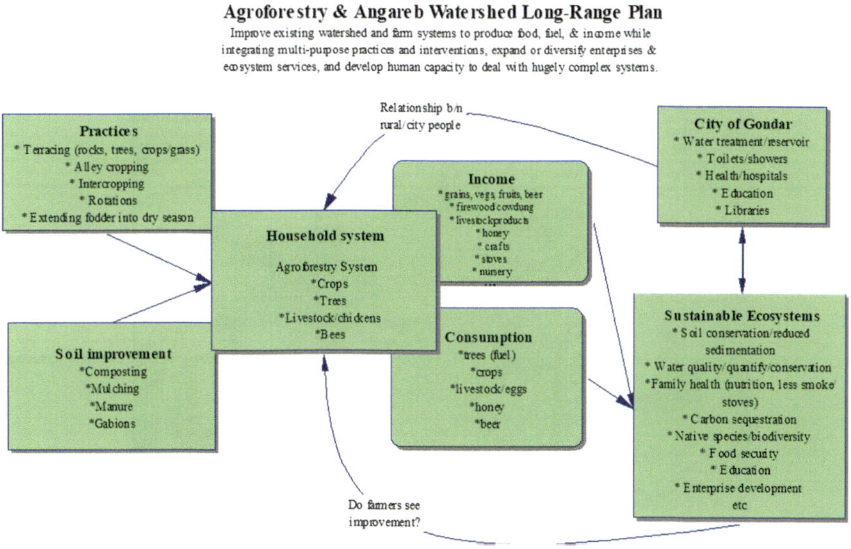

Fig. 18.6 Angereb watershed long-range plan

and Sesbania (*Sesbania sesban*). Women were hired to plant seeds and take care of the seedlings at the nursery and created jobs for women in the neighborhoods. Figures 18.7, 18.8, 18.9, and 18.10 show the seedling preparation and reforestation process.

Farmers were involved in tree planting and agroforestry practices, such as, establishing riparian buffers along the streams, alley cropping for soil and water conservation by planting trees on terraces and agroforests/home gardens to address food security. Men prepared the site for field planting and take care of the seedlings by weeding, cultivation, and protecting from grazing animals. The first tree planting started in 2008 and 67,000 seedlings were planted on 20 ha in the immediate vicinity of the reservoir. During the 2009 planting season, an additional 250,000 seedlings planted around the reservoir. In 2010, tree-planting efforts shifted from the immediate area of the reservoir to various sub-watersheds within the Angereb watershed. In 2013, Gondar citizens committee recommended a more comprehensive approach, and the Bukaya sheleko sub-watershed (Fig. 18.9) selected as a demonstration area for more integrated and comprehensive approach to watershed restoration.

In addition to planting trees, income generation projects were included to provide residents with alternatives to gathering and selling firewood and grazing the hillsides. The tree planting continued from 2015 to 2019 and the CSCA-Gondar

Fig. 18.7 Sowing multipurpose seedlings Weleka Nursery

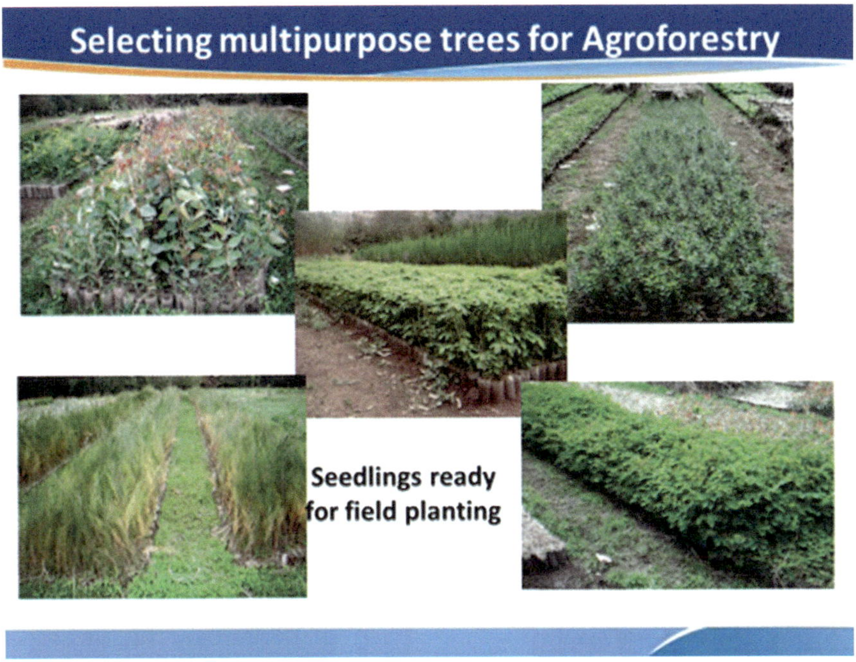

Fig. 18.8 Seedlings of multipurpose trees ready for planting

provided funding to purchase seeds and raise the necessary seedlings in partnership with the Gondar Department of Agriculture and Gondar-based nonprofit Bridge of Hope. Through this project, 2.5 million tree seedlings planted on 560 hectares in the Angereb watershed in ten years (Table 18.1). These trees planted on terraces for soil and water conservation as alley cropping provided riparian buffers to reduce siltation of the reservoir. The planted multipurpose trees in home gardens address food security and alleviate poverty. In Table 18.1, it is assumed, 1.5 × 1.5 m spacing (2.25 m^2/seedling), 4450 seedlings/ha, and dividing 2,497,220 million seedlings by 4450 gives 561 ha.

Agroforestry for Food Security

Mixed farming, which integrates trees, crops, and livestock, is common practice in the Ethiopian highlands. This farming practice planting patterns include sequential and intercropped designs that optimize land use and labor, conserve resources, and provide multiple benefits. Alley cropping which involves planting trees closely within rows, but rows are widely spaced on the contour to form natural terraces when coppiced at 25 cm to collect soil, while tree residues are used as fuel,

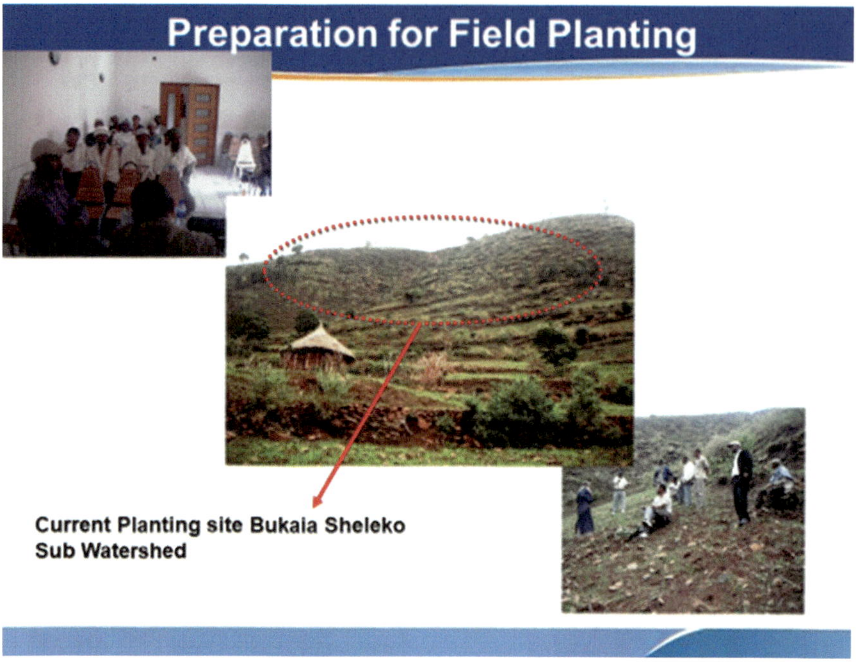

Fig. 18.9 Site preparation for field planting

fertilizer, or fodder. On steep hillsides, this is a common practice to produce wheat, barley, and teff and conserve soil and water. One framer in Sabia sub-watershed planted gesho shrubs on the contour in a limited form of alley cropping giving promise that careful selection of tree species may make this practice feasible on steep sites. Gesho is a cash crop used for local brewing. Farmers in the Angereb watershed should also be encouraged to plant nitrogen fixing trees and shrubs to improve soil fertility and increase crop yields and income.

Agroforestry for Soil and Water Conservation

Farmers expressed interest to improve or practice conservation that include terraces using rocks and planting trees and interplant grasses in upland fields. Trees and shrubs planted on terraces reduce soil erosion, create physical barrier, and enhance water infiltration. Eucalyptus trees planted in woodlots, on terraces or rock piles, and along riparian streams to reduce bank collapse. Farmers also expressed interest in planting multipurpose trees along 5 m of the stream, inter planting trees and grasses to upland cropland, and adding fruit (low-chill apple, mango, citrus, etc.) and gesho shrubs to their farming system. The fruit trees and gesho can offer

Fig. 18.10 Agroforestry tree panting project

Table 18.1 Seedling produced and planted in the Angereb watershed 2008–2019

Year	# Seedlings
2008	67,000
2009	250,000
2010	350,000
2011	291,165
2012	228,347
2013	253,160
2014	262,548
2015	200,000
2016	125,000
2017	125,000
2018	125,000
2019	100,000
Total	2,497,220

additional income sources but need additional study perhaps by the University of Gondar. Farmers might consider grass mixes to optimize fodder choices for livestock while contributing to ecosystem services and conservation practices.

Agroforestry for Biodiversity

Home gardens agroforestry is a more complex multi-stratum than other agroforestry systems (Asfaw et al. 2015). In home garden agroforestry, cultivation of planned and intensively managed trees, crops, and livestock exists. In the Angereb watershed, farmers by growing food crops, shrubs, herbs, and medicinal plants practice home garden. Traditional medicinal shrubs and herbs grow in the understory. Home gardens are diverse, provide ecosystem balance and sustainability, household food security, and rural development in Ethiopia. The multipurpose trees and shrubs planted in agroforestry systems increase vegetation diversity. The diverse nature of agroforestry helps increase biodiversity, reduce crop failure, and economic risks.

In addition to tree planting and agroforestry practices, the CSCA-Gondar constructed 19 water sources including drilled wells, hand-dug wells, and developed springs to provide clean drinking water for 1,700 households in Gondar city and the surrounding villages (CSCA-Gondar 2020).

Agroforestry for Mitigating Climate

Managing sustainable development in the face of a drastically changing climate can be very difficult. Knowing and deciding, which actions taken in the present will be sustainable in the future, are even harder decision to make. Management for future conditions is even more difficult given that some tree species may not even be present in the future. Although the best and most current science and literature can be applied to ecosystem-based approaches, the uncertainties about future climates make management a difficult task (Bishaw et al. 2013). Although there are many pessimistic outlooks regarding future resource management and climate change, carbon offset sales programs may be a positive policy that could relate to more sustainable development. These carbon offset management actions would be looking at additional values beyond just the board foot value of agroforestry.

Conclusions and Recommendations

Agroforestry is a dynamic, ecological-based natural resources management system that provides many environmental services and production systems and address the social and economic benefits that society needs. In addition to the environmental services and improving production systems, agroforestry plays great role in climate change adaptation and mitigation (Kuyah et al. 2019; Tamirat and Mekides 2020). Climate change is one of the challenges of subsistence farmers, and farmers in the Angereb watershed are no exception. Because of its diverse nature, agroforestry serves as tools for climate change adaptation and mitigation. Farmers by using drought resilient crop varieties such as sorghum, millet, and yam, they can adapt to climate change. Farmers should also change their farming strategies by adapting shorter agricultural calendar. Regarding livestock, farmers should switch to animals, which can live more easily in dry conditions such as goats that do not need as much water as sheep and cows (Bishaw et al. 2013). Agroforestry also sequester more carbon from the atmosphere, trees, and shrubs on farms can sequester carbon and mitigate climate change. Agroforestry systems store significant amount of carbon belowground. Carbon stored in agroforestry systems could be sold in carbon credit markets and generate additional income to the farm household.

To promote sustainable integrated watershed management and agroforestry practices in the Angereb watershed, we recommend the following biophysical, social, cultural, and policy frameworks. (1) Control soil erosion by constructing check dams, terraces, and soil improvement activities. (2) Focus on multipurpose trees, shrubs, and fruit trees to improve agricultural production and increase incomes. (3) Use participatory approaches, build on existing culture, and address local development issues. (4) Build capacity and awareness of local people through training and workshops. (5) Promote education and training in environmental health and climate change at elementary, high schools, and universities. (6) Increase the scientific knowledge on ecosystem-based approach by engaging universities, national, and international research institutions. (7) Increase interagency collaborations among ministries, such as agriculture, forestry, environment, water, health, and education, etc. These are similar to the findings and recommendations given to improve watershed management practices in Ethiopia (Negasa 2020).

We recommend scaling up the ecosystem-based approaches and sustainability concepts using lessons learned from the integrated watershed management and agroforestry practices on tree planting, soil and water conservation, and food security in the Angereb watershed to the larger Blue Nile River Basin. Despite the many opportunities agroforestry provides, there are several challenges to scale up agroforestry practices. These are issues regarding social (demographic factors, land ownership, unavailability of markets, infrastructure), economic (financial incentives, economic benefits), and environmental (soil erosion, water quality, global climate change) (Tamirat and Mekides 2020).

Lake Tana, one of the biggest Lake in Ethiopia, is affected by an invasive weed, water hyacinth. This weed has affected the biological, physical, and social life in and around the lake. Integrated control strategy, such as manual, mechanical, and biological control can be used to remove this weed. It is also proposed to improve the riparian buffers along the lake to filter sediment, nutrients, and pesticides. We recommend integrated watershed management and agroforestry practices to reduce siltation of the Lake and address environment services and food security.

Concerning the Grand Ethiopian Renascence Dam (GERD), while completion of the first phase is a great success, there is great concern about siltation of the dam (Deribe and Tirusew 2020) and the dam being affected by water hyacinth. We recommend treating the whole Blue Nile River Basin with integrated watershed management and agroforestry practices to address the ecological, social, and economic issues and promote sustainable development. We hope this will address the environmental degradation, soil erosion, and food security, which will have great impacts on survival of Lake Tana and extend the longevity of the Grand Ethiopian Renaissance Dam (GERD) by reducing siltation.

References

Amare A (2005) Study of sediment yield from the watershed of the angereb reservoir. M.Sc. Thesis, Alemaya University, December 2005, p 114

Asfaw Z, Linger E, Zewudie S (2015) Plant species richness and structure of homegarden agroforestry in Jabithenan District, Northwestern Ethiopia. Int J Environ Sci 4(2):52–58

Bishaw B (2001) Deforestation and land degradation in Ethiopian highlands: a strategy for physical recovery, vol 8, issue 1. Oregon State University, Corvallis. North East Africa Sturdies (ISSN 0740–9133), pp 7–26

Bishaw B, Neufeild H, Mowo J, Abdulkadir A, Muriuki J, Dalle G, Assefa T, Guillozet K, Kassa H, Dawson IK, Luedeling E, Mbow C (2013) In: Davis CM, Bernart B, Dmitriev A (eds) Farmers' strategies for adapting to and mitigating climate variabilty and change through agroforestry in Ethiopia and Kenya. Forestry Communications Group, Oregon State University, Corvallis, Oregon, pp 15–20

Brundtland, Gro Harlem (1987) Report of the World Commission on Environment and Development: Our Common Future. http://www.un-documents.net/our-common-future.pdf. Accessed 12 Jan 2021

Convention on Biological Diversity (CBD), Ecosystem Approach. https://www.cbd.int/ecosystem/. Accessed 31 Dec 2020

Corvallis Sister Cities Association-Gondar, Ethiopia (CSCA-Gondar) (2020) Restoring the angereb reservoir and watershed. Accessed 31 Dec 2020. https://sistercities.corvallis.or.us/angereb-watershed-restoration/

Gashaw T, Bantider A, G/Silassie G (2014) Land degradation in Ethiopia: causes, impacts and rehabilitation techniques. J Environ Earth Sci 4(9). ISSN 2224–3216 (paper) ISSN 2225–0948 (online)

Giddings B, Hopwood B, O'Brien G (2002) Environment, economy and society: fitting them together into sustainable development. Sustain Dev 10(4):187–196

Guillozet P (2010) Angereb watershed targeted revegetation strategy. Gondar, Ethiopia. Corvallis Sister Cities Association Gondar, Ethiopia, Report April 2010. p19

Haileslassie A, Hagos F, Mapedza E, Sadoff C, Awulachew SB, Gebreselassie S, Peden D (2008) Institutional settings and livelihood strategies in the Blue Nile Basin: implications for upstream/downstream linkages. Colombo, Sri Lanka: International Water Management Institute, p 81 (IWMI Working Paper 132)

Haileslassie A, Hagos F, Awulachew SB, Peden D, Gebreselassie S, Negash F (2008) Indicators of environmental degradation in the Blue Nile Basin: exploring prospects for payment for environmental services. Paper presented at the Second Nile Development Forum, Khartoum, Sudan, 17–19 November 2008, 37p. https://hdl.handle.net/10568/38113

Kidane D, Alemu B (2015) The effect of upstream land-use practices on soil erosion and sedimentation in the Upper Blue Nile Basin. Res J Agric Environ Manag 4(2):055–068. Available online at http://www.apexjournal.org

Kuhlman T, Farrington J (2010) What is sustainability? Sustainability 2(11):3436–3448

Kuyah S, Whitney C, Jonsson M, Sileshi G, Oborn I, Muthuri C, Luedeling E (2019) Can agroforestry enhance ecosystem services provision without reducing productivity. Agroforestry 2019, Regular Talk

Negasa DJ (2020) Major constraints of watershed management practices in ethiopia and ways forward (review article). Int J Environ Protect Policy 8(4):70–76. https://doi.org/10.11648/j.ijepp.20200804.11

Tamirat W, Mekides A (2020) Opportunities and challenges of scaling up agroforestry practices in Sub-Saharan Africa: a review. Agric Rev 41(3):216–226. Submitted: 09–06–2020 Accepted: 22–08–2020 Published: 11–09–2020

Deribe M, Tirusew A (2020) Beyond the normalization of climate change, Ethiopia's Green Legacy is a must have insurance for the GERD. Accessed May 18, 2021https://www.weaspire.info/.

Chapter 19
Rainfall-Runoff and Sediment Yield Modeling in Headwater Catchments of Lake Tana Sub-Basin, Ethiopia

Sileshie Mesfin, Arega Mullu, and Assefa M. Melesse

Abstract Tana sub-basin receives rainfall during summer as a primary source of stream flow. This research aimed to relate the catchment characteristics with hydrology of headwater catchments. Rainfall-runoff and sediment yield modeling in relation to catchment characteristics was conducted. Land use, soil, digital elevation model (DEM), precipitation, water quality and quantity were used in the analysis. ERDAS, GIS and HYDATA software were used to evaluate the physical catchment characteristics and meteorological data. Model parameters were calibrated using observed data of 2009–2011 and validated over the period of 2012. Particle swarm optimization was used to determine optimal model parameters. The model parameters were related to catchment characteristics using linear regression in a statistical software. Normalized difference vegetation index (NDVI) and hypsometric integral are directly proportional to direct runoff parameter C. Likewise, the baseflow model parameter b is positively affected by elongation ratio and average slope, and NDVI shows negative effect. Finally, regional model for the hydrology of the headwater catchments was developed.

Keywords Model parameter · Calibrated · Validated · Particle swarm optimization · SPSS

S. Mesfin (✉)
Amhara Design and Supervision Works Enterprise, Bahir Dar, Ethiopia

A. Mullu
Hydraulic and Water Resources Engineering, Debre Markos University, Debre Markos, Ethiopia

A. M. Melesse
Department of Earth and Environment, Institute of Environment, Florida International University, Miami, USA
e-mail: melessea@fiu.edu

A. M. Melesse et al. (eds.), *Nile and Grand Ethiopian Renaissance Dam*,
Springer Geography, https://doi.org/10.1007/978-3-030-76437-1_19

Introduction

Geology, topography, meteorology and biological factors affect the quantity and quality of surface and groundwater. Water quality is highly variable, and human activities as well as watershed processes play important roles.

Even though water quality of many rivers is deteriorating, restorative efforts are proven to improve the quality of these rivers by reducing sediment and pollutants from reaching receiving water bodies. Strengthening and synthesizing appropriate evidence regarding the interrelation between the benefits of human being from rivers and improved river health is very important (Parker and Oates 2016).

Lake Tana sub-basin is home of over 2 million people making it one of the most densely populated upper part of Blue Nile Basin. However, it is one of the most affected sub-basins with high soil erosion, sediment loss, land degradation and water quality deteriorations affecting the livelihood of its inhabitants (Setegn et al. 2008). The area of Lake Tana sub-basin is 15,069 km^2 and receives mean annual rainfall of about 1280 mm (Asalf 2018), and its mean annual evapotranspiration is 773 mm (Wale et al. 2008). Gilgel Abbay, Gummara, Rib and Megech Rivers are the main tributaries which contribute over 93% of the runoff water to the lake (Asalf 2018).

In many developing countries, many important rivers are ungaged; however, the science of understanding river flows becomes crucial and important. To analyze flow events, focuses were given as an emerging field of study of eco-hydrological science. As a result of this, obtaining well recorded, reliable and organized data of runoff and baseflows of rivers is a challenge.

Lake Tana sub-basin is part of the upper Blue Nile Rivers Basin constituting the major part of the Nile River Basin. Hydrology of the Nile River Basin has been studied focusing on rainfall-runoff processes, stream flow modeling, sediment dynamics, teleconnections and river flow, land use dynamics, climate change impact, groundwater flow modeling, hydrodynamics of Lake Tana, water allocation and demand analysis (Melesse et al. 2009a, b, 2011a, 2014; Abtew et al. 2009a, b; Abtew and Melesse 2014a, b, c; Yitayew and Melesse 2011; Chebud and Melesse 2013, 2009a, b; Setegn et al. 2009, 2010a, b, 2011; Melesse 2011; Dessu and Melesse 2012, 2013; Dessu et al. 2014).

Soil erosion and sediment transport and deposition are major water resources and watershed management problem facing most developing countries. Deforested and degraded watersheds accelerate runoff process and hence sediment detachment and transport. Various studies (Aga et al. 2018, 2019; Defersha and Melesse 2012a, b; Defersha et al. 2010, 2011, 2012; Maalim and Melesse 2013; Maalim et al. 2013; Setegn et al. 2010a, b; Melesse et al. 2011a, b; Msagahaa et al. 2014; Wang et al. 2008a, b, c; Mekonnen and Melesse 2011; Setegn et al. 2009; Yesuf et al. 2015) used in situ data, laboratory, field and watershed scale studies to understand soil erosion processes and hence estimate sediment production.

Researchers have used models to estimate runoff for watersheds that have limited observed flow data. Models are proven to be useful tools capable of representing essential features and processes of watersheds. One of the most commonly used hydrological model is the Soil Water Assessment Tool (SWAT) . Researchers have used SWAT not only for estimating runoff but also to evaluate the effect of rainfall variability, land use change and also climate change on flow, soil water, evapotranspiration and groundwater flow (Dessu and Melesse 2012, 2013; Dessu et al. 2014; Wang et al. 2006, 2008a, b, c; Wang and Melesse 2005, 2006; Behulu et al. 2013, 2014; Setegn et al. 2009, 2010a, b, 2011, 2014; Mango et al. 2011a, b; Getachew and Melesse 2012; Assefa et al. 2014; Grey et al. 2013; Yesuf et al. 2015).

The purpose of this study is to show catchment characteristics of headwater watersheds and study how the runoff is affected by these catchment characteristics by developing a simple rainfall-runoff relationship. The study also tries to address the relationship of runoff in rivers and sediment concentration. An attempt is also made to regionalize the model parameters based on the best-fitted catchment characteristics.

Methodology

Description of the Study Area

The area of Lake Tana sub-basin, which is one of the sub-basins of the Blue Nile River Basin, is 15,096 km^2. It is located in Ethiopia's North-West highlands, in Amhara region between 10°56′ to 12°45′N latitude and 36°44′–38°14′E longitude. The sub-basin is one of the areas in Ethiopia that significantly supports the livelihood of 2,940,102 people in the upper Nile River Basin (Commission 2008). Eight micro-catchments were selected for the study, which are found in Tana Basin. These eight sites are led by Tana Beles Integrated Water Resource Development Program (TBIWRDP) Watershed Monitoring and Evaluation (WME). These sites are found in Gilgel Abbay, Gumara and Ribb sub-catchments of Lake Tana sub-basin; see Table 19.1 and Fig. 19.1.

Table 19.1 Geographic location of study sub-catchments

No	Study catchments	Main catchment	Area (km^2)	Latitude (°N)	Longitude (°E)
1	Agar	Gumara	5.20	11.62	37.72
2	Bosa	Gilgel Abbay	7.08	11.00	37.28
3	Dabzute	Ribb	4.90	11.83	38.08
4	Enkulal	Gumara	3.68	11.62	37.78
5	Genamechawecha	Gumara	5.20	11.72	38.11
6	Guali	Ribb	1.44	11.82	38.09
7	Tikurwuha	Ribb	5.00	11.77	38.14
8	Toma	Gilgel Abbay	6.20	11.23	37.18

Data Collection and Analysis

For this study, daily river staff gage records, land use, soil information and meteorological data were collected. Water level records, rainfall records and flow velocity records were collected from Tana Beles Integrated Water Resource Development Program (TBIWRDP) Watershed Monitoring Program (WMP).

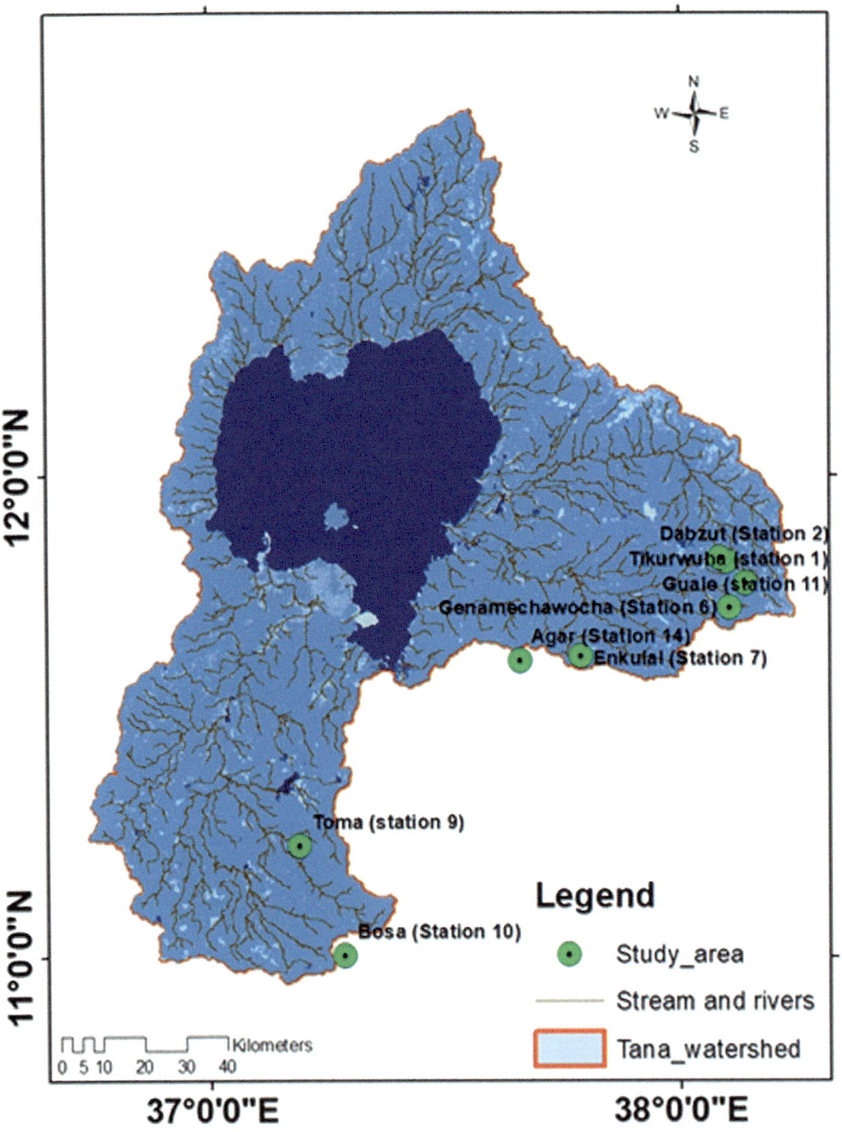

Fig. 19.1 Figure location map of study area

Shuttle Radar Topographic Mission Digital Elevation Model (SRTM-DEM), Landsat image from Landsat archives and soil map were collected from Landsat websites and Ethiopian Ministry of Water, Irrigation and Electric (MoWIE), respectively. Figs. 19.2 and 19.3 show the river gauge and general area of the study site, respectively.

Fig. 19.2 Gauging station of Guali river

Fig. 19.3 Photo of Gena Mechawocha watershed

Field data were collected, different land uses were identified, and coordinates of training samples were taken by GPS. Total 120 GPS points were taken from each of the main types of land uses: cultivated lands, forests, bushlands, grazing and degraded lands. These training points were verified through Google Earth. Using these training points, land uses were prepared using Earth Resources and Data Analysis System (ERDAS) Imagine, an image processing software.

Missing data were filled by regression analysis from neighboring stations, and consistency of the data is seen by visual inspection. Since the data are from a short period monitoring program, (less than four years), double mass analysis was not possible.

Rainfall and Flow Measurement

Rainfall data were collected two times a day at 7:00 and 19:00 h by site managers available in the area and appointed by TBIWRDP. To account rainfall distribution throughout the watershed, there were 20 rain gage stations in these eight catchments averaging 2.5 rain gages per catchment. Rain gages were installed in the farmers' farm plots, and additional records were collected during any time of the day when there is rain since rainfall recording type could not catch all daily rains. The analysis and conversion of original rainfall data to daily monthly and annual were conducted by using HYDATA software. Since in each catchment 1–5 rainfall measurement stations were installed, arithmetic mean and Thiessen polygon methods were used to get the average rainfall of the catchment.

Similar to rainfall, discharge measurements were conducted by TBIWRDP. Staff gages were observed twice a day at 7:00 and 19:00, and additional records were conducted when there is high discharge following high rain. For small flows, current meter and floating method discharge measurements were conducted. Discharge estimation using manning and crest stage was also held. Discharge obtained using current meter, floating method, manning estimation and crest stage estimation were used to develop rating curves for each catchment. Different discharge captured methods are used to account all stage gage heights.

Baseflow Separation

Two parameter digital filters baseflow separation method was used (Eckhardt 2005), which is a baseflow separation method used to analyze continuous hydrograph over a long time such as years, which were used to separate the baseflow. Eight catchments of the Lake Tana basin were separated into their respective component of direct runoff and baseflow. The approach used a numerical algorithm (a digital filter) to partition the stream flow hydrograph into direct runoff and baseflow. The two parameters baseflow index (BFI) and Alpha are used in the algorithm.

The alpha parameter (Alpha) for the basin has been determined by iterating the parameter between 0.925 and 0.995, until it gives good result (Asmerom 2008). Then the Alpha parameter, filter parameter value of 0.995 was used since it gives the best estimate in the Tana Basin study. And the second parameter, BFI, value was calculated in HYDATA software for each study catchments.

Sediment Measurement and Analysis

Suspended sediment samples were taken when the Secchi turbidity is less than eight numerical reading (for Secchi 8 and above, no sample was taken; flow being considered as sediment free) twice a day together with staff gage reading at 7:00 and 19:00, and extra samples were taken when excess discharge occurred. The suspended sediment was taken to laboratory with one litter plastic bottle, and its sediment concentration was analyzed. The sediment load was calculated by using hourly interpolated suspended sediment and water level by using HYDATA. Suspended sediment and water level were computed in spreadsheet to get sediment load, and this sediment load is processed via HYDATA to get daily, monthly and annual sediment load.

Catchment Characteristics

Soil data of major soil types of study catchments were collected from MoWIE, GIS Department. Land use maps were prepared from Landsat images with 30-m resolution for September 2011 (path/row 169/52, 169/53 and 170/52). Image analyses were conducted using ERDAS Imagine and GIS software. Supervised land use classifications were done using training points taken in the field, and Google earth and signatures of each land use were prepared. Using these training point signatures, the land use types were identified. Physiographic, geology, soil, land use and cover conditions and normalized differential vegetation index (NDVI) were retrieved for all study catchments.

Development of Regional Rainfall-Runoff Model for Small Catchments

For rainfall-runoff modeling through regionalization, calibration of the model from gaged catchments is required to get the model parameters. These model parameters were related to catchment characteristics to get regional model. Parameter values were calibrated automatically using the particles swarm optimization (PSO) technique.

To define the model parameters and see the relationships for gaged study areas and formulate models, the initial values of parameters were calibrated with observed discharge. The performance and level of goodness of fit were determined by the objective functions of root mean square error, coefficient of determination, Nash–Sutcliff efficiency and percent bias.

Validation has been often used using "split sample" process, whereby the period of observed data (say the first two-thirds of available record which is 2009–2011) is used for calibration and the remaining one-third (2012 data) used for validation. The model that was calibrated using the calibration data set is run for validation period without change in the model parameter and goodness of fit statistics is computed for the validation process.

Since the main purpose of the study is to formulate the simple rainfall-runoff model for small catchments of Tana sub-basin, based on the responses of catchment characteristics, there is a need to formulate a relationship between model parameter and catchment characteristics values of the gaged catchments. Therefore, linear regression is used for the study. Regression analysis is a statistical tool for the computation of relationships between one or more other variables (usually called independent variables) and a given variable (usually called dependent variable) (Wale et al. 2008).

Results and Discussion

River Discharge in the Upper Catchment

The rating curve was developed taking two scenarios of power and polynomial equations. The power relation was used when the water level is less than one meter because the values of the polynomial relation become less than zero which indicated in the first column rating curve equation. Polynomial relation was taken when the water level reading is greater than one meter, which is indicated by the second column rating curve equation because the values of the power trend have extremely exaggerated flow values. River flow estimates were derived using hourly interpolated water level and derived rating curve formula. Rating curves for both flow scenarios are shown in Table 19.2. HYDATA software was used to get daily, decadal and monthly data and Excel spreadsheet showed the study catchments flows are either perennial or seasonal. Even in some years, when there is more than normal rainfall, seasonal rivers would have perennials flows.

Table 19.2 Rating curves study catchments

No.	Catchment name	Rating curve (H <=1 m)	Rating curve (H > 1 m)
1	Agar	$Q = 0.088\ H^{8.62}$	$Q = 5.342\ H^2 - 8.035\ H + 2.817$
2	Bosa	$Q = 1.190\ H^{2.59}$	$Q = 3.071\ H^2 - 1.048\ H + 0.074$
3	Dabzute	$Q = 2.791\ H^{4.30}$	$Q = 5.368\ H^2 - 2.819\ H + 0.361$
4	Enkulal	$Q = 0.980\ H^{3.96}$	$Q = 1.666\ H^2 + 1.174\ H - 0.598$
5	Genamechawocha	$Q = 1.459\ H^{2.57}$	$Q = 3.298\ H^2 - 2.823\ H + 0.400$
6	Guali	$Q = 0.443\ H^{4.48}$	$Q = 2.016\ H^2 + 0.391\ H - 0.649$
7	Tikurwuha	$Q = 1.127\ H^{3.23}$	$Q = 2.091\ H^2 + 0.771\ H - 0.589$
8	Toma	$Q = 0.253\ H^{5.33}$	$Q = 4.987\ H^2 - 7.843\ H + 3.197$

where H is staff gage height in meter, and Q is discharge in m^3/s

Baseflow Index

Baseflow index was computed using HYDATA software from total runoff analysis. The baseflow index is used to separate the baseflow and direct runoff from total runoff to obtain the baseflow model parameter. The minimum value of the baseflow index was 0.2076 in Enkulal catchment, and the maximum value was 0.8358 in Genamechawocha catchment (Table 19.3).

Model Parameters

Optimized values using particle swarm optimization of the model parameters based on the observed discharge and rainfall are indicated in Table 19.4. The optimized model parameters are on daily basis. Parameter C in the second column of Table 19.4 which is the total flow parameter ranges 0.3370–0.6876, and parameter b, the baseflow parameter ranges 0.0076–0.2197. Parameter C was computed independently after baseflow separation has been applied as direct runoff and baseflow, and the direct runoff parameter C becomes in the range of 0.2342–0.5550.

Table 19.3 Baseflow index of study catchments

Catchments	Baseflow index
Agare	0.3452
Bosa	0.4649
Dabzute	0.2954
Enkulal	0.2076
Genamechawocha	0.8358
Guale	0.3859
Tikurwuha	0.5266
Toma	0.3308

Table 19.4 Model parameters of study catchments before and after separation of baseflow

Catchment	Parameter *C*	Parameter *b*	*C*-value after baseflow separation
Agar	0.3370	0.0076	0.3188
Bosa	0.3377	0.1327	0.3488
Dabzut	0.6876	0.0159	0.6044
Enkulal	0.3370	0.0076	0.4012
Genamechawocha	0.3194	0.2197	0.2342
Guali	0.6876	0.0159	0.5550
Tikurwuha	0.4074	0.1175	0.4176
Toma	0.3370	0.0076	0.3052

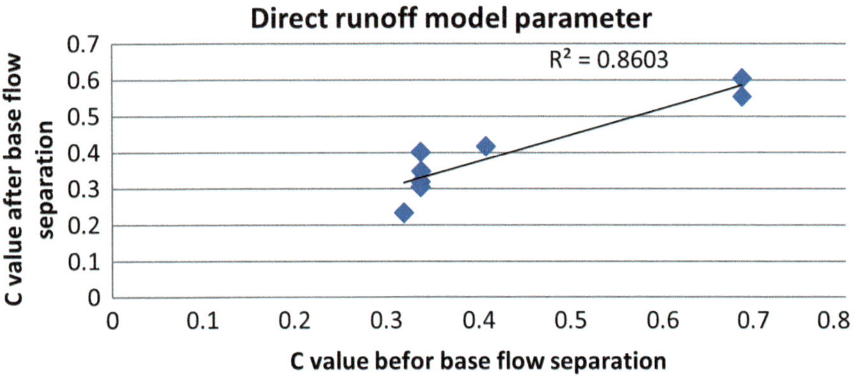

Fig. 19.4 The relation between the two *C* values

The values of total runoff parameter and direct runoff parameter (columns 2 and 4) were plotted as shown in Fig. 19.4. The result shows that R^2 is 0.86, and hence, there is good relation between the *C* values so that we can take the *C* after baseflow separation value as direct runoff response.

Calibration

Calibration results indicate that the root mean square error is around 0, coefficient of determination is greater than 0.6, and the Nash–Sutcliff coefficient is greater than 0.6 that means the parameters are estimated showing good agreement with observed data. The negative value of the percent bias (PBIAS) indicated there was an over-estimation of the simulation results, and the positive indicates under estimation of results; hence, Tikurwuha, Guale, Bosa and Dabzute results show over-estimated values, whereas Genamechawocha, Enkulal, Toma and Agare show under estimation. Generally, the values of PBIAS are less than 25 which are acceptable (Table 19.5).

Table 19.5 Calibration of model parameters

Catchment	Calibration			
	RMSE	R^2	NSE	PBIAS
Tikurwuha	0.138	0.609	0.602	−8.25
Guale	0.058	0.738	0.702	−13.19
Genamechawocha	0.090	0.648	0.637	0.131
Enkulal	0.130	0.639	0.610	7.69
Toma	0.171	0.680	0.674	9.7
Bosa	0.200	0.675	0.661	−13.19
Dabzut	0.212	0.685	0.680	−5.97
Agar	0.152	0.604	0.688	3.85

Validation

In order to test the model parameter outside the period of calibration and to see the consistency of the data, validation of this parameter was conducted. Validation is conducted using the 2012 data which is outside the period of calibration and tested by objective functions. The root mean square error shows good fit to the data, and the Nash–Sutcliff and coefficient of determination show there is good agreement between the simulated and observed data. Tikurwuha, Guale, Genamechawocha, Toma, Bosa and Dabzut catchments simulation results show over-estimation that indicated there is a decrease of flow, whereas Enkulal and Agar show an increasing in the flow. Except Enkulal catchment, the rest had similar trend in calibration and validation periods (Table 19.6).

Area of land use proportions, soil type proportions and NDVI, circularity ratio, elongation ratio, drainage density, average slope in percent and hypsometric integral of the study catchments were set as an independent parameter in the analysis as seen in Table 19.7. Forest land, grazing land and cultivated land were in the range of 0–18%, 0–34.7% and 49.6–100%, respectively. The study catchments are dominated by Luvisols, Leptosols and Nitosols and found in the range of 5.33–100%, 0–94.67% and 0–82.17%, respectively. NDVI, circularity ratio (CR), elongation ratio (ER), drainage density (DD), average slope in percent and

Table 19.6 Validation of model parameters

Catchment	Validation			
	RMSE	R^2	NSE	PBIAS
Tikurwuha	0.125	0.847	0.61	−83.86
Guale	0.059	0.692	0.676	−14.45
Genamechawocha	0.059	0.816	0.645	−14.59
Enkulal	0.107	0.649	0.623	6.29
Toma	0.1207	0.808	0.791	−4.44
Bosa	0.201	0.670	0.621	−19.79
Dabzut	0.202	0.737	0.733	−4.31
Agar	0.160	0.646	0.608	28.23

Table 19.7 Land cover and soil types by catchment

Catchment	Forest (%)	Grazing (%)	Cultivated (%)	Cr-Luvisol (%)	Len-Leptosol (%)	Hu-Nit osol (%)
Agar	17.70	15.90	66.40	100.00	0.00	0.00
Bosa	15.70	34.70	49.60	100.00	0.00	0.00
Dabzut	21.40	4.00	74.60	100.00	0.00	0.00
Enkulal	6.69	29.38	63.92	100.00	0.00	0.00
Genamechawocha	18.00	25.40	56.60	41.59	58.41	0.00
Guali	0.00	0.00	100.00	100.00	0.00	0.00
Tikurwuha	0.50	1.00	98.50	5.33	94.67	0.00
Toma	0.00	12.40	87.60	17.83	0.00	82.17

Table 19.8 Catchment characteristics

Catchment	NDVI (Average)	CR	ER	DD	Slope (Average)	HI
Agar	0.33	0.300	1.955	1.164	10.85	0.439
Bosa	0.52	0.400	1.318	3.263	27.38	0.398
Dabzut	0.41	0.382	1.922	1.358	12.16	0.564
Enkulal	0.32	0.364	1.548	1.651	14.94	0.385
Genamechawocha	0.38	0.310	2.207	2.412	19.61	0.444
Guali	0.39	0.271	1.891	1.281	14.27	0.515
Tikurwuha	0.43	0.711	1.687	1.960	19.72	0.454
Toma	0.55	0.283	2.004	1.788	18.54	0.352

hypsometric integral (HI) range 0.32–0.55, 0.271–0.4, 1.318–2.207, 1.164–3.263, 10.85–27.38 and 0.352–0.564, respectively (Table 19.8).

Regionalization of Model Parameters

Two model parameters for direct runoff (*C*) are developed since grazing land and cultivated land are multi-collinear negatively (−0.89); they give good regression value interchangeably. The R^2 for both C-values is 0.997, and the b-value is 0.910. The F tests for all model parameters are significant at 95% level of significance. The individual *t* tests also are significant for all independent parameters included in the model parameter equations. Based on the analysis model developed in relation to rainfall, runoff model parameter and catchment area, the following daily regional formula is developed as seen in Eq. 19.1, and direct runoff model parameters are synthesized in best-fitted catchment characteristics as seen below.

$$Q = \frac{C * R * A}{86.4} + b \tag{19.1}$$

where Q = daily mean discharge (m^3/s), R = daily rainfall (mm), A = catchment area (km^2), C = direct runoff-derived model parameter, and b = baseflow (delayed runoff)-derived model parameter. To consider the daily time scale, we divide the result by 86.4. Two formulations for C value are developed for cultivated land and grazing land.

For cultivated land, C is given by

$$C = -0.321 - 0.025 * \mathrm{Fo} - 0.011 * \mathrm{Cu} + 1.342 * \mathrm{NDVI} - 0.021 * \mathrm{S} + 3.748 * \mathrm{HI} \tag{19.2}$$

For grazing land, C is

$$C = -1.468 - 0.014 * \mathrm{Fo} + 0.011 * \mathrm{Gr} + 1.342 * \mathrm{NDVI} - 0.021 * \mathrm{S} + 3.748 * \mathrm{HI} \tag{19.3}$$

$$b = -0.358 + 0.184 * \mathrm{ER} + 0.023 * \mathrm{S} - 0.714 * \mathrm{NDVI} \tag{19.4}$$

where Fo = forest (% of total land), Cu = cultivated land (% of total land), Gr = grazing land (% of total land), S = average slope, NDVI = normalized differential vegetation index, ER = elongation ratio, HI = hypsometric integral and C and b =model parameters.

The equation of model parameters indicates percentage increase of forest land, cultivated land and slope decrease the direct runoff whereas NDVI, grazing land and HI increase it.

Evaluation of the Model

For applicability of the model, evaluation is essential in gaged catchments. Anjeni catchment is selected for evaluation of the model, the physical catchment characteristics are computed, and the catchment has 0.342 direct runoff model parameter and 0.005 delayed runoff model parameter. The observed and simulated runoff of the catchment is tested by objective functions, and the result shows RMSE, 0.039 and PBIAS, 4.9 are in acceptable limit but coefficient of determination and NSE are low, which are 0.465 and 0.4, respectively. The result of NSE and coefficient of determination indicates that the model has some overfitting problem (Table 19.9).

Table 19.9 Model parameter evaluation value of Anjeni watershed

Data type	Anjeni watershed	Data type	Anjeni watershed
Period of data	1/1/86–12/31/1993	NDVI	0.4
Area (km^2)	1.565	Elongation Ratio	1.9
Longest path (km)	2.646	*C*-value	0.342
Elevation (Max)	2505	*b*-value	0.005
Elevation (Min)	2405	R^2	0.465
Elevation (mean)	2448	RMSE	0.039
Hypsometric integral	0.4	PBIAS	4.9
Slope (%)	13%	NSE	0.4
Land use	Cultivated 100%		

Sediment Yield

The relations of discharge to suspended sediment load for eight catchments were developed in power relation (Table 19.10). Based on the correlation between discharge and sediment and sediment yield equation (Table 19.10), simulated sediment yield was obtained. The relation shows it has good agreement between the observed and simulated suspended sediment since the R^2 is more than 0.6 for all catchments. When we compare the land use and the sediment Bosa, Enkulal and Guale catchments are covered with cultivated and grazing lands (including degraded lands), and the sediment rate is very high in these catchments. This may indicate that cultivated and grazing lands are sediment source areas.

Different catchments have different characteristics which influence the hydrologic flow regime. It is important to know that this difference can help in planning action on variables which need management measures in the catchments. To facilitate this, it is helpful to classify and recognize the basin based on catchment characteristics which are easily computed. Recent development in hydrology seeks to find a way of understanding the interrelations of hydrological processes at the landscape scale by analyzing catchment characteristics (Sivapalan 2006).

This study addressed modeling of rainfall-runoff and runoff and sediment yield based on the catchment characteristics. TBIWRDP study of selected catchments

Table 19.10 Sediment rating curve of the study area

Catchment	Sediment yield equation	R^2-value
Agar	$Q_s = 181.1\ Q^{1.13}$	0.776
Bosa	$Q_s = 530.6\ Q^{1.59}$	0.796
Dabzut	$Q_s = 174.3\ Q^{1.16}$	0.760
Enkulal	$Q_s = 476.4\ Q^{1.38}$	0.788
Genamechawoch	$Q_s = 221.1\ Q^{1.27}$	0.804
Guale	$Q_s = 464.7\ Q^{1.36}$	0.838
Tikurwuha	$Q_s = 251.1\ Q^{1.21}$	0.801
Toma	$Q_s = 197.8\ Q^{1.18}$	0.681

was used for this study. The models in this study showed that forests and cultivated lands and slopes have inversely proportional to direct runoff, and grazing lands which include degraded areas and foothill grass lands, NDVI and hypsometric integral are directly proportional to direct runoff. Besides, the baseflow is directly proportional to elongation ratio and average slope of the area and inversely proportional to NDVI.

Some researchers recently conducted studies in different areas and reached similar conclusions. Teketay and Bekele referenced in Mugasha et al. (2004) indicated that grazing and degraded bushlands characterized by land degradation in the basin increase rapid runoff. Engda and others (Engda et al. 2011) indicated surface runoff source areas are degraded areas with less soil cover, and foot hill areas are also related to saturation excess runoff. Hill slope areas are recharge areas so that they transmit water as interflow. Study by Abebe and Foerch (2006) also found that baseflow index was negatively related to NDVI, this could be related to higher uptake of water by riparian forest during dry season. The model can be applicable to determine the water resource of micro-catchments in ungaged watersheds of Lake Tana Basin.

The model has some limitations such as overfitting, and this can be because of the small number of samples (number of catchments) used for the regression. This model will best fit if the number of samples is large and covers different topography. There is also some limitation of the model in that for gentler slopes catchment, the baseflow parameter is not well estimated, and this is revealed as slopes relate directly to direct runoff.

The sediment model corresponding to observed discharge is shown in power regression relation that shows significant relation since the R^2 is more than 0.6 for all catchments which is in good agreement with the observed one. The sediment flow may depend on land uses of the catchment. Catchments dominated by grazing and cultivated land show high sediment yield.

Conclusion

All catchment characteristics (independent variables) were equally important in predicting and explaining the hydrologic process in the catchment. The individual variables which were selected based on the statistical significance from all groups of independent variables, in general, were better in explaining the variation. The predictor variables also are supported by the research findings that are reasonably significant. Hill slopes are interflow areas, and degraded bushlands and grazing lands are saturation excess areas so that they are direct runoff source areas. This idea is supported by the model developed in this research.

Hydrological characterization based on the catchment characteristics could be used as a source of baseline data for the implementation of catchment management in the basin. It could be especially useful for the management of ungaged catchments. The study showed that by determining catchment characteristics, the ungaged catchments water resource can be determined so that water resource allocation is possible.

Acknowledgements Authors acknowledge assistance in the time of data collection, organization and analysis period given by Wase-Tana project staff (VLIR OI project Belgium-Ethiopia), Blue Nile Institute (Bahir Dar University) Bureau of Water, Irrigation and Energy (BoWIE), and Tana Beles Project staffs.

References

Abebe A, Foerch G (2006) Catchment characteristics as predictors of base flow index (BFI) in Wabi Shebele river basin, East Africa. Paper presented at the proceedings of conference on international agricultural research for development

Abtew W, Melesse AM (2014a) Nile River basin hydrology. In: Melesse AM, Abtew W, Setegn S (eds) Nile River basin: ecohydrological challenges, climate change and hydropolitics, pp 7–22

Abtew W, Melesse AM (2014b) Climate teleconnections and water management. In: Nile River Basin. Springer International Publishing, pp 685–705

Abtew W, Melesse AM (2014c) Transboundary Rivers and the Nile. In: Nile River Basin. Springer International Publishing, pp 565–579

Abtew W, Melesse A, Desalegn T (2009a) Spatial, inter and intra-annual variability of the Blue Nile River Basin Rainfall. Hydrol Process 23(21):3075–3082

Abtew W, Melesse A, Desalegn T (2009b) El Niño Southern Oscillation link to the Blue Nile River Basin hydrology. Hydrol Process Special Issue Nile Hydrol 23(26):3653–3660

Aga AO, Chane B, Melesse AM (2018) Soil erosion modelling and risk assessment in data Scarce Rift Valley Lake regions, Ethiopia. Water 10:1684. https://doi.org/10.3390/w10111684

Aga AO, Melesse AM, Chane B (2019) Estimating the sediment flux and budget for a data limited Rift Valley Lake in Ethiopia. Hydrology 6(1). https://doi.org/10.3390/hydrology6010001

Asalf S (2018) Evaluate the effects of watershed characteristics on stream flow, the case of Blue Nile River Basin, Beshilo Watershed. Addis Ababa University, Ethiopia

Asmerom GH (2008) Groundwater contribution and recharge estimation in the Upper Blue Nile flows, Ethiopia

Assefa A, Melesse AM, Admasu S (2014) Climate change in Upper Gilgel Abay River catchment, Blue Nile Basin Ethiopia, In: Melesse AM, Abtew W, Setegn S (eds) Nile River Basin: ecohydrological challenges, climate change and hydropolitics, pp 363–388

Behulu F, Setegn S, Melesse AM, Fiori A (2013) Hydrological analysis of the Upper Tiber Basin: a watershed modeling approach. Hydrol Process 27(16):2339–2351

Behulu F, Setegn S, Melesse AM, Romano E, Fiori A (2014) Impact of climate change on the hydrology of Upper Tiber River Basin using bias corrected regional climate model. Water Res Manag 1–17

Chebud YA, Melesse AM (2009a) Numerical modeling of the groundwater flow system of the Gumera Sub-Basin in Lake Tana Basin, Ethiopia. Hydrol Process Special Issue Nile Hydrol 23 (26):3694–3704

Chebud YA, Melesse AM (2009b) Modeling lake stage and water balance of Lake Tana, Ethiopia. Hydrol Process 23(25):3534–3544

Chebud Y, Melesse AM (2013) Stage level, volume, and time-frequency change information content of Lake Tana using stochastic approaches. Hydrol Process 27(10):1475–1483. https://doi.org/10.1002/hyp.9291

Commission FDR o EPC (2008) Summary and statistical report of the 2007 population and housing census–population size by age and sex. Addis Ababa

Defersha MB, Melesse AM (2012a) Effect of rainfall intensity, slope and antecedent moisture content on sediment concentration and sediment enrichment ratio. CATENA 90:47–52
Defersha MB, Melesse AM (2012b) Field-scale investigation of the effect of land use on sediment yield and surface runoff using runoff plot data and models in the Mara River basin, Kenya. CATENA 89:54–64
Defersha MB, Quraishi S, Melesse AM (2010) Interrill erosion, runoff and sediment size distribution as affected by slope steepness and antecedent moisture content. Hydrol Earth Syst Sci Discuss 7(6447–6489):2010
Defersha MB, Quraishi S, Melesse AM (2011) The effect of slope steepness and antecedent moisture content on interrill erosion, runoff and sediment size distribution in the highlands of Ethiopia. Hydrol Earth Syst Sci 15:2367–2375. https://doi.org/10.5194/hess-15-2367-2011
Defersha MB, Melesse AM, McClain M (2012) Watershed scale application of WEPP and EROSION 3D models for assessment of potential sediment source areas and runoff flux in the Mara River Basin, Kenya. CATENA 95:63–72
Dessu SB, Melesse AM (2012) Modeling the rainfall-runoff process of the Mara River Basin using SWAT. Hydrol Process 26(26):4038–4049
Dessu SB, Melesse AM (2013) Impact and uncertainties of climate change on the hydrology of the Mara River Basin. Hydrol Process 27(20):2973–2986
Dessu SB, Melesse AM, Bhat M, McClain M (2014) Assessment of water resources availability and demand in the Mara River Basin. CATENA 115:104–114
Eckhardt K (2005) How to construct recursive digital filters for baseflow separation. Hydrol Process Int J 19(2):507–515
Engda TA, Bayabil HK, Legesse ES, Ayana EK, Tilahun SA, Collick AS et al (2011) Watershed hydrology of the (semi) humid Ethiopian Highlands. In: Nile River Basin. Springer, pp 145–162
Getachew HE, Melesse AM (2012) Impact of land use/land cover change on the hydrology of angereb watershed, Ethiopia. Int J Water Sci 1(4):1–7. https://doi.org/10.5772/56266
Grey OP, Webber Dale G, Setegn SG, Melesse AM (2013) Application of the soil and water assessment tool (SWAT Model) on a small tropical Island State (Great River Watershed, Jamaica) as a tool in Integrated Watershed and Coastal Zone Management. Int J Trop Biol Conserv 62(3):293–305
Maalim FK, Melesse AM (2013) Modeling the impacts of subsurface drainage systems on runoff and sediment yield in the Le Sueur Watershed, Minnesota. Hydrol Sci J 58(3):1–17
Maalim FK, Melesse AM, Belmont P, Gran K (2013) Modeling the impact of land use changes on runoff and sediment yield in the Le Sueur Watershed, Minnesota using GeoWEPP. Catena 107, 35–45
Mango L, Melesse AM, McClain ME, Gann D, Setegn SG (2011a) Land use and climate change impacts on the hydrology of the upper Mara River Basin, Kenya: Results of a modeling study to support better resource management, special issue: climate, weather and hydrology of East African Highlands. Hydrol. Earth Syst. Sci. 15:2245–2258. https://doi.org/10.5194/hess-15-2245-2011
Mango L, Melesse AM, McClain ME, Gann D, Setegn SG (2011b) Hydro-meteorology and water budget of Mara River basin, Kenya: a land use change scenarios analysis, In: Melesse A (ed) Nile River Basin: hydrology, climate and water use. Springer Science Publisher, Chapter 2, pp 39–68. https://doi.org/10.1007/978-94-007-0689-7_2
Mekonnen M, Melesse A (2011) Soil erosion mapping and hotspot area identification using GIS and remote sensing in northwest Ethiopian highlands, near Lake Tana. In: Melesse A (ed) Nile River Basin: hydrology, climate and water use (chapter 10). Springer Science Publisher, pp 207–224. https://doi.org/10.1007/978-94-007-0689-7_10
Melesse AM (2011) Nile River Basin: hydrology, climate and water use. Springer Science & Business Media, Berlin
Melesse AM, Loukas Athanasios G, Senay Gabriel, Yitayew Muluneh (2009a) Climate change, land-cover dynamics and ecohydrology of the Nile River Basin. Hydrol Process Special Issue Nile Hydrol 23(26):3651–3652

Melesse AM, Abtew W, Desalegne T, Wang X (2009b) Low and high flow analysis and wavelet application for characterization of the Blue Nile River system. Hydrol Process 24(3):241–252
Melesse AM, Abtew W, Setegn S, Dessalegne T (2011a) Hydrological variability and climate of the Upper Blue Nile River Basin. In: Melesse A (ed) Nile River Basin: hydrology, climate and water use (chapter 1). Springer Science Publisher, pp 3–37. https://doi.org/10.1007/978-94-007-0689-7_1
Melesse AM, Ahmad S, McClain M, Wang X, Lim H (2011b) Sediment load prediction in large rivers: ANN approach. Agric Water Manag 98:855–866
Melesse A, Abtew W, Setegn SG (2014) Nile River Basin: ecohydrological challenges, climate change and hydropolitics. Springer Science & Business Media, Berlin
Yesuf HM, Alamirew T, Assen M, Melesse AM (2015) Modeling of sediment yield in Maybar gauged watershed using SWAT, northeast Ethiopia. CATENA 127:191–205
Msagahaa J, Ndomba, Melesse AM (2014) Modeling sediment dynamics: effect of land use, topography and land management. In: Melesse AM, Abtew W, Setegn S (eds) Nile River Basin: ecohydrological challenges, climate change and hydropolitics, pp 165–192
Mugasha A, Chamshama S, Gerald V (2004) Indicators and tools for restoration and sustainable management of forests in east Africa. State For For Res Tanzania Paper 3:66
Parker H, Oates N (2016) How do healthy rivers benefit society. A review of the evidence. ODI and WWF, London
Setegn SG, Srinivasan R, Dargahi B (2008) Hydrological modelling in the Lake Tana Basin, Ethiopia using SWAT model. Open Hydrol J 2(1)
Setegn SG, Srinivasan R, Dargahil B, Melesse AM (2009) Spatial delineation of soil erosion prone areas: application of SWAT and MCE approaches in the Lake Tana Basin, Ethiopia. Hydrol Process 23(26):3738–3750
Setegn SG, Dargahi B, Srinivasan R, Melesse AM (2010a) Modelling of sediment yield from Anjeni Gauged watershed, Ethiopia using SWAT. J Am Water Res Assoc 46(3):514–526
Setegn SG, Srinivasan R, Melesse AM, Dargahi B (2010b) SWAT model application and prediction uncertainty analysis in the Lake Tana Basin, Ethiopia. Hydrol Process 24(3):357–367
Setegn S, David R, Melesse AM, Bijan D, Ragahavan S (2011) Impact of climate change on the hydro-climatology of Lake Tana basin, Ethiopia. Water Resour Res 47, W04511:13. https://doi.org/10.1029/2010wr009248
Setegn SG, Melesse AM, Haiduk A, Webber D, Wang X, McClain M (2014) Spatiotemporal distribution of fresh water availability in the Rio Cobre Watershed, Jamaica. CATENA 120:81–90
Sivapalan M (2006) Pattern, process and function: elements of a unified theory of hydrology at the catchment scale. Encyclopedia of hydrological sciences
Wale A, Rientjes T, Dost R, Gieske A (2008) Hydrological balance of Lake Tana Upper Blue Nile Basin, Ethiopia
Wang X, Melesse AM (2005) Evaluations of the SWAT Model's Snowmelt hydrology in a Northwestern Minnesota Watershed. Trans ASAE 48(4):1359–1376
Wang X, Melesse AM (2006) Effects of STATSGO and SSURGO as inputs on SWAT model's snowmelt simulation. J Am Water Res Assoc 42(5):1217–1236
Wang X, Melesse AM, Yang W (2006) Influences of potential evapotranspiration estimation methods on SWAT's hydrologic simulation in a Northwestern Minnesota Watershed. Trans ASAE 49(6):1755–1771
Wang X, Shang S, Yang W, Melesse AM (2008a) Simulation of an agricultural watershed using an improved curve number method in SWAT. Tans Am Soc Agri Bio Eng 51(4):1323–1339
Wang X, Yang W, Melesse AM (2008b) Using hydrologic equivalent wetland concept within SWAT to estimate streamflow in watersheds with numerous wetlands. Tans Am Soc Agri Bio Eng 51(1):55–72

Wang X, Garza J, Whitney M, Melesse AM, Yang W (2008c) Prediction of sediment source areas within watersheds as affected by soil data resolution. In: Findley PN (ed) Environmental modelling: new research (Ch 7). Nova Science Publishers, Inc., Hauppauge, NY 11788, pp 151–185. ISBN: 978-1-60692-034-3

Yitayew M, Melesse AM (2011) Critical water resources management issues in Nile River Basin. In: Melesse A (ed) Nile River basin: hydrology, climate and water use (chapter 20). Springer Science Publisher, pp 401–416. https://doi.org/10.1007/978-94-007-0689-7_20

Chapter 20
Water Conservation Through Decentralized Rainwater Harvesting Under Climate Uncertainty

Mengistu A. Jemberie and Assefa M. Melesse

Abstract Water is essential to sustain life, and adequate, safe and accessible supply must be available to all. Millions of people throughout the world do not have access to clean water for domestic purposes. In many parts of the world, conventional piped water is either not available, unreliable or too expensive. One of the biggest challenges of the twenty-first century is to overcome the growing water shortage. Rainwater harvesting (RWH) has thus regained its importance as a valuable alternative or supplementary water resource, along with more conventional water supply technologies. Much actual or potential water shortages can be relieved if rainwater harvesting is practiced more widely. Severe water shortages and extremely fragile ecological conditions necessitate careful attention to water resources conservation and management. Nowadays, cumulative effects of climate change, population increase, development, and industrialization are leading to increased water demand which seeks careful and strategic management of the available resource. In the Ethiopian context, the average annual population growth is about 2.8% and twice of water demand increase. People collect and store rainwater in buckets, tanks, ponds, and wells. This is commonly referred to as rainwater harvesting and has been practiced for centuries. Rainwater can be used for multiple purposes ranging from irrigating crops to washing, cooking and drinking. In this paper, assessment and review are done on effects of climate change, population growth and development on water demand increase. Long-term average annual rainfall was accessed from meteorological data, and minimum numbers of houses data were taken from Ethiopian 2007 census data. Finally, possible amount of water harvested is estimated for domestic and other water uses.

Keywords Climate change · Rainwater harvesting · Ethiopia · Population growth · Water demand

M. A. Jemberie (✉)
Adama Science and Technology University, Adama 1888, Ethiopia

A. M. Melesse
Department of Earth and Environment, Institute of Environment, Florida International University, Miami FL 33199, USA
e-mail: Melessea@fiu.edu

A. M. Melesse et al. (eds.), *Nile and Grand Ethiopian Renaissance Dam*,
Springer Geography, https://doi.org/10.1007/978-3-030-76437-1_20

Introduction

Background

Water is increasingly recognized as a major component in economic development and poverty reduction. Several recent papers considering the importance of water in meeting the development goals have highlighted water's direct and indirect contributions, rather than just focusing on its central role in achieving the aim on environmental sustainability and the accompanying target on water supply and sanitation. This analysis illustrates the fact that water's interaction in the lives of the poor is complex in character and operates through multiple dimensions: improved livelihoods, security, reduced health risks, reduced vulnerability, and pro-poor economic growth. Further, as well as its significance in poverty alleviation, investment in water infrastructure and management has a major impact on national economies. Finally, there is increasing attention to the multiple values of water for society, including not only its importance in terms of ecosystem sustainability but also its cultural and social components (Dowa et al. 2010).

Ethiopia boasts a population of over 100 million people and which only 38% have access to safe drinking water sources and only 12% of the population use improved sanitation facilities. Ethiopia has found itself in an extreme water crisis situation, brought on mainly by severe drought, little governmental funding and assistance, and lack of water management and sanitation resources. The shortage of water and drought has brought an overwhelming spread of famine and food shortage (Hendrix 2012).

Even though there is sufficient water resource, due to the lack of water conservation and management system, millions of Ethiopians are forced to walk up to six hours in order to fetch water that is often source of disease. This is causing women and children to put water collection above all else, including school. It also is creating water-borne illnesses that infect and claim thousands of Ethiopian lives. Around 7.5 million Ethiopians suffer problems related to high fluoride levels. This number is only the tip of the iceberg when it comes to water-related disease in Ethiopia and without improving sanitation within the country; the numbers are set to only increase (World Bank 2012).

Like much of Africa, Ethiopia has become warmer over the past century and human-induced climate change will bring further warming over the next century at unprecedented rates. For the IPCC mid-range (A1B) emission scenario, the mean annual temperature will increase in the range of 0.9–1.1 °C by 2030, in the range of 1.7–2.1 °C by 2050, and in the range of 2.7–3.4 °C by 2080 over Ethiopia (UNDP 2007).

Climate Change and Variability

Currently, climate change is serious global challenge where developed countries contribute most to the climate impact and others becoming victims. According to the IPCC, Africa is one of the most vulnerable continents to climate change and climate variability (CLUVA 2013).

Total reliance on surface water from major rivers makes the urban water supply highly vulnerable to changes in the flow and water quality of these rivers and subject to water pollution incidents. The National Assessment Report on Climate Change, 2007, indicated decreasing trends in runoff in the last 40 years in the main rivers due to climate change. The watershed of major rivers and tributaries has been experiencing higher frequencies of consecutive years of drought. In addition, water quality in major rivers is deteriorating due to urban developments in the catchments. Especially rift valley lakes and rivers water qualities are deteriorated to great extent due to climate change and liquid waste disposals from development and agricultural areas. In some cases, water is no longer potable. Both volume and quality of available surface water resource are adversely affected as a result of climate change and development.

Flooding and drought, the two extreme events which are caused by climate change, are the main natural hazards that require water conservation and management considerations. Flooding, the flow of excess water over areas which are habitually dry, is one of the major natural hazards which impact prosperity and safety of human settlements. Drought is a deficiency of precipitation from an expected or "normal" amount, and when extended over a season or longer period of time, available water is insufficient to meet demands. It can be meteorological drought, a deficiency of precipitation from an expected or "normal" amount over an extended period of time; agricultural drought, a deficiency in water availability for crop or plant growth; hydrological drought, a deficiency in surface and subsurface water supplies that lead to a shortage of water to meet normal and specific water demands.

Climate change is a primary contributor to hydrological drought through the increase in heating of the earth's surface, resulting in a reduction of rainfall and cloud cover and, consequently, in greater evaporation rates. The resultant effects of drought are exacerbated by human activities such as deforestation, overgrazing, and poor cropping methods, which reduce the water retention of the soil and improper soil conservation techniques, which lead to soil degradation.

Surface temperature is expected to rise over the twenty-first century under all assessed emission scenarios. It is very likely that heat waves will occur more often and last longer and that extreme precipitation events will become more intense and frequent in many regions (IPCC 2014).

For more than three decades, Ethiopia has experienced recurrent deadly droughts including those of the 1972/73, 1984 and 2002/03. Communities whose livestock is often the most devastated by drought are located in the Afar, Oromia, Somali, and Southern regions. Desertification, brought on by human land-use pressure and frequent drought, has consumed significant land area and continues to threaten arable land (NAPA 2007). According to the report, there is increasing trend of temperature and some decrease in precipitation trend. For the IPCC mid-range (A1B) emission scenario, the mean annual temperature will increase in the range of 0.9–1.1 °C by 2030, in the range of 1.7–2.1 °C by 2050, and in the range of 2.7–3.4 °C by 2080 over Ethiopia compared to the 1961–1990 reference condition.

Water Supply and Demand

Water Stress and Scarcity

Water scarcity and stress are becoming common phenomenon caused by climate change. "Water Security" is herein described as the nexus between the availability, accessibility, and use of water. The concept is defined as availability of and access to water in sufficient quantity and quality to meet livelihood needs of all households throughout the year, without considering the needs of other users. "Water stress" is shortage of water, whereas "scarcity" is defined from the perspective of individual water users who lack secure access to safe and affordable water to consistently satisfy their need for food production, drinking, washing, or livelihoods (Molden 2007).

Water scarcity is first and foremost a poverty issue. About 1.2 billion people live in areas of physical water scarcity, and up to one in three people in the world face water shortages. In 2025, about 1.8 billion people will live in regions with absolute water scarcity and about two-thirds of the world's population in areas of water stress (FAO 2007).

Water Supply and Demand Imbalances

In the year 2000, about 7 of the 50 million people in rural areas (13%) and 8 of 11 million people in urban areas (73%) had access to safe drinking water. During the same period, access to basic services in sanitation was only 6% for rural and 62% for urban communities (Ndaruzaniye 2011). Water demand far exceeds supply by about 50%, and the water and sanitation distribution network reach only about 55% of the capital city, Addis Ababa, service area. Approximately half the population living in Addis Ababa has less than 12 h per day of water service, and a quarter have no service at all. Most people without water supply live in precarious conditions in Addis Ababa peripheral poor areas, which are now part of the city's administrative area for the last decade.

Addis Ababa water supply is 50 L per capita per day (lpcd) or 35 lpcd taking into account 35% of water losses or waste. When compared to other African countries, most comparable cities in Africa are provided with at least 100 lpcd with a 3.5% population growth rate by 2012, whereas most of Ethiopian cities are getting average of 20 lpcd, and average annual population growth is above 4% (Ndaruzaniye 2011).

About 12% of households in Addis Ababa have flush toilets that discharge to sewers or septic tanks, 63% of households use individual or shared pit latrines, and 25% do not have access to sanitation facilities. The occurrences of diseases are highest in densely populated areas where water supply, sanitation, and nutrition are inadequate (World Bank 2007).

Decentralized Rainwater Harvesting

Water conservation means preservation or restoration of the available resource of water in different forms. Water conservation involves keeping, preserving, and restoring of surface, subsurface, and groundwater sources. In this chapter, we will present surface water conservation through rainwater harvesting. Since the rainwater will be harvested at household levels, it is said to be decentralized.

Where there is no surface water, where groundwater is deep or inaccessible due to hard ground features, or where it is too salty, acidic, or otherwise unpleasant or unfit to drink, another source must be sought. In areas that have regular rainfall, the most appropriate alternative is the collection of rainwater, called "rainwater harvesting."

Falling rain can provide some of the cleanest naturally occurring water that is available anywhere. This is the result of a natural evapotranspiration process with risk of contamination only from airborne particles and from man-made pollution caused by the smoke and ash of fires and industrial processes, particularly those that burn fossil fuels.

Most modern technologies for obtaining drinking water are related to the exploitation of surface water from rivers, streams and lakes, and groundwater from wells and boreholes. However, these sources account for only 40% of total precipitation. It is evident, therefore, that there is considerable opportunity for the collection of rainwater when it falls, before huge losses occur due to evaporation and transpiration and before it becomes contaminated by natural means or man-made activities.

The term "Rainwater harvesting" is usually taken to mean the immediate collection of rainwater running off surfaces upon which it has fallen directly. This definition excludes runoff from land watersheds into streams, rivers, lakes, etc. Rainwater can be collected from roofs, and only to a lesser extent where it is collected from small ground, or rock (impervious), catchments (WaterAid 2013).

Rainwater can be collected from most forms of roof. Tiled roofs or roofs sheeted with corrugated mild steel and concrete slabs, etc., are preferable. Collected rainwater can be used for domestic, public, industrial, and other purposes. In addition, it can be used for groundwater recharge. The rainwater is then collected in gutters placed around the eaves of the building. The capacity of storage tanks is determined based on expected yield from the collection surface area and expected rainfall.

Description of Study Area

Climate of Ethiopia

Ethiopia has diverse climate influenced by latitude and altitude. There are five climatic zones, Wurch (cold to moist), Dega (cool to humid), Weynadega (cool sub-humid), Kola (warm semiarid), and Berha (hot arid) as defined by altitude, rainfall, length of growing season, and mean annual temperature (Berhanu et al. 2014). Figure 20.1a shows Ethiopia's maximum air temperature spatial variation. Ethiopia's spatial rainfall distribution is very high having annual rainfall less than 200mm in the east, southeast, and northeast; and most of western highlands receiving mean annual rainfall of over 1200 mm (Berhanu et al. 2014). Figure 20.1b shows Ethiopia's annual rainfall spatial variation.

Population Growth

Ethiopia is one of the countries having rapid increasing trend of population. The current population is above 110 million based on the United Nations estimates. The country's population is equivalent to 1.35% of the total world population and ranks number 13 in the list of countries by population. The population density is 102 per km^2 in which 19.4% of the total population is living in urban areas. Figure 20.2a shows spatial distribution of Ethiopia's population density. Figure 20.2b shows urban, rural, and total population changes since 1980 with projection to 2025.

Water Demand

Even though there is an economic water scarcity in the country, there is great increase in water demand for domestic, industrial, commercial, institutional, and public uses due to the following major factors. According to WHO (2014), the minimum per capita water demand is around 20 lpcd which is the lowest marginal demand. The demand will increase with population growth, development and

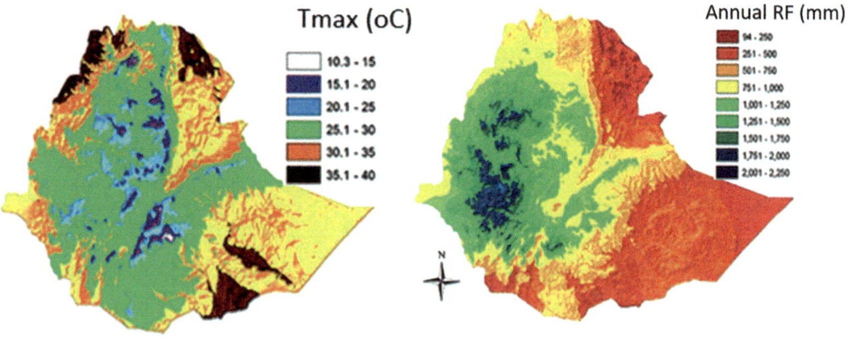

Fig. 20.1 Ethiopia's maximum air temperature (**a**) and annual rainfall (mm) (**b**)

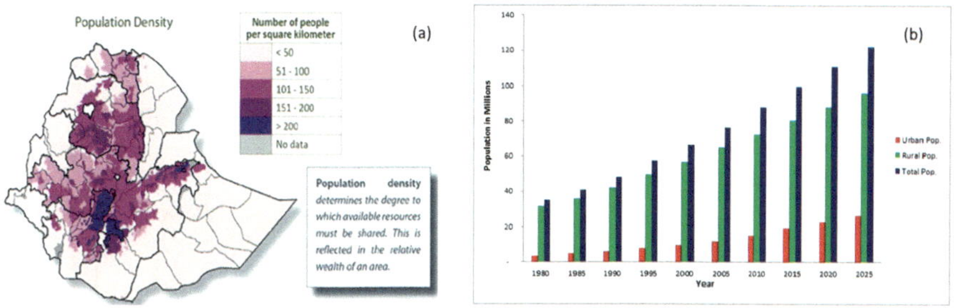

Fig. 20.2 Ethiopia's population density spatial distribution (Rettberg et al. 2017) (**a**) and urban, rural, and total historical population growth (**b**)

industrialization, increased life standard, and climate change with increasing in temperature.

The Blue Nile Basin

The basin accounts about 23% of country's land area and is the habitat of 37.6% of the total population. The basin has relatively higher population density compared to other parts of the country. The basin gets relatively higher amount of annual precipitation ranging from 800 to 2220 mm with an average amount of 1535 mm which indicates that there is high potential of rainfall harvesting for future integrated water use management (Johnston and McCartney 2010). Figure 20.3 shows the Blue Nile Basin with climatic zones.

Methodology and Analysis

The population and household data are collected from Ethiopian 2007 census and from United Nations world population report. The climate data such as precipitation and temperature are obtained from Ethiopian meteorological agency. The demand and supply data and standards are obtained from Ethiopian Ministry of Water, Energy and Electricity, from World Health Organization (WHO 2014) and literatures. Accordingly, the available data for computation and analysis are the following:

From 2007 Ethiopian Census data:

- The average number of individuals per household is 5.6 and taking max value as 7 for rural areas and 6 for urban areas.
- Ratio of households to housing units in rural areas is 1.05, whereas in urban areas is 1.1 and percent of current urban population is 19.8.

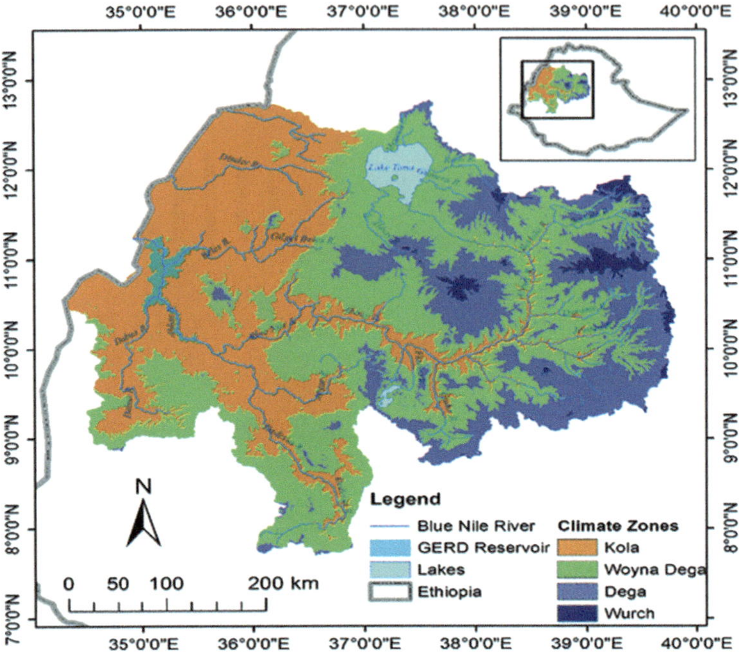

Fig. 20.3 The Blue Nile Basin and climatic zones (Abtew and Dessu 2019)

- In rural areas, the average area for housing is 40 m^2 and for those in urban areas is 36 m^2.

Based on the data obtained from Ethiopian Meteorological Agency, mean annual rainfall of 850 mm is taken for analysis even though more than 60% of the population lives in areas getting mean annual rainfall exceeding 1200 mm. In the Blue Nile Basin, 1300 mm annual rainfall is used.

According to WHO standard, minimum daily domestic per capita water demand for developing countries is 20 l/day in which only 10–15% is for drinking purpose. For analysis purpose, we have taken 30 lpcd for rural areas and 40 lpcd for urban areas which are above the standard. From total daily water demand, more than 26% is lost for toilets, 23% for washing cloths and dishes, 18.5% for showering and the rest for gardening purpose and losses.

In most of rural areas, there is additional shelter with minimum roof area 10 m^2 for cattle sheltering usually made from iron sheets, plastic, or other materials. Due to increasing life standard and climate change, annual domestic water demand increases by 1.5%. Current annual population growth rate is 2.48% for the total population in which 5.12% and 2.44% are growth rates of urban and rural areas, respectively. Table 20.1 shows Ethiopia's population growth between 1980 and 2015.

Table 20.1 Population and its corresponding growth rate data

Year	Total pop.	Urban pop.	Rural pop.	UGR	RGR	UAGR (%)	RAGR (%)
1980	35,239,974	3,668,755	31,571,219	0.273	0.1436	5.4604	2.8725
1985	40,775,997	4,670,398	36,105,599	0.2983	0.1631	5.9658	3.2615
1990	48,057,094	6,063,524	41,993,570	0.3004	0.1752	6.0076	3.5047
1995	57,237,226	7,884,886	49,352,340	0.2342	0.1491	4.6843	2.9825
2000	66,443,603	9,731,656	56,711,947	0.2288	0.14	4.5764	2.7994
2005	76,608,431	11,958,476	64,649,955	0.2614	0.1211	5.2272	2.4216
2010	87,561,814	15,083,947	72,477,867	0.2772	0.1055	5.5449	2.1102
2015	99,390,750	19,265,898	80,124,852	0.2676	0.1425	5.3524	2.8503

Pop population, *UGR* urban growth rate, *RGR* rural growth rate, *UAGR* urban annual growth rate, and *RAGR* rural annual growth rate

To estimate volume of harvested rainwater from rooftops, the rational formula is employed (Eq. 20.1).

$$Q = R * P * A \tag{20.1}$$

where

Q Mean annual harvested rainwater volume (m^3)
R Runoff coefficient (unit less)
P Mean annual precipitation (mm)
A Surface area of rooftops (m^2)

The runoff coefficient is considered to account potential losses due to splashing, evaporation, leakage, and overflow of rooftops and minimum value is taken as 0.8 (20% loss).

Results and Discussion

Ethiopia's Potential of Rainwater Harvesting

As it is known that Ethiopia is the country of "Water Tower" and getting annual mean rain fall of 1100 mm, there is significantly high potential of rainwater harvesting from rooftops in both rural and urban areas of the country. As the census data indicates, there is rapid increases of population which will also contribute to increase in housing developments. Living standard of rural community can be improved from potential rainwater harvesting due to rooftop type and area changes and increased yield. Using hydrological analysis from the data collected, the following results are obtained. If appropriately managed and awareness is created, there is a potential to store about half billion m^3 water from residential rooftops only. Table 20.2 shows total rural population (Pop), number of household (HH),

Table 20.2 Estimated contribution of harvested rainwater from total rural domestic water demand

Year	Pop	No. of HH	No. of HHU	RWH (m^3)	Per capita demand (lpcd)	Total demand (m^3)	RWH as % of total
1980	31,571,219	4,510,174	4,295,404	116,834,987	19.5	224,708,151	52
1985	36,105,599	5,157,943	4,912,326	133,615,278	21	276,749,416	48.3
1990	41,993,570	5,999,081	5,713,411	155,404,776	22.5	344,872,194	45.1
1995	49,352,340	7,050,334	6,714,604	182,637,231	24	432,326,498	42.2
2000	56,711,947	8,101,707	7,715,911	209,872,783	25.5	527,846,447	39.8
2005	64,649,955	9,235,708	8,795,912	239,248,813	27	637,125,307	37.6
2010	72,477,867	10,353,981	9,860,934	268,217,413	28.5	753,951,011	35.6
2015	80,124,852	11,446,407	10,901,340	296,516,459	30	877,367,129	33.8

number of housing units (HU), estimated annual rainwater volume (RWH), per capita water demand, total water demand and rainwater as percent of total water demand for from 1980 to 2015. Table 20.3 shows similar data for urban domestic water demand.

Tables 20.2 and 20.3 show that huge amount of water can be conserved and harvested from the rainfall. The magnitude of harvested rainfall depends on the number of housing units and the available areas of the rooftops. As the population increases, the available number of housing units will also increase even though the rate is not similar to the population and demand growth rates. The result and analysis did not incorporate the amount of rainwater harvested from rooftops of areas used for industrial, commercial, public, and other institutions. Only residential areas are considered. From Figs. 20.4 and 20.5, it shows that if potential available rainwater is harvested appropriately, it can fulfill above 40% of total domestic water demand in rural areas and 25% for urban community. In urban areas, harvested rainwater can satisfy the toilet water demand and water demands used for gardening and cleaning purposes which account above 35% of total water demand.

If appropriately collected and managed, rainwater that can be harvested from rooftops of industrial, institutional, commercial and public areas and can fulfill the demands for toilet, gardening, and cleaning purposes. In addition to this, municipal water stress and direct runoff magnitude from paved areas can be reduced through proper water harvesting mechanisms.

Blue Nile Basin

As it is shown in the figure below, there is increasing water demand for both rural and urban communities. Figures 20.6 and 20.7 show that more than 50% of total water demand can be fulfilled through rainwater harvesting in rural areas, whereas above 40% of urban domestic demand can be met through decentralized rainwater harvesting.

Table 20.3 Share of harvested rainwater from total urban domestic water demand

Year	Pop	No. of HH	No. of HHU	RWH (m^3)	Per capita demand (lpcd)	Total demand (m^3)	RWH as % of total
1980	3,668,755	611,459	555,872	13,607,746	29.4	39,369,410	34.6
1985	4,670,398	778,400	707,636	17,322,931	31.2	53,186,492	32.6
1990	6,063,524	1,010,587	918,716	22,490,162	32	70,821,960	31.8
1995	7,884,886	1,314,148	1,194,680	29,245,759	33.8	97,275,839	30.1
2000	9,731,656	1,621,943	1,474,493	36,095,597	35.4	125,742,727	28.7
2005	11,958,476	1,993,079	1,811,890	44,355,075	37	161,499,218	27.5
2010	15,083,947	2,513,991	2,285,447	55,947,731	38.3	210,866,037	26.5
2015	19,265,898	3,210,983	2,919,075	71,458,967	40	281,282,111	25.4

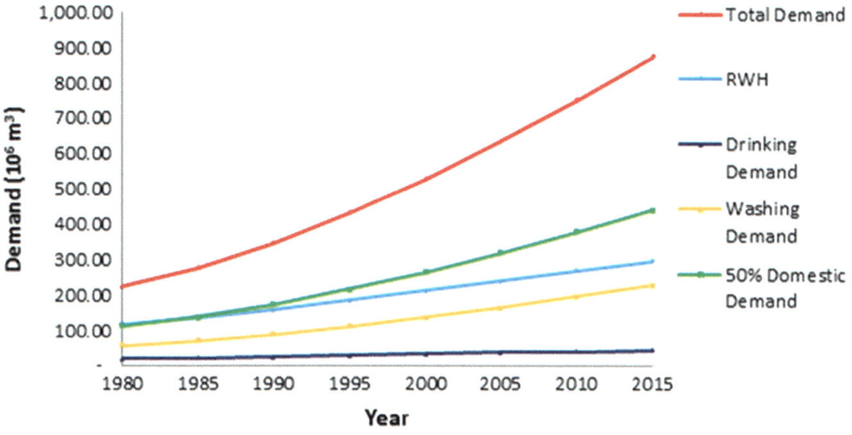

Fig. 20.4 Available rainwater harvest (RWH) for rural water use compared to total water demand

Even though there is increasing gap with water demand, harvested rainwater has an adequate amount to fulfill drinking water demand and demand for cloth washing in rural areas if it is collected timely and properly. The result indicates that unsafe and non-healthy rural water supply from rivers, dug wells, and springs can be solved through integrated and well-managed water harvesting technologies.

Summary

Decentralized rainwater harvesting can significantly reduce the stress of water supply and unmet demands in rural and urban areas. In urban areas, collected rainwater can be used for toilets (sanitary use), cleaning and washing, and for

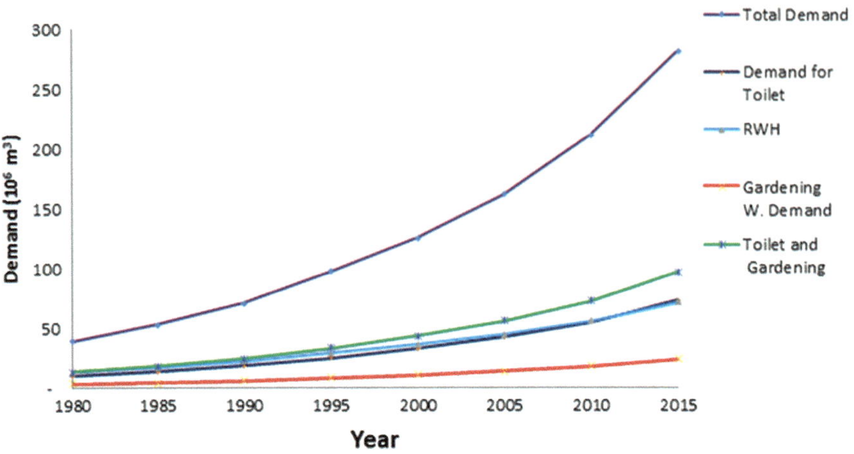

Fig. 20.5 Available rainwater harvest (RWH) for urban water use compared to total water demand

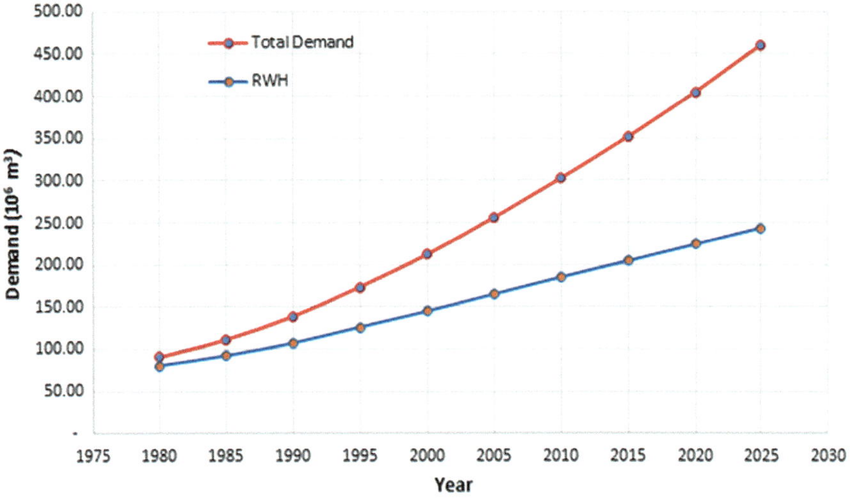

Fig. 20.6 Potential rainwater harvest for rural water use compared to total water demand

gardening purposes which consumes above 50% of total domestic water demand. In addition, it can be used as potable water for drinking and cooking if it is captured properly and neatly in those rural regions who are using unsafe surface water use. Furthermore, urban runoff problems due to increased impervious areas including more than 30% rooftops can be significantly reduced through rainwater collection. Decentralized rainwater harvesting through rooftops has great advantage to store

Fig. 20.7 Potential rainwater harvest (RWH) for urban water use compared to total water demand

potential rainfall using closed vessels (storages) without any loss to evaporation. Above ground rainwater harvesting is an alternative water storage in addition to ground water storage, especially at lowland areas with high evaporation.

References

Abtew W, Dessu SB (2019) Hydrology of the Blue Nile Basin overview. In: The grand Ethiopian Renaissance Dam on the Blue Nile. Springer Geography. Springer, Cham. https://doi.org/10.1007/978-3-319-97094-3_4

Berhanu B, Seleshi Y, Melesse A (2014) Chapter 6: Surface water and ground water resources of Ethiopia: potentials and challenges of water resources development. In: Melesse et al (eds) Nile River Basin. Springer, New York

CLUVA (2013) Climate change and vulnerability of African Cities. Italy, Napoli

Dowa A, Noel S, Shone G, Barron J, Soussan J (2010) Water and poverty linkages in Africa: Ethiopia case study. Stockholm Environment Institute, Stockholm

FAO (2007) Coping with water scarcity. Challenge of the twenty-first century. FAO, Rome

Hendrix M (2012) Water in Ethiopia: drought, disease and death. Global Majority E-Journal

Johnston R, McCartney M (2010) Inventory of water storage types in the Blue Nile and Volta river basins. International Water Management Institute, Colombo, Sri Lanka. 48 p. (IWMI Working Paper 140). https://doi.org/10.5337/2010.214

Molden D (ed) (2007) Water for food, water for life: a comprehensive assessment of water management in agriculture. International Water Management Institute, London

NAPA (2007) Climate change, national adaptation programme of action (NAPA) of Ethiopia. Addis Ababa

Ndaruzaniye V (2011) Water security in Ethiopia: risks and vulnerabilities' assessment. Global Water Institute for Africa Climate

Pachauri RK, Meyer LA (2014) Climate Change 2014: Synthesis Report. Contribution of Working Groups I, II and III to the Fifth Assessment Report of the Intergovernmental Panel on Climate Change. (p. 151). IPCC, Geneva, Switzerland

Rettberg S, Beckmann G, Minah M, Schelchen A (2017) Ethiopia's arid and semi-arid lowlands: towards inclusive and sustainable rural transformation country study. Centre for Rural Development (SLE), Burlin

UNDP (2007) Climate change national adaptation programme of action (NAPA) of Ethiopia. National Meteorological Agency, Ethiopia, Addis Ababa

WaterAid (2013) Rainwater harvesting. www.wateraid.org/Technologies. Accessed 23 Dec 2020

WHO (2014) Progress on drinking water and sanitation. World Health Organization and UNICEF, Luxembourg

World Bank (2007) Urban water supply and sanitation project. World Bank, Ethiopia

World Bank (2012) Ethiopia: climate risk factsheet. The World Bank, Washington, DC

https://www.nationalparks-worldwide.com/eaf/ethiopia/ethiopia-weather.html

Chapter 21
Gully Erosion and Effectiveness of Its Treatment Measures, Upper Abay Basin, in the Northwest Highlands of Ethiopia

Negesse Adimas, Mulatie Mekonnen, Derege Tsegaye, and Abebe Senamaw

Abstract In Ethiopia, gully erosion is a priority problem and identifying solution is a main concern for governmental and non-governmental organizations. The investigation of gully erosion and effectiveness of treatment practices is vital for the decision making. Therefore, this research was conducted with the objectives of assessing gully erosion and evaluating the impacts of gully treatment practices on gully erosion control and to address the research question: What are the major causes of gully formation and development in the watershed? How much soil was lost as a result of gully erosion? Are the implemented gulley treatment practices effective? Field measurement from four selected experimental gullies was conducted, (i) gully treated with physical practices, (ii) gully treated with biological/vegetative practices, (iii) gully treated with biophysical practices, and (iv) gully without any treatment as a control. Interview and focus group discussion, transect walk, soil sampling, satellite imagery, and field measurements were used for data collection. Statistical Package for Social Science (SPSS), Excel, Arc GIS, and Google Earth Software were used for data analysis. The result shows that the major causes of gully formation and development in the order of importance were local footpaths, surface runoff, free grazing, poor agricultural practices, and absence of proper waterway. Due to gully erosion, ~3.66 ha of land and ~55,198 t of soil were lost from the four identified gullies. The long-term gully erosion rate was found to be 15 t ha^{-1} yr^{-1}. Large amount of sediment was trapped by the gully treatment measures and help to reduce downstream sedimentation of water bodies. The integration of physical and biological (biophysical) practices was found to be more effective than individual practices. Land loss due to gully erosion also played a vital negative role on farmers' income in addition to the social influence by damaging roads, causing cattle deaths and flooding. In general, in Timet watershed,

N. Adimas
Mecha, Natural Resource Conservation and Management, TVET College, Merawi, Ethiopia

M. Mekonnen (✉) · D. Tsegaye · A. Senamaw
Natural Resources Management Department, College of Agriculture and Environmental Science, Bahir Dar University, Bahir Dar, Ethiopia

A. M. Melesse et al. (eds.), *Nile and Grand Ethiopian Renaissance Dam*, Springer Geography, https://doi.org/10.1007/978-3-030-76437-1_21

in the upper Blue Nile (Abay) basin of Ethiopia, there is high rate of gully erosion causing downstream reservoirs sedimentation, and biophysical gully treatment measures were found to be more effective in reducing gully erosion and downstream sedimentation.

Keywords Gully erosion · Gully treatment measures · Upper Abay Basin · Northwest highland of Ethiopia

Introduction

Gully erosion is a major cause of land degradation (Mekonnen et al. 2016; Addisu and Mekonnen 2019); an important source of sediment for water bodies that reduces water quality and degrades aquatic ecosystems (Valentin et al. 2005; Keesstra et al. 2009; Hughes and Prosser 2012); reduces effective soil depth and available water holding capacity, thus hindering the free movement of moisture, air, and nutrients within the soil system (Belay and Bewket 2012) and decreases the sizes of farmlands, and thus adversely affects farming activities (Yazie et al. 2020). Gully expansion also has economic-related negative influences by reducing the amount of land available for production (Zegeye et al. 2017; Yazie et al. 2020; Mekonnen 2020) and can contribute more than 90% of catchment sediment yield (Tebebu et al. 2010; Zegeye et al. 2014; Mekonnen et al. 2016). Gullies are the most destructive forms of soil erosion and difficult to reverse when matured (Moges and Holden 2008; Wijdenes Oostwoud and Bryan 2001; Mekonnen et al. 2020). Gullies can be formed either by surface runoff (Valentin et al. 2005; Rijkee et al. 2015) or by saturation of the soil profile (Tebebu et al. 2010; Mekonnen et al. 2016).

Gulley erosion is a global threat that aggravates sediment detachment, transport, and delivery. Unless the mechanisms are understood and measures are implemented, it can lead to land degradation and hence loss of arable land. A number of studies used laboratory, field scales, and modeling to understand soil erosion and sediment dynamics in various regions (Aga et al. 2018, 2019; Defersha and Melesse 2012a, b; Defersha et al. 2010; 2012; Maalim and Melesse 2013; Maalim et al. 2013b; Setegn et al. 2010; Melesse et al. 2011; Msagahaa et al. 2014; Wang et al. 2008; Mekonnen and Melesse 2011; Setegn et al. 2009; Mohammed et al. 2015).

Land degradation due to gully erosion by water is the main threat in Ethiopia, which is manifested by the presence of a lot of gully affected lands. For example, in the Eastern Abay (Nile) Basin, from 85,026 ha of land, ~2% (1370 ha) is affected by gully erosion (Desta and Adugna, 2012) and ~300,000 ha of the Amhara National Regional State with a total area of 157,077 km^2 are affected by gully erosion (Desta and Adugna 2012).

In the study area, Timet watershed, gully erosion is one of the major influential factors that affect negatively the economic and social activities of communities. Loss of soil; sediment deposition on the grazing lands, roads, and streams; and damaging of roads was major problems. It was also a cause for environmental pollution (like water and air pollution) in the lower side of the watershed. Loss of important soil nutrients and degradation of soil fertility has an impact on the production and productivity of crops. Improper design and layout of gully treatment practices, less attention given to the effect of gully erosion, and inappropriate selection of treatment practices were also major gaps. Gully expansion is also affecting the economic feasibility of farmlands by reducing the quantity of land available to farming.

To reverse the problems, gully erosion treatment practices have been implemented for the last three decades by the community and non-governmental organizations; however, such practices do not seem to be sustained and gully erosion is not reduced as desired (Mekonnen 2020). The investigation of gully erosion and effectiveness of treatment practices was vital for the decision making and sustainable planning processes of gully erosion prevention. Understanding the controlling factors of gully head migration and lateral expansion of gullies is also crucial to design appropriate gully control measures (Zegeye et al. 2017). Gully erosion processes are three dimensional (length, width, and depth) in nature (Valentin et al. 2005) and less known. Moreover, the processes of gully formation are not well understood specifically in the Nitosols-dominated humid highlands of Ethiopia (Rijkee et al. 2015). Gully erosion treatment structures that are effective in preventing gulling by overland flow in semiarid regions (Nyssen et al. 2002, 2006) may not be effective in the humid Ethiopian highlands, where interflow elevates ground water tables in the valley bottom that promote gully formation and expansion (Tebebu et al. 2010).

Therefore, the objectives of this study, in Timet watershed, in the northwest highlands of Ethiopia were to (i) assess gully erosion and the effectiveness of gully erosion treatment practices (biological, physical, and biophysical); (ii) identify the major causes of gully formation and development; (iii) assess gully morphological development since its initiation; (iv) quantify soil loss as a result of gully erosion; (v) evaluate the impacts of different gully management practices on gully erosion reduction and enhancing sediment deposition/trapping; and (vi) evaluate the economic impact of gully erosion due to land loss.

Negative Impacts of Gully Erosion

Gully erosion is one of the major global problems affecting agricultural production and sustainable use of natural resources. Although the problem is as old as the agricultural practices, its extent and adverse impact on crop production, land competition, ground water depletion, etc., is getting worse year after year. Gully erosion is the most prevalent form of soil erosion in Ethiopia, which dissects farmlands, impedes tillage operations, damages agricultural and residential areas, and restricts free movement of animals and humans.

Gullies are common features throughout the northwest Ethiopian highlands. Gullies cause not only the loss of soil volume and arable land, but also off-site sedimentation and the creation of efficient links between uplands and valley bottoms (Mekonnen et al. 2016). Gullies enhance dissection of the land and decreased crop yield and contribute to the lowering of the ground water table. Gully erosion strongly affects the general biological mass and the physical and chemical soil properties on cultivated and non-cultivated fields. It reduces effective soil depth and available water capacity, thus hindering the free movement of moisture, air, and nutrients within the soil system (Mekonnen et al. 2020). It shrinks the sizes of farm fields, alters movement of people and animals and thus adversely affects farming activities.

Gully Erosion Treatment Practices

Vegetation Cover

Tree planting and natural regeneration by area enclosure are common practices that can be managed by farmers in Ethiopia. Physical and vegetative soil and water conservation measures could also be constructed to regulate runoff. Such simple structures would promote retention storage, reduce erosion potential, and make water available to support the vegetation growth for longer period (Mekonnen and Getahun 2020; Moges and Holden 2008) .

Gully formation and development is often controlled by the inherent strength of the tree root mat that binds the surface soils until the undercut trees finally collapse. The increasing effects of plant roots on soil resistance to concentrated flow erosion mainly depend on the characteristics of effective roots (fibrils less than 1 mm in diameter) distributed densely in the depth 0–30 cm (Li et al. 2003). Plant roots reduce gully erosion in addition to improving soil physical properties such as structural stability and infiltration (Fig. 21.1). It was also recently reported that an increase in tiller density of different grass species results in a decrease of concentrated flow erosion rates (Mekonnen et al. 2016). In the Chinese loess plateau, an increase in grass land and forest land by 42% and a corresponding decrease in farmland by 46% reduced sediment production mainly due to gully erosion by 31% in the catchment (Li et al. 2003). From a land management perspective, the success of tree plantings, to mitigate gully erosion, depends on the stage of gully development and particularly on whether or not mass movement erosion has begun. Where mass movement assisted by excessive groundwater pressure is the main process leading to uncontrollable gully expansion, a particular attention must be paid to the stabilization of eroding riparian areas and swales, especially on the lower slopes of agricultural fields (Mekonnen et al. 2015).

Fig. 21.1 Role of trees in reducing the head gully erosion, northwest highlands of Ethiopia (Photograph by Mekonnen et al. 2016)

The principle of "prevention is better than cure" is highly relevant for gullies. Preventing the formation of a gully is much easier than controlling it once it has formed. If incipient gullies are not stabilized, they become longer, larger, and deeper practices (Desta and Adugna 2012).

Physical Measures

Stabilization of gullies involves the use of appropriate structural and vegetative measures in the head, floor, and sides of the gully. Once gullies have begun to form, they must be treated as soon as possible to minimize further damage and restore stability. There are a multitude of physical and biological techniques which can be applied for effective gully treatment. The combination of the two measures (bio-physical approach) is the best solution for effective gully control and for productive use of the gully area (Yazie et al. 2020).

Gully Reshaping and Filling

Gully wall reshaping is cutting off steep slopes of active gully flanks into gentle slope (Minimum at 45% slope), up to two-third of the total depth of the gully and constructing small trenches along contours for re-vegetating slanted part of the gully walls and beds. The reshaped gully area can be planted even be used for cultivation (Desta and Adugna 2012). At the very least, cultivation close to the gully edges should be discouraged as should channeling of water from the surrounding farms to the gully. Barriers, in conjunction with reduced flow, should eventually lead to a buildup, further reduce flow, and decrease sediment transport by positive feedback. Once the gully walls and floor stabilization takes place, the gully floor could be put to protective but beneficial uses like growing fodder for animals, rather than cultivation (Moges and Holden 2008; Addisu and Mekonnen 2019).

Brushwood Check Dams

The main objective of brushwood check dams is to hold soil material carried by flowing water in the gully. Small gully heads, no deeper than one meter, can also be stabilized by brushwood check dams. Brushwood check dams are temporary structures and should not be used to treat deep and ongoing gullies.

Stone Check Dam

Stone check dams are commonly used for gully treatment all over Ethiopia. Yet during implementation, three common mistakes have been occurring: The base of the check dam is often built to narrow, and therefore, the final structure will lack the recommended trapezoidal cross-section. The flanks or the anchorage into the sides of the gully are not cut in deep enough and when building up the check dam larger stones on the outsides are not properly combined with smaller stones on the inside. Spillways should be built having a parabolic/concave shape instead of the window shape and the width of the apron needs to be at least 1.5 times the effective height of the spillway (MOARD 2005).

Gabion Check Dam

Gabions are rectangular boxes of varying sizes and are mostly made of galvanized steel wire woven into mesh. The boxes are tied together with wire and then field with either stone or soil material and placed as building blocks. Small stones can be used as the wire mesh will prevent them being washed away. If large stones are used, they must be placed carefully with small stones filling the spaces between them otherwise water may jet through the gabion and undermine the ground beneath. Gabions are filled in situ, and as they are very heavy, they will not be washed away provided they have been correctly installed (MOARD 2005). Gabion check dams are important measures in trapping large amount of sediment and re-filling deep gullies with the sediment trapping efficiency of 74% (Mekonnen et al. 2015; Mekonnen and Getahun 2020).

Study Area and Methods

Study Area

The study area, Inguti Kebele (Timit watershed), is located in the northwest Ethiopian highlands, head catchment of the Abay Basin. Geographically, it is located between 297,000–304,000 m E and 1,259,200–1,268,200 m N (Fig. 21.2). It covers an area of 361.74 ha. It is situated at an elevation ranging from 2000–2159 m.a.s.l. with the slope ranged from 0 to 43%. The minimum and maximum temperatures were 14 and 27 °C, and the minimum and maximum annual rainfalls

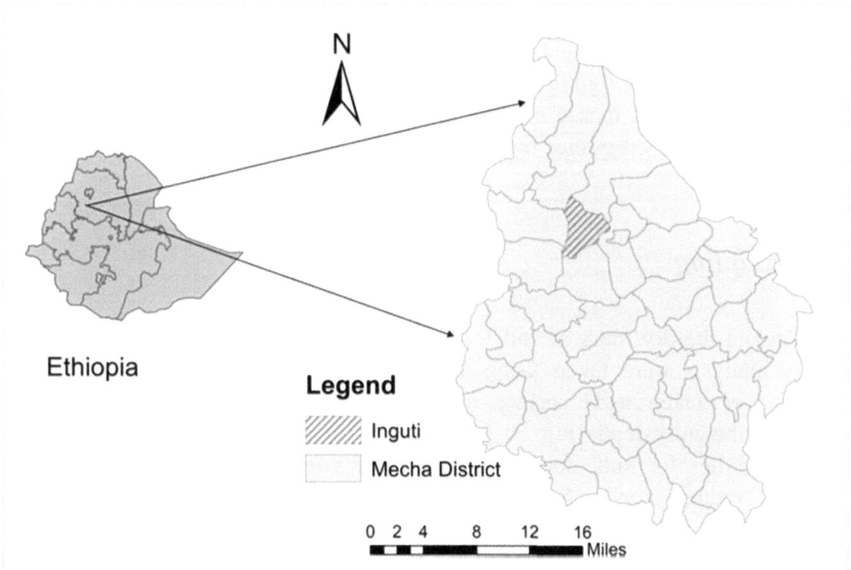

Fig. 21.2 Location map of the study area (Inguti Kebele, Timit watershed), northwest highlands of Ethiopia. Kebele is the lowest administrative level

were 1200 mm and 1700 mm, respectively. The soil in the watershed is red in color and dominated by Nitosols. The area is characterized by intensive agriculture with an average landholding of 1 ha per household. Land preparation is performed with the traditional *Maresha* plow pulled by a pair of oxen. Maize (*Zea mays*), wheat (*Triticum aestivum*), and finger millets (*Eleusine coracana*) are the major crops usually grown, and some of the cultivated lands are covered with Eucalyptus trees.

Experimental Gully Selection

In Inguti Kebele (Timit watershed), four gullies were selected: (i) gully treated with physical practices, (ii) gully treated with biological/vegetative practices, (iii) gully treated with both physical and biological practices (biophysical practices), and (iv) gully without any treatment as a control.

Types and Sources of Data

Topographic data (slope and elevation) were collected from the field using GPS. Soil samples were taken from the field using core sampler and soil auger to analyze soil bulk density and texture. Google Earth image was used to assess the change in gully dimensions and quantify the rate of gully development. Length, width, and depth measurements of the four gullies were collected using meter tape, nylon string, and pigs from the field. Soil loss volumes as a result of gully erosion were calculated from the four gullies on the field. The volumes of trapped sediment were estimated using field measurement. Surface area of eroded lands was measured from the four selected gullies by using measuring tape, nylon string, and pigs.

Soil Sampling

To calculate the dry mass of soil lost due to gully erosion, soil samples for bulk density analysis were taken using standard core sampler. That means the soil samples were taken at the surface, middle and bottom of each investigated gully down into the gully depth (Fig. 21.3a). Totally, 27 soil samples were collected from the four gullies for bulk density analysis. The core sampler has 5 cm diameter (2.5 cm radius) and 5 cm height with the volume calculated as ($V = \pi r^2 * h = 98$ cm^3). For texture analysis, nine soil samples were collected from the investigated four gullies that mean three samples at the head part, three samples at the middle, and three samples at the lower part of the gullies along the gully length (Fig. 21.3b). The soil auger was used to take soil samples at the depth of 20 cm.

Fig. 21.3 Soil sampling locations within the investigated gullies, soil sampling along the gully depth for bulk density analysis (**a**) and soil sampling along the gully length for soil texture analysis (**b**)

Satellite Imagery

To assess the gully morphological development, Google Earth was used. Google Earth Imagery was used to quantify gully surface area coverage and its changes from 2009 to 2017. The gully edges were digitized using Google Earth's polygon mapping tool. Then, the digitized polygons were used to calculate the length and width of the investigated gullies and the changes in size between the two investigated years.

Field Measurements

To assess the annual gully development rate, gully dimensions (depth, width and length) were measured before and after the rainy season of 2018. Measurement was done staring from the head of gully to down slope, which depends on gully condition. Gully cross-sectional geometry was surveyed by dividing the cross section into trapezoidal segments at abrupt changes in the bank and then measuring the width and depth of the gully at each segment. Gully dimensions were measured manually by crossing the gully into different sections based on the shape of gulley dimensions using tape meter, nylon string, and pigs in the beginning and ending of the rainy season to estimate the land losses and soil volume for each selected gullies.

Trapped Sediment Estimation

The trapped sediment dimensions (depth, width, and length) were measured manually using tape meter. To calculate the volume (V; m^3) of the sediment accumulated behind the check dams, the cross-sectional area (A; m^2) of the sedimentation times the length (*L*; m) from the check dams to the end of sedimentation upstream was calculated (Mekonnen et al. 2015).

Soil Texture and Bulk Density Analysis

Soil bulk density is one of the parameters that help to understand the causes of gully expansion. To analyze the soil bulk density, the collected samples were oven dried at 105 °C in the laboratory. Finally, bulk density was calculated by dividing the mass of the oven-dried soil to the volume of the core sampler. Soil textures were analyzed using hydrometric method from Amhara design and supervision works enterprise soil laboratory.

Soil Loss Analysis

To calculate the surface area occupied by the gully (*A*), volume of soil lost (*V*), long-term gully erosion (LTGE), and gully density (Gd), Eqs. 21.1–21.6 were used, respectively, (Tebebu et al. 2010; Zegeye et al. 2014; Mekonnen et al. 2015).

$$A = L * W \tag{21.1}$$

where *A* is surface area occupied by the gully, *L* is length of the gully, and *W* is width of gully

$$\text{Volume of soil lost } (V)\left(\text{m}^3\right) = \sum_{i=1}^{n} Vi \tag{21.2}$$

where V_i is the volume soil lost in a section gully (product of gully cross-sectional area to the gully section length).

$$\text{Long - term gully erosion rate}(\text{LTGE}) = \frac{VBd}{TA} \tag{21.3}$$

where *V* is volume of soil lost (m^3), Bd is bulk density of the soil (g/cm^3), *T* is age of the gully in years, *A* is area of the sub-watershed in hectares.

$$\text{Gully density} = \frac{\sum_{i=1}^{n} Li}{A} \tag{21.4}$$

where *Li* is length of unit gully in meter, *A* is area of the sub-watershed in hectares, and *n* is number of gullies

To calculate the volume (V; m^3) of the sediment accumulated behind the trapezoidal-shaped check dams (Eq. 21.5), the cross-sectional area (A; m^2) of the sedimentation times the length (L; m) from the check dam to the end of sedimentation upstream was calculated (Eq. 21.6). The cross-sectional area (A) of the trapped sediment is the average of the top and bottom widths ($b2$ and $b1$; m) of the sediment times its height (h; m) measured from the base of the dam to the sediment surface (Eq. 21.6). For rectangular shape dams, length * width * depth of the trapped sediment was used.

$$V = A * L \tag{21.5}$$

$$A = \frac{1}{2}(b1 + b2) * h \tag{21.6}$$

Results and Discussions

Causes of Gully Formation and Development

About 83% of the sampled informants strongly believed that topography (slope steepness and length) was the most important cause of gully erosion in the study area. Others like local roads (foot paths) along the slope (70%), deforestation (76%), surface runoff (60%), free grazing on the cultivated land (66%), poor agricultural practices (50%), absence of proper waterway (43%), soil type and traditional drainage channels (33%), respectively, were found to be the causes of gully formation and development (Table 21.1).

Intensive farming practice without appropriate natural resources management, deforestation, and overgrazing was found to trigger flooding and soil erosion by water. Mekonnen et al. (2015) reported that construction of roads has a great role on the formation and development of gullies than other factors. Deforestation through

Table 21.1 Farmers perception (30 interviewed farmers') on the causes of gully formation and development

Major causes of gully erosion	No. of responses	% of responses	Rank
Local foot paths	21	70	3
Deforestation	23	76.6	2
Topography	25	83	1
Poor agricultural practices	15	50	6
Surface runoff	18	60	5
Free grazing on the cultivated land	20	66	4
Absence of proper waterway	13	43	7
Traditional drainage channels	10	33	8

clearance of the surrounding woody biomass for the creation of new settlements and expansion of cultivated land resulted in the degradation of the land surface. Similarly, due to intensive farming practice, deforestation and overgrazing, the natural stability of the soil has been under constant degradation which is in line with the findings of (Mansur 2014). Valentin et al. (2005) also reported a similar finding that the formation and development of gullies was caused by overgrazing, road construction, urbanization, and improper farming. The result indicated that although some causes were natural the most important causes for gully formation and development were man-made or human activities, which could be controlled. Therefore, the local government with the integration of the community can at least reduce the causes of gully formation and development through awareness creation.

Soil Texture and Bulk Density

Soil texture is one of the physical soil characteristics that influence land use and management. As indicated in Table 21.2, soil texture in the watershed had heavy clay (69%). The implication of this finding is that there was low infiltration due high proportion of clay content (saturation). Therefore, there was a high concentration of surface runoff that might have increased gully formation and development, and the falling and slumping of gully walls. The mean bulk density was 1.22 g cm^{-3} (Table 21.2). This implies that the soil of the study area has low infiltration rate that creates a chance for surface flow. Therefore, soil texture and bulk density were found to be important causes for gully formation and development in addition to man-made causes mentioned, which is in line with the reports of Valentin et al (2005), Danladi and Ray (2014) and Zegeye et al. (2017).

Table 21.2 Soil laboratory analysis results of the average bulk density and textures under the four gullies

Soil properties	Physically treated gully	Biologically treated gully	Biophysically treated gully	Control (untreated) gully	Mean
Av. bulk density (kg/m^3)	1.2	1.2	1.22	1.24	1.22
Sand (%)	17	17	17	17	17
Silt (%)	17	15	14	11	14
Clay (%)	66	68	69	72	69

Topography

A GPS reading collected from the watershed outlet to the highest point of the watershed dividing line shows an elevation ranging between 2000 m and 2159 m above sea level. The slope gradient in the watershed ranged from 0 to 43% which means level to gentle in the outlet and steep toward the inlet (upper watershed dividing line). This difference in slope steepness accelerates the runoff flow and has been the cause for gully erosion. The longer the slope, the higher the runoff volume and concentration, which reduces runoff infiltration and increases runoff velocity and then gully erosion. Higher slope gradient in the upper part of the watershed leads to faster runoff velocity, which increases soil erosion and gully formation. Soil erodibility and gully formation are highly dependent on slope because erosion rates of all soils increase as slope angle and length increase (Meshesha et al. 2016; Mekonnen et al. 2016), and concentrated surface runoff from lengthy slope lands is an important cause of gully formation (Mekonnen and Melesse 2011). In this study, surface runoff was found to be one of the important causes of gully formation and development. This implies that soil and water conservation practices should be implemented within farmlands to reduce slope length and angle, which will reduce surface runoff concentration for the reduction of gully formation and development.

Land Use/Cover

Naturally, gully processes are accelerated by the intensification of farming systems. The depletion of the soil organic matter reduces the soil structural stability and enhances crusting, runoff production, and gully erosion (e.g., Valentin 2004). In Timet watershed higher gully distribution was observed on cultivated lands (70%) than other land uses, which shows that tillage has an impact for gully formation and development. The soil resistance to erosion is not only dependent on topographic position (slope steepness), but also highly affected by the extent of disturbance which occurs during tillage (Meshesha et al. 2016). Untilled soil is more compact (solid) and its aggregate stability and shear strength is higher while porosity and infiltration capacity is lower (Meshesha et al. 2016). Therefore, cultivated lands are more vulnerable than other land uses for gully formation and development. The map (Fig. 21.4) indicated that 70% of the watershed covered with cultivated lands. Therefore, land use/cover type has significant role for gully formation and development.

As indicated in Table 21.3, ~70% of the watershed was covered with cultivated lands for a long period of time. Therefore, during the rainy season, these cultivated lands were vulnerable for sheet and rill erosion. Concerning the surface area coverage of the investigated gullies, 3.12 ha (~85%) is in the cultivated land and the rest 0.54 ha (15%) was found in the grazing lands and forest lands.

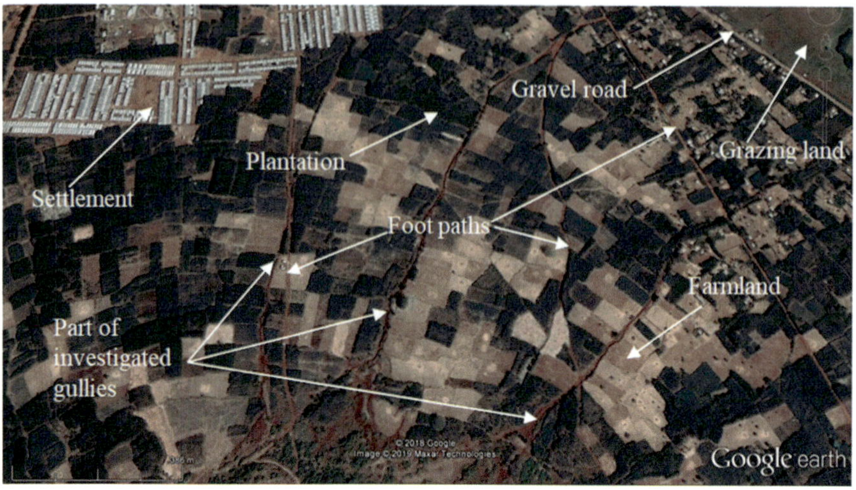

Fig. 21.4 Google Earth showing part of the study area land use/cover, gullied areas, gravel road and foot paths, in Inguti Kebele (Timit watershed), northwest highlands of Ethiopia

Table 21.3 Land use types and their surface area coverage in the study watershed

Land use type	2009		2017		Change (%)	Gully area (ha)
	Area (ha)	Area (%)	Area (ha)	Area (%)		
Plantation	25.5	7.0	33	9.15	+2.15	0.39
Cultivated lands	255	70.7	253	70	−0.7	3.12
Settlement	11.8	3.3	13	3.6	+0.3	–
Grazing land	64.9	18	59.4	16.5	−1.55	0.15
Road	3.5	1.0	2.91	0.8	−0.2	–
Total	360.7	100	360.7	100	0	3.66

Grazing land is the second largest cover in the watershed, which is ~16.5%. Grazing land cover reduced from 18% in 2009 to 16.5% in 2017 because of expansion of the small nearby town, Merawi. The expansion of Merawi town into the watershed has an impact on the watershed environment due to accumulation of wastes. Farmland (cultivated land) cover was the largest cover (70%) in the watershed that decreased by 0.7% during the investigation periods. Plantation cover is the 3rd largest cover (~9%), which shows an increasing trend by 2.15%, which was because of Eucalyptus and *Grevillia* trees plantation. Planting such tree species played an important role in controlling the expansion of gullies head cut. Roads (footpaths and gravel roads) and over grazing of the grazing lands are facilitation gully formation and development.

Gully Morphological Development

Based on interview results, gully incision started in 1980s from the upper part of Timet watershed due to sheet and rill erosion, which corresponds to when the indigenous vegetation on the hillsides was decreased gradually due to agricultural land expansion and other human factors. After gullies have been formed, slumping of gully walls was most severe in the middle part of the gullies as a result of surface flow moving through the cracks and causing local saturation near the banks and the property of clay soil. Gully expansion was severe (mostly at the top and middle part of the watershed) due to alternate swelling and shrinking of clays, which agreed with a study conducted in the nearest watershed by Zegeye et al (2017) in that the bottom lands of the watershed with Vertisols dominant, gully formation was severe due to alternate swelling and shrinking of expanding clays resulting in deep cracks in the dry season.

The form or shapes of investigated gullies (from top to bottom) in the watershed vary. At the top, the shape of the gullies had V-shape whereas in the middle part had trapezoidal. The trapezoidal shaped gully was formed as a result of concentrated runoff expanding the bottom side of the gully. But at the top, the runoff was expanding the side and head of the gully due to surface flow. The depth of the four gullies decreased at the bottom due to sediment deposition when the slope is flat and had U shape. To control the expansion of the gullies, the district office of agriculture in cooperation with the community started to construct soil and water conservation structures in the year 2015 consisting of graded soil bunds on the farmland, planting forage trees along the gullies and on the graded soil bunds, building stone and gabion check dams inside the gullies.

Gully dimension measurements (using Google earth version 2018) showed that in 2009 the total gully surface area was 2.82 ha but in 2017 it becomes 3.55 ha. From 2009 to 2017, in 9 years, the loss of land by the four gullies was 0.73 ha (0.08 ha per year). Gullies in the past 9 years increased by 20.56% (8% per year) (Table 21.4). Based on the result, gully dimension (depth and widths) has increased in the last 9 years. In year 2009, the length of the four identified gullies (G1, G2, G3, and G4) was 710 m, 790 m, 1440 m, and 1550 m, respectively. While in 2017, the gullies did not continuously increase (Table 21.4) because the gullies had reached the upper and lower ends of the watershed. However, there is continuous increase in depth and width in the middle part of identified gullies because the middle part of the gullies was reduced by the local road formation and cultivation near to the side of the gullies. Thus, it should be easier to restore already formed gullies by using the integration of physical and biological control methods.

Table 21.4 Change in gully dimension because of gully erosion (2009–2017)

Year	Treatment gullies	Gully length (km)	Gully surface area (ha)
2009	Physically treated gully (G1)	0.71	0.37
	Biologically treated gully (G2)	0.79	0.35
	Biophysically treated gully (G3)	1.44	1.03
	Untreated gully, control (G4)	1.55	1.07
	Total	4.49	2.82
2017	Physically treated gully (G1)	0.73	0.54
	Biologically treated gully (G2)	0.79	0.41
	Biophysically treated gully (G3)	1.51	1.25
	Untreated gully, control (G4)	1.68	1.35
	Total	4.71	3.55
Relative change in %		4.6	20.5

Land and Soil Loss

From the direct field measurement in May 2018, the total surface area of the four investigated gullies was 3.63 ha, which increased to 3.66 ha in November 2018 (Table 21.5). Therefore, 0.03 ha of land was lost within a single rainy season, and totally, 3.66 ha of land became out of production. The total volume of soil lost from the four gullies in the Timet watershed was 45,244.26 m^3 (Table 21.5). From the four investigated gullies, the longest and the shortest gullies were 1681 m and

Table 21.5 Annual development of the four investigated gullies in 2018 rainy season

Evaluation periods	Gully dimension	G1	G2	G3	G4	Total
Beginning of the rainy season	Length (m)	732	838	1377.9	1681	4629
	Average depth (m)	1.26	1.64	2.01	2.04	
	Average width (m)	5.76	3.9	5.34	6.37	
	Volume (m^3) (V1)	5316	5360	12,482	20,155	43,313
	Mass of loss soil (t)	6486	6539	15228	24589	52,842
	Surface area coverage (m^2)	5,502	4,287	12,606	13,896	36,291
End of the rainy season	Length (m)	732	838	1378	1681	4284
	Average depth (m)	1.3	1.7	2.03	2.1	
	Average width (m)	5.8	4	5.44	6.5	
	Volume (m^3) (V2)	5532	5698	12,842	21,171	45,244
	Mass of soil lost (t)	6750	6952	15667	25829	55197.9
	Surface area coverage (m^2)	5,530	4,384	12,656	13,995	36,565
Volume change and its percentage		4	6.3	2.9	5	
Long-term gully erosion rate t ha^{-1} yr^{-1}		0.6	0.6	1.4	2.4	5.1

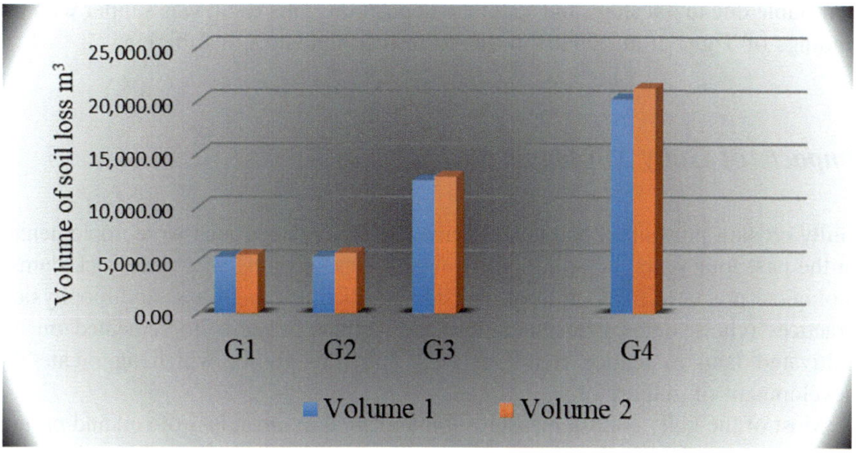

Fig. 21.5 Volume of the four experimental gullies that was measured before and after the rainy season. Volume 1 and volume 2 are the volume of soil loss before and after the rainy season, respectively

732.5 m, respectively. As indicated in Table 21.5, the untreated gully (G4) was the widest and the longest than others. Within a rainy season the average gully depth increased by 0.04 m, 0.06 m, and 0.02 m in G1, G2, and G3, respectively (Fig. 21.5), while in untreated gullies (G4) there is deposition of sediments in most part of the gully bed, which cause the depth to be reduced (−0.06 m).

Gully to plot area ratio was found to be 0.01, and percent coverage from the total watershed area was 1%. This shows that for every 100 units of land 1-unit area of land was damaged in the watershed. The total amount of soil removed from four gullies was 45,244.26 m^3 or 55,198 t (Table 21.5). Soil loss due to gully erosion within the watershed was found to be 153 t ha^{-1}, and the average amount of soil loss from each gully was 38.25 t ha^{-1}. Based on results of the interview of farmers, the age of the investigated gullies is about 30 years. Based on this information, the long-term rate of gully erosion calculated as 15.1 t ha^{-1} yr^{-1}. According to Pathak et al. (2006), gullies can be classified as small gully (<1 m), medium (between 1 and 5 m) and large gully (>5 m). Therefore, gullies in Timet watershed were medium in size.

Based on gully continuation, all gullies in the watershed are discontinuous and their depth increases from the upstream to its mid-point and then decreases gradually. Perhaps this is because of high accumulation of runoff and other human factors like roads, plowing near the gully bank at the middle section. Referring to the shape of gully cross section, those formed due to road drainage are trapezoidal, deep and wide but seem to be stable at the downstream section. The mid-sections of these gullies were trapezoidal shaped with steep-sided walls and are very wide. Head ward development of gullies is not due to reaching the upper most of the catchment divide. On the other hand, the lower parts of all gullies are U in shape and stable due to flat slope and sediment deposition. This result was similar with the findings of Tigist et al. (2010) in the northwest highlands of Ethiopia.

Impacts of Gully Management Practices

Gully erosion prevention and control practices in the study area were implemented in the past four years including physical practices (gabion and stone check dams), biological practices (plantation on the side and bed of the gullies) and biophysical practices (check dams with plantation). Soil bunds are also implemented on the cultivated land to reduce runoff entering into the gullies which aggravate the development of gullies.

Most of the gully control measures implemented to arrest loss of soil and protect from development of new gullies were ineffective. As depicted in Table 21.6, the average spacing between consecutive check dams is almost similar to the national recommended spacing. However, during implementation, common mistakes have been occurred: The base of the check dam often built is narrow; thus, the final structure would lack the recommended trapezoidal cross section. Good functioning check dams have a spillway, an apron, a concave plan form when looking down slope and are built at vertical intervals with heights that result in a negative slope gradient of the line connecting the spillway and the foot of the upstream dam. As shown in Fig. 21.6, the check dam was damaged due to the absence of apron, improper spillway and side key, which will give at least the following advantages: (i) reduced runoff erosivity, (ii) storage of large runoff and sediment volumes, (iii) ponding of water at the foot of the dams during rain events, in which the flow energy of runoff overtopping the dam is dissipated.

Table 21.6 Trapped sediment measured behind the stone check dam (physically treated gully, G1)

No. of Sections	Average length (m)	Average width (m)	Cross-sectional bed area (m^2)	Average depth (m)	Volume of trapped sediment (m^3)	Spacing b/n check dams (m)
1	8.7	4.67	40.63	0.46	18.69	17
2	7.7	4	30.8	0.53	16.32	15
3	11	3	33	0.68	22.44	21
4	21	4	84	0.48	40.32	29
5	9	4.17	37.53	0.25	9.38	14
6	10	2	20	0.3	6	18
Total			308.96		113.15	

Fig. 21.6 Ineffective stone and gabion check dams on large gullies. A lack of apron (**a**), ineffective side key (**b**), and damaged stone check dam (**c**)

Spillways should be built having a parabolic/concave shape instead of the window shape, and the width of the apron needs to be at least 1.5 times the effective height of the spillway (Asrat et al. 2005). The integration of physical and biological gully treatment practice is better than the individual practices because in the study area the integrated practices (gabion check dam with plantation) reduces bed and bank erosion by trapping more sediment. Mostly, the failures of check dams were poor design and specifications selection. Check dams will not be effective for large size gullies. Mekonnen et al. (2015) evaluated sediment storage dams (SSDs) constructed from gabion and stone and found as effective for large size and deep gully treatments, and therefore, SSDs will be alternative solutions to treat large size gullies in Timet watershed.

Based on the result, physical gully treatment practice, stone check dam, trapped $\sim$113.15 m^3 of sediment, whereas the biophysically treated gully (G3) trapped $\sim$357.7 m^3 sediments (Table 21.7). This means the integration of physical and biological (biophysical) practices trapped more sediment than the physical gully treatment measures. Moreover, physical practices (G1) requires continues maintenance to sustain the gullies. Rather the biophysical practices (G3) sustain the gully bed and side by increasing soil shear strength through their roots, slowing down the storm runoff, and trapping sediments. Biophysical practices have many purposes, not only used for gully treatment practices but also used for livestock feed as dual function. Therefore, biophysical gully treatment practices are better than physical treatment measures. Planting of soil and water resistant or tolerant species (e.g., *Grevilla robusta, Eucalyptus comaldunesis, Sesbania sesban*) on the gully bed and side breaks the flow and velocity of runoff, reduces bank failure, traps sediment, and protects the gully bed from erosion.

The biological practices (G2) reduce gully expansion by increasing soil shear strength through their roots, slowing down the storm runoff, and trapping sediments, which was also described by researchers, like Poesen et al. (2003). Therefore, planting suitable species on the gully face and around the boundary reduces or slows down bank failure and water-induced erosion, especially for fairly deep gullies. Gully measurement and observations indicated that the growth of vegetation on gully beds and banks, in addition to the building of check dams,

Table 21.7 Trapped sediment which was measured from biophysically treated gully (G3)

No. of sections	Average length (m)	Average. width (m)	Cross-sectional area (m^2)	Average depth (m)	Volume of trapped sediment (m^3)	Spacing b/n check dams (m)
1	13.7	2.1	28.77	0.49	14.097	16
2	21	2.32	48.65	0.45	21.89	25
3	26	1.37	35.53	0.32	11.37	27
4	11.25	2	22.5	0.35	7.87	13
5	18.8	1.9	35.72	0.48	16.97	20
6	15	1.98	29.63	0.5	14.8	15
7	12	1.7	20.4	0.48	9.78	12
8	14.75	2.5	36.25	0.32	11.48	15
9	16	2.25	36	0.33	11.88	17.5
10	12	2.37	28.4	0.5	14.2	13
11	9.3	2	18.6	0.4	7.44	11
12	22	1.95	42.9	0.6	25.74	22
13	20.25	2	40.5	0.4	16.2	21
14	16	2.1	33.6	2.1	10.08	16
15	17.43	2.77	48.29	0.28	13.52	18.5
16	20	3.5	70	0.21	14.7	24
17	12	1.75	21	0.33	6.93	12
18	20	2.9	58	0.53	30.74	20
19	28	2.57	71.87	0.33	23.72	33
20	31	2.13	66.13	0.38	25.13	38.5
21	7.4	2.5	18.5	0.27	5	8
22	12.7	2.6	33	0.5	16.51	12.7
23	18	3	55.2	0.4	22.08	18
24	8	2.9	23.2	0.24	5.57	8
Total			922.64		357.697	

rapidly stabilizes gullies. Through vegetation growth on the gully bottom, gully bed roughness is increased, reducing runoff velocity and favoring sediment deposition. The biomass produced in the gully area can be used as cattle fodder, which decreases grazing pressure in other areas. Biophysical gully control practices and integration of physical and biological practices appear to be a sustainable method of gully rehabilitation. At village level, it can be decided that gully bottoms and banks, which make up only a small part of the rangeland, are as far as possible made into enclosures. Rapidly recovering vegetation will lead to dam stabilization and increased sediment deposition.

Gully erosion can rapidly change landscapes as can be seen for instance G4 that have expanded laterally into the cultivated land through erosion of the right and left banks (west and east banks), whereas G1, G2, and G3 most lateral erosion of the left and right banks were rather limited due to the treatment practices. Therefore, untreated gully (G4) was the largest (in depth, width and length) and the most sever type of gully (Table 21.7). About 1.4 ha of land and 25,828.8 t of soil mass was reduced by this gully. It covers 38% of the four selected gullies and 0.4% of the total watershed. The roads and grazing lands were damaged by its flooding and sediment deposition.

Economic Impacts of Gully Erosion Due to Land Loss

Gully erosion in the study area had an adverse impact on the socioeconomic development of the community. From the field measurement, ~3.66 ha land had lost from the four gullies. Maize *(Zea mays)* and finger millet (*Eleusine coracana*) were the dominant crops in the watershed with an average productivity of 2000 kg ha^{-1} yr^{-1}. However, because of loss of productive land through gully erosion, the yield of maize was reduced by 183 t yr^{-1} in four gullies. In support of this finding based on the information gained from the field measurement, interview, and direct field observation, gully erosion in the watershed has the following major economic and social adverse impacts: (i) reduction of agricultural land and its resources including obliteration of crops by flooding and erosion that means gully erosion cuts farmlands causing very severe damage to agricultural lands and hence crop production; (ii) due to blocking of roads and flooding, the community is highly affected from their socioeconomic development and interaction with each other. As informed from local farmers, a large number of cattle were destroyed by the runoff flow collected inside the gullies during the study rainy season, and (iii) loss of soil fertility and reduction of land productivity, which requires large amount of artificial fertilizer which demoralized local farmers due to the expensive cost of the chemical fertilizer, (iv) deposition of sediment on the grazing lands, roads, and streams in the lower part of the watershed. Especially, G3 and G4 have a greater contribution of sediment yield to the grazing lands. Therefore, the fodder production from grazing lands declined.

There was also an environmental pollution due to sediment deposition. Chemicals (herbicides, pesticides, fertilizer, etc.), produced from the farm, may reach to the river by runoff and form a toxic compound on the river water. Although the water quality treatment requires another research, the people believed that the water quality decreased due to runoff and other domestic wastes from time to time because they used this river for domestic and other purposes. Similar studies in Debre Mewi watershed Tebebu et al. (2010) and Mekonnen et al. (2016) in Koga watershed reported as gully erosion are contributing significant sediment load to the downstream river network. In order to prevent siltation of the newly constructed reservoirs in Ethiopia, stopping gully erosion number is one solution. In general, this study indicates that gully erosion causes considerable on-site and off-site effects in the watershed, which is line with Mekonnen (2020).

Future Consideration

Proper design of check dams is very crucial at least to reduce the development of constructed gullies. All check dams should have properly constructed apron on the downstream side to protect the dam from undercutting. The width and depth of the spillway shall be determined by the width of the gully and the discharge rate. The check dam should be properly keyed to the floor and sides of the gully to improve stability, which involves excavation of foundation about 0.5 m deep and wide across the gully floor and 0.5 m into the gully sides. Check dams are better when they are integrated with plantations and gully reshaping on both sides is relevant for reducing the bank failure and widening of gully.

Weak points in gully head cut areas could be protected from overland flow by implementing cut off drains. Improving proper drainage structure could also reduce the rate of gully expansion. Runoff generated along the road should be diverted to nearby natural waterways or make proper artificial waterways to minimize development of the gullies. Appropriate biological and physical soil and water conservation practices must be practiced on the cultivated land to rehabilitate the gullies and to protect new formation of gullies. Farmers, development agents, and decision makers could promote the integrated application of biological and physical gully control approach since it is better than other control methods. The community should participate from the planning to implementation and be part of the evaluation process to prevent and control gully formation and development effectively. The government should also develop land use and administration polices and strategies with the proper participation of landholders and/or farmers with the responsibility implementation. It is important for the sustainable land management practices and ecological balance of the environment.

Conclusions

The major causes of gully formation and development were local footpaths, surface runoff, free grazing, poor agricultural practices, absence of proper waterway, and soil type, respectively. Gully dimension measurements showed that gullies are expanding at an alarmingly fast rate. Even though physical and biological gully treatment practices are playing a vital role in trapping sediment, their integration was found to be more effective. Blocking of roads and flooding, loss of cattle by runoff flow, deposition of sediment on the grazing lands, and roads were also other social and economic impacts on the community as a result of gully erosion. Generally, the study plays a great role for the decision making of developing natural resources management policies and strategies. Specifically, it could be used for sustainable planning processes of gully erosion prevention and control for the community and the district agricultural offices.

Acknowledgements The research was supported by Merawi Technical Vocational Education and Training Center and Bahir Dar University, which is highly appreciated and acknowledged. The authors appreciate Inguti kebele Timet watershed communities for their kind response to any requested information.

References

Addisu S, Mekonnen M (2019) Check dams and storages beyond trapping sediment, carbon sequestration for climate change mitigation, Northwest Ethiopia. Geoenviron Disasters 6(4). https://doi.org/10.1186/s40677-019-0120-1

Aga AO, Chane B, Melesse AM (2018) Soil erosion modelling and risk assessment in data scarce Rift Valley Lake regions, Ethiopia. Water 10:1684. https://doi.org/10.3390/w10111684

Aga AO, Melesse AM, Chane B (2019) Estimating the sediment flux and budget for a data limited Rift Valley Lake in Ethiopia. Hydrology 6(1). https://doi.org/10.3390/hydrology6010001

Asrat WA, Carucci V, Desta L, Yitayew A (2005) Community-based participatory watershed development: a guideline part 2

Belay M, Bewket W (2012) Assessment of gully erosion and practices for its control in north-western highlands of Ethiopia. Int J Environ Stud 69(5):714–728. https://doi.org/10.1080/00207233.2012.702478

Danladi A, Ray HH (2014) An analysis of some soil properties along gully erosion sites under different land use areas of Gombe Metropolis, Gombe State, Nigeria. J Geogr Reg Plann 7(5): 86–96

Defersha MB, Melesse AM (2012a) Effect of rainfall intensity, slope and antecedent moisture content on sediment concentration and sediment enrichment ratio. CATENA 90:47–52

Defersha MB, Melesse AM (2012b) Field-scale investigation of the effect of land use on sediment yield and surface runoff using runoff plot data and models in the Mara River basin, Kenya. CATENA 89:54–64

Defersha MB, Quraishi S, Melesse AM (2010) Interrill erosion, runoff and sediment size distribution as affected by slope steepness and antecedent moisture content, Hydrol. Earth Syst Sci Discuss 7:6447–6489

Defersha MB, Melesse AM, McClain M (2012) Watershed scale application of WEPP and EROSION 3D models for assessment of potential sediment source areas and runoff flux in the Mara River Basin, Kenya. CATENA 95:63–72

Desta L, Adugna B (2012) A field guide on gully prevention and control. Nile Basin initiative Eastern Nile subsidiary action program (ENS AP)

Hughes AO, Prosser IP (2012) Gully erosion prediction across a large region: Murray-Darling Basin, Australia. Soil Res 50(4):267–277

Keesstra SD, Van Dam O, Verstraeten GV, Van Huissteden J (2009) Changing sediment dynamics due to natural reforestation in the Dragonja catchment, SW Slovenia. CATENA 78(1):60–71

Li Y, Poesen J, Yang JC, Fu B, Zhang J, H (2003) Evaluating gully erosion using 137 Cs and 210 Pb/ 137CS ratio in a reservoir catchment. Soil Tillage Res 69:107–115

Maalim FK, Melesse AM (2013) Modeling the impacts of subsurface drainage systems on runoff and sediment yield in the Le Sueur Watershed, Minnesota. Hydrol Sci J 58(3):1–17

Maalim FK, Melesse AM, Belmont P, Gran K (2013) Modeling the impact of land use changes on runoff and sediment yield in the Le Sueur Watershed, Minnesota using GeoWEPP, Catena 107, 35–45

Mansur A (2014) An assessment of gully erosion in Dutse Sahelian Zone of Jigawa state, Nigeria, and its adverse consequences on the socio-economic development of the state. J Agric Environ Sci 3(3):17–25

Mekonnen M (2020) Impacts of soil and water conservation practices after half of a generation age, northwest highlands of Ethiopia. Soil Tillage Res. https://doi.org/10.1016/j.still.2020.104755

Mekonnen M, Getahun M (2020) Soil conservation practices contribution in trapping sediment and soil organic carbon, Minizr watershed, northwest highlands of Ethiopia. https://doi.org/10.1007/s11368-020-02611-5

Mekonnen M, Melesse AM (2011) Soil erosion mapping and hotspot area identification using GIS and remote sensing in northwest Ethiopian highlands, near Lake Tana. In: Nile River Basin. Springer, pp 207–224

Mekonnen M, Keesstra SD, Baartman JE, Ritsema CJ, Melesse AM (2015) Evaluating sediment storage dams: structural off-site sediment trapping measures in northwest Ethiopia. Cuadernos de investigación geográfica 41(1):7–22

Mekonnen M, Keesstra SD, Baartman JE, Stroosnijder L, Maroulis J (2016) Reducing sediment connectivity through man-made and natural sediment sinks in the Minizr catchment, Northwest Ethiopia. Land Degrad Dev 28(2):708–717

Mekonnen M, Abeje T, Addisu S (2020) Integrated watershed management on soil quality, crop productivity and climate change adaptation, dry highland of Northeast Ethiopia. Agric Syst J. https://doi.org/10.1016/j.agsy.2020.102964

Melesse AM, Ahmad S, McClain M, Wang X, Lim H (2011) Sediment load prediction in large rivers: ANN approach. Agric Water Manag 98:855–866

Meshesha DT, Tsunekawa A, Haregeweyn N (2016) Determination of soil erodibility using fluid energy method and measurement of the eroded mass. Geoderma 284:13–21

MOARD (2005) Community Based Participatory Watershed Development guideline part 1. Ministry of Agriculture and Rural Development. Addis Ababa, Ethiopia

Moges A, Holden NM (2008) Estimating the rate and consequences of gully development, a case study of Umbulo catchment in southern Ethiopia. Land Degrad Dev 19(5):574–586

Mohammed H, Alamirew T, Assen M, Melesse AM (2015) Modeling of sediment yield in Maybar gauged watershed using SWAT, northeast Ethiopia. CATENA 127:191–205

Msagahaa J, Ndomba, Melesse AM (2014) Modeling sediment dynamics: effect of land use, topography and land management. In: Melesse AM, Abtew W, Setegn S (eds) Nile River Basin: ecohydrological challenges, climate change and hydropolitics, pp 165–192

Nyssen J, Poesen J, Moeyersons J, Luyten E, Veyret-Picot M, Deckers J, Govers G (2002) Impact of road building on gully erosion risk: a case study from the northern Ethiopian highlands. J Brit Geomorphol Res Group 27(12):1267–1283

Nyssen J, Poesen J, Veyret-Picot M, Moeyersons J, Haile M, Deckers J, Govers G (2006) Assessment of gully erosion rates through interviews and measurements: a case study from northern Ethiopia. Earth Surf Process Land 31(2):167–185

Pathak P, Wani SP, Sudi R (2006) Gully control in SAT watersheds. J SAT Agric Res 2(1):1–23

Poesen J, Nachtergaele J, Verstraeten G, Valentin C (2003) Gully erosion and environmental change: importance and research needs. CATENA 50(2–4):91–133

Rijkee P, Keesstra SD, Mekonnen M (2015) Low-land gully formation in the Amhara region, Ethiopia. In: Geophysical research abstracts 17:EGU2015-15827

Setegn SG, Srinivasan R, Dargahi B, Melesse AM (2009) Spatial delineation of soil erosion prone areas: application of SWAT and MCE approaches in the Lake Tana Basin, Ethiopia. Hydrol Process Spec Issue Nile Hydrol 23(26):3738–3750

Setegn SG, Bijan Dargahi B, Srinivasan R, Melesse AM (2010) Modelling of sediment yield from Anjeni Gauged Watershed, Ethiopia using SWAT. JAWRA 46(3):514–526

Tebebu TY, Abiy AZ, Zegeye AD, Dahlke HE, Easton ZM, Tilahun SA, Steenhuis TS (2010) Surface and subsurface flow effect on permanent gully formation and upland erosion near Lake Tana in the northern highlands of Ethiopia. Hydrol Earth Syst Sci 14(11):2207–2217

Tigist YT, Anteneh ZA, Assafa DZ, Tilashwork CA (2010) Assessment of hydrological controls on gully formation and upland erosion near Lake Tana, northern highlands of Ethiopia. Cornell University, 2nd Joint Federal Interagency Conference, Las Vegas

Valentin C (2004) Overland flow, erosion and associated sediment and biogeochemical transports. Vegetation, water, humans and the climate: a new perspective on an interactive system. Global Change Series–The IGBP Series. Springer Verlag, Berlin, pp 317–322

Valentin C, Poesen J, Li Y (2005) Gully erosion: impacts, factors and control. CATENA 63(2–3):132–153

Wang X, Garza J, Whitney M, Melesse AM, Yang W (2008) Prediction of sediment source areas within watersheds as affected by soil data resolution. In: Findley PN (ed) Environmental modelling: new research (Ch 7), Nova Science Publishers, Inc., Hauppauge, NY 11788, pp 151–185. ISBN: 978-1-60692-034-3

Wijdenes Oostwoud DJ, Bryan R (2001) Gully-head erosion processes on a semi-arid valley floor in Kenya: a case study into temporal variation and sediment budgeting. Earth Surf Process Land 26(9):911–933

Yazie T, Mekonnen M, Derebe A (2020) Gully erosion and its impacts on soil loss and crop yield in three decades, northwest Ethiopia. Model Earth Syst Environ. https://doi.org/10.1007/s40808-020-01018-y

Zegeye AD, Damtew S, Tilahun AS, Langendoen E, Dagnew D, Guzman C, Steenhuis TS (2014) Gully development processes in the Ethiopian highlands. In: Proceedings of the second International Conference on the advancements in science and technology (ICAST), Bahir Dar University, pp 220–229

Zegeye AD, Steenhuis TS, Mekuria W, Dagnaw DC, Addisse MB, Tilahun SA, Kasse TA (2017) Effect of gully headcut treatment on sediment load and gully expansion in the sub humid Ethiopian Highlands. Environ Ecol Res 5(2):138–144

Chapter 22
Restoring Lake Tana Through Reduction of Outflow and Compensation of the Power Gap with an Alternative Energy Source

Yirgalem Assegid and Assefa M. Melesse

Abstract Since the beginning of the Tana Beles hydropower generation, Lake Tana had a half meter drop from its annual average because of inflow and outflow changes. The outflow pattern has changed because of the Beles hydropower project which demanded consistent flow of water, releasing 7 times more than the dry season's historical flow. On the other hand, inflow to the lake has decreased 1.29 BCM (35% of historical flow) due to irrigation abstractions on the tributaries of Lake Tana. Irrigation developments on Gumera, Rib, Megeche and Gilgel-Abay tributaries consume to 2890 m^3/ha annually to meet irrigation demand over 100,000+ ha. Finally, Lake Tana had to support navigation, meet the ecological requirement of the Blue Nile River riparian, and supply to TIS ESAT Falls for tourism. The highest hydrological pressure on Lake Tana seems to have emanated from hydropower generation. Accordingly, this study reviews the possibility of scaling down Beles hydropower by decreasing outflow from the lake and supplementing the reduced power production from alternative energy sources. A country-level solar energy survey indicates that the Western escarpments of the rift valley, specifically from Afar triangle all the way to southern Wollo and North Shoa, as well as North Western part of the country, surrounding Lake Tana have irradiation and photovoltaic (PV) values of 6 kWh/m^2 and 5.2 kWh/kW-p, respectively. The contribution of solar power to the energy sector of Ethiopia is only 1%, despite its plans to tap the priciest concentrated solar power (CSP) that produces 15,000 GW from 1 km^2. This study suggests that photovoltaic power is cheaper, and it could compensate even the maximum energy planned from the Beles hydropower station (400 MW). Producing 350 MW from solar energy needs 200 ha, which can make use of degraded and irrecoverable lands. The study concludes that solar energy harvesting evens out against hydropower generation on initial investment, economic lifetime, maintenance costs and abundance of energy

Y. Assegid (✉)
South Florida Water Management District, West Palm Beach, FL, USA

A. M. Melesse
Department of Earth and Environment, Florida International University, Miami, FL, USA
e-mail: melessea@fiu.edu

A. M. Melesse et al. (eds.), *Nile and Grand Ethiopian Renaissance Dam*,
Springer Geography, https://doi.org/10.1007/978-3-030-76437-1_22

sources. In fact, because of the erratic behavior of rainfall, Beles hydropower could be unreliable. The return from tourism sector is 12 times more than the maintenance cost of the solar energy technologies. Finally, the study recommends making use of an existing tunnel to fill a series of dams for irrigation purposes, which are already planned on the project design. The filling can take place during the wet season in which Lake Tana has the peak outflow.

Keywords Lake Tana · Tana Beles HP project · Alternative energy resources · Solar energy · Photovoltaic

Introduction

Economic and Ecohydrological Services of Lake Tana

Lake Tana is the largest freshwater body in Ethiopia that had an estimated area of 3156 km^2 and an average depth of 9 m. It is situated in the North Western highlands at an altitude of 1786 m.a.s.l. (Alemayehu et al. 2010). The lake gets inflow from contributory rivers such as Gumera, Rib, Megeche, and Gilgel Abay whose total watershed area adds up to 16,000 km^2. The lake is also the source of the Blue Nile River with annual contribution of 3.5BCM (7% of the Blue Nile River flow) (Belete 2013).

Lake Tana maintains hydrological and ecological balances, supports economic activities, and hosts sociocultural heritages. The lake serves to maintain microclimatic conditions of the surrounding locality through forcing of convective rainfall system (Haile et al. 2008). The hourly maximum rainfall happens to be within 20 km radius from the lake, and it is attributed to the convective rainfall pattern forced by the lake (Haile et al. 2008). The microclimatic stability enhances the agricultural activity which is the livelihood of people living around the lake. Similarly, Lake Tana maintains the wetland system surrounding the lake which in turn serves as a genetic pool of indigenous flora and fauna (Alemayehu et al. 2010).

Lake Tana contributes an estimated $200 million to the macroeconomy through hydropower generation of 560 MW (11 MW at Abay I, 73 MW at Abay II, and 460 MW at Beles I). Locally, it contributes $5.1 million through tourism at Bahir Dar city only (Woldu 2018); it contributes $15 million from fishery (18,000 ton of fish) (Amare et al. 2018), and supports 100,000 ha for irrigation (Worqlul et al. 2015) of which 54,000 ha is under implementation process (SMEC 2000; Alemayehu et al. 2010).

Water Demands from Lake Tana Basin for Hydropower and Irrigation

Out of the 100,000 ha irrigable potential, 50% is being under planning and implementation (Worqlul et al 2015). Current reservoir construction on the tributaries Gumera, Rib, Megeche, and Gilgel Abay and withdrawal from the lake for irrigation purposes are shown in Table 22.1. The design estimate shows demand for irrigation is 2890 m^3/ha (Worqlul et al. 2015) totaling 0.78 BCM, matching the demand estimate by Belete (2013). The four storage dams have an active storage volume of 1.2 BCM. In other words, 1.2 BCM will be reduced from inflow to the lake to satisfy irrigation demand and evaporation losses from the dams. Comparing against the total inflow of 3.5 BCM (Chebud and Melesse 2009), the irrigation demand cuts 35% of the inflow.

Beles I hydropower and Abay I and II hydropower demand an outflow of 2.9 BCM/yr and 2.4 BCM, respectively, with a constant flow (supply). Last but not least, the ecological requirement demands a constant flow of 17 m^3/s (Gebre et al. 2008) totaling 0.52 BCM. Currently, the Abay I and II hydropower is suspended (Mulat et al. 2018) due to supply and demand imbalance. Belete (2013) indicated the monthly demand still shows imbalances (higher demand than supply) for the months January to May. The same analysis shows the demand could only be met without affecting lake level if the water regulating weir at its natural outlet is raised to 1.5 m (Belete 2013). In other words, the dry season supply should be withdrawn increasing the storage (raising lake level by 1.5 high). This however causes flooding on Bahir Dar city and other rural areas (Belete 2013).

The mean annual lake level shows a negative trend and dropping by half a meter with the current demand (Worqlul et al. 2015; Alemayehu et al. 2010). Consequently, the lake recedes by 30 km^2 and its bed turn into farmland

Table 22.1 Costs of PV power plant production

Item	Cost (USD)	
PV panel cost	182,000,000	
Inverter	1,000,000	Every 5 years
Feasibility study	150,000	
Development cost	100,000	
Engineering cost	200,000	
Power plant cost	1,149,875	If existing hydropower substation is not used
Staff Training	600,000	
Miscellaneous	500,000	
Maintenance	350,000	Annual
Total initial cost	**186,049,875**	
Revenue	58,223,340	Annual

(Alemayehu et al. 2010). Lake Tana has been to its lowest level in 2003 (below 1785 m amsl) merely because of drought (Chebud and Melesse 2009), and water transits have been halted for months (Alemayehu et al. 2010). In short, Lake Tana is under pressure due to increased demand for hydropower generation and irrigation. To tackle vulnerability of the lake assessing alternative energy sources that fills the gap from reduction of hydropower generation is crucial while maintaining the irrigation practices.

Alternative Source of Energy and Availability for Ethiopia

Studies have shown that Ethiopia can generate up to 13,000 GW of hydropower from Eastern Blue Nile basin (Mulat et al. 2018). Accordingly, several projects have been operational, and some others are in the process of study and design. While this is commendable, increased pressure on key resource and causing ecological disruption cannot be traded off for any monetary value. In such circumstances, it is imperative to understand the risks and looking for alternative sources of energy.

Globally, about 9000 PV stations are built to produce 40 GW. In Africa, Libya has built 600 MW PV station to preserve oil and fossil fuel (Rehman et al. 2007). Saudi Arabia shared the same intentions of preserving its natural resources and conducted pilot-level grid-based solar power production and performance test (Rehman et al. 2007). The contribution of solar energy in Ethiopia is off-grid and limited to 1% (EEPCO 2015). Given the increase in experience, tapping solar power to reduce the pressure on Lake Tana is key. This is in line with the policy to grow the contribution at least to 15%. Apparently, some experience exists with NGOs and the private sector. For the rural areas, experimenting solar energy use for irrigation and other farming practices brings dynamism in the energy sector.

A country-level solar energy survey (World bank 2017) indicates that the Western escarpments of the rift valley, specifically from Afar triangle all the way to southern Wollo and North Shoa, as well as North Western part of the country, surrounding Lake Tana have irradiation and photovoltaic values of 6 kWh/m^2 and 5.2 kWh/kW-p, respectively (Figs. 22.1 and 22.2). For this study, photovoltaic technologies (PV) is chosen over concentrated solar power (CSP) for the reason that PV power plant is cheaper.

Theoretically, hydropower takes half the initial investment, 2 times higher economic lifetime, and 75% less maintenance cost compared to the solar power technologies (Timilsina 2011). Practically, because of the high sedimentation rate and erratic behavior of rainfall, the analysis could not hold at least for the case of Ethiopia. The initial investment per kWh of the Grand Ethiopian Renaissance Dam (GERD) with full production mode of 6000 MW is $833/kW which is double from the assumed cost and adds up 100$/kW more than the initial investment cost of the solar energy technology. If GERD produces 1800 MW, its initial cost is ($1400/kWh) which is twice more than the power cost (790$/kWh). In the case of

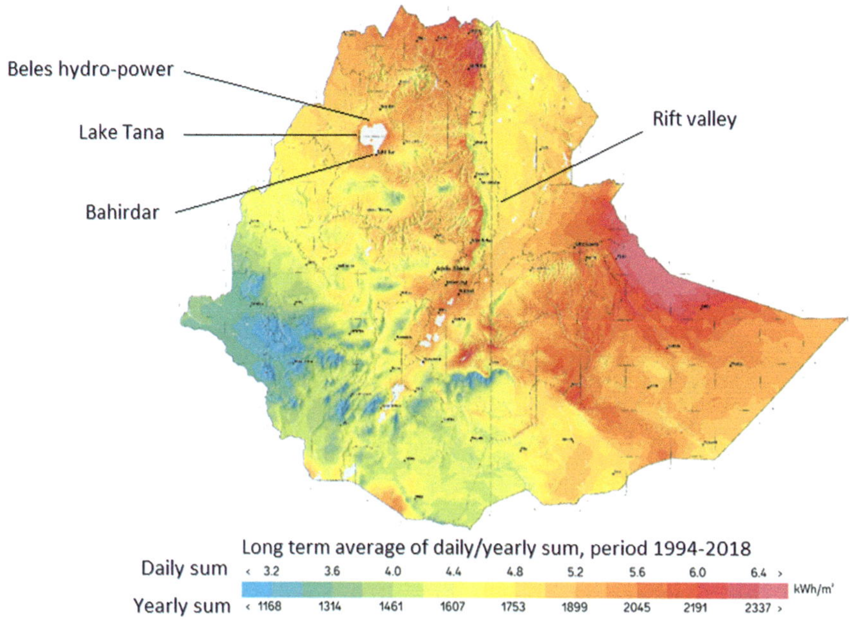

Fig. 22.1 Long-term average irradiance values (1994–2018)

economic lifetime, average soil loss from the highlands of Ethiopia ranges from 34 tons ha^{-1} y^{-1} (Zimale et al. 2016) and the estimate goes up to 526 tons ha^{-1} y^{-1} in gully erosion areas. With this sediment transport rate that was reported since 1974 (40+ years), and if only 10% of the catchment contributes to such rate of soil loss, the economic lifetime is barely 25 years which evens out with the PV technologies.

Methods

In this study, RETScreen (Government of Canada 2020) is used to analyze the solar energy production and operational costs, the type of product used to minimize cost, cash flows, financial viability, and risks. The tool is used globally and tested based on the performance tests of built solar powers. The model performs with the global database of photovoltaic products and hence gives access to get information on the attribute of each product. The first module of the model is the benchmark analysis which uses virtual energy analyzer based on location and planned type of facility (Fig. 22.3). A second module of the model is the feasibility analysis that includes the smart project identifier and analyzes the energy demand, cost, alternative energy sources, and associated financial viability and risks.

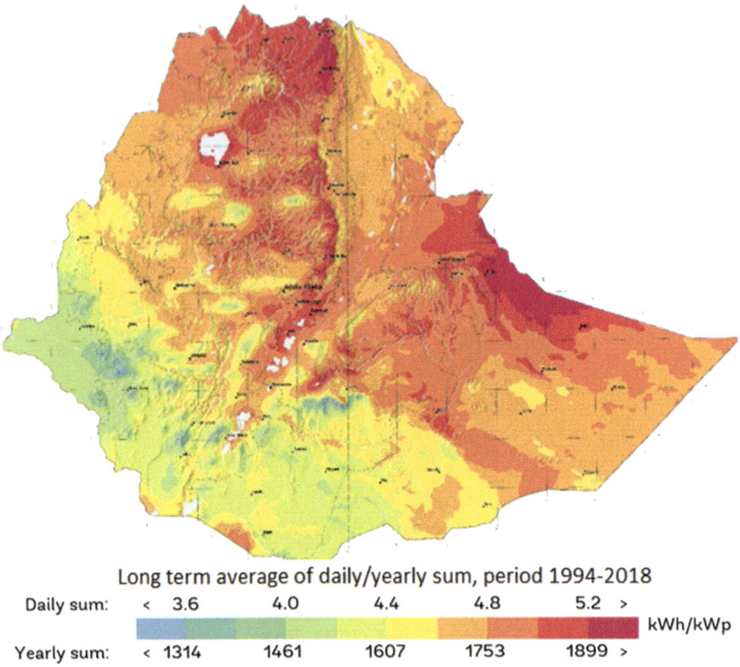

Fig. 22.2 Long-term average photovoltaic values (1994–2018)

Study Area

In this study, Bahir Dar is chosen for to do a benchmark survey for the availability of climatic data. The solar irradiance and photovoltaic at this station is considered to be similar to Beles Hydropower Station. Accordingly, a station at 11.6 N and 37.4 E named Bahir Dar Zuriya is used (Fig. 22.4). The facility is at 18,811 m.a.s.l and categorized as hot–humid climatic zone. The temporal solar irradiation shows a range of 5.5–6.4 kWh/m^2/d (Fig. 22.5) which is just 1kwh/m^2/d less compared to the maximum at Saudi Arabia.

Analysis

According to Mulat et al. (2018), the 84 MW Abay I and II HEP generation was already abandoned in a trade-off with Beles I. In this analysis, Abay I and II power plants are assumed to restart the production getting extra 84 MW from the current plan of production as it has a dual advantage of satisfying the ecological requirement flows and sustain tourism. Beles I is suggested to be reduced to 110 MW,

Fig. 22.3 Modules for the solar energy prefeasibility, feasibility, and performance analysis

Fig. 22.4 Location of Bahir Dar station on the map of Ethiopia

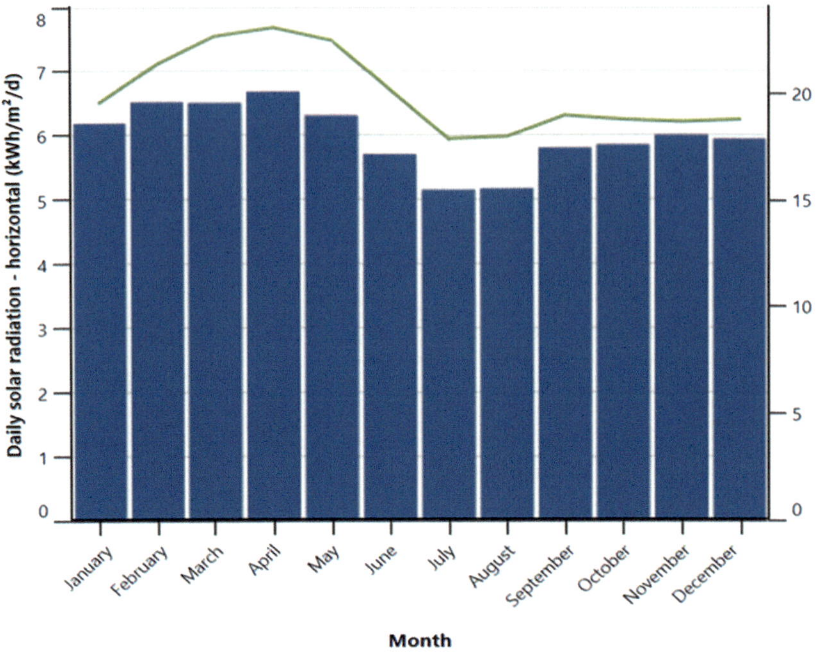

Fig. 22.5 Monthly solar irradiation (line graph) and air temperature (bar chart) Bahir Dar station

whereas the remaining 350 MW could be the gap that should be compensated based on the current plan.

The selection of specific PV products was made based on the cost, availability, maintenance costs, capacity of power per unit, and technical easiness. Accordingly, "mono-Si–LR6-60PH—310 W" manufactured by LONGi-Solar is selected. As a silicon-based material, it is the cheapest ($\sim$161/unit) and the cost gets 10% reduction at decadal scale. To produce 350 MW with 18.96% efficiency (vertical to horizontal irradiation conversion), it requires 1,129,033 units of panels. The area of each panel is 1.635 m^2, and it requires 200 ha (184 ha for panels and 16 ha for access and other accessories).

About 97% of the investment cost goes to the PV technologies (Table 22.1). The future of tapping solar energy is inevitable as cost of the PV technologies gets cheaper. The cost of the technology has reduced from $64/W in 1976 to $1.4/W in 2010 (Timilsina et al. 2012). The cost of PV for this study (mono-Si–LR6-60PH—310 W) is $0.5/W.

The current cost of production of PV power is competitive compared to other power production systems (Fig. 22.6). The low cost of production clearly gives an increased margin for financial viability. For this project, although production cost of PV could go as small as $0.025 kW/h, the analysis was made at a cost $0.1 kW/h (4 times the minimum).

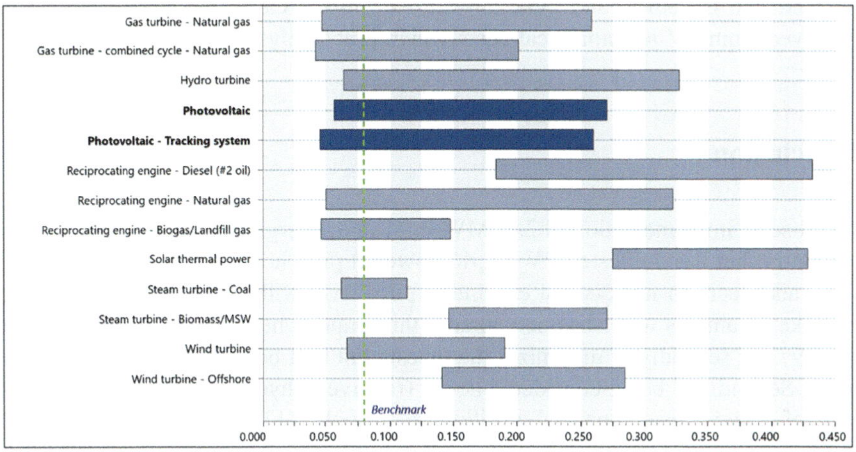

Fig. 22.6 Energy production cost—central grid range ($kW/h)

A more conservative approach is followed in this project, and a debt period of 10 years unlike the global experience of 15 years is used. Also, the rate of escalation of the power is set at 2% compared to the 4%. The project lifetime is set at 25 years as is in all parts of the world. The results show that the internal rate of return (IRR) 0.75 is greater than the reduced rate of return (0.09). The simple payback 3.2 years and equity pay back of 1.2 years show a positive cash flow could be gained in a very short period. Last, the average inflation rate of 2% is used, and it turns a net present value (NPV) cumulative cash flow of 1.8 billion USD. If the inflation rate is 5%, the NPV cumulative cash flow would be 3 billion USD. For a 10% inflation rate (Ethiopian context), the NPV cumulative cash flow reaches 6 billion USD. Finally, the benefit–const ratio of 10.1 seals the feasibility of the PV project.

The risk associated with financial viability is conducted with sensitivity analysis of 25% in all input values of the itemized costs shown in Table 22.2, interest rate,

Table 22.2 Sensitivity and risk analysis: Sensitivity range at 25%

Parameter	Value	Range(-/+)	Minimum	Maximum
Initial costs	186,049,75 USD	25%	139,537,406 USD	232,562,344 USD
Operation and maintenance	350,000 USD	25%	262,500 USD	437,500 USD
Electricity exported to grid	582,233.4 MWH	25%	436,675 MWH	727,791.75 MWH
Electricity export rate	100 USD/MWH	25%	75 USD/MWH	125 USD/MWH
Debt ratio	70%	25%	2.50%	87.50%
Debt interest rate	7%	25%	25%	8.75%
Debt term	15 yr	25%	11 Yr	19

reduced rate of return, and inflation rates. In all the cases, the maximum risk is observed from PV technological costs which practically is declining since 1976.

Conclusion

A full-scale implementation of 350 MW from solar energy may seem unrealistic for a country that has not tested PV in grid power at pilot level. A flexibility would be vital, and decision makers can execute at pilot level with the vision to expand till the lake regains its water balance. Firstly, this strategy helps to test performance of the PV, and secondly, it minimizes the investment cost benefiting from the existing trend. Second, PV being considered newer (relative to hydropower), there would be lack of interest from financial institutions at such a bigger scale. Third, setting institutional arrangement and skilled labor needs time, and hence, a pilot scale could serve as experimentation time to develop the institutional framework.

In summary, Lake Tana has for long sustained civilizations and livelihoods of millions. So, the cost to scale back Beles I hydropower by replacing solar PV to fill the power gap must be viewed from such perspective. Comparing internal rate of return or the discount rate against any projects is not a level field given the importance of lake, while the PV shows viability in all economic parameters.

References

Alemayehu T, McCartney M, Kebede S (2010) Modelling to evaluate the water resources implications of the planned infrastructure development in the Lake Tana sub-basin, Ethiopia. Ecohydrol Hydrobiol 10(s2–4):211–221. https://doi.org/10.2478/v10104-011-0023-6

Amare D, Endalew M, Debas T, Demissew A, Temesgen K, et al. (2018) Fishing Condition and Fishers Income: The case of Lake Tana, Ethiopia. Int J Aquac Fish Sci 4(1):006–009. https://doi.org/10.17352/2455-8400.000035

Belete MA (2013) Modeling and analysis of Lake Tana sub-basin water resources system, PhD thesis. Agricultural and Environmental Sciences Faculty of the University of Rostockm Germany

Chebud YA, Melesse AM (2009) Modelling lake stage and water balance of Lake Tana, Ethiopia. Hydrol Process 23:1534–1544

EEPCO (2015) Ethiopian Electric Power Energy sector–investment opportunities, UK-ETHIOPIA Trade and Investment Forum 2015, London, UK. https://www.developingmarkets.com/sites/default/files/AA%20Eth%20Energy%20Sector%20Presentation%20London.pdf

Gebre A, Getachew D, McCartney M (2008) Stakeholder analysis of the Chara weir, Lake Tana, a study report submitted to The International Water Management Institute (IWMI). Addis Ababa, Ethiopia

Government of Canada (2020) RETScreen: https://www.nrcan.gc.ca/maps-tools-publications/tools/data-analysis-software-modelling/retscreen/7465

Haile AT, Rientjes T, Gieske A, Gebremichael M (2008) Rainfall variability over mountainous and adjacent lake areas: the case of lake Tana at the source of Blue Nile River. J Appl Meteorol Climatol 48:1696–1708

Mulat AG, Moges SA, Moges MA (2018) Evaluation of multi storage hydro power development in the Blue Nile basin, regional perspective. J Hydrol Reg Stud 16(2018):1–14
Rehman S, Bader MA, Al-Moallem SA (2007) Cost of solar energy generated using PV panels. Renew Sustain Energy Rev 11:1843–1857
SMEC (2000) Hydrological study of the Tana-Beles sub-basins. Main report
Timilsina GR, Kurdgelashvilib L, Narbelc PA (2012) Solar energy: markets, economics and policies. Renew Sustain Energy Rev 16:449–465
Woldu MG (2018) Community based tourism in Lake Tana growth corridor of the Amhara region of Ethiopia: the missing link among stakeholders and implications to tourism industry. Cogent Soc Sci 4(1):1450335
World Bank (2017) Global Soar Atlas, https://www.worldbank.org/en/topic/energy/publication/solar-photovoltaic-power-potential-by-country
Worqlul A, Collick AS, Rossiter DG, Langan S, Steenhius TS (2015) Assessment of surface water irrigation potential in the highlands of Ethiopia, the Lake Tana Basin. CATENA 129:76–85
Zimale FA, Mogesm MA, Alemu ML, Ayana, Demissie SS, Tilahun SA, Steenhuis TS (2016) Calculating the sediment budget of a tropical lake in the Blue Nile basin: Lake Tana. J Soil. https://doi.org/10.5194/soil-2015-84

Chapter 23
Blue Nile River Course Morphology Change Response to Watershed Treatments: Lake Tana to Grand Ethiopian Renaissance Dam

Mebruk Mohammed, Belete Berhanu, and Agizew Nigussie

Abstract Within the Blue Nile River (BNR) system, large amounts of riverine deposits and influx of hill slope materials provide the sediment for the channel to transport. The amount of material that the river system can transport is directly related to the discharge and slope, which in turn are driven by watershed management, climate and river cross section lithology. In this study, the channel morphology change is particularly described as the change in elevation of the riverbed level. The possible variations in sediment supply under different scenarios of watershed management practices and the corresponding changes in the riverbed level were computed. Three different watershed management (agronomic practices, mechanical measures and biological measures) scenarios were tested for their possible effect on the BNR morphology. The analysis for each of the scenarios effect on the riverbed level change was obtained from a calibrated finite element model for sediment and diffusive wave transport models. The current sediment supply and river sediment transport capacity were first evaluated. This was achieved by first calibrating the model with observed average monthly flows and sediment concentration data of two stations located at Kessie Bridge and Ethiopia-Sudan border. The performance of the model was measured by NSE, PBIAS and RSR values for which the model has resulted in acceptable performance level. The morphologic characters of the river (riverbed elevation, cross section and slope) used in the river hydraulics and sediment transport analysis were obtained from 30 m digital elevation model. The results indicated a decrease in the average riverbed level during flood rising time (June to July) and an increase during flood recession time (August to September). The decrease in the riverbed level was due to

M. Mohammed (✉) · B. Berhanu · A. Nigussie
School of Civil and Environmental Engineering, Addis Ababa Institute of Technology, Addis Ababa University, Addis Ababa, Ethiopia
e-mail: mebruk.mohammed@aait.edu.et

B. Berhanu
e-mail: belete.berhanu@aait.edu.et

A. Nigussie
e-mail: agizew.nigussie@aait.edu.et

A. M. Melesse et al. (eds.), *Nile and Grand Ethiopian Renaissance Dam*, Springer Geography, https://doi.org/10.1007/978-3-030-76437-1_23

the reduction in the sediment concentration which gives the flow a higher erosive power than the current situation. The scoured bed in the flood rising period has created a larger cross section for the flow in the recession time, which resulted in a decrease in the river competence that ultimately creates deposition of sediments. The result has shown a relatively high change in the riverbed level between 300 and 550 km downstream of Lake Tana. In between these two distances, from Lake Tana along BNR, Muger, Guder and Fincha Rivers join the main BNR course. However, the effect is becoming smaller while moving downstream. The river morphology around the location of the Grand Ethiopian Renaissance Dam is also significantly affected by the watershed treatments; this is mainly due to the treatments made on the Dabus River catchments, rather than the other main rivers upstream. Among the scenarios developed for improving the watershed, scenario 3 (biological measures) has resulted in the highest riverbed level change in both flood peaking and recession periods. Thus, this scenario is expected to bring about lesser sediment for the river to transport and the greatest change in the river morphology when compared to the other scenarios.

Keywords Blue Nile · Grand Ethiopian Renaissance Dam · River morphology · Watershed treatment

Introduction

Many factors influence the morphology of a river system. Long time scale processes (e.g., tectonic changes) through short-term drivers (watershed management practices, climatic changes) ultimately determine first-order channel characteristics such as slope, discharge and capacity. The amount of material that the river system can transport is directly related to the discharge and slope, which are in turn driven by watershed management, climate and bedrock lithology. Among these drivers behind the morphology, the one that is caused by change in watershed management practice of a fluvial system is explored in this paper. Due to the construction of the Grand Ethiopian Renaissance Dam (GERD), Ethiopia has embarked on restoring the catchment condition of the Blue Nile River. Such intense and wide catchment restoration work will have a direct impact on the river's morphological character. The study will show watershed management practices impact on the morphology of the Blue Nile River (BNR) upstream of GERD site.

The BNR is part of the larger Nile River watershed and is situated along the North Western flank of the Mountains of Ethiopia and flows northeast out of Lake Tana (Fig. 23.1). The Mountains are a complex assemblage of 56–250 m thick volcanic and basement rocks. The volcanic rocks exist for the vast portion of the upstream part, while basement rocks constitute downstream part of the river (MacAlister et al. 2012). Downstream of Lake Tana, the BNR flows through a mixture of bedrock and highland outwash deposits. Ahmed and Ismail (2008) specified that most of the sediment in the Nile flows from the Ethiopian Highlands

Fig. 23.1 BNR basin drainage systems

through the Blue Nile and Atbara Rivers 95% of which is during the flood season (July–Sept.).

Methodology

Most of the sediment transport models used in river engineering are one dimensional, especially those used for long-term simulation of a long river reach (Yang et al. 2015). A longitudinal profile shown below (Fig. 23.2) is developed from 30 m digital elevation model (DEM) for the BNR. The long profile is created by taking about 597 straight reaches along the BNR. The longitudinal profile of a stream shows its slope or gradient. Since a river channel or river system is generally steepest in its upper regions, most river profiles are concave upward. Shulits (1941) provided an equation for the concave horizontal profile in terms of distance along the stream (Eq. 23.1)

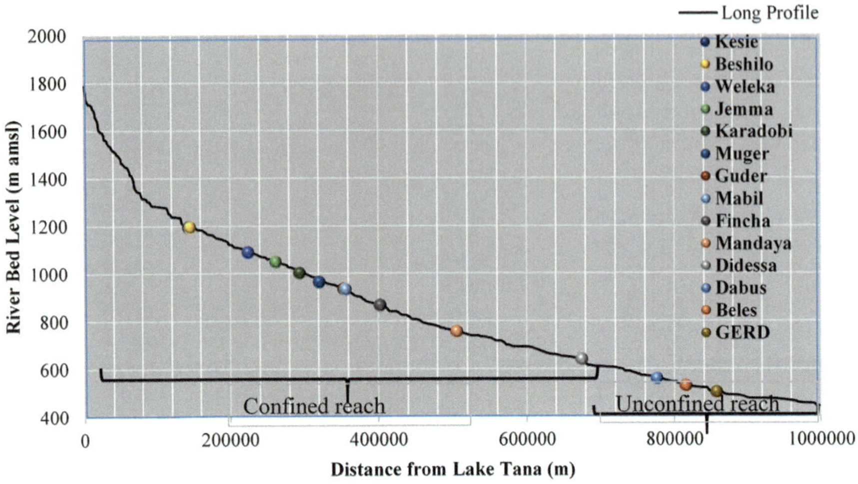

Fig. 23.2 Long profile of BNR with locations of the major tributaries

$$S_x = S_0 e^{-ax} \tag{23.1}$$

where S_x is the slope at any station a distance x downstream of reference station in feet, S_0 is the slope at a reference station, and a is a coefficient of slope change.

According to the equation suggested by Shulits (1941), the value of a and S_0 that would best fit the collected slope data is -2×10^{-6} and 0.0022, respectively. The coefficient of determination, R^2, obtained between the predicted and existing river slope is 0.2354 which is far from perfect. The general curve is disturbed by the presence of bends. Hence, the longitudinal slope of the thalweg will never completely follow a simple functional relationship; however, such poor relationship between the existing riverbed profile and the ideal case suggests that the riverbed slope is still active, and hence, there is high possibility of riverbed profile change to occur in future.

River Cross Section

Cross section across the river is taken along a line perpendicular to the direction of flow. The cross section length is extended till the left and right banks are 20 m above the corresponding minimum riverbed level. The area of cross section (m^2), wetted perimeter (m) and top width (m) for any given water depth (between 0 and 20 m at 0.25 m interval) was calculated. Thus, for a given roughness coefficient, a rating curve was drawn for the 597 sections.

Between Lake Tana and the confluence with the Didessa River, the BNR system flows through reaches of volcanic rocks (MacAlister et al. 2012). Where the river has incised through the thin soil deposits into the underlying bedrock, its slope has attuned such that through the bedrock reaches the river has enough stream power to erode, while some 15 km downstream of the Didessa River confluence point its slope decreases and the system is depositing sediment. These bedrock and alluvial reaches are denoted here as "confined" and "unconfined" channel reaches in Fig. 23.2, respectively. Downstream of these reaches, the channel width increases, slope decreases, and hence velocity decreases leading to deposition of sediment and formation of bars and small islands.

A conceptual model for identifying river morphology change due to possible watershed management scenario is presented, within the context of the BNR system, as shown in Fig. 23.3. Among the main drivers of this model (tectonics, watershed management and lithology), watershed management practice is considered here for its impact on the BNR morphology change as the other two are not expected to change in the coming few years.

In this study, the channel morphology change is particularly described as the change in elevation of the riverbed level. The focus is on predicting the vulnerability of the river's morphology due to any intervention made on the watershed through understanding of the links between intermediate processes and channel morphology.

Watershed Management Scenarios

Watershed management scenarios were developed based on land capability classes and their respective limiting factors as recommended by Tulu (2011) and Escobedo (1998). Three conservation practices (agronomic, mechanical and biological) were

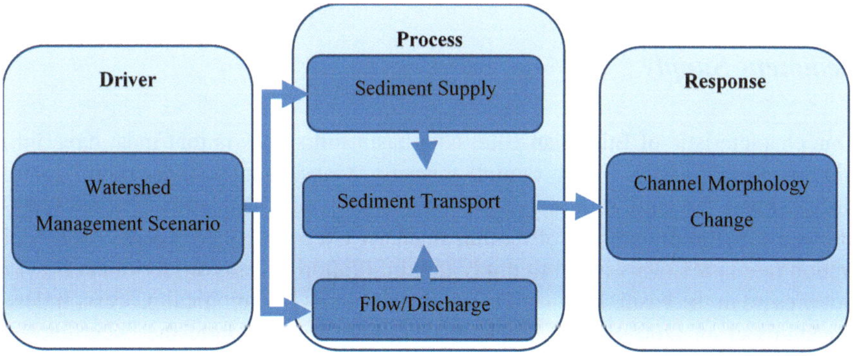

Fig. 23.3 Conceptual model outlining the driver, processes and outcomes behind river morphology in a sediment-dominated system such as the BNR

developed as scenarios for each land capability class unit (LCU). Land capability classes are units of land that relatively have the same degree of limitation. Land capability classes are usually grouped into two: land suitable for cultivation (capability class signified as I, II, III, IV) and land not suitable for cultivation but suitable for permanent vegetation like pastures, orchard and forests (V—wet land suitable for temporary grazing or rice cultivation; VI—land suitable for perennial crops or grazing; VII—land suitable for forest; VIII—land not suitable for agriculture). The degree of the risk to soil erosion increases from I to IV and VI to VIII.

The capability classes are further classified into sub-classes based on erosion-associated factors like slope (L), soil depth (D), past erosion (E), water logging (W), infiltration (I), texture (T) and stoniness or rockiness (S). The symbols L, D, E, W, I, T or S are added to the land capability classes to indicate the specific erosion-associated factor. The land capability classification of Ethiopia is presented in Table 23.1.

For land capability classification of Blue Nile Basin, ArcGIS and TEFASES algorithm after Berhanu and Aray (2005) was adopted. TEFASES algorithm was used to compare the cell values of the land feature layers to get the land capability classification of the basin (Fig. 23.4).

Agronomic practices scenario is related to management options of crop cultivation which include strip cropping, contour cropping, change of crop species, etc. Mechanical measure scenario refers to watershed treatments including engineering-based physical conservation measures like soil/stone bund, terracing, cutoff drain, waterways, etc. Biological measure scenario refers to watershed management options related to treatment with vegetation cover by grassland improvement, tree planting, etc. Scenarios for a land capability unit were also developed from one or two of the three categories. For instance, three scenarios were developed from only biological and mechanical measures when land capability units are VI or greater as these land capability units are not suitable for agricultural activities and in turn unable to recommend agronomic practices. The proposed scenarios are shown in Table 23.2.

Sediment Supply

One characteristic of Ethiopian Blue Nile basin hill slopes is that most have infiltration rates in excess of the rainfall intensity. Consequently, most runoff is produced when the soil saturates or from shallow, degraded soils. Engda (2009) showed that the probability of rainfall intensity exceeding the measured soil infiltration rate is 8%. This is not to imply that infiltration excess, or Hortoman flow, is not present in the basin, but that it is not the dominant hydrological process. Indeed, Steenhuis et al. (2009) and Collick et al. (2009) not only note the occurrence of infiltration excess runoff but also state that it is predominantly found in areas with exposed bedrock or in extremely shallow and degraded soils.

Table 23.1 Land capability classification of Ethiopia with limiting factors

No	LCC	Description	Limiting factor
1	I	L1, D1, E0, W0, I0, T3-5, S0	Nil
2	IID	L1, D2, E0, W0, I0, T3-6, S0-1	Soil depth (100–150 cm)
3	IIT	L1, D1, E0, W0, I0, T6, S0-1	Soil texture (silty clay loam)
4	IIS	L1, D1, E0, W0, I0, T3-5, S1	Stoniness (15–30%)
5	IIL	L2, D1-2, E0, W0, I0, T3-6, S0-1	Slope (3–8%)
6	IIIE	L2, D1-2, E1, W0-1, I0-1, T3-7, S0-2	Past erosion (slight erosion)
7	IIIW	L2, D1-2, E0, W1, I0-1, T3-7, S0-2	Water logging (intermittently water logged)
8	IIII	L2, D1-2, E0, W0, I1, T3-7, S0-2	Infiltration (moderate)
9	IIIT	L2, D1-2, E0, W0, I0, T7, S0-2	Soil texture (heavy clay)
10	IIIs	L2, D1-2, E0, W0, I0, T7, S2	Stoniness (30–50%)
11	IIIL	L3, D1-2, E0-1, W0-1, I0-1, T3-7, S0-2	Slope (8–15%)
12	IVD	L3, D3, E0-2, W0-2, I0-2, T2-7, S0-2	Soil depth (50–100 cm)
13	IVE	L3, D1-2, E2, W0-2, I0-2, T2-7, S0-2	Past erosion (moderate)
14	IVW	L3, D1-2, E0-2, W2, I0-2, T2-7, S0-2	Water logging (regular waterlogged)
15	IVI	L3, D1-2, E0-1, W0-1, I2, T2-7, S0-2	Infiltration (poor)
16	IVT	L3, D1-2, E0-1, W0-1, I0-1, T2, S0-2	Soil texture (sandy loam)
17	IVL	L4, D1-3, E0-2, W0-2, I0-2, T2-7, S0-2	Slope (15–30%)
18	VID	L1-4, D4, E0-2, W0-2, I0-2, T2-7, S0-3	Soil depth (25–50 cm)
19	VIS	L1-4, D1-3, E0-2, W0-2, I0-2, T2-7, S3	Stoniness (50–90%)
20	VIL	L5, D1-4, E0-2, W0-2, I0-2, T2-7, S0-3	Slope (30–50%)
21	VIIE	L5, D1-4, E3, W0-2, I0-2, T2-7, S0-3	Past erosion (severe)
22	VIIL	L6, D1-4, E0-3, W0-2, I0-2, T2-7, S0-3	Slope (>50%)
23	VIIID	L1-6, D5, E0-4, W0-2, I0-2, T1-7, S0-4	Soil depth (<25 cm)
24	VIIIE	L1-6, D1-4, E4, W0-2, I0-2, T1-7, S0-4	Past erosion (very severe)
25	VIIID	L1-6, D1-4, E0-3, W0-2, I0-2, T1, S0-4	Soil texture (sand)
26	VIIID	L1-6, D1-4, E0-3, W0-2, I0-2, T2-7, S4	Stoniness (>90%)
27	VW	L1-6, D1-5, E0-4, W3, I0-2, T1-7, S0-4	Water logging (swamp)

Source Tana-Beles WME (2012)

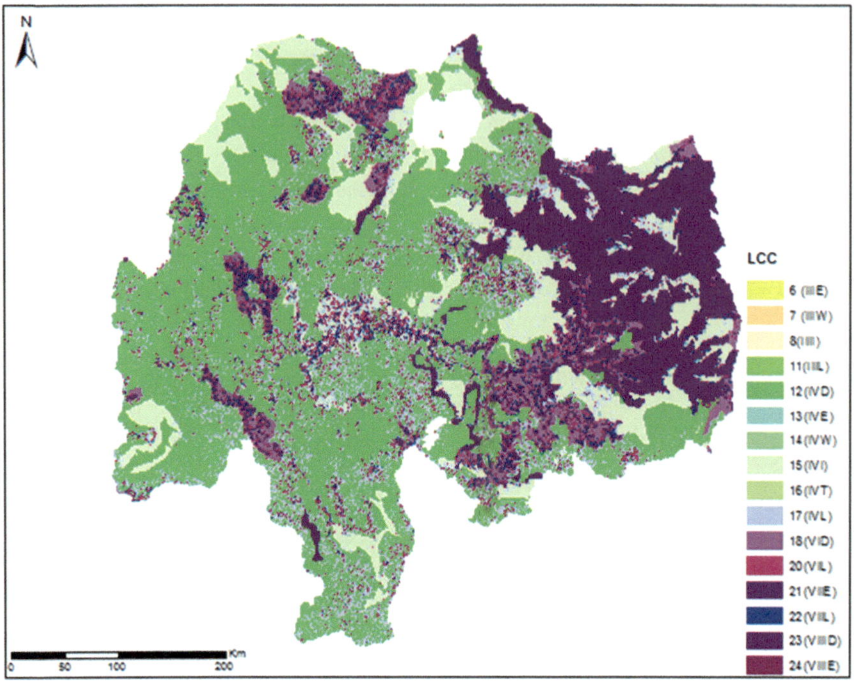

Fig. 23.4 Land capability classification of Blue Nile Basin

The sediment supply to BNR is computed on the basis of the soil loss obtained across its basins for each land capability classes. Revised Universal Soil Loss Equation (RUSLE) is used to quantify this loss for two scenarios, viz. with and without watershed management practices. The RUSLE equation is defined by Eq. 23.2.

$$A = R \times K \times L \times S \times C \times P \tag{23.2}$$

where:

- A = Average annual soil loss per unit area (tons/ha/year).
- R = Rainfall–runoff erosivity factor—the rainfall erosion index. The R-factor is calculated as total storm energy (E) times the maximum 30 min intensity (I_{30}), or EI, and is expressed as the rainfall erosion index. Alternatively, the R-factor of a given area can be computed using a monthly rainfall data as $R = \sum_{i=1}^{12} \frac{P_i^2}{P}$ where R = rainfall erosivity factor, p_i = total amount of precipitation in ith month of the year and P = total amount of precipitation during the year (mm).
- K = Soil erodibility factor: the soil loss rate per erosion index unit for a specified soil.

Table 23.2 Proposed watershed management scenarios

LCU	Scenario 1 (agronomic)	Scenario 2 (mechanical)	Scenario 3 (biological)
IIIE	Contour cropping	Cultivation + cutoff drain + waterways	Controlled grazing
IIIW	Alley cropping	Bed and furrow + cultivation	Convert to water logging resistance species
IIII	Convert to deep-rooted species	Deep plowing	Controlled grazing
IIIL	Alley cropping	Fanya Juu + cutoff drain + swaterways	Cultivation with grass strip
IVD	Convert to shallow root crop + strip cropping	Convert to shallow root crop + cutoff drain + waterways	Controlled grazing
IVE	Cultivation + hedgerow with fruit trees	Fanya Juu + cutoff drain + waterways	Controlled grazing
IVW	Convert to water logging resistance species	Bed and furrow + waterways	Fodder trees
IVI	Convert to deep-rooted species	Deep plowing + grass strip	Controlled grazing + grass development
IVT	Selective cropping	Fanya Juu + grass strip	Controlled grazing + grass development
IVL	Cultivation + hedgerow with fruit trees	Stone/soil bund	Grass production + cutoff drain
VID	Perennial crops	Fruit + contour bund	Apiculture development
VIL	Perennial crops	Fruit + trench	Apiculture development
VIIE	Area closure	Gully control + trench	Tree plantation + trench
VIIL	Area closure	Hillside terrace	Grassland development + cut and carry
VIIID	Area closure	Forest + trench	Forest development
VIIIE	Area closure	Forest + gully control	Forest development

- L = Slope length factor: the ratio of soil loss from the field slope length to soil loss from a 22.1 m length under identical conditions.
- S = Slope steepness factor: the ratio of soil loss from the field slope gradient to soil loss from a 9% slope under otherwise identical conditions.
- The L and S factors are commonly computed as topographic LS factor. In GIS environment, unit stream power erosion and deposition model are recommended

for *LS* calculations (Pelton et al. 2014). The major steps to compute the *LS* factor in the GIS environment involve:

- Calculate flow direction from clipped Watershed DEM layer using flow direction tool
- Calculate flow accumulation with flow accumulation tool
- Calculate slope of watershed in degrees using slope tool
- Compute the LS factor using Eq. 23.3:

$$\begin{aligned} &\text{Power}(\text{``flowacc''} \times [\text{cell resolution}]/22.1, 0.4) \\ &\quad \times \text{Power}(\sin(\text{``sloperasterdeg''} \times 0.01745)/0.09, 1.4) \times 1.4 \end{aligned} \tag{23.3}$$

- *C* = Cover management factor is computed as the soil loss ratio from a field under a given cover compared to the loss from a field under continuous and bare fallow conditions with up—down slope tillage (Wischmeier and Smith 1978). The *C*-factors suggested by Hurni (1987) were used as a base for computing *C* values for average cover conditions (Table 23.3).
- *P* = Support practice factor is the ratio of soil loss with a support practice such as contouring, strip cropping or terracing to soil loss with straight-row farming up and down the slope. The *P* values vary from 0 to 1 (Table 23.4) where the highest values correspond to a bare land without any support practices.

Table 23.3 *C* values suggested by Hurni (1987)

Cover type	*C* value	Cover type	*C* value
Bad land hard	0.050	Sorghum–maize	0.10
Bad land soft	0.400	Cereals and pulses	0.15
Dense grass	0.010	*Teff*	0.25
Degraded grass	0.050	Fallow hard	0.05
Dense forest	0.001	Fallow plowed	0.60
Other forest (modest ground cover)	0.010	Continuous fallow (without cover)	1.00

Table 23.4 P values for combined support practices

Land slope (%)	Contour factor	Strip crop factor	Terrace factor	
			Closed outlet	Open outlet
1–2	0.6	0.3	0.5	0.7
3–8	0.5	0.25	0.6	0.8
9–12	0.6	0.3	0.7	0.8
13–16	0.7	0.35	0.8	0.9
17–20	0.8	0.4	0.9	0.9
21–25	0.9	0.45	1.0	1.0

*Source*Wischmeier and Smith (1978)

Soil erosion models, such as RUSLE estimate gross soil erosion rate at plot-scale (Wischmeier and Smith 1978; Renard et al. 1997). Erosion rates estimated by RUSLE are often higher than those measured at catchment outlets. Sediment delivery ratio (SDR) is used to correct such reduction effect. Mathematically, it is expressed by Eq. 23.4.

$$\mathrm{SDR} = Y/E \tag{23.4}$$

where Y is average annual sediment yield per unit area and E represent average annual erosion over the same area. Factors that influence the values of SDR includes: the hydrological input (mainly rainfall), landscape properties (vegetation, topography, soil properties) and their complex interaction. These factors make it difficult to compute or measure SDR for a given catchment. Researchers have developed different relationship for sediment deliver ratio using different input parameters. Some of them use event storm (Williams and Berndt 1972) or rainfall–runoff factors (Arnold et al. 1996) or drainage area and annual runoff depth (Dency and Bolten 1976)-based relationships for the estimation of SDR or sediment yield directly. The input parameters for these relationships by themselves need complicated hydrological models. Others use drainage area only for the development of SDR equation (Renfro 1975; Vanoni 1975; USDA SCS 1979). Although the SDR-area-based lumped methods are easy and direct for applications, most of them underestimate the sediment yield since they do not consider the effects of various physiographic attributes of a basin (Khanbilvardi and Rogwski 1984). In this research, GIS supported distributed slope, gradient, relief-length ratio and drainage area-based estimation of SDR as explained in Mutua and Klik (2006), Williams and Berndt (1972) is employed.

Once the SDR is obtained, the weight of the sediment yield is computed at the confluence of each BNR major tributary, and it is converted to volume by specific weight ratio (Eq. 23.5)

$$V = W/\gamma \tag{23.5}$$

where V is the annual volume of sediment (million m^3/year), W is the annual yield of sediment (million ton/year) computed from the soil loss analysis and γ is the unit weight (ton/m^3) of the soil, which is derived from the percentage distribution of sand, silt and clay in the geo-soil database attributes of the basin (Berhanu et al. 2013).

Sediment Transport

This study opts to identify the BNR reaches that are more vulnerable to changes in sediment input to the river system. Thus, width average sediment transport analysis is used. For this proposed width average model, the most applicable continuity equation is Eq. 23.6a.

$$\frac{\partial z}{\partial t} = \frac{T_E}{1 - p_o} \frac{\partial q_s}{\partial x} \quad (23.6a)$$

where

- p_o is the porosity; here it is assumed constant throughout the whole simulation and is calculated by Eq. 23.6b.

$$p_o = 1 - \frac{\gamma_{md}}{\gamma_s} \quad (23.6b)$$

where γ_s and γ_{md} are the unit weight and dry unit weight of the transported material, respectively, and q_s is computed by Eq. 23.6c.

$$q_s = \frac{Q_s}{w} \quad (23.6c)$$

where q_s is the volumetric sediment discharge per unit width (w), Q_s is the total volumetric sediment discharge. ∂x is the reach distance and T_E, is trap efficiency which describes the percentage of sediment that settles out over a given distance Δx, and computed by Eq. 23.6d.

$$T_E = 1 - \mathrm{e}^{-\frac{\omega \Delta x}{q}} \quad (23.6d)$$

where q is the discharge per unit width within the reach, and ω is the particle fall velocity. According to Julien (2010), without re-suspension, 99% of the sediment in suspension settles within a distance $\Delta x > \Delta x_c$ computed by Eq. 23.6e.

$$\Delta x_c = 4.6 \frac{hV}{\omega} \quad (23.6e)$$

- **Particle fall velocity** (ω): For mean particle size of ds ω is given by Julien (2010), Eq. 23.6f.

$$\omega = \frac{8 v_m}{d_s} \left(\left(1 + 0.0139 d_*^3\right)^{0.5} - 1 \right) \quad (23.6f)$$

where v_m is the viscosity of the river water, d_* is the dimensionless grain diameter and is calculated by Eq. 23.6g.

$$d_* = d_s \left(\frac{(G-1)g}{v_m^2} \right)^{1/3} \quad (23.6g)$$

where G is the specific gravity of the sediment and g is acceleration due to gravity.

Aggradation and degradation of a channel bed were estimated over time by finite difference schemes to approximate both the spatial and the temporal derivatives of the sediment continuity equation. A spatial grid that shall satisfy the criteria, $T_E \approx 1$ ($\Delta x > \Delta x_c$), is first created. The width averaged volumetric sediment flow rate was calculated on this grid, and its derivative was estimated using a first-order backward difference scheme (Eq. 23.6h).

$$\frac{\partial q_{s(i)}}{\partial x} \approx \frac{q_{s(i)} - q_{s(i-1)}}{\Delta x} \quad (23.6h)$$

The governing partial differential equation is then marched forward in time using forward Euler method, which is first-order method and is shown by Eq. 23.6i.

$$\frac{\partial z_{(i)}}{\partial t} \approx \frac{z_{(i+1)} - z_{(i)}}{\Delta t} \quad (23.6i)$$

Combining these two difference approximations yields the final scheme (Eq. 23.6j).

$$\begin{aligned} \frac{z_{(i+1,j)} - z_{(i,j)}}{\Delta t} &= -\frac{T_E}{1-p_o} \left(\frac{q_{s(i-1,j)} - q_{s(i-1,j-1)}}{\Delta x} \right) \Rightarrow z_{(i+1,j)} \\ &= z_{(i,j)} + \frac{\Delta t}{\Delta x} \left(-\frac{T_{Ej}}{1-p_o} \right) \left(q_{s(i-1,j)} - q_{s(i-1,j-1)} \right) \end{aligned} \quad (23.6j)$$

which make the change in riverbed level represented by Eq. 23.6k.

$$\Delta z_{(i+1,j)} = \frac{\Delta t}{\Delta x} \left(-\frac{T_{Ej}}{1-p_o} \right) \left(q_{s(i-1,j)} - q_{s(i-1,j-1)} \right) \quad (23.6k)$$

Using such a simplistic scheme creates a number of stability issue. To enforce stability of this scheme, Julien (2010) suggest a distribution factor α ($0 \le \alpha \le 1$) to be implemented; this will distribute changes in bed level between the different ends of the reach (Eq. 23.6l).

$$\Delta z_{(i+1,j)} = \alpha \left(\Delta z_{(i+1,j)} \right); \quad \Delta z_{(i+1,j+1)} = (1-\alpha) \left(\Delta z_{(i+1,j)} \right) \quad (23.6l)$$

which makes the final change in elevation (Eq. 23.6m)

$$\Delta z_{(i+1,j\text{New})} = \alpha\left(\Delta z_{(i+1,j)}\right) + (1-\alpha)\left(\Delta z_{(i+1,j-1)}\right) \tag{23.6m}$$

Reach distance Δx: Within the model, Δx was taken as the straight reach along the river. River cross sections were used as topographic raw data. Along the Lake Tana GERD course of the BNR, a total of 598 cross sections that form 597 reach distances were formed. However, to make use of the flow and sediment data at the Ethiopian Sudan border, the reach was extended. Each cross section is separated in straight gaps of maximum 5,052 m and a minimum of 363 m. The average reach distance is about 1,463 m. The cross section data defining the channel geometry are generated with the aid of a digital elevation model (DEM).

Average particle size: Ahmed and Ismail (2008) describe that at Roseries reservoir, just downstream of Ethiopia—Sudan border, the suspended sediment load is mainly distributed as 30% clay (<0.002 mm grain size diameter), 40% silt (0.002–0.02 mm) and 30% fine sand (0.02–0.2 mm). From the suspended load and bed load distribution at Roseries, the average particle size that will enter the Blue Nile upstream shall be larger in size. This is because of faster flows that are expected at the upper reaches of the Blue Nile. Thus, the average particle size over the whole BNR reach upstream is taken as 0.2 mm.

Particle density: Proper density of the transported sediment is necessary for estimating the particle fall velocity. It is usually 1.1–1.5 tones/m^3 and depends on the deposit's consolidations. Based on a limited scattered sediment data, ENTRO (2007) uses the density of sediment deposits as 1.5 tones/m^3.

Discharge per unit width q: The discharge per unit width shall be related to the flood capable of changing the channel morphology, which in this study case is the effective discharge. The discharge carrying the maximum sediment load has been called the effective discharge/channel forming discharge (Leopold 1992). In most cases, the effective discharge was found to be approximately equal to the bank full discharge (Leopold 1992; Andrews 1984). For BNR, the bank full discharge occurs in the months of **July, August and September**. These flood waves experience attenuation, and Fan and Li (2004) suggest that such floods are better approximated by diffusion waves rather than kinematic waves. The governing equation for diffusive flood wave is written as Eq. 23.6n (Fan and Li 2004).

$$\frac{\partial Q}{\partial t} + c\frac{\partial Q}{\partial x} - \frac{q_o}{2S_o}\frac{\partial^2 Q}{\partial x^2} = f(x,t) \tag{23.6n}$$

where S_o is the riverbed slope, q_o is an average discharge per unit channel width (m^2/s), Q is volume flow rate (Discharge) and c is the celerity of diffusion wave and is calculated by Eq. 23.6o.

$$c = \frac{\mathrm{d}Q}{\mathrm{d}A} = \frac{\mathrm{d}Q/\mathrm{d}h}{\mathrm{d}A/\mathrm{d}h} = \frac{1}{w}\frac{\mathrm{d}Q}{\mathrm{d}h} \tag{23.6o}$$

In Eq. 23.6o, w is the width of water surface and h is the depth of water.

From the Manning equation, the celerity speed, c, in a wide channel is $c = (5/3)$ v in which v ($=Q/A$) is velocity of flow. This c and the coefficient $q_o/(2S_o)$ in the above equation were multiplied by factor α and β, respectively, so that the non-uniform flow nature and rough cross-sectional data use effect on the model are handled.

Thus, the final model used in this study has the form of Eq. 23.6p.

$$\frac{\partial Q}{\partial t} + \alpha c \frac{\partial Q}{\partial x} - \beta \frac{q_o}{2S_o} \frac{\partial^2 Q}{\partial x^2} = f(x, t) \tag{23.6p}$$

where β and α are constants to be found by trial and error so that the measured and simulated discharge values are correlated within acceptable range.

Diffusion coefficient, D, is computed by Eq. 23.6q

$$D = \beta \frac{q_o}{2S_o} \tag{23.6q}$$

q_o is a reference flow per unit width ($m^3/s/m$). The flood wave is considered to be a perturbation with respect to this reference discharge. The inhomogeneous term $f(x, t)$ is defined by Eq. 23.6r.

$$f(x, t) = \alpha c q_1 - \beta D \frac{\partial q_1}{\partial x} \tag{23.6r}$$

where q_l denotes the lateral inflow per unit width of the confluent region.

According to Fan and Li (2004), optimal results are obtained if c is close to and not greater than unity and if the sum of c and D is greater than 1. In addition, t_r/t should be greater than 5. Besides, a diffusion wave should satisfy the inequality in Eq. 23.6s.

$$t_r S_o \left(\frac{g}{d_o}\right) \geq 15 \tag{23.6s}$$

where t_r (s) is the time-to-peak of the inflow hydrograph, S_o the channel slope, d_o the average flow depth in meters and g the gravitational constant equal to 9.81 m/s^2.

Diffusion wave routing is accomplished by combining and solving the above equations through careful satisfactions of the stated criteria. Since Manning's equation is used to estimate the friction slope S_f, RNB was described as a series of cross sections that adequately describe the hydraulic character of the entire flow path, including slope, geometry and roughness. Besides, finite element method was used to solve the above governing partial differential equation. In the diffusion wave equation, the time derivative is first order, and therefore, for solving it, an initial condition is needed. An initial discharge that uniformly increases from Lake Tana to the border is assumed.

Solving the governing equations also need boundary conditions to be specified at some points. The hydrological data are the most important boundary condition of the model. Hydrographs and/or water levels at known locations provide the necessary information. The hydrological data are needed as a sequence of water level (m) or discharge (m^3/s) in time (s). Hydrographs can either be implemented in the model as an inflow boundary condition or an external source or sink. The inflow boundary condition is used to simulate discharge from upstream that enters the study reach. Additional discharge coming from a tributary is handled in the model as a source.

A continuous hydrograph was used for the inflow boundary condition at the out let of Lake Tana. Major tributaries (e.g., Beshilo, Weleka, Jemma, Guder, Muger, Fincha, Didessa, Dabus and Beles) that drain into the Blue Nile in the study reach are taken as sources described by a hydrograph. Other minor tributaries have small water contribution to the Blue Nile flow, and thus, a uniform discharge per unit width of the boundary is given over such reach. This uniform discharge per unit width to the river could also account for any subsurface flow contribution that might exist along the river reach. A discharge per unit width could be positive or negative; a negative value implies the river acts as a recharge to the groundwater and vice versa for the positive value. This uniform discharge is fixed by trial and error, and its value is expected to vary monthly in accordance with the basin hydrograph as shown in Fig. 23.5.

In this study, two effective roughness values were used: representing the unconfined and confined sections of the BNR (Fig. 23.2). For these two sets of effective roughness coefficients, existing riverbed slope and cross sections at the 598 locations, the diffusion wave equation was solved. The roughness coefficient of the river, the diffusion coefficient and advective coefficients were adjusted by trial

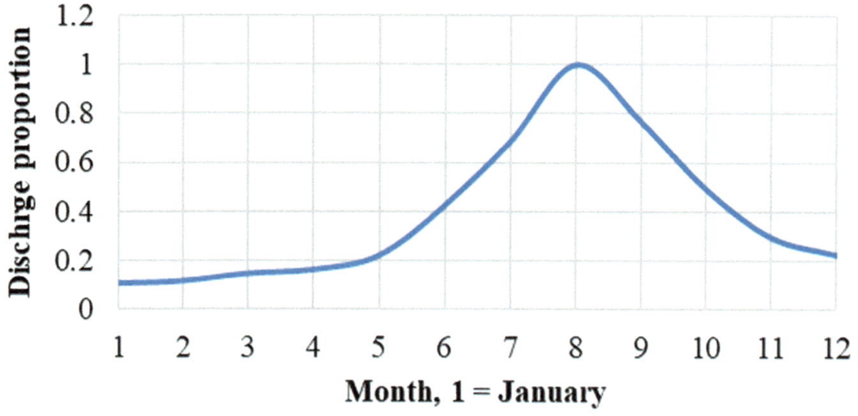

Fig. 23.5 Relative uniform discharges per unit length in proportion to the maximum

and error to fit the hydrograph of the BNR at Ethiopia–Sudan border and at Kessie Bridge. The calibrated model is used in determining the discharge per unit width q.

Sediment Advection–Diffusion Equation

The general relationship describing conservation of sediment mass for incompressible dilute suspensions subjected to diffusion, mixing, dispersion and advection with point sediment sources is obtained from Julien (2010), Eq. 23.7a

$$\frac{\partial q_s}{\partial t} + V\frac{\partial q_s}{\partial x} + (D + \varepsilon_x)\frac{\partial^2 q_s}{\partial x^2} = f(x, t) \tag{23.7a}$$

where q_s is the sediment concentration and V is the average velocity of flow.

In laminar flow, the turbulent mixing and dispersive coefficient vanish ($\varepsilon_x = 0$). Conversely, in turbulent flows like the case of BNR, the molecular diffusion coefficient D is negligible compared to the turbulent mixing and dispersion coefficients ($D \ll \varepsilon_x$). Thus, the equation is reduced to Eq. 23.7b.

$$\frac{\partial q_s}{\partial t} + V\frac{\partial q_s}{\partial x} + \varepsilon_x\frac{\partial^2 q_s}{\partial x^2} = f(x, t) \tag{23.7b}$$

in which the inhomogeneous term $f(x, t)$ is defined by Eq. 23.7c.

$$f(x, t) = Vc_1 - \varepsilon_x\frac{\partial c_1}{\partial x} \tag{23.7c}$$

where c_1 denotes the lateral inflow sediment concentration per unit width of the confluent section.

Equation 23.7d is similar to the equation given for flow except that the advective and diffusion coefficients are replaced by average velocity and turbulent mixing and dispersion coefficients, respectively. The specific numerical solution to this equation needs appropriate initial and boundary conditions. The initial condition is set as zero concentration of sediment in the entire river reach. This is because the initial time step starts in January where the sediment flow is nil.

$$\frac{\partial q_s}{\partial t} + \alpha V\frac{\partial q_s}{\partial x} + \beta\varepsilon_x\frac{\partial^2 q_s}{\partial x^2} = \alpha Vq_{s1} - \beta\varepsilon_x\frac{\partial q_{s1}}{\partial x} \tag{23.7d}$$

The sediment inflow boundary condition is used to simulate sediment from upstream that enters the study reach. Additional sediment load coming from a tributary is described with a source. Major tributaries (e.g., Beshilo, Weleka, Jemma, Guder, Muger, Fincha, Didessa, Dabus and Beles) that drain into the Blue

Table 23.5 Sediment concentration observed at El Diem station

Month	Decade	Sed. conc. (ppm)*	Average sed. conc. (ppm)	% of the total
June	2	1956	461.45	12
	3	–		
July	1	3361	1419.691	35
	2	3895		
	3	4335		
August	1	5660	1143.5	29
	2	3095		
	3	2948		
September	1	3589	764.9	19
	2	2305		
	3	1755		
October	1	1294	217.3182	5
	2	591		
	3	317		

**Source* Awulachew et al. (2008)

Nile in the study reach are taken as sources described by sediment load curves. Other minor tributaries have small sediment contribution to the Blue Nile flow, and thus, a uniform sediment discharge per unit width of the boundary is given over the study reach. This uniform sediment discharge per unit width to the river is fixed by trial and error, and its value is expected to vary monthly from (June to July) in accordance with the basin sediment load curve observed at El diem station (Table 23.5).

Model Performance Evaluation

Model evaluation is an essential measure to verify the robustness of the model. In this study, three model evaluation methods were used following Moriasi et al. (2007) model evaluation guideline. These methods are (i) Nash–Sutcliffe efficiency (NSE), (ii) percent bias (PBIAS) and (iii) ratio of the root mean square error to the standard deviation of measured data (RSR). The Nash–Sutcliffe efficiency (NSE) is computed as the ratio of residual variance to measured data variances. The Nash–Sutcliffe is calculated using Eq. 23.8.

$$\mathrm{NSE} = 1 - \left[\frac{\sum_1^n \left(X_i^{\mathrm{Obs}} - X_i^{\mathrm{Sim}}\right)^2}{\sum_1^n \left(X_i^{\mathrm{Obs}} - X^{\mathrm{Mean}}\right)^2}\right] \tag{23.8}$$

where X^{obs} is observed and X^{sim} *is simulated* variables (flow in $m^3\ s^{-1}$ or sediment concentration in mg l^{-1}), X^{mean} is mean of n observed values and n = number of observations.

The percent bias (PBIAS) on average measures the simulated data tendency to be larger or smaller than their observed values. The PBIAS is calculated with Eq. 23.9

$$\mathrm{PBIAS} = \frac{\sum_{1}^{n}\left(X_i^{\mathrm{Obs}} - X_i^{\mathrm{Sim}}\right)}{\sum_{1}^{n}\left(X_i^{\mathrm{Obs}}\right)} \times 100 \tag{23.9}$$

The ratio of root means square error to the standard deviation of measured data (RSR) is shown in Eq. 23.10.

$$RSR = \frac{\sqrt{\sum_{1}^{n}\left(X_i^{\mathrm{Obs}} - X_i^{\mathrm{Sim}}\right)^2}}{\sqrt{\sum_{1}^{n}\left(X_i^{\mathrm{Obs}} - X^{\mathrm{Mean}}\right)^2}} \tag{23.10}$$

According to Moriasi et al. (2007) stream flow model simulation judged as satisfactory for monthly time step, if NSE is greater than 0.5, RSR is less than 0.70 and PBIAS is within ±25% for flow and NSE greater than 0.5, RSR less than 0.70 and PBIAS within ±55% for sediment.

Results and Discussion

In this study, the channel morphology change is particularly described as the change in elevation of the riverbed level. Under different scenarios of watershed management practices and the corresponding variation in the sediment supply, the change in the riverbed level is computed. The current sediment supply and river sediment transport capacity was first evaluated. This was achieved by first calibrating the model for observed average monthly flows and sediment concentration data at Kessie Bridge and Ethiopia-Sudan border.

In the model, average monthly point discharges from main tributaries were first obtained from the hydrologic analysis of the basin. In addition, trial values of discharges per unit length that vary by location and time are input to the model. In the model, water is assumed to recharge the groundwater from the river in any time of the year in the unconfined reach of the river (Fig. 23.2). This is because groundwater is assumed to discharge toward the BNR along the confined reaches of the BNR and groundwater level is expected to be lower than the river water level and higher evaporation from the river at unconfined reaches of BNR.

The model flow predictions were tested against average monthly flow at Kessie Bridge and at Ethiopia–Sudan border, as shown in Fig. 23.6. The simulated average monthly flow matched the observed values with NSE, RSR and PBIAS equal to

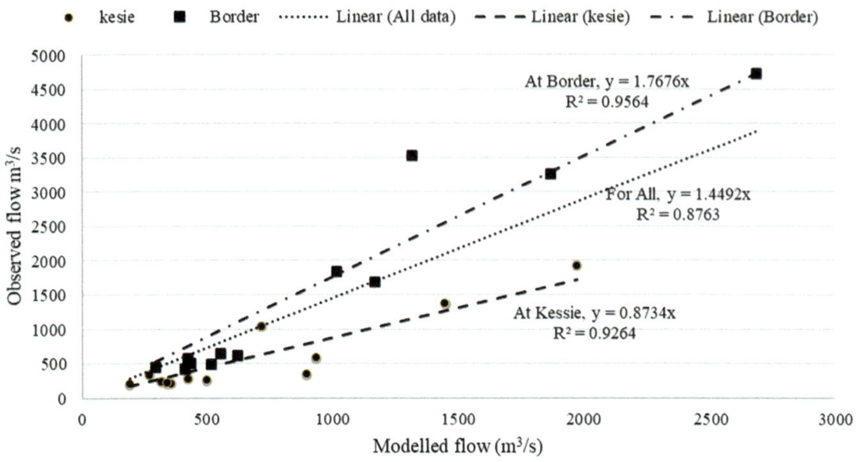

Fig. 23.6 Model and measured monthly discharges comparison

0.80, 0.45, and − 22%, respectively, for flow at Kessie Bridge and 0.51, 0.7 and 39%, respectively, for flow at Ethiopia–Sudan border. For the flow observed at both locations, the NSE, RSR and PBIAS values obtained were 0.55, 0.7 and 23%, respectively. From the results, it can be seen that except the PBIAS for flow at border, all are within the permissible limits.

The model sediment predictions were also tested against average monthly sediment data at Kessie Bridge and at Ethiopia–Sudan border as shown in Fig. 23.7. The simulated average monthly sediment load matched the observed values with NSE, RSR and PBIAS equal to 0.69, 0.56 and −25%, respectively, at Kessie Bridge and 0.81, 0.43 and 11%, respectively, for sediment load at Ethiopia–Sudan border. For the sediment load observed at both locations, the NSE, RSR and PBIAS values

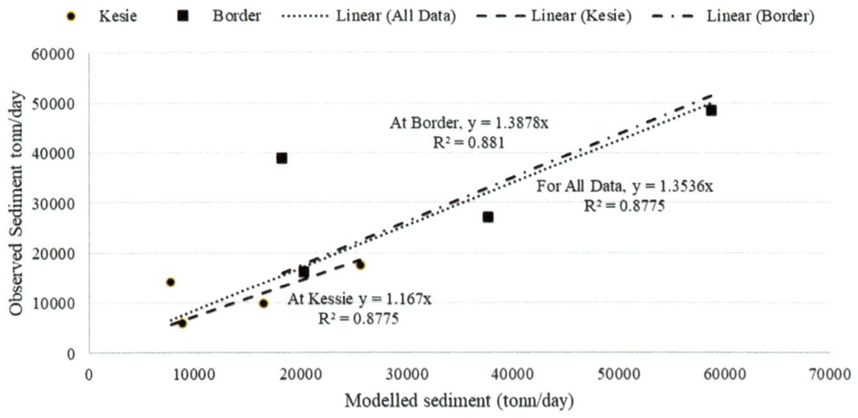

Fig. 23.7 Model and measured monthly average sediment flow comparison

Table 23.6 Model parameters obtained after calibration

	Hydrograph/reach		Sediment trans.	
	Confined	Unconfined	Confined	Unconfined
Manning roughness	0.0035	0.0017	0.0035	0.0017
Lateral flow/sediment	0.1	-0.1	0.1	0.1
Coefficients				
• α	0.03	0.025	80	80
• β	10	20	1.5×10^6	1.5×10^6

obtained were 0.80, 0.45 and 6%, respectively. The simulated average monthly sediment which matched the observed values at both locations has respected the recommended NSE, RSR and PBIAS values.

According to the performance criteria suggested by Moriasi et al. (2007), the above model fit statistics for both flow and sediment models are acceptable. These results indicate that the flow and sediment model parameters and discharge per unit length values are satisfactory. Thus, the model can be applied to describe the sediment and flow transport along the BNR. The models were run with the parameters shown in Table 23.6.

Channel Morphology Change

The channel bed elevation change equation discussed above is shown by Eq. 23.11.

$$\Delta z_{(i+1,j)} = \frac{\Delta t}{\Delta x}\left(-\frac{T_{Ej}}{1-p_o}\right)\left(q_{s(i-1,j)} - q_{s(i-1,j-1)}\right); \quad \text{where } T_{Ej} = 1 - e^{-\frac{\omega \Delta x}{q}} \tag{23.11}$$

All inputs to this equation can be obtained from the calibrated flow and sediment transport models. For the different scenarios of watershed development, the change in the riverbed level is shown in Fig. 23.8. The figure shows the relative riverbed level change across the BNR course to GERD dam site if the two watershed practice scenarios (scenarios 2 and 3) are executed.

From Fig. 23.8, it is clear that the average riverbed level change varies from location to location. The maximum change in the river morphology will occur between 320 and 550 km distances. This is mainly due to the fact that the main degraded catchment in the basin occur upstream of this location, where the interventions has reduced sediment input to BNR significantly. Muger, Guder and Fincha contribute their water and sediment yield at these locations in short distance interval. However, the effect is becoming smaller while moving downstream. The river morphology around the location of the GERD is also significantly affected by the watershed treatments; this is mainly due to the treatments made on the Dabus

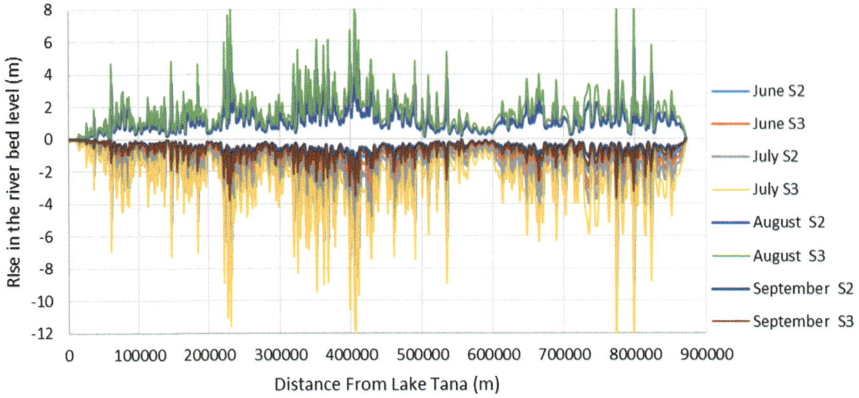

Fig. 23.8 Riverbed level variation for proposed soil and water conservation measures

River catchments, rather than the other main rivers upstream. The soil and water conservation measures done in both scenarios for the catchment have resulted in degradation of the riverbed in June, July and September, while aggradation is observed in August for both scenarios.

Among the two scenarios, scenario 3 is found to have relatively larger impact on the river morphology change (Fig. 23.9). Figure 23.9 shows the riverbed variation at 600 km distance. The riverbed level decreases during the flood rising time and increases during flood recession time. The decrease in the riverbed level has arisen due to the reduction in the sediment concentration which gives the existing flow a higher erosive power than the base scenario. The rise in the riverbed level is due to the fact that scoured bed has created a larger cross section for the flow which resulted in decrease in velocity which ultimately creates deposition. The yearly average riverbed variation is negative implying that riverbed degradation is the dominant process formed.

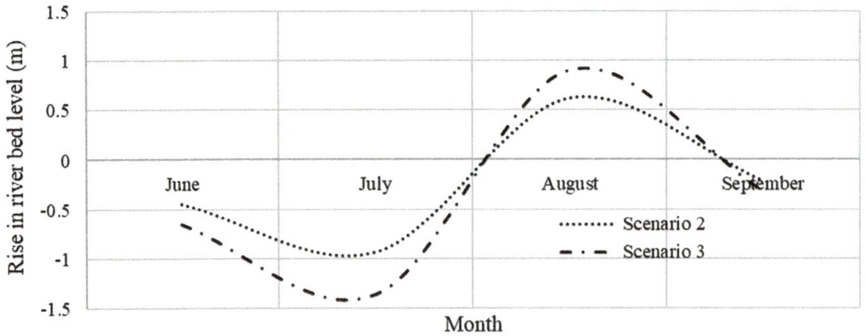

Fig. 23.9 Relative rise in riverbed level selected at 600 km distance from lake Tana

Among the scenarios developed for improving the watershed management, scenario 3 has resulted in the highest riverbed level change in both flood peaking and recession periods. Thus, this scenario is expected to bring about the most change in the river morphology than the other scenarios.

Conclusion

With respect to the BNR system, large amounts of riverine deposits and influx of hill slope material provide the sediment for the channel to transport. The amount of material that the river system can transport is directly related to the discharge and slope, which are in turn driven by watershed management, climate and bedrock lithology. Watershed management practice in addition to runoff change will bring about sediment supply change. Three different watershed management (agronomic practices, mechanical measures and biological measures) scenarios were tested for their possible effect on the BNR morphology. Agronomic practices (scenario 1) are related to management of crop cultivation (strip cropping, contour cropping, change of crop species, etc.). Mechanical measures (scenario 2) include engineering-based physical conservation measures like soil/stone bund, terracing, cutoff drain, waterways, etc. Biological measures (scenario 3) are treatment options related to vegetation. The analysis for each of the scenarios effect on the riverbed level change was obtained from a calibrated finite element model for sediment and diffusive wave transport models. The current sediment supply and river sediment transport capacity were first evaluated. This was achieved by first calibrating the model with observed average monthly flows and sediment concentration data of two stations located at Kessie Bridge and Ethiopia–Sudan border. The performance of the model was measured by NSE, PBIAS and RSR values for which the model has resulted in acceptable performance level. However, the morphologic characters of the river (riverbed elevation, cross section and slope) used in the river flow and sediment transport analysis were obtained from 30 m digital elevation model which make the result to be taken cautiously. Besides, the scenarios developed has bring about the sediment yield change with small change in the annual runoff generation.

The results of this study for all scenarios showed reductions in sediment input for the river to transport. The results also indicated a decrease in the average riverbed level during flood rising time (June to end of July) and an increase during flood recession time (August to September). The decrease in the riverbed level was due to the reduction in the sediment concentration which gives the flow a higher erosive power than the base scenario. The scoured bed in the flood rising period has created a larger cross section for the flow in the recession time, which resulted in a decrease in the river competence that ultimately creates deposition of sediments.

The result has shown a relatively high change in the riverbed level between 300 and 550 km downstream of Lake Tana. In between these two distances, from Lake Tana along BNR, Muger, Guder and Fincha rivers join the main BNR course. However, the effect is becoming smaller while moving downstream. The river morphology around the location of the GERD is also significantly affected by the watershed treatments; this is mainly due to the treatments made on the Dabus River catchments, rather than the other main rivers upstream. The proposed watershed management measures in the catchments of these rivers have resulted in larger sediment reduction. Among the scenarios developed for improving the watershed, scenario 3 (Biological measures) has resulted in the highest riverbed level change in both flood peaking and recession periods. Thus, this scenario is expected to bring about lesser sediment for the river to transport and the greatest change in the river morphology when compared to the other scenarios.

References

Ahmed AA, Ismail AE (2008) Sediment in the Nile River System. UNESCO-IHP-International Sediment Initiative

Andrews ED (1984) Bed material entrainment and hydraulic geometry of gravel-bed rivers in Colorado. Geol SOC Am Bull 95:371–378

Arnold JG, Williams JR, Srinivasan R, King KW (1996) The soil and water assessment tool (SWAT) user's manual. Temple, TX

Awulachew SB, McCartney M, Steenhuis TS, Ahmed AA (2008) A review of hydrology, sediment and water resource use in the Blue Nile Basin. International Water Management Institute, Colombo, Sri Lanka, 87 p (IWMI Working Paper 131)

Berhanu BK, Aray DS (2005) TEFASES–A SDSS for land capability classification for Ethiopian watersheds. In: Proceedings of the international conference on hydrological perspectives for sustainable development 2005, vol 1, pp 32–44

Berhanu BK, Assefa MM, Seleshi Y (2013) GIS-based hydrological zones and soil geo-database of Ethiopia, Catena, vol 104. Elsevier Publisher, pp 21–31, 0341–8162

Collick AS, Easton ZM, Engda TA, Ashagrie BB, Adgo E, Awulachew SB, Zeleke G, Steenhuis TS (2009) Application of physically based water balance model on four watersheds throughout the upper Nile basin in Ethiopia. Hydrol Process 23:3718–3727

Dency, Bolten S (1976) Ecological issues in floodplains and riparian corridors, white paper. Center for Streamside Studies, University of Washington

Eastern Nile Technical Regional Office (ENTRO) (2007) Eastern Nile Power Trade Program study, ENPT 001 2009

Engda TA (2009) Modeling rainfall-runoff-soil loss relationships of North-eastern Highlands of Ethiopia, Andit Tid watershed, Ethiopia. Master Thesis, Integrated Watershed Management and Hydrology Program, Cornell University, Bahir Dar Ethiopia and Ithaca, NY

Escobedo J (1998) Assistance to soil and water conservation program, Phase III, Field Document No. 19 ETH/85/016. CFSCDD, MOA, FAO, October 1998, Addis Ababa, Ethiopia

Fan P, Li JC (2004) Diffusive wave solutions for open channel flow with uniform and concentrated lateral inflow. Adv Water Resour 29(2006):1000–1019

Hurni H (1987) Soil conservation in Ethiopia (Amharic version), CFSCDD, Ministry of Agriculture, Addis Ababa, 103 pp

Julien PY (2010) Erosion and sedimentation. Colorado State University, Cambridge University Press

Khanbilvardi RM, Rogowski AS (1984) Quantitative evaluation of sediment delivery ratios. Water Resour Bull 20(6):865–874

Leopold LB (1992) Dynamics of gravel-bed rivers (ed. by Billi P, Hey RD, Thorne CR, Tacconi P). Wiley

MacAlister C, Pavelic P, Tindimugaya C, Ayenew T, Ibrahim ME, Abdel Meguid M (2012) The Nile River Basin Water, Agriculture, Governance and Livelihoods, IWMI (ed by Awulachew, Smakhtin, Molden and Peden)
Moriasi DN, Arnold JG, Van liew MW, Bingner RL, Harmel RD, Veith TL (2007) Model evaluation guidelines for systematic quantification of accuracy in watershed simulations. Am Soc Agricult Biol Eng. 50(3):885−900. ISSN 0001−2351
Mutua BM, Klik A (2006) Estimating spatial sediment delivery ratio on a large rural catchment. J Spat Hydrol 6(1):64–80
Pelton J, Frazier E, Pickilingis E (2014) Calculating slope length factor (LS) in the revised universal soil loss equation (RUSLE). Accessed Feb 3 2020 from http://gis4geomorphology.com/ls-factor-in-rusle/LS-Factor-in-RUSLE-with-ArcGIS-10.x_Pelton_Frazier_Pikcilingis_2014
Renard KG, Foster GR, Weesies GA, McCool DK, Yoder DC (1997) Predicting soil erosion by water: a guide to conservation planning with the revised universal soil loss equation (RUSLE), Agriculture Handbook No. 703, Agricultural Research Service, Washington, District of Colum, 404 pp
Renfro GW (1975) Use of erosion equations and sediment delivery ratios for predicting sediment yield. In: Present and prospective technology for predicting sediment yields and sources, agricultural resources services, ARS-S-40. US Dept. Agric., Washington, D.C., pp 33–45
Shulits S (1941) Rational equation of river-bed profile. American Geophysical Union
Steenhuis T, Collick A, Easton Z, Leggesse E, Bayabil H, White E, Awulachew S, Adgo E, Ahmed A (2009) Predicting discharge and sediment for the Abay (Blue Nile) with a simple model. Hydrol Process 23:3728–3737
Tana-Beles WME (2012) Training manual on GIS—land capability classification for purpose of SWC, Kombolcha (Wello), Ethiopia
Tulu T (2011) Soil and water conservation for sustainable agriculture. Mega Publishing and Distribution P.L.C, Addis Ababa
USDA-SCS (1979) United States Department of Agriculture—Soil Conservation Service. National Engineering Handbook, Sec. 4. Hydrology
Vanoni VA (ed) (1975) Sedimentation engineering, ASCE manuals and reports on engineering practice No.54. American Society of Civil Engineers, New York
Williams JR, Brendt AD (1972) Sediment yield computed with the universal equation. Proc Am Soc Civil Eng 98(HY12):2087–2098
Wischmeier WH, Smith DD (1978) Predicting rainfall erosion losses—a guide to conservation planning. USDA Handbook 537, Washington, D.C.
Yang CT, Simões FJM, Huang J, Greimann B (2015) Generalized sediment transport models for alluvial rivers and reservoirs, US-China workshop on advanced computational modelling in hydroscience & engineering September 19–21. Mississippi, USA, Oxford

Chapter 24
Comparison of Spatial Interpolation Techniques of Rainfall for Hydrological Applications in a Complex Mountainous Region of the Upper Blue Nile Basin

Haileyesus Belay Lakew and Semu A. Moges

Abstract The quality of measured rainfall data and the network coverage of the gauge stations have a significant impact on the hydrological simulation. This chapter aims to present and evaluate different spatial interpolation techniques to produce a daily temporal scale of gridded rainfall datasets (0.10°). The interpolated gridded rainfall datasets are evaluated for the hydrological simulation in a complex mountainous region of the upper Blue Nile Basin. Five interpolation techniques of cubic, linear, natural, nearest, and spline are applied to interpolate and grid observed rainfall data using MATLAB. Fully distributed hydrological model of Coupled Routing and Excess STorage (CREST) is calibrated from 2005 to 2012 and validated from 2013 to 2017. Independent calibration is carried out for respective interpolated rainfall data for the hydrological simulation and comparison. The result indicates that among the interpolation techniques, the nearest method of interpolation reveals the best performance for the hydrological simulation of the daily gauged streamflow data. The NSCE and bias show 0.89 and −5.8% as well as 0.86 and −8.1% performance metrics for the calibration and validation periods, respectively. Therefore, for the complex mountainous region of the upper Blue Nile Basin, the nearest interpolation scheme is more applicable to generate high-resolution rainfall data for various water resources applications.

Keywords CREST · Spatial interpolation of rainfall · Interpolation techniques · Hydrologic simulation · Upper Blue Nile

H. B. Lakew (✉)
Ethiopian Space Science and Technology Institute, Addis Ababa, Ethiopia

S. A. Moges
Civil and Environmental Engineering, University of Connecticut, Storrs, CT, USA

A. M. Melesse et al. (eds.), *Nile and Grand Ethiopian Renaissance Dam*,
Springer Geography, https://doi.org/10.1007/978-3-030-76437-1_24

Introduction

Accurate assessment of spatial and temporal precipitation is crucial for simulating hydrological processes for water resource management, hydrologic modeling, recharge assessment, and irrigation scheduling (Cheng et al. 2017; Yang et al. 2015). Because precipitation influences the spatial organization of runoff, evaporation, and soil moisture, consequently affecting the ecological process (Taylor et al. 1997; Wang et al. 2017). The spatial and temporal resolutions of precipitation are vital for water resources applications. The precipitation data are mainly sourced from ground-based measurements taken from rain gauges. However, obtaining accurate precipitation data with better spatial coverage in Africa remains a challenge, especially in mountainous regions, mainly due to insufficient rain gauges. The situation in Ethiopia and the upper Blue Nile are not different from this real picture of gauging stations in Africa (Cristiano et al. 2017; Lakew et al. 2019). To overcome this challenge, different satellite-based and reanalysis precipitation products have emerged globally for the last two decades.

These global products are evaluated for local case studies through direct comparison with gauged rainfall. Besides, global precipitation products are also evaluated for the hydrological simulations to estimate the observed streamflow globally (Beck et al. 2017; Lakew et al. 2017, 2020). The performance of MSWEP product is evaluated using hydrological modeling for 9,011 catchments (<50,000 km^2) across the globe. The quality of MSWEP is compared against four state-of-the-art gauge-adjusted precipitation (P) datasets (WFDEI-CRU, GPCP-1DD, TMPA 3B42, and CPC Unified). The multi-source precipitation product of MSWEP obtained a median calibration NSCE of 0.52 versus 0.29–0.39 for the other P datasets (Beck et al. 2017). The result reveals that the MSWEP precipitation shows better performance from the rest of the products. However, MSWEP cannot capture and simulate the magnitude of the observed flow for the 9011 catchments across the globe with a median performance of 0.52.

Multiple globally available precipitation products are evaluated in the data scarce region of the upper Blue Nile Basin over multi-scales (1,656–199,812 km^2) focusing on multi-year (2000–2012) for daily simulation (Lakew et al. 2020) applying CREST hydrological model. The evaluation compares five precipitation products of gauge-adjusted Climate Prediction Center Morphing Technique (CMORPH), Tropical Rainfall Measuring Mission (TRMM) Multi-Satellite Precipitation Analysis 3B42 version 7 (TMPA), ERA-Interim (ERAI), Global Precipitation Climatology Centre (GPCC), and Multi-Source Weighted Ensemble Precipitation (MSWEP). Performance results indicate that the MSWEP precipitation shows consistently better performance. However, the product needs improvement to estimate the total annual flow volume of the observed flow for the three nested watersheds to perform close to the gauged rainfall. In the same basin of upper Blue Nile Basin (Gilgel Abbay Basin watershed) both satellite and reanalysis precipitation products are evaluated for the hydrological simulation performance (Lakew et al. 2017). Three precipitation products of gauge-adjusted (corrected)

CMORPH, (TRMM) TMPA 3B42v7, and ECMWF reanalysis products are evaluated using CREST model. The result indicates that all products under independent and RainFact calibration mode reproduce daily streamflow better than the simulation mode with NSCE performance above 70% in calibration and 60% in validation periods. These all evaluation results reveal that both satellite and reanalysis precipitation products cannot capture the gauged rainfall magnitude and need correction due to the indirect nature of the measurement (Bitew et al. 2012; Hirpa et al. 2010; Lakew et al. 2020; Sahlu et al. 2017).

Therefore, spatial interpolation using the available main source of gauged rainfall data with finer spatial scale can narrow this gap. Different types of spatial interpolation techniques are available and used to obtain gridded precipitation data based on gauge observations (Tabios III and Salas 1985). However, different spatial interpolation techniques have different performance to capture the hydrograph shape and magnitude of the observed streamflow in the hydrological simulation of a basin. In the complex mountainous region of the upper Blue Nile Basin, from different types of spatial interpolation techniques, the best fitting interpolation technique is not evaluated and identified to use the grid rainfall data for different water resources applications.

This research mainly aims to evaluate and identify the accurate spatial interpolation technique of the gauged rainfall dataset from cubic, linear, natural, nearest, and spline for the complex mountainous region of the upper Blue Nile Basin, Ethiopia. A framework of spatial interpolation and gridding scheme is developed by MATLAB using the available gauged rainfall data. Fully distributed hydrological model CREST is calibrated independently from 2005 to 2012 and validated from 2013 to 2017 for each spatially interpolated rainfall. The comparison is made by the hydrological simulation performance of the gauged rainfall data applying different spatial interpolation techniques using CREST model in the upper Blue Nile.

Materials and Methods

Study Area

The Nile proper is formed by two major tributaries the White Nile and the Blue Nile. The Upper Blue Nile (locally known as Abbay Basin) originates from the Ethiopian plateau and contributes 60% of the flow to the Nile River at Aswan, Egypt (Conway and Hulme 1993). The upper Blue Nile Basin is the largest river basin in terms of volume of discharge and second largest in terms of area in Ethiopia. It comprises 199,812 km^2 between 34°24′E–39°48′E and 7°42′N–12°30′N 20 percent of the area of Ethiopia and has a mean annual discharge of 48.5 cubic kilometers (1912–1997; 1,536 m^{3}s^{-1}) (Waterbury 1988). The basin drains a large portion of the central and southwestern Ethiopian Highlands. The river has cut a deep and circuitous course through the central Ethiopian Highlands and in some places its gorge is one kilometer deep. Its course flows 900 km from Lake Tana until it leaves Ethiopia and crosses into the vast plains of Sudan. There is only one significant waterfall, at Tis Isat (Tis Abbay), roughly 25 km from Lake Tana where the river drops 50 m into the Blue Nile Gorge. Much of the highland plateau is above 1,500 m and consists of rolling ridges and flat grassland meadows with meandering streams that waterfall over vertical sides of canyons. The river basin is composed mainly of volcanic and Pre-Cambrian basement rocks with small areas of sedimentary rocks. The soils generally consist of latosols on gentle slopes and deep vertisols in flatter areas subject to waterlogging (Conway 2000).

Hydrometeorological Datasets

Obtaining long-term historical gauged rainfall data is a challenging task globally, especially in Africa, Ethiopia, in which rain gauge stations are sparsely located. Besides, the rain gauge stations have missing data with short records. The available historical gauged daily rainfall data is obtained from the data records of the Ethiopian Meteorological Agency (EMA). In the upper Blue Nile Basin, there are 153 gauging stations from the EMA data records. A total of 128 in situ rainfall measuring stations (more than 84% from the total) that have good quality and record length (2005–2017) used for the study (Fig. 24.1). The daily based streamflow data at the same time length of the gauged rainfall (2005–2017) is obtained from the Sudan Ministry of Water Resources, Irrigation and Electricity. This streamflow data used as a reference to tune the parameters during the independent calibration at the outlet of the basin of the upper Blue Nile Basin at the Eldiem streamflow station. The Eldiem streamflow station is located at the border of Sudan and Ethiopia that is managed by the Sudanese government near to the Great Ethiopian Renaissance Dam (GERD).

In addition to rainfall and streamflow datasets, the potential evapotranspiration (PET) is one of the compulsory time series datasets that used to set up the distributed hydrological model CREST. The PET data is derived from the Penman–Monteith method (mm/day) (Allen et al. 1998) with the same daily temporal resolution of the rainfall and streamflow datasets. The potential evapotranspiration of a given crop is defined as soil evaporation and plant transpiration under unlimited soil water supply and actual meteorological conditions. It is an important agrometeorological parameter for climatological and hydrological studies, as well as for irrigation planning and management (Labedzki et al. 2011). The grid PET data that is available with 0.5° spatial and a daily temporal resolution accessed from the earthH2Observe data portal (https://wci.earth2observe.eu/portal/).

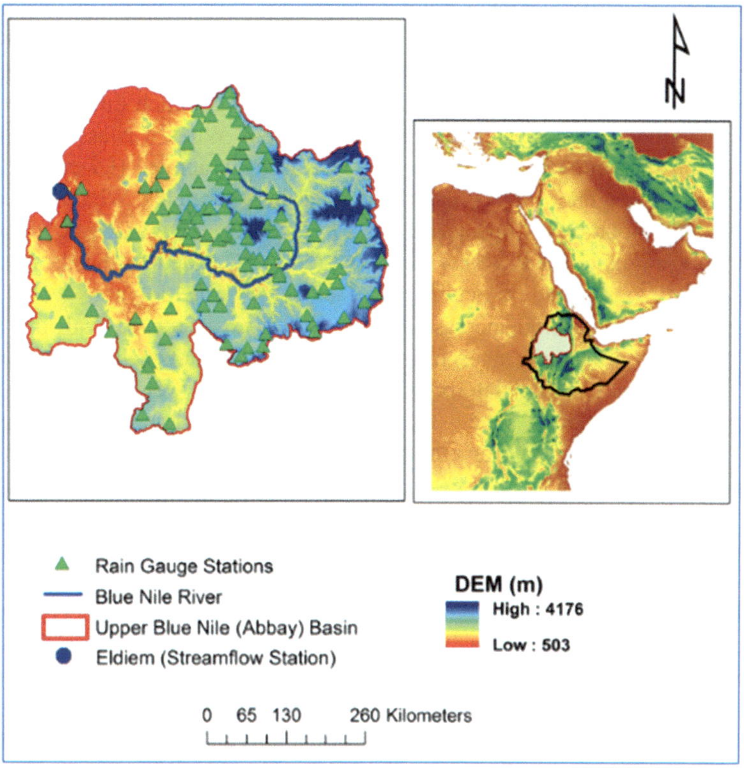

Fig. 24.1 Location of rain gauge and streamflow stations in the upper Blue Nile Basin relative to *East Africa*

Interpolation Techniques

Five different interpolation methods (cubic, linear, natural, nearest, and spline) are tested in this study to interpolate the available rain gauge data to a finer resolution of daily and 0.10° temporal and spatial resolutions, respectively. These methods are chosen due to that they are representative of available interpolation procedures, widely used in rainfall interpolation, and easy to implement in MATLAB (for hydrological simulation using CREST hydrological model).

Cubic interpolation, the interpolated value at a query point is based on a cubic interpolation of the values at neighboring grid points in each respective dimension (Kalyan and Rao 2013). The interpolation is based on a cubic convolution. Grid must have uniform spacing in each dimension, but the spacing does not have to be the same for all dimensions. It requires at least four points in each dimension and more memory and computation time than "linear."

Simple kriging (linear) is one of the interpolation techniques of kriging. In kriging, the value of the interest variable is estimated for a particular point using a weighted sum of the available point observations. The weights of the data are chosen so that the interpolation is unbiased, and the variance is minimized. Kriging is the first method of interpolation to take into account the spatial dependence structure of the data. There are several types of kriging, which differ according to the form applied to the mean of the interest variable: When it is assumed that the mean is constant and known, simple kriging (linear) is applied (Ly et al. 2013).

The natural neighbor method is quite popular in some fields. What is the natural neighbor interpolation? Consider a set of Thiessen polygons (the dual of a Delaunay triangulation). If a new point (target) were added to the dataset, these Thiessen polygons would be modified. In fact, some of the polygons would shrink in size, while none would increase in size. The area associated with the target's Thiessen polygon that was taken from an existing polygon is called the "borrowed area." The natural neighbor interpolation algorithm uses a weighted average of the neighboring observations, where the weights are proportional to the "borrowed area.` The natural neighbor method does not extrapolate contours beyond the convex hull of the data locations (i.e., the outline of the Thiessen polygons) (Yang et al. 2004).

The nearest neighbor method is also known as Thiessen polygon method. The Thiessen polygon (nearest) method assumes that the estimated values can take on the observed values of the closest station (Nalder and Wein 1998). The method requires the construction of a Thiessen polygon network. These polygons are formed by the mediators of segments joining the nearby stations to other related stations. The surface of each polygon is determined and used to balance the rain quantity of the station at the center of the polygon. The polygon must be changed every time a station is added or deleted from the network. The deletion of a station is referred to as "missing rainfall." This method, although more popular than taking the simple average of the number of stations, is not suitable for mountainous regions, because of the orographic influence of the rain (Goovaerts 2003; Ly et al. 2013).

The spline interpolation method is based on a mathematical model for surface estimation that fits a minimum-curvature surface through the input points. The method fits a mathematical function to a specified number of the nearest input points, while passing through the sample points. This method is not appropriate if there are large changes in the surface within a short distance, because it can overshoot estimated values (Ruelland et al. 2008).

Hydrological Model

The Coupled Routing and Excess STorage (CREST) distributed hydrological model simulates the spatio-temporal variation of water and energy fluxes and storages on a regular grid with the grid cell resolution being user-defined, thereby enabling global and regional scale applications (Wang et al. 2011). CREST's distinguishing characteristics include: (1) distributed rainfall-runoff generation and cell-to-cell routing;

(2) coupled runoff generation and routing via three feedback mechanisms (including rainfall-runoff generation, evapotranspiration, and sub-grid cell routing); and (3) representation of sub-grid cell variability of soil moisture storage capacity and sub-grid cell routing (via linear reservoirs). The coupling between the runoff generation and routing mechanisms allows detailed and realistic treatment of hydrological variables such as soil moisture. Furthermore, the representation of soil moisture variability and routing processes at the sub-grid scale enables the CREST model to be readily scalable to multi-scale modeling research. In this study, CREST V.2.1, which involves coupled routing scheme based on a fully distributed linear reservoir (FDLRR) scheme (Shen and Hong 2014), has been applied for the calibration and simulation of the precipitation product (Lakew 2020).

CREST Model Setup

The requirements of data to setup the CREST hydrological model are discussed below. Two types of datasets are required (i) the geospatial input data which are static in nature and (ii) the hydrometeorological input data which are dynamic time series. The geospatial data with the resolution of 1 km by 1 km resolution digital elevation model (DEM) is extracted from the United States Geological Survey (USGS) National Elevation Dataset (NED) (https://www.usgs.gov). The DEM is used to derive the geospatial data for CREST model such as slope, flow direction, and flow accumulation and drainage area of the case study. The second input consists of the dynamic hydrometeorological datasets (Lakew et al. 2020). The meteorological datasets (rainfall and PET) are geo-referenced with the geographic location of the upper Blue Nile. The process of geo-referencing and gridding is carried out using MATLAB. Daily time series observed streamflow (hydrological) dataset is used at the outlet of the basin to calibrate model parameters.

Performance Evaluation

The gauged rainfall datasets gridded applying different interpolation techniques are independently calibrated, and the simulated flows from interpolated techniques are evaluated using the reference gauged streamflow data. The performance of the simulated flow of the interpolated rainfall datasets with the observed streamflow data is evaluated using three different evaluation metrics and shown in Eqs. 24.1–24.3.

First, for statistical goodness of fit of simulated streamflow, the Nash–Sutcliffe coefficient of efficiency (Nash and Sutcliffe 1970) is utilized.

$$\mathrm{NSCE} = 1 - \frac{\sum (Q_{i,o} - Q_{i,c})^2}{\sum (Q_{i,o} - \overline{Q_o})^2} \tag{24.1}$$

where $Q_{i,o}$ is the observed streamflow of the *i*th day; $Q_{i,c}$ is the simulated streamflow of ith day; and $\overline{Q_o}$ is the average of all the daily observed streamflow values. If NSCE $\leq$ 0, then the model provides no skill in relation to using the observed mean as a predictor and values greater than zero indicating better agreement.

Second, the correlation coefficient (hereafter CC) is used to assess the agreement between the simulated flows and the observed streamflow. The value of CC is such that $-1 < \text{CC} < +1$.

$$\text{CC} = \frac{\sum (Q_{i,o} - \overline{Q_o})(Q_{i,c} - \overline{Q_c})}{\sum (Q_{i,o} - \overline{Q_o})^2 (Q_{i,c} - \overline{Q_c})^2} \tag{24.2}$$

where $\overline{Q_c}$ is the average of daily simulated streamflow values.

Third, relative bias ratio assesses the systematic bias of the simulated discharge:

Bias is defined as the average difference between the simulated flows and the observed streamflow. The value of bias can be positive as well as negative. A positive bias indicates overestimation of the observed streamflow amount while a negative bias indicates underestimation of it.

$$\text{Bias} = \frac{\sum Q_{i,c} - \sum Q_{i,o}}{\sum Q_{i,o}} * 100\% \tag{24.3}$$

The best skill occurs with NSCE = 1, CC = 1, and bias = 0%.

Results and Discussion

The simulated flows from the independently calibrated interpolated rainfall are evaluated with the observed streamflow of the upper Blue Nile Basin measured at the Eldiem station which is near to the border of Sudan and Ethiopia as well as close to the Great Ethiopian Renaissance Dam (GERD). The performance of the simulated flow from the five interpolation techniques is evaluated using NSCE, bias and CC metrics.

The NSCE performance manifests that the whole interpolation techniques reveal good performance between 0.72 and 0.89 range for both calibration and validation periods. Especially the nearest interpolation technique is more applicable to generate high resolution (0.10°, daily) interpolated rainfall data for the hydrological simulation in the complex mountainous region of the upper Blue Nile Basin. The nearest interpolation technique shows 0.89 and 0.86 of the NSCE performance metric for calibration and simulation periods, respectively, as shown in Table 24.1. This indicates that based on the NSCE performance evaluation metric, the nearest interpolation technique is more applicable to generate 0.10° spatial and daily temporal resolutions for the hydrological simulation in the upper Blue Nile Basin

Table 24.1 Types of interpolation techniques and the performance metrics

Interpolation techniques		Calibration (2000–2006)	Validation (2007–12)
Cubic	NSCE	0.77	0.72
	Bias (%)	−9.1	−15.4
	CC	0.88	0.85
Linear (Simple Kriging)	NSCE	0.76	0.72
	Bias (%)	−11.6	−17.2
	CC	0.88	0.85
Natural	NSCE	0.77	0.74
	Bias (%)	−13	−17.7
	CC	0.88	0.85
Nearest (Thiessen polygon)	NSCE	0.89	0.86
	Bias (%)	−5.8	−8.1
	CC	0.93	0.89
Spline	NSCE	0.82	0.74
	Bias (%)	−6.5	−13.34
	CC	0.90	0.86

that has complex mountainous region. Among the five interpolation techniques, even though linear reveals good performance of NSCE (>0.72) for both calibration and validation periods, but it shows low performance relative to the other rainfall interpolation techniques. The linear interpolation technique shows 0.76 and 0.72 NSCE performance to simulate the streamflow for calibration and validation periods, respectively.

A positive bias indicates an overestimation of the observed streamflow amount, while a negative bias indicates an underestimation of it. Therefore, the bias results manifest that the whole interpolation techniques show limitation to estimate the magnitude of the observed flow with negative values which is an indication of underestimation. The underestimation is due to that during simulation flow applying CREST model, the simulated flow from different products reveals limitation to estimate and to capture the peaks of the observed flow (Lakew et al. 2020). However, similar to the NSCE performance, the nearest interpolation technique reveals the best bias performance with the minimum bias value from the rest interpolation techniques for both calibration and validation periods. The bias values are −5.8% and −8.1% for calibration and validation periods, respectively, to simulate the upper Blue Nile flow using the nearest technique of gauged rainfall interpolation. The natural interpolation technique shows the maximum bias values of −13% and −17.7% for both calibration and validation periods, respectively, as shown in Table 24.1.

The CC results indicate that the simulated flow from interpolated rainfall data using the five techniques of interpolation have high performance of correlation with the observed streamflow of the upper Blue Nile Basin. The whole interpolated techniques score above 0.85 of the CC performance metric for both calibration and validation periods to simulate the observed streamflow of the basin. Especially, like the performance of NSCE and bias evaluation metrics, the CC performance metric of the nearest interpolation technique outperforms from the rest interpolated techniques to simulate the streamflow. The result of the nearest interpolation technique shows 0.93 and 0.89 of CC performance for calibration and validation periods, respectively.

The whole evaluation metrics of NSCE, bias, and CC results show that the nearest outperforms the rest interpolation techniques to interpolate the gauged rainfall and simulate the observed streamflow of the upper Blue Nile Basin.

Quantile–Quantile Evaluation

The quantile–quantile (Q-Q) plot is a graphical technique for determining if two datasets come from populations with a common distribution. A 45-degree reference line is also plotted. If the two sets come from a population with the same distribution, the points should fall approximately along this reference line (Wilk and Gnanadesikan 1968). The greater the departure from this reference line, the greater the evidence for the conclusion that the two datasets come from populations with different distributions or with the same distribution but biased or shifted with the mean.

The simulated flows of the upper Blue Nile Basin using different interpolation techniques of the gauged rainfall are plotted using Q-Q and shown in Fig. 24.2. The simulated flows were plotted for three temporal resolutions of daily, monthly, and wet season of the basin. The wet season of the basin is between June and September and the same in which the rainfall magnitude is a peak (JJAS). Notably, rainfall and the river streamflow magnitude generally peak between July and August, its inter-annual variability is greatest at the beginning and end of the rainy season, in June and September. Therefore, like the daily and monthly temporal resolutions, the wet season (JJAS) of the simulated flows of the basin using different techniques of interpolation for the gauged rainfall are plotted using the Q-Q plot. The figure of the three temporal resolutions of the Q-Q plot shows that the interpolated gauged rainfall applying different interpolation techniques try to capture the reference of the observed flow of the upper Blue Nile Basin. The performances of the interpolation techniques reveal two significant sets of quantile performances that overestimate (nearest and spline) and underestimate (cubic, linear, and natural) the observed flow for the daily, monthly, and wet season temporal resolutions. Even though the nearest and spline techniques of interpolation overestimate the observed quantile values, but it shows better performance to capture the reference line to estimate the quantile of the observed streamflow magnitude. Especially, the interpolated gauged rainfall dataset using nearest interpolation techniques shows improved performance to simulate the quantile magnitude of the observed flow simulated flow of the upper Blue Nile Basin for the three temporal resolutions. Particularly, the nearest interpolation technique shows enhanced performance to capture the peak quantiles of the observed flows for daily, monthly, and wet season temporal resolutions from the rest interpolation techniques as shown in Fig. 24.2.

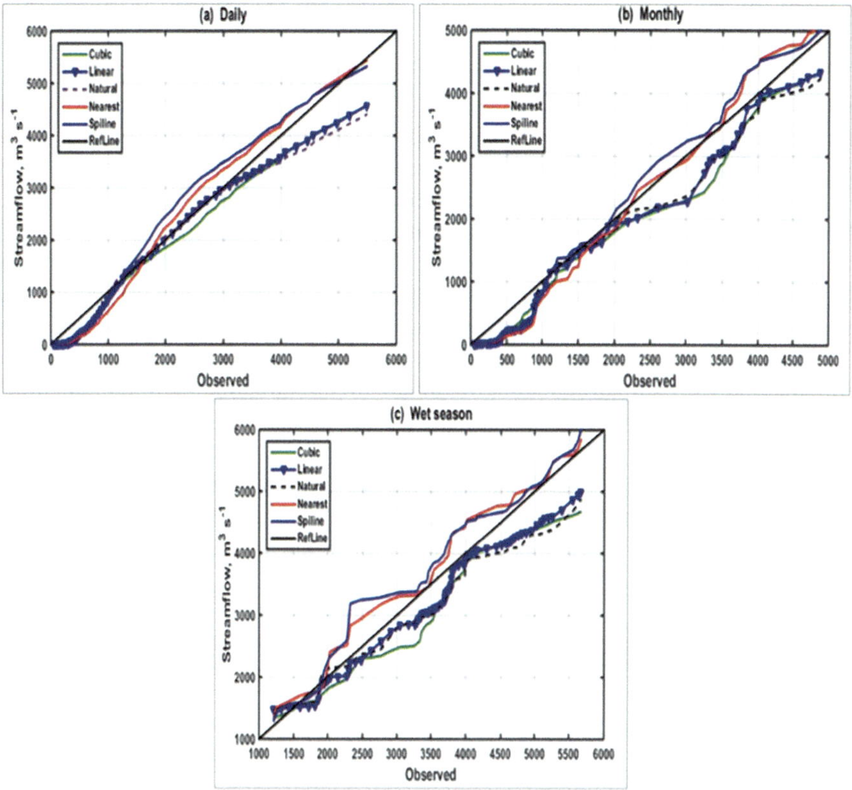

Fig. 24.2 Q-Q plot of different interpolation techniques for daily, monthly, and wet season temporal resolutions

Conclusions

The purpose of this study is to evaluate different spatial interpolation techniques to produce a daily temporal scale of gridded rainfall data (0.10°) for the hydrological simulation in a complex mountainous region of the upper Blue Nile Basin. Five interpolation techniques of cubic, linear, natural, nearest, and spline are applied to interpolate and grid observed rainfall data using MATLAB. The interpolated rainfall datasets were independently calibrated using a fully distributed hydrological model of Coupled Routing and Excess STorage (CREST). The model is calibrated from 2005 to 2012 and validated from 2013 to 2017. The performance of the interpolation techniques is evaluated using NSCE, bias, and CC metrics, as well as using Q-Q plot of graphical evaluation technique.

The NSCE result reveals that the five interpolation techniques perform between 0.72 and 0.89 range for calibration and validation periods. Especially, the nearest rainfall interpolation technique is more applicable to simulate the streamflow magnitude of the upper Blue Nile Basin.

The bias values for the five interpolation techniques are negative, which is an indication of underestimation. However, the nearest interpolation technique reveals the minimum negative bias value and the best bias performance to estimate the total volume of the observed streamflow of the upper Blue Nile Basin.

The CC result reveals that the five interpolation techniques score above 0.85 of the CC performance metric for calibration and validation periods.

The NSCE, bias, and CC results reveal that the five interpolation techniques have the potential to generate high resolution gridded rainfall data. Especially, the nearest interpolation technique outperforms the rest of the interpolation techniques.

The Q-Q plot reveals that nearest interpolation technique shows the better performance to the estimate the peak quantiles of the observed flow for the whole temporal resolution of daily, monthly, and wet seasons of the upper Blue Nile Basin.

The general result of NSCE, bias, CC, and Q-Q plot shows that the nearest interpolation technique of the gauged rainfall to generate daily high spatial resolution (0.10°) data is more applicable to simulate the streamflow in the complex mountainous region of the upper Blue Nile Basin.

Acknowledgements The author would like to thank the Ethiopian Meteorological Agency (EMA) and the Sudan Ministry of Water Resources, Irrigation, and Electricity for the updated observed rainfall and streamflow data, respectively.

References

Allen RG, Pereira LS, Raes D, Smith M (1998) Crop evapotranspiration—guidelines for computing crop water requirements—FAO irrigation and drainage paper 56. FAO Irrig Drain Pap 56. Food Agric Organ, Rome, Italy. United Nations ISBN 92–5–104219–5

Beck HE, van Dijk AIJM, Levizzani V, Schellekens J, Miralles DG, Martens B, de Roo A (2017) MSWEP: 3-hourly 0.25° global gridded precipitation (1979–2015) by merging gauge, satellite, and reanalysis data. Hydrol Earth Syst Sci 21:589–615. https://doi.org/10.5194/hess-21-589-2017

Bitew MM, Gebremichael M, Ghebremichael LT, Bayissa YA (2012) Evaluation of high-resolution satellite rainfall products through streamflow simulation in a hydrological modeling of a small mountainous watershed in Ethiopia. J Hydrometeorol 13:338–350. https://doi.org/10.1175/2011JHM1292.1

Cheng M, Wang Y, Engel B, Zhang W, Peng H, Chen X, Xia H (2017) Performance assessment of spatial interpolation of precipitation for hydrological process simulation in the three Gorges basin. Water. https://doi.org/10.3390/w9110838

Conway D (2000) The climate and hydrology of the Upper Blue Nile River. Geogr J 166:49–62. https://doi.org/10.1111/j.1475-4959.2000.tb00006.x

Conway D, Hulme M (1993) Recent fluctuations in precipitation and runoff over the Nile sub-basins and their impact on main Nile discharge. Clim Change 25:127–151. https://doi.org/10.1007/BF01661202

Cristiano E, ten Veldhuis M-C, van de Giesen N (2017) Spatial and temporal variability of rainfall and their effects on hydrological response in urban areas—a review. Hydrol Earth Syst Sci 21:3859–3878. https://doi.org/10.5194/hess-21-3859-2017

Goovaerts P (2003) Using elevation to aid the geostatistical mapping of rainfall erosivity. CATENA 34:227–242. https://doi.org/10.1016/S0341-8162(98)00116-7

Hirpa FA, Gebremichael M, Hopson T (2010) Evaluation of high-resolution satellite precipitation products over very complex Terrain in Ethiopia. J Appl Meteorol Climatol 49:1044–1051. https://doi.org/10.1175/2009jamc2298.1

Kalyan P, Rao PSRC (2013) Cubic spline interpolation with new conditions on M0 and Mn. Int J Eng Res Technol 2

Labedzki L, Kanecka-Geszke E, Bąk B, Słowińska S (2011) Estimation of reference evapotranspiration using the FAO Penman-Monteith method for climatic conditions of Poland. https://doi.org/10.13140/2.1.5080.4802

Lakew HB (2020) Investigating the effectiveness of bias correction and merging MSWEP with gauged rainfall for the hydrological simulation of the upper Blue Nile basin. J Hydrol Reg Stud 32:100741. https://doi.org/10.1016/j.ejrh.2020.100741

Lakew HB, Moges SA, Asfaw DH (2017) Hydrological evaluation of satellite and reanalysis precipitation products in the Upper Blue Nile Basin: a case study of Gilgel Abbay. Hydrology 4:39. https://doi.org/10.3390/hydrology4030039

Lakew HB, Moges SA, Anagnostou EN, Nikolopoulos EI, Asfaw DH (2019) Evaluation of global water resources reanalysis runoff products for local water resources applications: case study-Upper Blue Nile Basin of Ethiopia. Water Resour Manag https://doi.org/10.1007/s11269-019-2190-y

Lakew HB, Moges SA, Asfaw DH (2020) Hydrological performance evaluation of multiple satellite precipitation products in the Upper Blue Nile Basin, Ethiopia. J Hydrol Reg Stud 27:100664. https://doi.org/10.1016/j.ejrh.2020.100664

Ly S, Charles C, Degré A (2013) Different methods for spatial interpolation of rainfall data for operational hydrology and hydrological modeling at watershed scale. BASE [En ligne].

Nalder IA, Wein RW (1998) Spatial interpolation of climatic normals: test of a new method in the Canadian Boreal Forest. Agric for Meteorol 92:211–225

Nash JE, Sutcliffe JV (1970) River flow forecasting through conceptual models part I—a discussion of principles. J Hydrol 10:282–290. https://doi.org/10.1016/0022-1694(70)90255-6

Ruelland D, Ardoin-Bardin S, Billen G, Servat E (2008) Sensitivity of a lumped and semi-distributed hydrological model to several methods of rainfall interpolation on a large basin in West Africa. J Hydrol 361:96–117. https://doi.org/10.1016/j.jhydrol.2008.07.049

Sahlu D, Moges SA, Nikolopoulos EI, Anagnostou EN, Hailu D (2017) Evaluation of high-resolution multisatellite and reanalysis rainfall products over East Africa. Adv Meteorol 2017:14. https://doi.org/10.1155/2017/4957960

Shen X, Hong Y (2014) CREST coupled routing and excess STorage CREST User Manual v2.1. Univ. Oklahoma HyDROS Lab (http://hydro.ou.edu).

Tabios GQ III, Salas JD (1985) A comparative analysis of techniques for spatial interpolation of precipitation1. JAWRA J Am Water Resour Assoc 21:365–380. https://doi.org/10.1111/j.1752-1688.1985.tb00147.x

Taylor CM, Harding RJ, Thorpe AJ, Bessemoulin P (1997) A mesoscale simulation of land surface heterogeneity from HAPEX-Sahel. J Hydrol 188–189, 1040–1066. https://doi.org/10.1016/S0022-1694(96)03182-4

Wang J, Hong Y, Li L, Gourley JJ, Khan SI, Yilmaz KK, Adler RF, Policelli FS, Habib S, Irwn D, Limaye AS, Korme T, Okello L (2011) The coupled routing and excess storage (CREST) distributed hydrological model. Hydrol Sci J 56:84–98. https://doi.org/10.1080/02626667.2010.543087

Wang Y, Yang H, Yang D, Qin Y, Gao B, Cong Z (2017) Spatial interpolation of daily precipitation in a high mountainous watershed based on gauge observations and a regional climate model simulation. J Hydrometeorol 18:845–862. https://doi.org/10.1175/JHM-D-16-0089.1

Waterbury J (1988) Water use and demand in the Nile Basin. In: Howell PP, Lock M, Cobb S (eds) The/onglei canal impact oppor. CUP, Cambridge, pp 64–84.
Wilk MB, Gnanadesikan R (1968) Probability plotting methods for the analysis for the analysis of data. Biometrika 55:1–17. https://doi.org/10.1093/biomet/55.1.1
Yang C, Kao S, Lee F, Hung P (2004) Twelve different interpolation methods: a case study of Surfer 8.0 (2004). SPRS 778–783
Yang X, Xie X, Liu DL, Ji F, Wang L (2015) Spatial interpolation of daily rainfall data for local climate impact assessment over greater Sydney region. Adv. Meteorol. 2015:563629. https://doi.org/10.1155/2015/563629

Chapter 25
Historical Trend Analysis of Rainfall in Amhara National Regional State

Sileshie Mesfin, Anwar Assefa Adem, Arega Mullu, and Assefa M. Melesse

Abstract Ethiopian agriculture is mostly rainfed and is threatened by interannual and seasonal rainfall variability. This study aimed to assess rainfall in the Amhara Region, Ethiopia. Two rainfall datasets: observed and Climate Forecasting System Reanalysis (CFSR) were used. CFSR data was bias corrected using rain gauge data. In order to analyze the rainfall characteristics, Mann-Kendal trend test, Sen's slope estimator, connected component (CC)-based homogeneity test and seasonality index were used. The results showed that Amhara Region received mean annual rainfall of 1,150 mm. Western part of Amhara Region showed significant increasing trend. Decreasing trend was observed in the remaining part of the region but the trend was not significant at 95% of significance level except for some northeastern part of the region. Amhara Region has three homogeneous rainfall regions: Region 1, 2 and 3. The region also has four major groups of seasons. Area of Region A received 75–80% of rain over five to six months while Region B received this amount of rain in four months. Region C received similar amount of rainfall in three months and Region D received same amount of rainfall in two or less months. In 79% of Amhara Region, rainfall is unimodal and received in wet season. The rest part of the region, 21% and mainly in the eastern part, has bimodal rainfall. From the result, we can conclude that Amhara Region has ample amount of rain but concentrated in few months with considerable variability and trend. Therefore, it is important to adjust the cropping system with the variability of rainfall.

S. Mesfin (✉)
Amhara Design & Supervision Works Enterprise, P.O. Box 1921, Bahir Dar, Ethiopia

A. A. Adem
Department of Natural Resource Management, Faculty of Civil and Water Resource Engineering, Bahir Dar University, Bahir Dar, Ethiopia

A. Mullu
Hydraulic and Water Resources Engineering, Debre Markos University, Debre Markos, Ethiopia

A. M. Melesse
Department of Earth and Environment, Institute of Environment, Florida International University, Miami, USA
e-mail: melessea@fiu.edu

A. M. Melesse et al. (eds.), *Nile and Grand Ethiopian Renaissance Dam*, Springer Geography, https://doi.org/10.1007/978-3-030-76437-1_25

Keywords Rainfall · CFSR · Amhara Region · Unimodal · Bimodal

Introduction

Rainfall distribution in the Nile Basin is varied and the upper stream part of the basin receives large amount of rainfall compared to the lower section of the basin in Sudan and Egypt. Over the last few decades, the rainfall in the basin has been highly variable where some parts receive lower than long-term average rainfall. Some studies attribute these changes to climate change. These changes in rainfall can bring about changes on water resources variability, especially on river flows, soil moisture, evapotranspiration and groundwater flow (Mango et al. 2011a, b; Behulu et al. 2014; Assefa et al. 2014; Melesse et al. 2009a, 2011a; Dessu and Melesse 2013; Grey et al. 2013). The major part of the upper Blue Nile River Basin is located in the Amhara Regional State. The Nile River system receives over 85% of the annual flow from Ethiopia, mainly the Blue Nile River Basin. A number of studies in the Nile Basin encompassing the hydrology, soil erosion, sediment transport, climate change and others were studied by a number of researchers (Melesse et al. 2009b, 2011b, 2014; Abtew et al. 2009a, b; Abtew and Melesse 2014a, b, c; Yitayew and Melesse 2011; Chebud and Melesse 2013, 2009a, b; Melesse 2011; Dessu and Melesse 2012, 2013; Dessu et al. 2014). These studies indicated that the basin is highly vulnerable to hydrological variability and flow fluctuations.

There are different types of seasonality in Ethiopia. The eastern Somali region and South-Central region of Ethiopia receive rainfall twice a year during spring (March to June) and summer (June to September), whereas western part of the country receives rainfall in the summer (June to September). Many farmers grow high yielding but slowly maturing crops that are growing in both summer and spring seasons.

Based on observation in quality-controlled selected stations, between mid-1970s to late 2000s, spring and summer rainfall decreased by 15–20% across parts of southeastern, southern and southwestern Ethiopia. Because of anticipated rainfall and temperature changes in standardized indices, dryness frequencies for selected crop growing areas and contraction of areas receiving adequate rainfall for viable agricultural livelihood occurred (Funk et al. 2012a).

Rainfall analyzed isohyets in 1960–1989, 1990–2009 and as projected forward for 2010–2039, climate changes observed in pastoralists areas, show that there is northwest ward retreat of the 250 mm spring season rainfall from March–June. More frequent and severe droughts are experiencing in these areas and making it difficult to support livestock raring to recover from poor rainy season. Approximately, the area of contraction affects 7 million people (Funk et al. 2012b).

The impact of rainfall on crop production can be related to its intra-seasonal distribution or its total seasonal amount. With very low total seasonal amounts, crop production suffers the most in the extreme cases of droughts. But more understated intra-seasonal variations in rainfall distribution during crop growing periods can also cause considerable reductions in yields without a change in total seasonal amount. Bewket (2009) indicated that the total amount of rainfall and number of rain days are very important as the seasonal total rainfall volume.

The aim of this study was to analyze rainfall seasonality and variability at annual timescales and examine homogeneity by using the Amhara Region as a case study site. The specific objectives of the study reported in this chapters were to assess the magnitude of rainfall in the region and examine local scale rainfall variability, seasonality and trend by using data from a relatively adequate network of stations.

Study Area and Methods

Description of the Study Area

The research was conducted in Amhara National Regional State (ANRS) (Fig. 25.1) which is located between 9°21′ to 14°0′ North and 36°20′ and 40°20′ East. The region has varied topographic of very diverse in nature covering lowland, midland and highland plains, mountains, rugged lands, undulating landforms, chains of plateaus that range from less than 500 to 4,620 masl at Ras-Dashen (the highest peak in Ethiopia). The various topographic feature of the region along with other factors has contributed to the variable rainfall in the region.

Methods

Data Collection

First base map of the region was prepared and different types of data were collected. More than 182 meteorological stations having data periods between the year 1980 and 2016 were collected from National Meteorological Agency (NMA) of Ethiopia. All these stations recorded rainfall and many of them also record temperature but limited number of stations record meteorological variables such as relative humidity, sunshine hours and wind speed. Based on our meteorological data period (1987–2013) and after managing the missed data, only 23 stations were used for the analysis. Since we have limited meteorological stations with good spatial distribution and short period of rainfall records, this study used additional meteorological data from climate forecast system reanalysis (CFSR) (https://globalweather.tamu.edu/). This study, therefore, include 152 rainfall grid points covering the extent of the region. This CFSR data was used after bias correction using ground observations. From the 152 grid points, 119 were selected for the analysis after managing

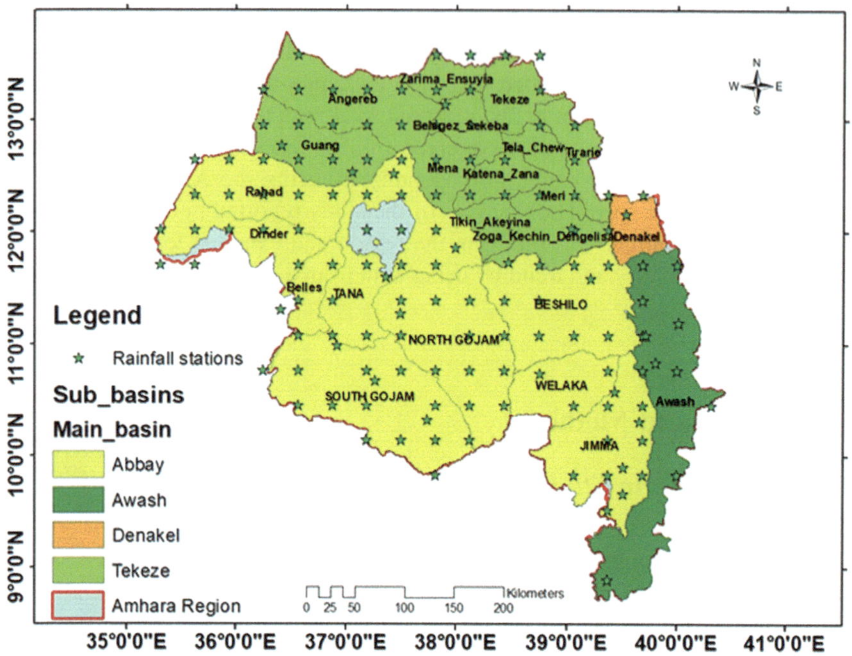

Fig. 25.1 Location map of the study area

goodness of fit of the bias corrected versus the observed data. Therefore, this study used 23 stations data from the observed and 119 grid points from CFSR.

Data Analysis

Filling Missing Data

This study used the method of arithmetic mean, which is a simple method for filling missing rainfall data. The missed data was computed from records of several stations at the same time summed and then divided by the number of stations. This method is applied if the magnitude of the difference between the stations' average annual rainfall and annual rainfall the station having missed data is less than 10%. The formula of missing data in station (p) is given by

$$P = \frac{P_1 + P_2 + P_3 + \cdots P_n}{N} \tag{25.1}$$

where p is missing rainfall data; P_1, P_2, P_3, …, P_n are rainfall data at the station 1, 2, 3, …, n; and n is number of stations.

The normal ratio method was used in cases when the normal annual precipitation at any of the index station is different by more than 10% of the station in question. In this method, the amounts at the index stations are weighted by the ratios of the normal annual precipitation values. That is, precipitation P_x at station X is:

$$P_x = \frac{1}{m}\sum_{i}^{m}\left[\frac{N_x}{N_i}\right]P_i \tag{25.2}$$

where P_x = Estimate for the ungauged station, P_i = Data values of station used for estimation, N_x = Normal annual data value of X station, N_i = Normal annual data value of surrounding stations and m = Number of surrounding stations.

Linear regression between test station (station A) and observed stations (station B, and C) is given by

$$P_x = a + bAP_{\mathrm{A}} + bBP_{\mathrm{B}} + bCP_{\mathrm{c}} \tag{25.3}$$

where a and b are regression constants.

Data Preparation

Observed meteorological data were not adequate to cover the region and the data time period is not similar. Some have long data record but some are very much recent, therefore, additional data were essential, and hence 152 sets of data were downloaded from CFSR archives from 1987–2013 for 27 years. The CFSR data was biased with underestimates or overestimates in the Amhara Region. These data were adjusted and bias corrected by specific locality of observed data. Specific locality was prepared by computing and dividing the monthly observed values into many classes by interpolating throughout the region. The monthly values were interpolated by kriging and the region was classified into 12 regions arbitrarily and converted to polygon. Then, the average values were taken to represent the observed monthly mean value. The raw data of CFSR were selected by clipping the point of raw CFSR data that is inside the classified region.

Linear scaling (LS) bias correction method was used to correct and adjust the CFSR data. The LS can match the monthly mean of corrected values with that of observed values (Lenderink et al. 2007). A multiplier and temperature with an additive term are used in the correction of the precipitation.

$$P_{\mathrm{c}} = P_{\mathrm{r,day}} * \frac{P_{\mathrm{o,month}}}{P_{\mathrm{r,month}}} \tag{25.4}$$

$$T_{\mathrm{c,day}} = T_{\mathrm{r,day}} + T_{\mathrm{o,month}} - T_{\mathrm{r,month}} \tag{25.5}$$

where P_{c}, P_{o} and P_{r} are corrected, observed and raw precipitation, respectively. Similarly, T_{c}, T_{o} and T_{r} are corrected, observed and raw temperature, respectively. The corrected CFSR data with nearby observed data were tested using performance evaluation criteria such as Nash–Sutcliffe efficiency (NSE), coefficient of determination (R^2), mean absolute error (MAE) and percent bias (PBIAS).

$$\mathrm{RMSE} = \sqrt{\frac{1}{N}\sum (P_o - P_c)^2} \tag{25.6}$$

$$\mathrm{NSE} = 1 - \frac{\sum (P_o - P_c)^2}{\sum \left(P_o - P_{o,\mathrm{mean}}\right)^2} \tag{25.7}$$

$$\mathrm{PBIAS} = \frac{\sum (P_o - P_c) * 100}{\sum P_o} \tag{25.8}$$

Mann–Kendal Trend Test

Mann–Kendall test is based on the difference ($X_i - X_K$) in years having rainfall in temporal periods ($X_i, \ldots, X_N$). S is calculated based on total signs (it is always between −1 and 1). In cases when $N > 10$, the standard Z is calculated as follows:

$$S = \sum_{i=1}^{n-1} \sum_{j=i+1}^{n} \mathrm{sgn}\left(X_j - X_i\right) \tag{25.9}$$

where the X_i is the actual time data for the time series of i = 1, 2, 3, to n, n is sample size. The S has approximately normal distribution.

The variance of S is

$$\mathrm{Var}(S) = \frac{n(n-1)(2n+5) - \sum_{i=1}^{m} t(t-1)(2t+5)}{18} \tag{25.10}$$

where t is the number of times (i.e., frequency) that the rank t appears. The MK test uses the following test statistic and positive value of Z shows the increasing trend and negative value shows decreasing trend.

$$Z_c = \begin{cases} (s-1)/\left(\sqrt{\mathrm{Var}(s)}\right), & s > 0 \\ 0, & S = 0 \\ \frac{s+1}{\sqrt{\mathrm{Var}(s)}}, & s < 0 \end{cases} \tag{25.11}$$

Sen's Slope Estimator

The magnitude of the drift of data points can be computed using Sen's slope (β) method, as described by (Sen 1968). Sen's method is a nonparametric method (Sen 1968; Theil 1950) for estimating the slopes of N pairs of data points. A positive value of Sen's slope indicates an 'upward' (increasing) trend and a negative value suggests a 'downward' (decreasing) trend in the time series.

Rainfall Homogeneity

Coefficient of variation was computed for each station first and classified in to three regions with 10%–20% (Region I), 20%–30% (Region II) and greater 30% (Region III) (Hare 2003). Then, homogeneity test was held to evaluate the homogeneity between stations in the region. Taking the coefficient of variation (CV) of each gauge and computing their CV's, stations were grouped according to the coefficient of variation in ascending order. Station which has low CV values are grouped together and those which have high CV and medium CV are grouped together similarly. Using the equations indicated from 25.12 through 25.15, the CV of their CV's were computed and connected component (CC) value was taken as the threshold of 0.3 and homogeneity test was conducted. The areas were tested for their significance as homogenous rainfall regions which would have CC value less than 0.3. There is less variability of annual rainfall in the station in those grouped under Region I and they are homogenous between stations. Similarly, the coefficient of variation of high value was grouped together that means there is high temporal variability of annual rainfall from year to year in those stations and this property makes them grouped together.

$$\mathrm{CC} = \frac{\mathrm{st\,dev\,CV}}{\mathrm{mean\,CV}} \tag{25.12}$$

$$\mathrm{mean\,CV} = \frac{\mathrm{CV}}{N} \tag{25.13}$$

$$\mathrm{st\,dev\,CV} = \sqrt{\frac{\sum (\mathrm{CV}_i - \mathrm{mean\,CV})^2}{N}} \tag{25.14}$$

$$\mathrm{CV} = \frac{\mathrm{STDEV\ of}\ X}{\mathrm{Mean\ of}\ X} \tag{25.15}$$

Rainfall Regime Classifications

The Seasonality Index (SI) shows the standardized timing and duration of rainfall without the influence of rainfall amount (Walsh and Lawler 1981). The SI is defined as the sum of the absolute deviation of mean monthly rainfall from the overall monthly mean and divided by the mean annual rainfall (R):

Table 25.1 Classification of the seasonality index (SI) based on Walsh and Lawler (1981) rainfall duration

Class code	Rainfall regime	Seasonality index (SI)	Rainfall duration in month
1	Very equable	≤ 0.19	≥ 270
2	Equable with a definite wetter season	0.20–0.39	180–269
3	Rather seasonal with a short drier season	0.40–0.59	150–179
4	Seasonal	0.60–0.79	120–149
5	Markedly seasonal with a long drier season	0.80–0.99	90–119
6	Most rain in 3 or less month	1.00–1.19	60–89
7	Extreme, almost all rain in 1 to 2 months	≥ 1.20	<60

$$SI = \frac{1}{R}\sum_{n=1}^{n=12} \text{abs}\left(X_n - \frac{R}{12}\right) \tag{25.16}$$

The value of the SI varies from 0 (when all months share the same amount of rainfall) to 1.83 (when all rainfall incidences occur in a single month). Walsh and Lawler (1981) also proposed rainfall regime classification based on SI values (Table 25.1).

Seasons of the Amhara Region were also identified as bimodal and unimodal in the region. Bimodal region received rainfall in two seasons which are in Belg (January to March) and Kiremt (June to September) and unimodal rainfall received only one season Kiremt (June to September).

Results and Discussion

Rainfall Characteristic of the Region

Observed Versus Bias Corrected Rainfall

It is shown that CFSR dataset exhibited underestimation and overestimation of the mean annual rainfall. However, the overall trend of the raw CFSR dataset overestimates the mean annual rainfall of the Amhara Region as it has been reviewed in Seasonality Index study of Ethiopia (Berhanu et al. 2016). However, corrected CFSR rainfall data has excellent goodness of fit with the observed rainfall data (Table 25.2) with most part having positive value of percentage bias (PBIAS) and Nash–Sutcliffe efficiency (NSE) value 0.98 and more. The coefficient of determination (R^2) was also more than 0.79. The Golina and Awash watershed stations were treated separately by observing the value with each month bias correction. Since there was a poor fit due to bimodal type of rainfall, the bias correction was done on monthly basis and that improved the fit.

Table 25.2 Performance of corrected CFSR rainfall for selected grid points as an example

Main basins	Station	R^2	NSE	PBIAS
Abbay	p123372	0.97	0.999	28.94
Abbay	p111372	0.87	0.999	9.96
Awash	p95397	0.79	0.999	9.31
Awash	p92397	0.82	0.999	20.32
Denakel	p120397	0.14	0.988	−21.48
Tekeze	p117384	0.98	0.998	32.66
Tekeze	p130378	0.96	0.998	−20.46

Mean Annual and Areal Rainfall

The mean annual rainfall of Amhara Region ranged from 586 to 2362 mm. These data are used to correct raw data of CFSR rainfall. Figure 25.2 showed the

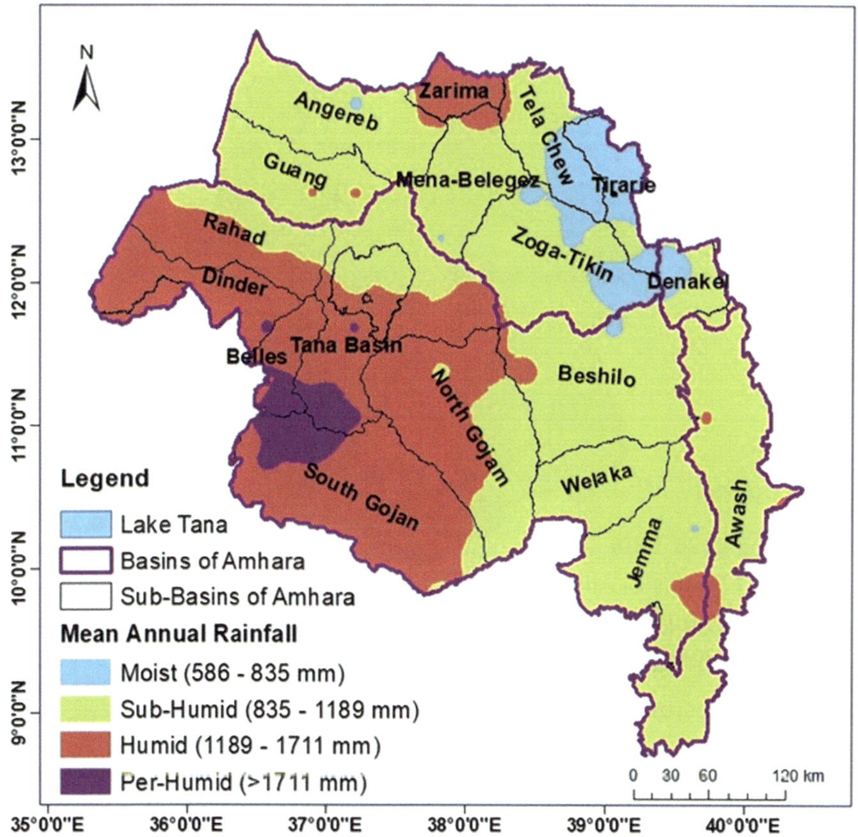

Fig. 25.2 Mean annual rainfall distribution of Amhara Region

interpolated mean annual rainfall of the region. Most of the region received mean annual rainfall between 910 and 1,400. Wag Himra Zone is the area receiving low amount of annual rainfall (~586 mm) while Awi Zone is the area receiving high amount of rainfall (~2,362 mm). The mean annual rainfall of Amhara Region is 1,150 mm with different spatial distribution and temporal variability. This average annual rainfall contributes 180.7 billion cubic meter (BCM) of rain water.

Rainfall Trend of Amhara Region

Analysis of Sen's slope indicated that most part of Amhara Region rainfall was decreasing during this study period, but a significant decreasing trend was observed around northwestern area of the region. The magnitude of the decrease, as indicated by the value given by Sen's slope method, was very high. A statistically

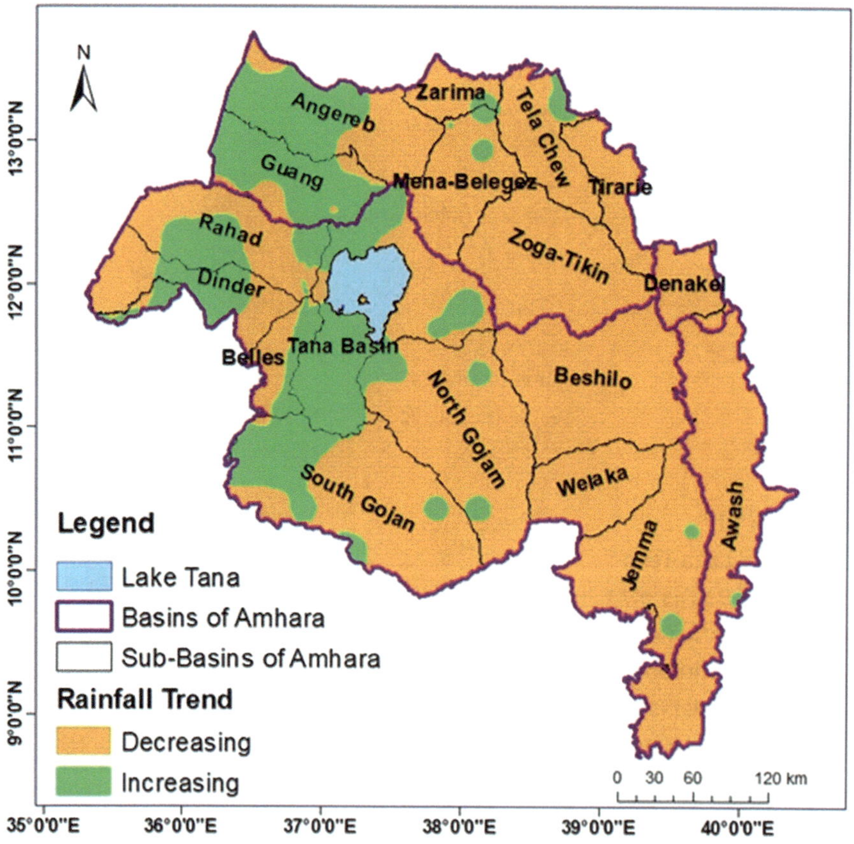

Fig. 25.3 Areas experiencing decreasing and increasing rainfall trends in Amhara Region based on Sen's slope

non-significant decreasing and non-significant increasing trend was observed for the most part of the region. In some places, significant increasing was observed. The value of Sen's slope ranged between −48 and 20. The interpolated value of slope shows decreasing and increasing trend of rainfall in the region (Figs. 25.3 and 25.4).

The trend significance and area proportions are indicated in Table 25.3 and showed that 63% of the region has no significant change at 95% significance level, even though there is a decrease in rain amount. Sekota, Gazgibla, Gidan, Kobo, Lasta, Guba Lfto, Delanta, Dawunt, Mekidela, Tenta, Saynit, Dessie, Leganbo, Kelela, Woredas show significantly decreasing rainfall trend.

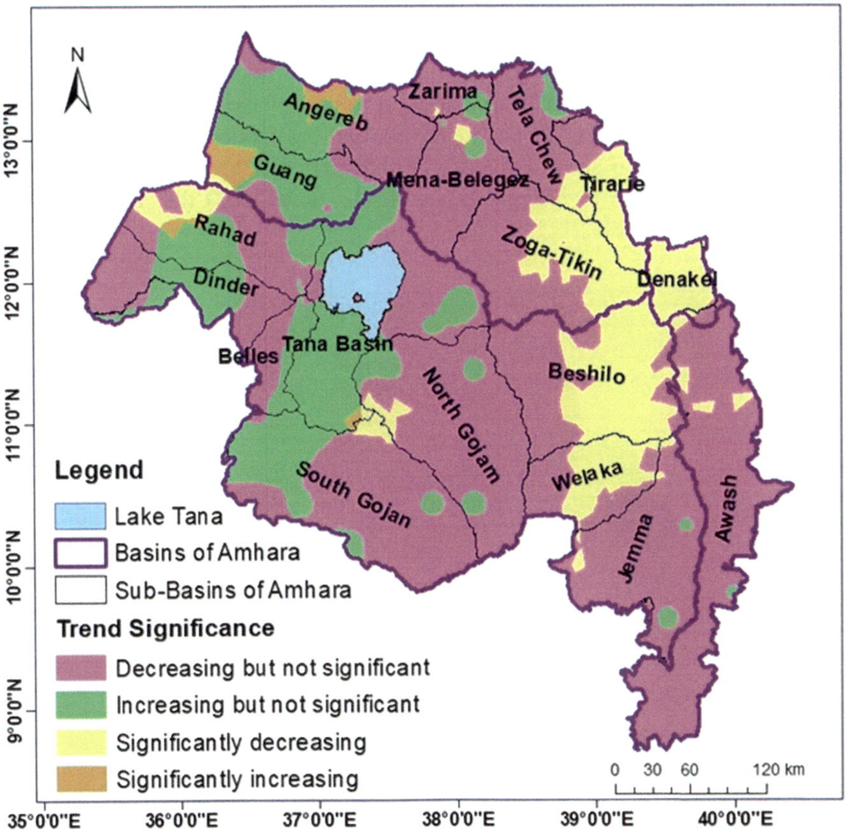

Fig. 25.4 Spatial representation of rainfall trend classes based on Mann–Kendal trend test

Table 25.3 Areas experiencing decreasing and increasing rainfall trends in Amhara Region based on their significance level

No	Significance	Area (km^2)	Area (%)
1	Significantly decreasing	24,064.1	15.4
2	Decreasing but not significant	98,580.2	63.0
3	Significantly increasing	2,245.02	1.4
4	Increasing but not significant	31,477.9	20.1
	Total	156,367.3	100

Rainfall Variability

The region has got three different homogenous regions based on rainfall variation. The western part of Amhara has less variability of rainfall which has minimum value of CV (10%) and maximum (20%) which is grouped under Region 1, covers total area of 35,386.68 km^2 (22.52%). The East Amhara part has high variability from year to year, (30%–63% CV) which is grouped under Region 3 and covers total area of 63,283 km^2 (40.28%). The central part of the regional state grouped under Region 2 covers 58,457 km^2 (37.20%) and has moderate variability (CV of 20%–30%). Areas that show a coefficient of variation (CV) greater than 30% are reported as vulnerable to drought (Haile 1988; Hare 2003) that means 40% of

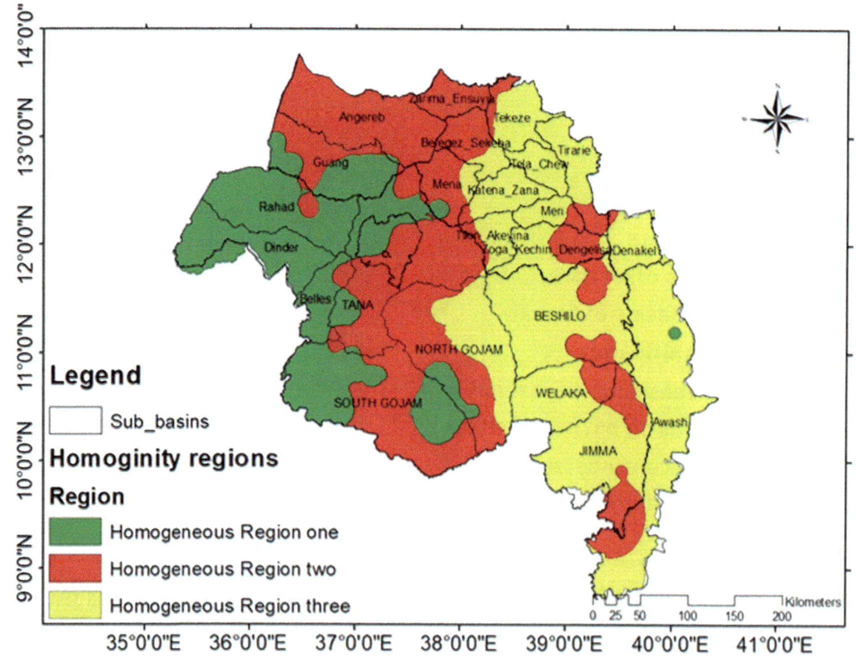

Fig. 25.5 Spatial representation of rainfall variability under three homogeneous regions

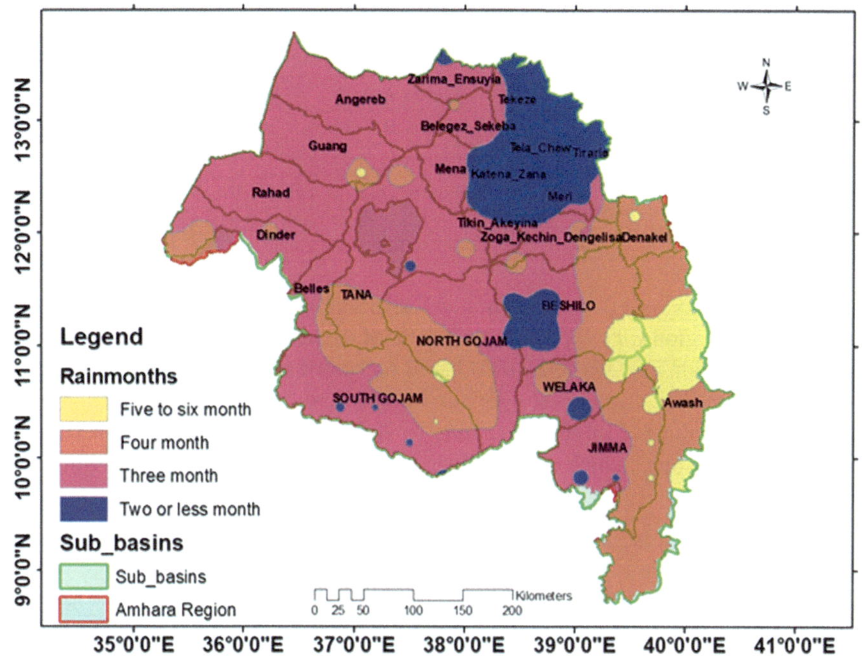

Fig. 25.6 Seasonality index of Amhara Region

Table 25.4 Area contribution of homogeneous rainfall variability regions in Amhara Region

Region	Area (km^2)	Area (%)
Less variable (Region 1)	35,386.68	22.52
Moderately variable (Region 2)	58,456.99	37.20
Highly variable (Region 3)	63,283.00	40.28
Total	157,126.67	100

Amhara Region is vulnerable to drought and is shown as homogeneous Region 3 (Fig. 25.5; Table 25.4)

Seasonality Index and Seasons of the Region

The value of SI ranged between 0.72 and 1.23 (Fig. 25.6). The SI value increases from the southwest to the northeast of the region. The amount of rainfall an area receives significantly affects its agricultural prospects. This study grouped the seasonality index into four categories. Area of Region A receives 75–80% of rain in 5–6 months and it covers 5.09% of the total region. Similarly, Region B, Region C

Table 25.5 Rainfall duration in areas of Amhara Region

Number of rainfalls receiving months	Region	Area (ha)	Proportion (%)
Five months	Region A	799,890.1	5.09
Four months	Region B	4,581,629.8	29.16
Three months	Region C	8,497,029.5	54.08
Two or less months	Region D	1,8341,29.6	11.67
Total		15,712,679.1	100.00

and Region D receive 70–80% of their annual rainfall in 4, 3 and 2 or less months, respectively. Region C covers more than 50% of the area in the region (Table 25.5).

Amhara Region also was classified based on seasonal rainfall distribution. There are two seasons in Amhara Region: unimodal and bimodal. Unimodal rainfall region receives rainfall in summer (June to August) and bimodal season has two distinct wet seasons, which are summer (most part of rainfall distribution) and

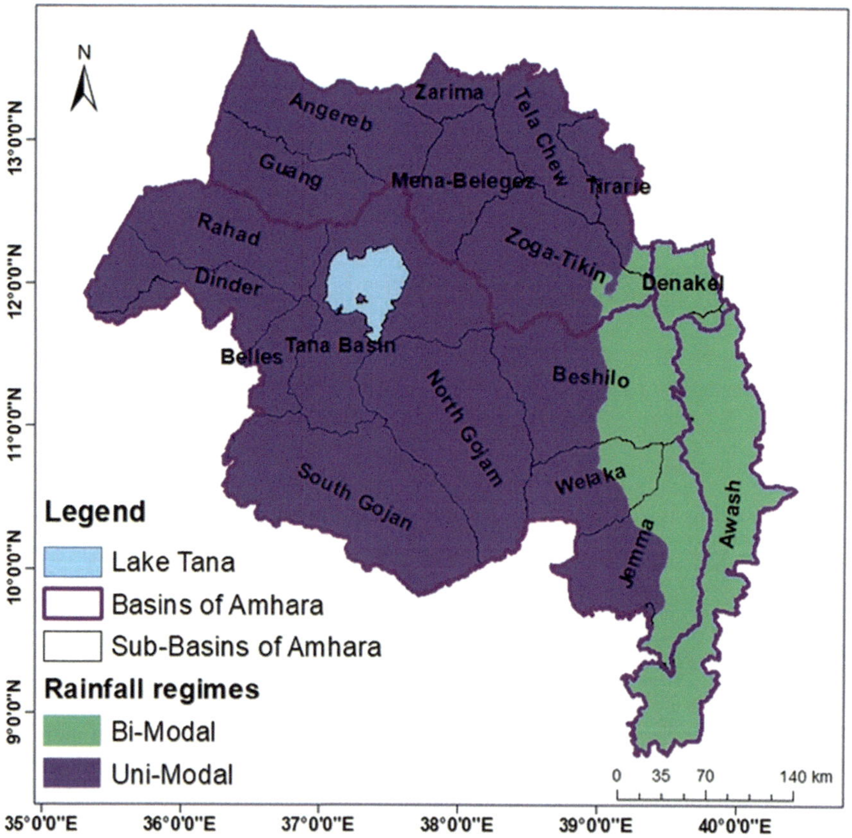

Fig. 25.7 Bimodal and unimodal seasonal coverage of Amhara Region

Table 25.6 Areal coverage of bimodal and unimodal seasons in Amhara Region

Season	Area (km^2)	Area in %
Un-modal	123,297.37	78.5
Bimodal	33,829.38	21.5
Total	157,126.75	100.0

spring (January to March). In areas having spring season, 40% of annual rainfall is received in this period. Bimodal seasonal distribution of rainfall covers 21.5% of the region and the remaining areas in the region receives unimodal rainfall. Figure 25.7 shows the map of seasons in Amhara Region (Table 25.6).

Summary

Rainfall pattern is decreasing in most part of the region, and in some parts, it is not significantly decreasing. As a result, in some places of the region, drought is intense, therefore, afforestation and some method of water conservation measures are necessary.

Acknowledgements We are grateful to Amhara Design and Supervision Works Enterprise (ADSWE), Land Use and Environmental Impact Assessment study division staffs and Abbay Basin Authority especially, to Habtamu Tamir. We thank the National Meteorological Agency of Ethiopia for the data provided.

References

Abtew W, Melesse AM (2014a) Nile River Basin hydology. In: Melesse AM, Abtew W, Setegn S (eds) Nile River Basin: ecohydrological challenges, climate change and hydropolitics, pp 7–22

Abtew W, Melesse AM (2014b) Climate teleconnections and water management. In: Nile River Basin. Springer International Publishing, Berlin, pp 685–705

Abtew W, Melesse AM (2014c) Transboundary rivers and the Nile. In: Nile River Basin. Springer International Publishing, Berlin, pp 565–579

Abtew W, Melesse A, Desalegn T (2009a) Spatial, inter and intra-annual variability of the Blue Nile River Basin Rainfall. Hydrol Process 23(21):3075–3082

Abtew W, Melesse A, Desalegn T (2009b) El Niño Southern oscillation link to the Blue Nile River Basin hydrology. Hydrol Process Special Issue Nile Hydrol 23(26):3653–3660

Assefa A, Melesse AM, Admasu S (2014) Climate change in Upper Gilgel Abay River Catchment, Blue Nile Basin Ethiopia. In: Melesse AM, Abtew W, Setegn S (eds) Nile River Basin: ecohydrological challenges, climate change and hydropolitics, pp 363–388

Behulu F, Setegn S, Melesse AM, Romano E, Fiori A (2014) Impact of climate change on the Hydrology of Upper Tiber River Basin using bias corrected regional climate model. Water Resour Manage 28:1327–1343

Berhanu B, Seleshi Y, Demisse SS, Melesse AM (2016) Bias correction and characterization of climate forecast system re-analysis daily precipitation in Ethiopia using fuzzy overlay. Meteorol Appl J Forecast Pract Appl Training Tech Model 23(2):230–243

Bewket W (2009) Rainfall variability and crop production in Ethiopia: case study in the Amhara region. Paper presented at the proceedings of the 16th international conference of Ethiopian studies.

Chebud Y, Melesse AM (2013) Stage level, volume, and time-frequency change information content of Lake Tana using stochastic approaches. Hydrol Process 27(10):1475–1483. https://doi.org/10.1002/hyp.9291

Chebud YA, Melesse AM (2009a) Numerical modeling of the groundwater flow system of the Gumera Sub-Basin in Lake Tana Basin Ethiopia. Hydrol Process Special Issue Nile Hydrol 23 (26):3694–3704

Chebud YA, Melesse AM (2009b) Modeling lake stage and water balance of Lake Tana Ethiopia. Hydrol Process 23(25):3534–3544

Dessu SB, Melesse AM (2012) Modeling the rainfall-runoff process of the Mara River Basin using SWAT. Hydrol Process 26(26):4038–4049

Dessu SB, Melesse AM (2013) Impact and uncertainties of climate change on the hydrology of the Mara River Basin. Hydrol Process 27(20):2973–2986

Dessu SB, Melesse AM, Bhat M, McClain M (2014) Assessment of water resources availability and demand in the Mara River Basin. CATENA 115:104–114

Funk C, Michaelsen J, Marshall M (2012a) Mapping recent decadal climate variations in precipitation and temperature across Eastern Africa and the Sahel. In: Wardlow BD, Anderson MC, Verdin JP (eds) Remote sensing of drought: innovative monitoring approaches. CRC Press, Boca Raton, FL, pp 331–358

Funk C, Rowland J, Eilerts G, Kebebe E, Biru N, White L, Galu G (2012b) A climate trend analysis of Ethiopia. US Geological Survey, Fact Sheet, 3053

Grey OP, Webber DG, Setegn SG, Melesse AM (2013) Application of the soil and water assessment tool (SWAT Model) on a small tropical island state (Great River Watershed, Jamaica) as a tool in integrated watershed and coastal zone management. Int J Tropical Biol Conserv 62(3):293–305

Haile T (1988) Causes and characteristics of drought in Ethiopia. Ethiopian J Agric Sci

Hare W (2003) Assessment of knowledge on impacts of climate change-contribution to the specification of art. 2 of the UNFCCC: impacts on ecosystems, food production, water and socio-economic systems

Lenderink G, Buishand A, Deursen Wv (2007) Estimates of future discharges of the river Rhine using two scenario methodologies: direct versus delta approach. Hydrol Earth Syst Sci 11 (3):1145–1159

Mango L, Melesse AM, McClain ME, Gann D, Setegn SG (2011a) Land use and climate change impacts on the hydrology of the upper Mara River Basin, Kenya: results of a modeling study to support better resource management. Special issue: climate, weather and hydrology of East African Highlands. Hydrol Earth Syst Sci 15:2245–2258. https://doi.org/10.5194/hess-15-2245-2011.

Mango L, Melesse AM, McClain ME, Gann D, Setegn SG (2011b) Hydro-meteorology and water budget of Mara River basin, Kenya: a land use change scenarios analysis. In: Melesse A (ed) Nile River Basin: hydrology, climate and water use, chapter 2. Springer Science Publisher, pp 39–68. https://doi.org/10.1007/978-94-007-0689-7_2.

Melesse A, Abtew W, Setegn SG (2014) Nile River Basin: ecohydrological challenges, climate change and hydropolitics. Springer Science & Business Media

Melesse A, Bekele S, McCornick P (2011a) Hydrology of the Niles in the face of land-use and climate dynamics. In: Melesse A (ed) Nile River Basin: hydrology, climate and water use. Springer Science Publisher, pp vii–xvii. https://doi.org/10.1007/978-94-007-0689-7

Melesse AM (2011) Nile River Basin: hydrology, climate and water use. Springer Science & Business Media

Melesse AM, Abtew W, Setegn S, Dessalegne T (2011b) Hydrological variability and climate of the Upper Blue Nile River Basin In: Melesse A (ed) Nile River Basin: hydrology, climate and water use, chapter 1. Springer Science Publisher, pp 3–37. https://doi.org/10.1007/978-94-007-0689-7_1

Melesse AM, Loukas AG, Senay G, Yitayew M (2009a) Climate change, land-cover dynamics and ecohydrology of the Nile River Basin. Hydrol Process Special Issue Nile Hydrol 23(26):3651–3652

Melesse AM, Abtew W, Desalegne T, Wang X (2009b) Low and high flow analysis and wavelet application for characterization of the Blue Nile River System. Hydrol Process 24(3):241–252

Sen PK (1968) Estimates of the regression coefficient based on Kendall's tau. J Am Stat Assoc 63 (324):1379–1389

Theil H (1950) A rank-invariant method of linear and polynominal regression analysis (parts 1–3). Paper presented at the Ned. Akad. Wetensch. Proc Ser A

Walsh R, Lawler D (1981) Rainfall seasonality: description, spatial patterns and change through time. Weather 36(7):201–208

Yitayew M, Melesse AM (2011) Critical water resources management issues in Nile River Basin. In: Melesse A (ed) Nile River Basin: hydrology, climate and water use, chapter 20. Springer Science Publisher, pp 401–416. https://doi.org/10.1007/978-94-007-0689-7_20

Chapter 26
Micro-Watershed Hydrological Monitoring and Evaluation. A Case Study at Lake Tana Sub-Basin, Ethiopia

Sileshie Mesfin, Arega Mullu, and Kibret Kassie

Abstract The vicious circle of poverty in Ethiopia is the manifestation of degradation of natural resources. Ministry of Agriculture (MoA) has implemented watershed management activities to restore the soil and water resources. This study was conducted in hydrologic monitoring sites in Tana Lake sub-basin, Ethiopia with the main objective of evaluating the response of watersheds in water quality and quantity as a result of implemented watershed management practices. Rainfall, hydrology and sediment concentration data of the watersheds were used and HYDATA and excel spread sheet were used for the analysis. Paired *t*-test was used to evaluate the significance of the rainfall, water quality and quantity. The results of the analysis show that there is no significant change observed in rainfall time series. There is an increase of baseflow in stations 6, 7 and 9 from main or project stations and 11 from control stations. But there was no significant change observed in stations 1 and 2 from main and 10 from control stations. The paired *t*-test also shows that a significant change was observed on the sediment concentration between the year 2010 and 2014 but there was no change between the main and control stations. The average sediment load reduction was 50% on streams in these small watersheds ranging 14–70% within the study period. Therefore, it was concluded that selected watershed management practices have significant effect on the increase of baseflow and hence sediment loss reduction.

Keywords Lake Tana sub-basin · HYDATA · Baseflow · Sediment load

S. Mesfin (✉)
Amhara Design and Supervision Works Enterprise, P.O. Box 1921, Bahir Dar, Ethiopia

A. Mullu
Hydraulic and Water Resources Engineering, Debre Markos University, Debre Markos, Ethiopia

K. Kassie
Bureau of Water, Irrigation and Energy, Bahir Dar, Ethiopia

A. M. Melesse et al. (eds.), *Nile and Grand Ethiopian Renaissance Dam*, Springer Geography, https://doi.org/10.1007/978-3-030-76437-1_26

Introduction

Understanding hydrological processes of watersheds will necessitate acquisition of field data for evaluation as well as modeling the dynamics of water sediment and pollutants. Various researchers used different models for simulating hydrological variables including runoff, evapotranspiration, groundwater flow and others. The most commonly used hydrological model is Soil and Water Assessment Tool (SWAT). It was used to model major hydrologic process in various parts of the world (Dessu and Melesse 2012, 2013; Dessu et al. 2014; Wang et al. 2008a, b, c, 2006; Wang and Melesse 2005, 2006; Behulu et al. 2014, 2012; Setegn et al. 2009, 2010a, c, 2011, 2014; Mango et al. 2011a, b; Getachew and Melesse 2012; Assefa et al. 2014; Grey et al. 2013; Yesuf et al. 2015).

One of the most important landscape processes responsible for hydrological variability is land use and land cover. Land use/land cover dynamics is an important landscape process capable of altering the fluxes of water, sediment, contaminants and energy. Mainly caused by human, impact of land use on water resources availability is high. Various studies used land cover mapping tools and methods to understand land use changes, inventory of forest and natural resources as well as understand the changes in the hydrologic behavior of watersheds (Getachew and Melesse 2012; Mango et al. 2011a, b; Wondie et al. 2011, 2012; Melesse and Jordan 2002, 2003; Melesse et al. 2007; Yesuf et al. 2013).

Degraded watershed which is deforested tends to increase wet season flow and hence reduce recharge and baseflow. Sediment deposition in reservoirs and dams has been a critical problem for water resources managers. A number of studies used laboratory, field scales and modeling studies to understand soil erosion and sediment dynamics in various regions (Aga et al. 2018, 2019; Defersha and Melesse 2012a, b; Defersha et al. 2010, 2012; Maalim and Melesse 2013; Maalim et al. 2013; Setegn et al. 2010b; Melesse et al. 2011a, b; Msagahaa et al. 2014; Wang et al. 2008a; Mekonnen and Melesse 2011; Setegn et al. 2009; Yesuf et al. 2015).

Ethiopia is a home of high mountains and plateaus. The tallest mountain, Ras Dashen, the largest water body of Ethiopia, Lake Tana and the source of Blue Nile is found in Amhara Region. Thus, the country is known as "the Water Tower of Africa." But nowadays rivers are murky which carries mud, the agricultural areas are degraded and at risk of loss of productivity. This is mainly due to lack of appropriate natural resource management and a series problem which may affect the quality and quantity of precious natural resource Ethiopia has; water (FAO 2001).

Lake Tana is the source of the Blue Nile River. Hydrology of the Nile River Basin has been studied by various researchers. Research on stream flow modeling, sediment dynamics, teleconnections and river flow, land use dynamics, climate change impact, groundwater flow modeling, hydrodynamics of Lake Tana, water allocation and demand analysis was reported in various studies (Melesse et al. 2009, 2010, 2011a, b, 2014; Abtew et al. 2009a, b; Abtew and Melesse 2014a, b, c; Yitayew and Melesse 2011; Chebud and Melesse 2013, 2009a, b; Setegn et al. 2009, 2010a, c; Melesse 2011; Dessu and Melesse 2012, 2013; Dessu et al. 2014).

According to the report of Bureau of Agriculture (BoA), in Ethiopia, more than ninety woredas with more than two million household inhabitants are drought prone and at risk of water shortage. Even though, Ethiopia is rich in water resource and has many rivers, millions of people are do not have access to freshwater. Ethiopia receives more than 750 mm of rainfall in over 70% of the arable land while 110 billion cubic meters (BCM) of rainwater loss as surface runoff. Besides, the groundwater resource is estimated around 4.6 BCM (Rämi and OCHA 2003).

The Tana and Beles Integrated Water Resource Development Program (TBIWRDP) was implemented over a period of eight years, since the 2008/2009 fiscal year. It was undertaken in the Abbay River Basin and has focused on the Tana and Beles sub-basins which are located within the Amhara and Benishangul Gumuz Regions, respectively (Rojas et al. 2016).

Hydrological monitoring as part of the Tana-Beles watershed monitoring and evaluation program was designed and established in the study area. The micro-watershed monitoring program had been operational since 2009 as part of Micro-watershed Monitoring Program (MMP) of Bureau of Water Irrigation and Energy Development (BoWIED). The training and technical advice in collaboration with the Bureau of Agriculture was provided by the technical advisers (TA) from the support of the Finland Government. Tana Sub-basin Organization (TaSBO) was responsible for the continuation of the MMP activities. The monitoring network consists of 15 stations in small micro-watersheds. Data was collected twice a day recording rainfall, water flow, and sediment concentration (when the water is turbid, sediment samples were collected) at the monitoring stations (Kainulainen 2012).

It is recognized that a long time series and data of several sample watersheds are required to establish hydrological impact of watershed management. But because of data constraint, data are not collected in all stations. Eight out of 15 stations, which are stations 1, 2, 6, 7, 9, from main or project and 10, 11 and 14 from control micro-watersheds were selected for analysis from 2009 to 2015. The main objective of this study was to evaluate watershed management in headwater landscapes in respective of the response to baseflow and sediment yield and hence evaluate the effect of selected watershed management practices in reducing soil erosion and runoff. Comparisons were made between the main or project and the control micro-watersheds.

Materials and Methods

Description of the Study Area

Tana sub-basin is located within the upper Blue Nile Rivers Basin, between 10°56′ to 12°45′ N latitude and 36°44′–38°14′ E longitude (Fig. 26.1). The sub-basin has a total area of 14,845.5 km^2, of which Lake Tana occupies 3,057.4 km^2. The seasonal and perennial wetlands have an area of 382.1 and 11.9 km^2, respectively. Tana sub-basin is the main sources of the Blue Nile River and one of the sixteen

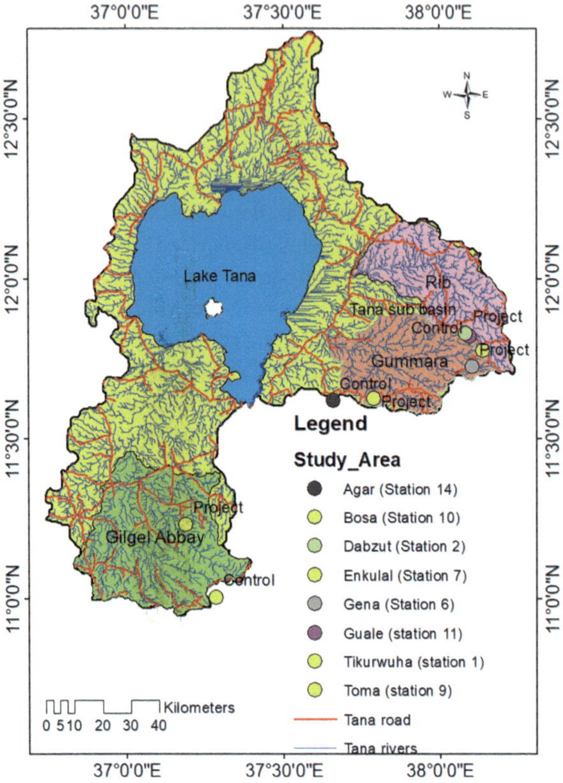

Fig. 26.1 Location map of the study watersheds

sub-basins in the Abbay (Blue Nile) Basin. The sub-basin is located in Amhara National Regional State (ANRS) encompassing four administrative zones (North Gondar zone in the north and north-west, South Gondar in the east & south-east, West Gojam and Awi zone in the south and south-west).

Tana sub-basin can be divided into different physiographic regions. These include plateau top, rugged mountainous, volcanic terrain, escarpment, low relief terrain in quaternary volcanic rocks and sediments and flat and gentle plains covered with lacustrine sediments. Topographic slope generally decreases toward the Lake Tana from all sides. The minimum and maximum elevations are 1,779 and 4,089 m, respectively.

The precipitation in Ethiopia falls almost entirely in the form of rainfall. The Ethiopian territory is divided into four zones marked as A, B, C and D, each of them with different precipitation patterns as per the seasonal classification and precipitation regime of Ethiopia (after NMSA 1996). Tana sub-basin belongs to Zone B (b2 & b3), which is characterized with dry as well as wet seasons. The rainfall is uni-modal and is mainly from April/May to October/November. The rainfall during summer originates from the Indian and Atlantic oceans when the Inter-Tropical Convergence Zone (ITCZ) moves to about 19°N, from June to

September just to the north of Lake Tana. From October to May, the ITCZ moves southwards and dry conditions exist in the region (Lamb et al. 2007).

Temperature varies in the study area and does not show distinct relationship with altitude. The mean annual temperature for these stations varies from the minimum of 15.8 °C for Debretabor meteorological station to the maximum of 20.6 °C at Gonder.

The study was undertaken with the micro-watersheds monitoring sites which are found in Gummara, Rib and Gilgel Abbay catchments of Tana sub-basin as part of sub-watersheds of Baskura (Rib), Kantai (Rib), Enkulal (Gumara) and Engule (Jema) watersheds.

There were 15 micro-watersheds within the hydrologic monitoring program (HMP) but because of data constraint, eight watersheds with better data availability were selected to conduct the research. These selected watersheds are indicated in Table 26.1.

Main or Project and Control Station

Main and Control Stations

There were five main or project stations selected for this study and since 2010, soil and water conservation structures were constructed and implemented including bunds, check dams, trenches and different biological conservation structures. There were three control stations which are inside or in the vicinity of project sub-watersheds that do not implement any soil and water conservation measures.

Table 26.1 Size, elevation and location of HMP stations

Sub-watershed	Micro-watershed	Area (km^2)	*X* (m)	*Y* (m)	Altitude (m)	Status
Kantai (Rib)	Tikurwuha (station 1)	5.1	406,275	1,301,889	2,870	Project
	Guale (station 11)	4.3	401,611	1,306,972	2,693	Control
	Dabzut (station 2)	1.2	400,159	1,307,844	2,640	Project
Enkulal (Gumara)	Enkulal (station 7)	4.1	367,967	1,285,429	2,300	Project
	Agar (station 14)	5.2	353,607	1,284,700	2,317	Control
Engule (Jemma)	Toma (station 9)	6.2	302,222	1,242,181	2,080	Project
	Bosa (station 10)	7.2	312,566	1,217,019	2,129	Control
Zefie (Rib)	Gena (station 6)	5.0	402,519	1,296,369	2,609	Project

Watershed Management

Both main and control watersheds have more or less similar land cover and land use types at the beginning of the project. There were few or no conservation works implemented in both watersheds. At the beginning of the program in 2009/2010, different conservation measures and watershed management activities were introduced in the main or project micro-watersheds. The introduced conservation practices implemented on the different monitoring watersheds were stone bunds, soil bunds, stone faced soil bund, fanya-juu, trenches, waterways and cutoff drains. Gully treatment and rehabilitation were also done on the main watersheds. According to Bureau of Agriculture of ANRS report, the total area that soil and water conservation practices introduced were 54,937.75 ha and treated gully with gully planting, check dam and the like were 1,061 ha and treated degraded lands of 19,646.2 ha in the period of 2009–2016. Even though the controlled watersheds did not have conservation practices by the project, due to the government mass movement of conservation practices, limited different types of conservation practices were introduced on the control monitoring watersheds. Figure 26.2 shows one of the study site before and after water conservation measures. Figures 26.3, 26.4, 26.5 and 26.6 show the gauge installation at the monitoring staions.

Data Source

The data used were found from Tana Sub-basin Organization (TaSBO). Micro-watersheds were established in Baskura, Kentay, Enkulal and Jema clusters

Fig. 26.2 Enkulal (station 7) shots before (left photo) and after (right photo) soil and water conservation practices

Fig. 26.3 Simple garden rain gauges were used at the MMP stations

Fig. 26.4 Current meter measurement with current meter at station 11

Fig. 26.5 Examples of staff levels, painted on natural rock, metal meter section mounted on eucalyptus poles and modern staff gauge

Fig. 26.6 Crest stage gauge installation at station 2 (left photo) and station 6

which are less than 10 km^2 monitored by Tana-Beles Integrated Watershed Management office. Staff gauge, rainfall and Secchi turbidity and bottling of turbid water were held by site managers in TBIWRMP. Suspended sediment was analyzed in Amhara Design and Supervision Works Enterprise (ADSWE) laboratory. This report was compiled based on data from 2009 to 2015 (Table 26.2).

Rainfall

Rainfall was measured from few numbers of rain gauges installed in the vicinity of farmers to collect data regularly twice a day at 7:00 and 19:00 h. In addition, records were made as often as high rain was observed. Hence, the data was processed by HYDATA software to change raw data to daily, monthly and annual for the study period.

Discharge Measurement and Staff Gauge Reading

Crest stage discharge measurements were conducted by TBIWRDP Watershed Monitoring Program (WMP) to capture peak floods. For small flows, current meter and floating method of discharge measurement were conducted. Staff gauges were observed twice a day at 7:00 and 19:00 and additional records were conducted when there is high discharge following high rain. Analysis of flow was made by HYDATA software and excel spread sheet.

Table 26.2 Number of observations and samples in MMP 2009–2015

Parameter	Year of data collection							
	2009	2010	2011	2012	2013	2014	2015	Total
Rainfall (mm)	3,116	>12,777	>15,000	>15,000	>15,000	>15,000	>15,000	
Secchi turbidity (no.)	3,131	12,069	12,469	12,340	12,360	12,350	12,030	
Staff gauge level (m)	3,132	11,812	12,409	12,125	12,360	12,300	12,800	
Flow measurement (m^3/s)								>500
Sediment sample (mg/l)	1,425	4,176	3,237	2,873	3,268			

Rating Curve Preparation

Rating curve was prepared using the current meter flow measurement, staff gauge reading and float method using indirect method (manning equation), since it was difficult to measure high flood during the field visit time, it was possible to include the high flood using computation. Rating curve was established using the recorded flow of current meter and floating methods, manning equation and the value was tested using goodness-of-fit and coefficient of determination (R^2).

Baseflow Separation

The flows of the rivers were estimated using hourly interpolated water level and derived rating curve formula using HYDATA software and excel spreadsheet. Interpolation of original data to hourly data was used to get hourly data of staff gauge and these hourly data were used to prepare the flow data using rating curve with excel spread sheet. Again, the hourly flow data were processed in the HYDATA software to get daily and monthly flow data. After flow of rivers was estimated, baseflow separation was conducted with local minimum method of BFI + hydro office software.

Sediment Measurement and Analysis

Suspended sediment was collected in field twice a day at 7:00 and 19:00 when Secchi turbidity is greater than 8 which means the river is sediment free at this reading and taken to laboratory by one letter plastic bottle. HYDATA analysis was done to change daily, annual and monthly bases. Sediment load analysis was done based on the flow and suspended sediment. The hourly flow of stream at the watershed was computed from stage discharge relation of rating curve and suspended sediment was obtained from laboratory analysis. Thus, sediment load in tons was computed by multiplying flow of hourly period and sediment concentration in mg/l. The following formula was used to compute sediment load.

$$\text{Sediment load}\left(\frac{\text{tonnes}}{\text{hr}}\right) = \left[\frac{\text{Flow}\left(\frac{\text{m3}}{\text{hr}}\right) * \text{Sed.concen}\left(\frac{\text{mg}}{\text{l}}\right) * 1000\left(\frac{1}{\text{m3}}\right)}{1000\left(\frac{\text{mg}}{\text{kg}}\right) * 1000\left(\frac{\text{kg}}{\text{tonne}}\right)}\right] \tag{26.1}$$

HYDATA Interpolation and Analysis

HYDATA software was used to store the field data. The water level and suspended sediment is changed to hourly time series and interpolated accordingly. Using the rating curve, the flow was computed for hourly time period using excel spread sheet and imported to HYDATA to convert to daily and monthly period. Similarly, sediment load is calculated using suspended sediment interpolated data with corresponding flow. Again, imported to HYADATA and convert to daily and monthly period. In turn, exporting the data in a preferable time series analysis were conducted in excel.

Significance Test

A paired *t*-test was used to compare two population means where we have two samples in which observations in one sample can be paired with observations in the other sample. A paired *t*-test was conducted to test the significance of temporal and spatial variations of baseflow and sediment changes.

Results and Discussion

Rainfall

Annual rainfall of the study area ranges from 916 mm in 2014 in station 1 to 1,857 mm in 2011 in station 7. Average rainfall of the area is 1,423 mm in 2010–2015. The smallest record was observed in Rib catchment and the highest record was observed in Gumara catchment. When we see the temporal distribution in 2014, lowest rainfall was observed except station 2 which has better record in 2014. Rainfall records were examined by paired *t*-test *i* between years of records to see whether the rainfall has an effect on the decrease or increase of baseflow and sediment concentration. In this study, paired *t*-test of rainfall in 2011 and 2014 months was examined, t-calculated is 0.9 which is less than t-tabulated, 1.79 and the null hypothesis has accepted; hence, there is no significant change in rainfall trend (Tables 26.3 and 26.4). Figure 26.7 also shows similar pattern of rainfall distribution. Table 26.3 shows the total amount of rain recorded from 2011 to 2015 in four stations.

Table 26.3 Annual rainfall of stations from 2010 to 2015

Year	Station 1	Station 2	Station 6	Station 7
2011	1,511.00	1,450.00	1,418.50	1,857.00
2012	1,276.00	1,462.00	1,358.00	1,419.00
2013	1,573.00	1,515.00	1,471.00	1,849.00
2014	916.00	1,658.33	932.00	1,068.00
2015	1,641.00	NA	1,131.00	1,548.00

NA not available.

Table 26.4 Rainfall paired *t*-test

t-Test: paired two samples for means		
	Variable 1	Variable 2
Mean	120.8333	138.1944
Variance	19,846.33	31,566.49
Observations	12	12
Pearson correlation	0.939515	
Hypothesized mean difference	0	
d*f*	11	
t stat	−0.90856	
P(*T* ≤ *tt* one-tail	0.191528	
t critical one-tail	1.795885	
P(*T* ≤ *t*) two-tail	0.383056	
t critical two-tail	2.200985	

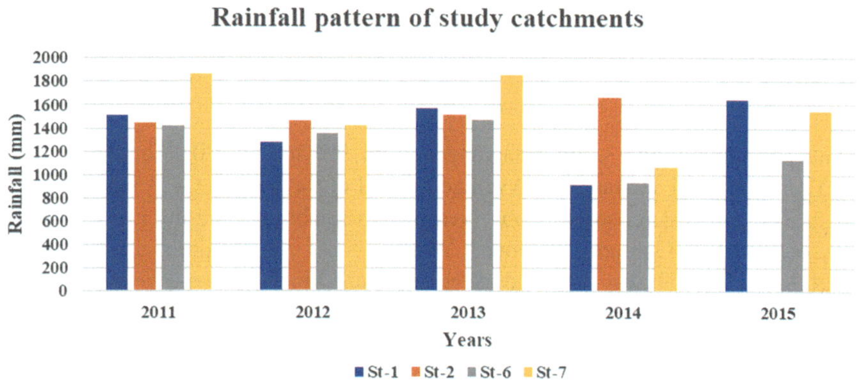

Fig. 26.7 Annual rainfall patterns of study watersheds

Flow

Rating Curve Equation

The accuracy of the rating curve is important to better estimate flows. The accuracy of the rating curve is evaluated by the goodness-of-fit and coefficient of determination. Generally, stations show better goodness-of-fit in the range of 70–97% and when R^2 is more than 0.80 (ranging 0.81–0.93). The lowest goodness-of-fit was observed in station 10 (<70%) and this station the farthest from Bahir Dar where monitoring office is located. This station is not visited regularly than other stations, therefore, the flow records used to prepare rating curve was poor. The other stations

Table 26.5 Rating curve

No	Station	Rating curve equations	Goodness-of-fit (%)	R^2
1	St 1	$Q = 0.641*(h + 0.26)^{4.667}$	77	0.88
2	St 2u	$Q = 0.515*(h + 0.20)^{7.203}$	90	0.88
3	St 6	$Q = 0.170*(h + 0.70)^{2.860}$	89	0.87
4	St 7	$Q = 0.385*(h + 0.25)^{3.422}$	97	0.81
5	St 9	$Q = 0.031*(h + 0.80)^{5.228}$	82	0.85
6	St 10	$Q = 0.199*(h + 0.80)^{2.817}$	70	0.93
7	St 11u	$Q = 0.290*(h + 0.25)^{4.580}$	85	0.86

have better estimates of goodness-of-fit; therefore, the rating curve can be used to estimate the discharge values (Q) from staff gauge readings (h) (Table 26.5).

Baseflow

For the baseflow analysis in response to watershed management, comparisons were done between main station, control station and temporal variation of records in watersheds upstream of the hydrological monitoring stations. The monthly baseflow of the main stations and control stations were tested by paired t-test. Between main and control stations, a one tailed t-test result showed t-calculated is 0.48 which is less than t-tabulated (2.13). Thus, the null hypothesis was accepted, and as a result, there is no significant change between the main stations and controls. See the significance in Tables 26.6 and 26.7.

Table 26.6 Baseflow paired t-test

t-Test: paired two samples for means between baseflow of control and project stations		
	Variable 1	Variable 2
Mean	0.17346	0.1724
Variance	0.00263	0.000527
Observations	5	5
Pearson correlation	0.015372	
Hypothesized mean difference	0	
df	4	
t stat	0.042427	
$P(T \leq t$ one-tail	0.484096	
t critical one-tail	2.131847	
$P(T \leq t)$ two-tail	0.968192	
t critical two-tail	2.776445	

Table 26.7 Baseflow of studied watersheds in m^3/s

Year	Main stations					Control stations		
	St 1	St 2	St 6	St 7	St 9	St 10	St 11	
2010	0.185	-	0.507	0.062	0.288	0.329	0.073	
2011	0.033	0.021	0.333	0.061	0.334	0.321	0.021	
2012	0.027	0.051	0.123	0.054	0.275	0.299	0.019	
2013	0.044	0.06	0.212	0.061	0.403	0.362	0.014	
2014	0.123	0.024	0.296	0.114	0.612	0.278	0.008	
2015	0.177		0.151	0.082				

Paired *t*-test also applied on temporal variations of the year 2010 and 2014 which is within the period of soil and water conservation was constructed and stabilized. According to the hypothesis station 1 and station 2 from main stations and station 10 from control station, null hypothesis was accepted but stations 6, 7 and 9 from the main and station 11 from control stations null hypothesis was rejected (Table 26.8). Thus, the baseflow of the former stations did not show changes but the latter stations have significant change on temporal bases. There is an increase of baseflow in these latter stations except station 11 which clearly shows a decrease of baseflow. At station 11, the average annual flow record in 2010 was 0.073 m^3/s, but in 2014, it decreased to 0.008 m^3/s (Table 26.9). The time series graph shows that in few years after start of conservation practices, baseflow decreased in main stations but after 2012 a general increase was observed which can be due to the stabilization of the constructed conservation measures and the response of the watershed after some years (Fig. 26.8). In the project stations, there was a clear baseflow increase than the control stations (Fig. 26.8). The moving average graph shows that the baseflow was decreasing in the beginning of construction years and gradually an increment is observed (Fig. 26.9).

Table 26.8 Paired *t*-test of baseflow for each station

Stations	*T* calculated	*T* tabulated	Null hypothesis
St 1	0.22	1.8	Accepted
St 2	1.75	1.8	Accepted
St 6	11.78	1.8	Rejected
St 7	6.94	1.8	Rejected
St 9	7.93	1.8	Rejected
St 10 (control)	1.74	1.8	Accepted
St 11 (control)	3.04	1.8	Rejected

Table 26.9 Suspended sediment of watersheds in mg/l

Year	Main stations						Control stations			
	St 1	St 2	St 6	St 7	St 9	Average	St 10	St 11	St 14	Average
2010	2.9	2.1	2.9	3.1	2.3	2.66	2.9	2.5	-	2.7
2011	2.3	1.5	1.9	1.5	1.5	1.74	2.1	1.2	1.5	1.6
2012	1.5	1.6	1.5	1.5	1.5	1.52	2.1	1.3	1.4	1.6
2013	1.3	1.3	1.3	1.5	1.4	1.36	1.7	1.2	1.3	1.4
2014	0.6	1.3	1.3	1.3	1.3	1.16	1.3	1.3	1.3	1.3
2015	1.5	1.5	1.3	1.4		1.425				

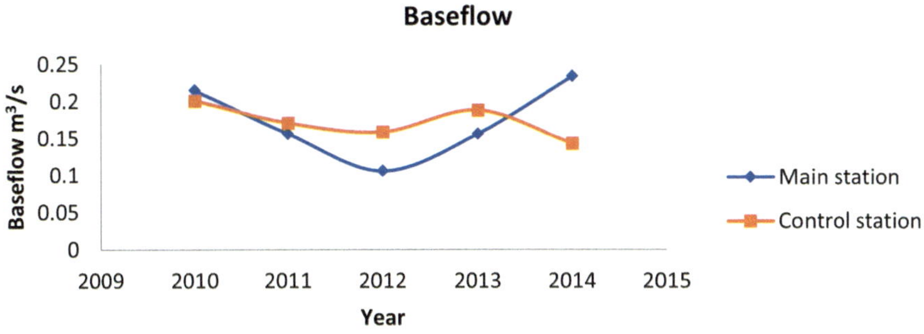

Fig. 26.8 Baseflow of main and control stations

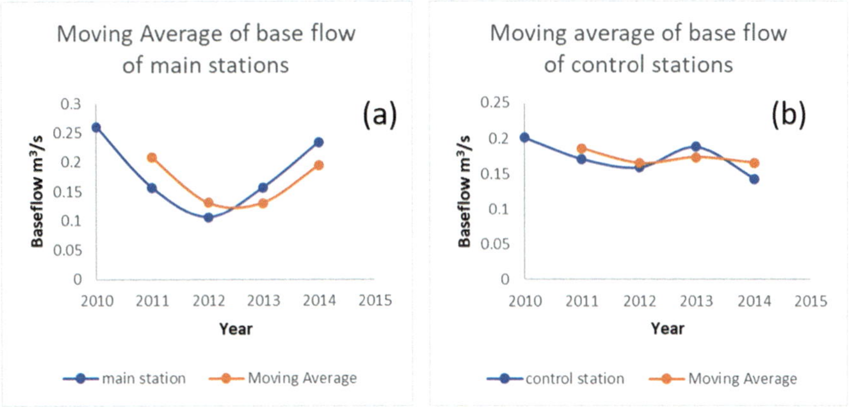

Fig. 26.9 Baseflow moving average of main stations (**a**) and control stations (**b**)

Sediment

Suspended Sediment

The average values of sediment in mg/l in main stations decreased from 2.66 to 1.425 mg/l from 2010 to 2015 years and from control stations it decreased from 2.7 to 1.3 mg/l from 2010 to 2014 years (Table 26.9). These showed that there was a decrease of suspended sediment in the main as well as control stations. Comparison of suspended sediment records between main and controls and temporal variations was also made using paired *t*-test and moving average (Figs. 26.10 and 26.11). The paired *t*-test was accomplished in suspended sediment between main and control stations, one tailed t-calculated is 0.02 which is less than t-tabulated 1.89 (Table 26.10). Thus, there is no significant change. The moving average shows that the temporal variation of stations as shown in Fig. 26.11. The annual values show a decrease trend in main as well as control stations. Generally, a decreasing of suspended sediment concentration from 2010 to 2015 was observed (Fig. 26.11).

Sediment Load

Most of the sediment load occurred early in the rainy season in June and July. Figure 26.12 shows the trend of discharge and sediment of study watersheds. Average sediment load in main station was recorded 2008 tons in 2010, but it decreased to 914 tons in 2015. Similarly, average sediment load of control stations record was 1895 tons in 2010 and it decreased to 246 tons in 2015. The monthly sediment load of watersheds at main and controls were tested by paired *t*-test. And the null hypothesis was accepted (t-calculated = 0.11 and t-tabulated = 2.13), and hence no significant change between the main and control stations. Figure 26.12 indicates that the flow increases, whereas the sediment load decreases starting in 2012.

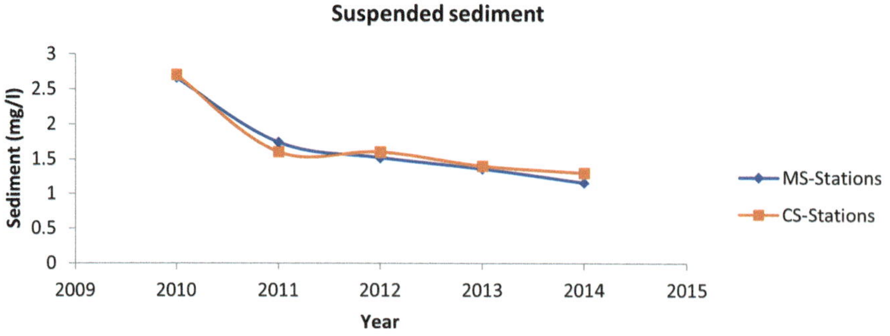

Fig. 26.10 Suspended sediment of main and control stations

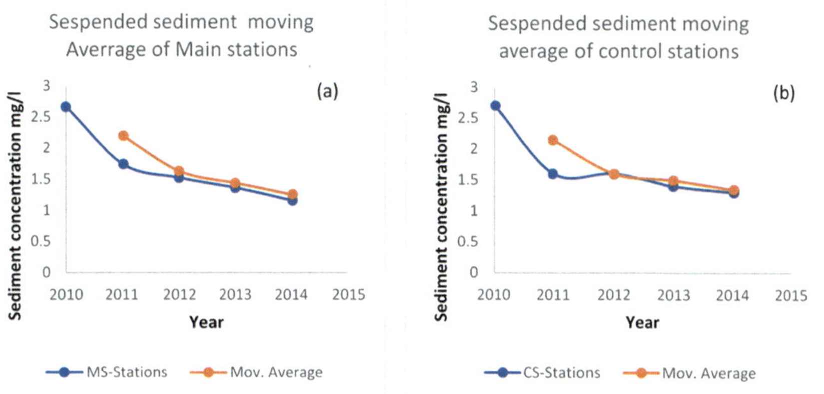

Fig. 26.11 Suspended sediment moving average of main (**a**) and control stations (**b**)

Table 26.10 Paired *t*-test of suspended sediment

t-Test: paired two samples for means		
	Variable 1	Variable 2
Mean	1.6875	1.2125
Variance	0.138393	0.06125
Observations	8	8
Pearson correlation	−0.66527	
Hypothesized mean difference	0	
d*f*	7	
t stat	2.367088	
P($T \leq t$) one-tail	0.02491	
t critical one-tail	1.894579	
P($T \leq t$) two-tail	0.049819	
t critical two-tail	2.364624	

But when we see the trend of the sediment load, the null hypothesis was rejected, the t-value of calculated (4.64) is greater than tabulated value (1.94) as it is shown in Table 26.13. Thus, there is significant decrease from starting period of conservation to 2015. There was a decreasing trend in sediment load in the control as well as project stations since the year the project started. The moving average of the sediment load also indicated that there was a decreasing trend since the conservation was started (Tables 26.11 and 26.12).

As shown in Table 26.14, the decreasing percentage of sediment load at each station was observed. The maximum sediment load decrease was 70% at station 1 and minimum decrease was observed (14%) at station 9. The analysis showed that sediment load decreased as a result of sustainable land management which was

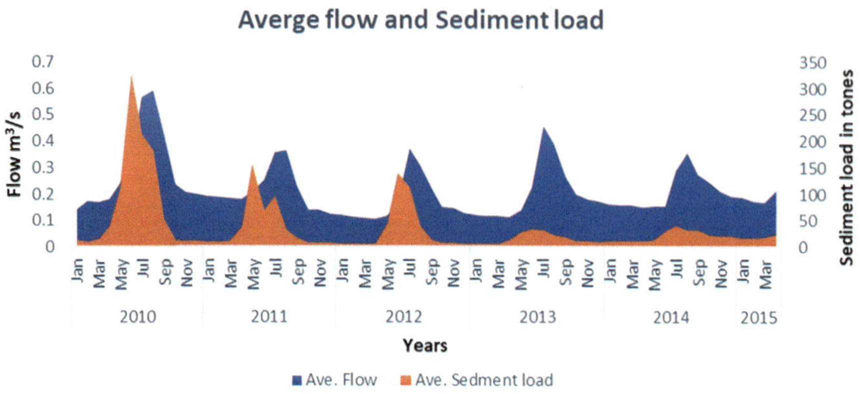

Fig. 26.12 Average annual flow and sediment yield of study watersheds

practiced for the study period. So, due to the impact of watershed management activities upstream of the hydrological stations, there is a reduction of more than 10% average suspended sediment load measured over the project life which is 50%.

Conclusion and Recommendation

Conclusion

It is shown that watershed management impacts water quality and quantity of the small watersheds. The rainfall has shown no significant change over the study period and among the project sites. There was an increase of baseflow in some stations such as station 6, 7, 9, in the main stations, and decreasing trend was observed in station 11 in control stations but in stations 1 and 2, which are main stations and station 10, a control station, there was no change observed. This may be because the short period of response for the management activities implemented which was not showing a clear change in the baseflow. Suspended sediment has decreased from 2010 to 2015. The sediment load was decreasing both in main and control stations and this may be because of large-scale watershed management activities that was implemented in control micro-watersheds by the government watershed development and management program. The overall project results show that the sediment load was decreasing with a range of 14–70% in monitoring stations, since the start of the watershed management program up to 2015; so, some of the results in some micro-watersheds is above the initial target set by the program. The total average, deducted from observations of the hydrological stations in the micro-watersheds of the headwater area, is 50%. The hydrological observations and watershed treatments at project areas show promising results that can be scaled up to the sub-basins in Abbay and/or other basins. It would be widely used as the

Table 26.11 Sediment load of watersheds in tons

Year	Main station						Control station		
	St 1	St 2	St 6	St 7	St 9	Average	St 10	St 11	Average
2010	2,715.24	2,477.46	2,070.66	1,036.47	1,740.69	2,008.104	2,942.63	848.87	1,895.75
2011	856.28	317.04	1,227.58	335.69	1,899.7	927.258	1,354.17	630.09	992.13
2012	583.41	1,060.55	389.76	416.09	1,550.76	800.114	1,460.04	71.15	765.595
2013	128.94	271.84	341.89	264.78	1,025.37	406.564	1,375.56	36.81	706.185
2014	141.42		395.56	231.13	1,304.82	518.2325	476.91	15.71	246.31
2015	536.26		330.88	1,875.71	-	914.2833	-	-	

Table 26.12 Paired *t*-test of sediment load between control and main/project stations

t-Test: paired two samples for means		
	Variable 1	Variable 2
Mean	911.3252	921.194
Variance	429,269.8	370,174.9
Observations	5	5
Pearson correlation	0.960214	
Hypothesized mean difference	0	
d*f*	4	
t stat	−0.11984	
$P(T \leq t)$ one-tail	0.455193	
t critical one-tail	2.131847	
$P(T \leq t)$ two-tail	0.910387	
t critical two-tail	2.776445	

Table 26.13 Paired *t*-test of temporal variation of sediment load

t-Test: paired two samples for means		
	Variable 1	Variable 2
Mean	1,976.003	366.5071
Variance	658,296.9	203,306.9
Observations	7	7
Pearson correlation	0.027825	
Hypothesized mean difference	0	
d*f*	6	
t stat	4.642775	
$P(T \leq t)$ one-tail	0.001765	
t critical one-tail	1.94318	
$P(T \leq t)$ two-tail	0.00353	
t critical two-tail	2.446912	

Table 26.14 Average sediment load decrease in percent of selected stations

Year	St 1	St 2	St 6	St 7	St 9	St 10	St 11
2010	0	0	0	0	0	0	0
2011	68	87	41	68	-9	54	26
2012	79	57	81	60	11	50	92
2013	95	89	83	74	41	53	96
2014	95		81	78	25	84	98
2015	80		84	81			
Station average (%)	70	58	62	33	14	48	62
Total watershed average (%)	50						

fundamental center of experience sharing and forefront of watershed management projects as the first unique project integrating the hydrologic monitoring with the integrated watershed management efforts to portray indicator results linked to project outcomes. The results found can be used to get the total sediment load derived from homogeneous watersheds in the sub-basin, viz., coupled with the hydrologic monitoring. In general, the project impacts have gradually led to intensifying and even in the support of the rural community engaged in additional watershed development works of adjacent watersheds. So, the flooding and negative effects on ecologically sensitive areas, owing to lagged flow and reduction of sediment loading in most downstream reaches, are believed to be diminished as a result of model watershed developments in the watersheds. This particular study result does not fully include peak hydrologic flow. The results are mainly derived from mid-flow and low flow hydrologic data collected from stations in the monitoring network in the headwater area.

Recommendation

The following recommendations are made based on the limited data and results reported in this study.

- To increase data quality, maintenance of the observed stations is required.
- Further, land cover-land use change assessments shall be coupled with the hydrologic monitoring results. Other impacts such as socio-economic conditions must also be evaluated; these may show benefits derived from changes of the watershed management in addition to hydrology of the project watersheds.
- Training is required to data observers (site managers) and technical staffs of Tana sub-basin organization so that the data collection and analysis results may have better quality outputs.
- The data need be further refined and organized into its own project database program and technical staff following the project work since this specific hydrological monitoring program need vast resource and immense volume of work.
- The flow data used to prepare the rating curves must be the focus of the measurement taking into account mainly the peak flows monitoring in the rainy season.
- The data entry needs to be done by trained and responsible person so that data errors on project results can be minimized.
- It is suggested that the measurement program shall continue for three to five years to show better correlated results (baseflow and sediment concentration) with growing watershed management efforts in the sub-basins located farther in the upstream mountainous areas.

Acknowledgements We thank the Tana Basin Organization staffs and Abbay Basin Authority, especially Ato Habtamu Tamir. Thanks also to Dr. Mikaela for the informal training on the use of HYDATA.

References

Abtew W, Melesse AM (2014a) Nile River Basin hydology. In: Melesse AM, Abtew W, Setegn S (eds) Nile River Basin: ecohydrological challenges, climate change and hydropolitics, pp 7–22

Abtew W, Melesse AM (2014b) Climate teleconnections and water management. In: Nile River basin. Springer International Publishing, Berlin, pp 685–705

Abtew W, Melesse AM (2014c) Transboundary rivers and the Nile. In: Nile River basin. Springer International Publishing, Berlin, pp 565–579

Abtew W, Melesse A, Desalegn T (2009a) Spatial, inter and intra-annual variability of the Blue Nile River Basin Rainfall. Hydrol Process 23(21):3075–3082

Abtew W, Melesse A, Desalegn T (2009b) El Niño Southern oscillation link to the Blue Nile River Basin hydrology. Hydrol Process Special Issue Nile Hydrol 23(26):3653–3660

Aga AO, Chane B, Melesse AM (2018) Soil erosion modelling and risk assessment in data scarce Rift Valley Lake Regions, Ethiopia. Water 10:1684. https://doi.org/10.3390/w10111684

Aga AO, Melesse AM, Chane B (2019) Estimating the sediment flux and budget for a data limited Rift Valley Lake in Ethiopia. Hydrology 6(1). https://doi.org/10.3390/hydrology6010001

Assefa A, Melesse AM, Admasu S (2014) Climate change in Upper Gilgel Abay River catchment, Blue Nile Basin Ethiopia. In: Melesse AM, Abtew W, Setegn S (eds) Nile River basin: ecohydrological challenges, climate change and hydropolitics, pp 363–388.

Behulu F, Setegn S, Melesse AM, Fiori A (2013) Hydrological analysis of the Upper Tiber Basin: a watershed modeling approach. Hydrol Process 27(16):2339–2351

Behulu F, Setegn S, Melesse AM, Romano E, Fiori A (2014) Impact of climate change on the hydrology of Upper Tiber River Basin using bias corrected regional climate model. Water Res Manage 1–17

Chebud Y, Melesse AM (2013) Stage level, volume, and time-frequency change information content of Lake Tana using stochastic approaches. Hydrol Process 27(10): 1475–1483. https://doi.org/10.1002/hyp.9291

Chebud YA, Melesse AM (2009a) Numerical modeling of the groundwater flow system of the Gumera Sub-Basin in Lake Tana Basin. Ethiopia Hydrol Process Special Issue Nile Hydrol 23 (26):3694–3704

Chebud YA, Melesse AM (2009b) Modeling lake stage and water balance of Lake Tana. Ethiopia Hydrol Process 23(25):3534–3544

Defersha MB, Melesse AM (2012) (2012a) Effect of rainfall intensity, slope and antecedent moisture content on sediment concentration and sediment enrichment ratio. CATENA 90:47–52

Defersha MB, Melesse AM (2011) (2012b) Field-scale investigation of the effect of land use on sediment yield and surface runoff using runoff plot data and models in the Mara River basin, Kenya. CATENA 89:54–64

Defersha MB, Quraishi S, Melesse AM (2010) Interrill erosion, runoff and sediment size distribution as affected by slope steepness and antecedent moisture content. Hydrol Earth Syst Sci Discuss 7(6447–6489):2010

Defersha MB, Quraishi S, Melesse AM (2011) The effect of slope steepness and antecedent moisture content on interrill erosion, runoff and sediment size distribution in the highlands of Ethiopia. Hydrol Earth Syst Sci 15:2367–2375. https://doi.org/10.5194/hess-15-2367-2011

Defersha MB, Melesse AM, McClain M (2012) Watershed scale application of WEPP and EROSION 3D models for assessment of potential sediment source areas and runoff flux in the Mara River Basin Kenya. CATENA 95:63–72

Dessu SB, Melesse AM (2012) Modeling the rainfall-runoff process of the Mara River Basin using SWAT. Hydrol Process 26(26):4038–4049

Dessu SB, Melesse AM (2013) Impact and Uncertainties of climate change on the hydrology of the Mara River Basin. Hydrol Process 27(20):2973–2986

Dessu SB, Melesse AM, Bhat M, McClain M (2014) Assessment of water resources availability and demand in the Mara River Basin. CATENA 115:104–114

FAO (2001) The state of food and agriculture 2001. Food & Agriculture Org

Getachew HE, Melesse AM (2012) Impact of land use/land cover change on the hydrology of Angereb Watershed, Ethiopia. Int J Water Sci 1(4):1–7. https://doi.org/10.5772/56266

Grey OP, Webber DG, Setegn SG, Melesse AM (2013) Application of the soil and water assessment tool (SWAT Model) on a small tropical island state (Great River Watershed, Jamaica) as a tool in integrated watershed and coastal zone management. Int J Tropical Biol Conserv 62(3):293–305

Kainulainen A (2012) Comparing the monitoring and evaluation systems of watershed management related development projects in Amhara, Ethiopia

Lamb HF et al (2007) Late Pleistocene desiccation of Lake Tana, source of the blue Nile. Quatern Sci Rev 26(3–4):287–299

Maalim FK, Melesse AM (2013) Modeling the impacts of subsurface drainage systems on runoff and sediment yield in the Le Sueur Watershed, Minnesota. Hydrol Sci J 58(3):1–17

Maalim FK, Melesse AM, Belmont P, Gran K (2013) Modeling the impact of land use changes on runoff and sediment yield in the Le Sueur Watershed, Minnesota Using GeoWEPP. CATENA 107:35–45

Mango L, Melesse AM, McClain ME, Gann D, Setegn SG (2011a) Land use and climate change impacts on the hydrology of the upper Mara River Basin, Kenya: results of a modeling study to support better resource management. Special issue: climate, weather and hydrology of East African Highlands. Hydrol Earth Syst Sci 15:2245–2258. https://doi.org/10.5194/hess-15-2245-2011.

Mango L, Melesse AM, McClain ME, Gann D, Setegn SG (2011b) Hydro-meteorology and water budget of Mara River basin, Kenya: a land use change scenarios analysis. In: Melesse A (ed) Nile River basin: hydrology, climate and water use, chapter 2. Springer Science Publisher, pp 39–68. https://doi.org/10.1007/978-94-007-0689-7_2.

Mekonnen M, Melesse A (2011) Soil erosion mapping and hotspot area identification using GIS and remote sensing in northwest Ethiopian highlands, near Lake Tana. In: Melesse A (ed) Nile River Basin: hydrology, climate and water use, chapter 10. Springer Science Publisher, pp 207–224. https://doi.org/10.1007/978-94-007-0689-7_10

Melesse A, Weng Q, Thenkabail P, Senay G (2007) Remote sensing sensors and applications in environmental resources mapping and modeling, special issue: remote sensing of natural resources and the environment. Sensors 7:3209–3241

Melesse A, Abtew W, Setegn SG (2014) Nile River Basin: ecohydrological challenges, climate change and hydropolitics. Springer Science & Business Media

Melesse AM (2011) Nile River Basin: hydrology, climate and water use. Springer Science & Business Media

Melesse AM, Jordan JD (2002) A comparison of fuzzy vs augmented-ISODATA classification algorithm for cloud and cloud-shadow discrimination in landsat imagery. Photogrammetric Eng Remote Sens 68(9):905–911

Melesse AM, Jordan JD (2003) Spatially distributed watershed mapping and modeling: land cover and microclimate mapping using landsat imagery part 1. J Spatial Hydrology (e-journal) 3(2)

Melesse AM, Loukas AG, Senay G, Yitayew M (2009) Climate change, land-cover dynamics and ecohydrology of the Nile River Basin. Hydrol Process Special Issue Nile Hydrol 23(26):3651–3652

Melesse AM, Abtew W, Desalegne T, Wang X (2010) Low and high flow analysis and wavelet application for characterization of the Blue Nile River System. Hydrol Process 24(3):241–252 (2009b)

Melesse AM, Abtew W, Setegn S, Dessalegne T (2011a) Hydrological variability and climate of the Upper Blue Nile River Basin. In: Melesse A (ed) Nile River Basin: hydrology, climate and water use, chapter 1. Springer Science Publisher, pp 3–37. https://doi.org/10.1007/978-94-007-0689-7_1

Melesse AM, Ahmad S, McClain M, Wang X, Lim H (2011b) Sediment load prediction in large rivers: ANN approach. Agric Water Manag 98:855–866

Msagahaa J, Ndomba, Melesse AM (2014) Modeling sediment dynamics: effect of land use, topography and land management. In: Melesse AM, Abtew W, Setegn S (eds) Nile River Basin: ecohydrological challenges, climate change and hydropolitics, pp 165–192

Rämi H, UN OCHA-Ethiopia (2003) Ponds filled with challenges. Water harvesting-experiences in Amhara and Tigray. UN OCHA-Ethiopia, pp 1–6.

Rojas RV et al (2016) Healthy soils: a prerequisite for sustainable food security. Environ Earth Sci 75(3):180

Setegn S, Rayner D, Melesse AM, Dargahi B, Srinivasan R (2011) Impact of climate change on the hydro-climatology of Lake Tana basin, Ethiopia. Water Resour Res 47(W04511):13. https://doi.org/10.1029/2010WR009248.

Setegn SG, Srinivasan R, Dargahil B, Melesse AM (2009) Spatial delineation of soil erosion prone areas: application of SWAT and MCE approaches in the Lake Tana Basin, Ethiopia. Hydrol Process 23(26):3738–3750

Setegn SG, Dargahi B, Srinivasan R, Melesse AM (2010a) Modelling of sediment yield from Anjeni Gauged Watershed, Ethiopia Using SWAT. J Am Water Res Assoc 46(3):514–526

Setegn SG, Dargahi B, Srinivasan R, Melesse AM (2010b) Modelling of sediment yield from Anjeni Gauged Watershed Ethiopia using SWAT. JAWRA 46(3):514–526

Setegn SG, Srinivasan R, Melesse AM, Dargahi B (2010c) SWAT model application and prediction uncertainty analysis in the Lake Tana Basin Ethiopia. Hydrol Process 24(3):357–367

Setegn SG, Melesse AM, Haiduk A, Webber D, Wang X, McClain M (2014) Spatiotemporal distribution of fresh water availability in the Rio Cobre Watershed, Jamaica. CATENA 120:81–90

Wang X, Melesse AM (2005) Evaluations of the SWAT model's snowmelt hydrology in a Northwestern Minnesota Watershed. Trans ASAE 48(4):1359–1376

Wang X, Melesse AM (2006) Effects of STATSGO and SSURGO as Inputs on SWAT model's snowmelt simulation. J Am Water Res Assoc 42(5):1217–1236

Wang X, Melesse AM, Yang W (2006) Influences of potential evapotranspiration estimation methods on SWAT's hydrologic simulation in a Northwestern Minnesota Watershed. Trans ASAE 49(6):1755–1771

Wang X, Shang S, Yang W, Melesse AM (2008a) Simulation of an agricultural watershed using an improved curve number method in SWAT. Trans Am Soc Agri Bio Eng 51(4):1323–1339

Wang X, Yang W, Melesse AM (2008b) Using hydrologic equivalent wetland concept within SWAT to estimate streamflow in watersheds with numerous wetlands. Trans Am Soc Agri Bio Eng 51(1):55–72

Wang X, Garza J, Whitney M, Melesse AM, Yang W (2008c) Prediction of sediment source areas within watersheds as affected by soil data resolution. In: Findley PN (ed) Environmental modelling: new research, chapter 7. Nova Science Publishers, Inc., Hauppauge, pp 151–185. ISBN: 978-1-60692-034-3

Wondie M, Schneider W, Melesse AM, Teketay D (2011) Spatial and temporal land cover changes in the Simen Mountains National Park, a world heritage site in Northwestern Ethiopia. Remote Sens 3:752–766. https://doi.org/10.3390/rs3040752

Wondie M, Teketay D, Melesse AM, Schneider W (2012) Relationship between topographic variables and land cover in the Simen Mountains National Park, a World Heritage Site in northern Ethiopia. Int J Remote Sens Appl 2(2):36–43

Yitayew M, Melesse AM (2011) Critical water resources management issues in Nile River Basin. In: Melesse A (ed) Nile River Basin: hydrology, climate and water use, chapter 20. Springer Science Publisher, pp 401–416. https://doi.org/10.1007/978-94-007-0689-7_20.

Yesuf HM, Alamirew A, Assen M, Melesse A (2013) Spatiotemporal mapping of land cover in Lake Hardibo Drainage Basin, Northeast Ethiopia: 1957–2007. Water conservation: practices, challenges and future implications. Nova Publishers, pp 147–164

Yesuf HM, Alamirew T, Assen M, Melesse AM (2015) Modeling of sediment yield in Maybar gauged watershed using SWAT, northeast Ethiopia. CATENA 127:191–205

Index

A. M. Melesse et al. (eds.), *Nile and Grand Ethiopian Renaissance Dam*, Springer Geography, https://doi.org/10.1007/978-3-030-76437-1

GPSR Compliance

The European Union's (EU) General Product Safety Regulation (GPSR) is a set of rules that requires consumer products to be safe and our obligations to ensure this.

If you have any concerns about our products, you can contact us on ProductSafety@springernature.com

In case Publisher is established outside the EU, the EU authorized representative is:

Springer Nature Customer Service Center GmbH
Europaplatz 3
69115 Heidelberg, Germany

Batch number: 10732076

Printed by Printforce, the Netherlands